The Masterpiece of Nature

Originally published in 1982, *The Masterpiece of Nature* examines sex as representative of the most important challenge to the modern theory of evolution. The book suggests that sex evolved, not as the result of normal Darwinian processes of natural selection, but through competition between populations or species - a hypothesis elsewhere almost universally discredited. The book also discusses the nature of sex and its consequences for the individual and for the population, as well as various other theories of sex. Since the value of these theories is held to reside wholly in their ability to predict the patterns of sexuality observed in nature, the book seeks to provide an extensive review of the circumstances in which sexuality is attenuated or lost throughout the animal kingdom, and these facts are then used to weigh up the merits of the rival theories. This book will be of interest to researchers in the area of genetics, ecology and evolutionary biology.

The Masterpiece of Nature

The Evolution and Genetics of Sexuality

by Graham Bell

Routledge
Taylor & Francis Group

First published in 1982
by Croom Helm Ltd.

This edition first published in 2019 by Routledge
2 Park Square, Milton Park, Abingdon, Oxon, OX14 4RN
and by Routledge
711 Third Avenue, New York, NY 10017

Routledge is an imprint of the Taylor & Francis Group, an informa business

© 1982 Graham Bell

Publisher's Note
The publisher has gone to great lengths to ensure the quality of this reprint but points out that some imperfections in the original copies may be apparent.

Disclaimer
The publisher has made every effort to trace copyright holders and welcomes correspondence from those they have been unable to contact.

A Library of Congress record exists under LCCN: 81016045

ISBN 13: 978-0-367-33925-8 (hbk)
ISBN 13: 978-0-429-32288-4 (ebk)
ISBN 13: 978-0-367-33927-2 (pbk)

THE MASTERPIECE OF NATURE

. . . the larger and more perfect animals are now propagated by sexual reproduction only, which seems to have been the chef d'oeuvre, or capital work of nature . . .

— Erasmus Darwin (1803) *The Temple of Nature; or, the Origin of Society.*
  J. Johnson, London. (Additional Notes, p. 36 of first edition.)

The formation of the organs of sexual generation . . . seems the chef d'oeuvre, the master-piece of nature . . .

— Erasmus Darwin (1803) *Zoonomia; or, the Laws of Organic Life.* Thomas and Andrews, Boston. (Sect. **XXXIX** 5.2, p. 404 of second American, from third London, edition.)

CROOM HELM APPLIED BIOLOGY SERIES
Edited by Peter J. Baron, BSc. Hons, PhD

NUTRITION AND DISEASE
Edited by R. J. Jarrett

THE NATURE OF ENZYMOLOGY
R. L. Foster

NUTRITION AND FOOD PROCESSING
H. G. Muller and G. Tobin

THE MASTERPIECE OF NATURE

THE EVOLUTION AND GENETICS OF SEXUALITY

GRAHAM BELL

CROOM HELM
London & Canberra

© 1982 Graham Bell
Croom Helm Ltd, 2-10 St John's Road, London SW11

British Library Cataloguing in Publication Data

Bell, Graham
    The masterpiece of nature.
    1. Sex (Biology)
    2. Evolution
    I. Title
    574.1'66   QH481

    ISBN 0-85664-753-5

Printed and bound in Great Britain by
Biddles Ltd, Guildford and King's Lynn

The Flight of the Sparrow

1862   'We do not even in the least know the final cause of sexuality;
        why new beings should be produced by the union of the two
        sexual elements, instead of by a process of parthenogenesis . . .
        The whole subject is as yet hidden in darkness'.

                                    — C. R. Darwin
                                    J. Proc. Linn. Soc. (Botany) **6**, 77-96

1889   ' . . . the part that amphigony has to play in nature . . . is
        not only important, but is of the very highest imaginable
        importance.'

                                    — A. Weismann
                                    Essays on Heredity, p. 281

1932   ' . . . genetics has finally solved the age-old problem of the
        reason for the existence . . . of sexuality and sex . . .'

                                    — H. J. Muller
                                    Amer. Natur. **66**, 118-38

1975   'The main work of providing a workable theoretical structure
        for understanding the enormous diversity of life cycles remains
        to be done.'

                                    — G. C. Williams
                                    Sex and Evolution, p. 119

1976   'One is left with the feeling that some essential feature of the
        situation is being overlooked.'

                                    — J. Maynard Smith
                                    J. Theoret. Biol. **63**, 245-58

CONTENTS

Contents

For my father

FIGURES

TABLES

Tables

PREFACE AND ACKNOWLEDGEMENTS

Three years ago I read George Williams' book on the evolution of sex. I found it both stimulating and exasperating: stimulating, because it revealed to me vast problems at the heart of biology whose existence I had before only dimly suspected; exasperating, because it provided a beginning and a middle without an end – and if Williams himself despaired of bringing his arguments to a successful conclusion, surely they must rest on a fallacy? After sketching out what seemed to me a sounder basis for theory, I later found a very similar opinion expressed in Michael Ghiselin's account of sexuality. But Ghiselin's book provided an end without beginning or middle; and his impressive edifice of facts and interpretations, which will one day receive the recognition it deserves, is flawed throughout by a failure to reduce his arguments to the formal propositions of population genetics (a failure, I should add, that he would not only admit, but vigorously defend). A little later John Maynard Smith's summary of his thinking on the subject appeared, and proved to be as polished and ingenious as one had expected, and as inconclusive as one had feared. It seemed to me that a major contribution could be made by someone who had not the mathematical skills of Maynard Smith nor the perceptiveness of Williams nor the biological insight of Ghiselin, but owned enough of all three of these to borrow the rest. So I wrote this book, in large part as a response to and a comment on the thinking of these three men. It would not have been completed without their criticisms of my early attempts to grapple with the problem: too just to permit complacency, too amiable to induce despondency.

I wrote the book as a member of the ecology group at McGill, and could not have chosen a better place. Since the group includes two convinced and (worse) logical anti-Darwinists I soon found that I could not get away with the deplorably loose reasoning of much evolutionary theory, and was forced to adopt a much more respectful attitude to facts and a much less respectful attitude to theories. The irritants, who must take most of the blame for the length of my middle chapter, are Rob Peters and Frank Rigler; the lubricants have been Don Kramer, Martin Lechowicz and Bill Leggett.

The colleagues who have commented on bits of manuscript or sketches of ideas are too numerous to mention individually, but Brian Charlesworth, Ric Charnov and Paul Handford have been particularly influential in moulding my views on genetic systems for better or worse.

Finally, I must mention my wife and children, since they have insisted. Without their close and constant support this book could not have been completed in less than half the time. But perhaps without them it would not have been worth completing at all.

Graham Bell
Montreal

1 THE PARADOX OF SEXUALITY

Introduction

Sex is the queen of problems in evolutionary biology. Perhaps no other natural phenomenon has aroused so much interest; certainly none has sowed as much confusion. The insights of Darwin and Mendel, which have illuminated so many mysteries, have so far failed to shed more than a dim and wavering light on the central mystery of sexuality, emphasizing its obscurity by its very isolation. No doubt the roots of this difficulty lie very deep. There are problems which are not excessively difficult to solve, but which are exceedingly difficult to see; not because they are obscure or trivial, but because they are painted so large in the foreground of the canvas that the eye glides over them, taking them as the givens which can be used to solve other and more important problems whilst not themselves requiring solution. After more than a century of Darwinism, during which time most of the conspicuous details in the background have yielded their secrets, we are too close to the canvas to appreciate that large areas in the foreground are still uncharted, still less explored. It seems that some of the most fundamental questions in evolutionary biology have scarcely ever been asked, and consequently still await an answer. Every student knows that homologous chromosomes usually segregate randomly during the division of the nucleus; no professor knows why. Every layman knows that all the familiar animals and plants have two sexes, but never more; few scientists have thought to ask, and none have succeeded in understanding, why there should not often be three or many sexes, as there are in some ciliates and fungi. The largest and least ignorable and most obdurate of these questions is, why sex? Or, to put this more technically, what is the functional significance of sexuality, which leads to its maintenance under natural selection in biological populations?

1.1 Sex, Gender and Reproduction

The earliest attempts to answer this question were doomed to failure, because the lack of a correct theory of genetics made it impossible to identify the consequences of sexuality with precision, or even to distinguish the concept of sexuality from that of gender on the one hand and from that of reproduction on the other. Although an understanding of the nature of sex hinges on an understanding of genetics, I shall have to assume an acquaintance with the elements of Mendelism and with the cytology of the cell cycle if I am to avoid an elaborate introduction to the topic; any interested reader who lacks this knowledge will find an excellent introductory account in Lewis and John (1972), and a more extensive treatment in White (1973). The Glossary on p. 501 gives short definitions of technical terms. Briefly, sex is a composite process in the course of which genomes are diversified by a type of nuclear division called *meiosis*, and by a type of nuclear fusion called *syngamy*, or *fertilization*. During meiosis genetic material may be exchanged between homologous chromosomes by the process of *crossing-over*, and this exchange is called *genetic recombination*. At the same time the total quantity of genetic material is reduced, since during meiosis the nucleus divides twice but the chromosomal material is replicated only once. In higher plants and animals the course of meiosis differs according to whether the meiotic products are macrogametes (ovules, eggs) or microgametes (pollen, sperm). During spermatogenesis, a diploid nucleus carrying two sets of chromosomes is replaced at the end of meiosis by four haploid nuclei, each of which carries a single set; during oogenesis the product of meiosis is a single haploid cell, the other three haploid nuclei having been eliminated as polar bodies. The main features of meiosis and its place in the life cycle are sketched in Figure 1.1.

The essential point is that meiosis shuffles the deck: the genome is rearranged by genetic recombination and by the random allocation of one member of each pair of homologous chromosomes to each of the meiotic products. The diploid nucleus which entered meiosis was made up of two haploid sets of chromosomes; the haploid nucleus or nuclei remaining at the end of the process are genetically unlike either of the original haploid sets. In a phrase, meiosis diversifies the haploid genome. The haploid products of meiosis may live for a long while independently of the diploid cells from which they were created, as do the gametophytes of many lower plants. In higher plants and animals the independent life of the haploid generation is usually brief or nonexistent, and meiosis is followed more or less immediately by syngamy, or fertilization. Syngamy involves the union of two haploid nuclei, and thus reconstitutes the diploid state. The diploid product of fusion, the zygote, has been formed from two meiotically diversified haploid nuclei, and is therefore itself different from any of the diploid nuclei from which it has descended. Syngamy thus diversifies the diploid genome. If we are concerned only with the diploid-to-diploid cycle, as we usually are when dealing with higher organisms, the obvious and immediate consequence of sexuality can be summarized in a single word: diversification.

'Sex' is therefore synonymous with 'mixis' (Ghiselin, 1974a): it is a process which changes the relationship between different elements (genes or linkage groups) of the genome, with or without the introduction of completely novel genetic material. Unfortunately, the derivation of the word 'sex' is from a Latin root meaning 'to cut', so that in common usage it means 'gender', a term which

---

Figure 1.1A (Plate): The Two Upper Photographs Show Two Small Aquatic Oligochaetes in the Family Naiadidae, *Chaetogaster* and *Stylaria*, both of which Normally Reproduce by Asexual Vegetative Fission (Section 4.24). Fission zones are clearly visible in *Chaetogaster*, at the upper right. *Hydra pseudooligactis*, in the lower photograph, also commonly reproduces by paratomical vegetative fission ('budding'), but is also often sexual (Section 4.4). The photograph shows a female carrying a fertilized egg.

Figure 1.1: Mixis and its Place in the Life Cycle of Some Common Freshwater Animals. All of the animals pictured here were isolated from a single tow sample taken in a macrophyte bed in a small lake in Southern Québec. (Photographs courtesy of Guy L'Heureux, McGill University, Canada.)

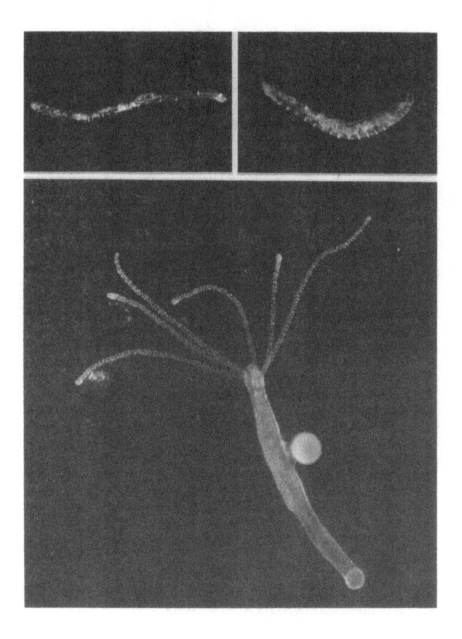

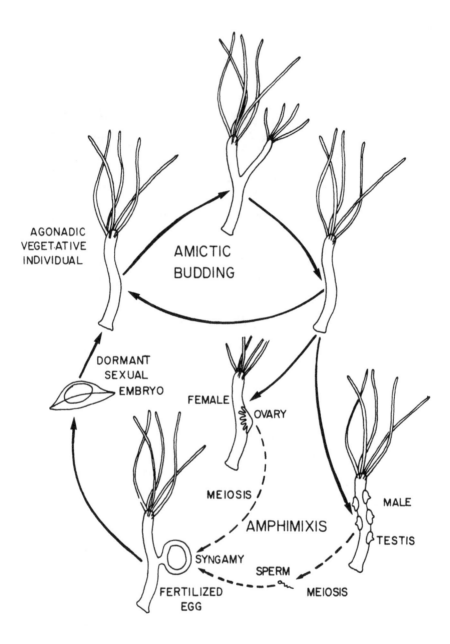

Figure 1.1B: The Life Cycle of *Hydra pseudooligactis*. This is a heterogonic cycle in which a period of asexual budding is followed by the production of dormant embryos by amphimixis. Some populations of *Hydra* are perennially asexual (Section 4.4).

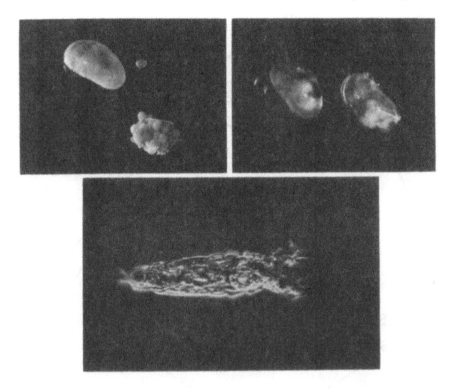

Figure 1.1C (Plate): Obligate Apomicts. The photograph at upper left shows the eggs, a juvenile and an adult of an unidentified parthenogenetic ostracod. Two adults of the parthenogenetic ostracod *Cypridopsis vidua* are shown at upper right (Section 4.30). The lower photograph shows a bdelloid rotifer (?*Philodina*), photographed in feeding position from above (Section 4.10).

is meaningful only in organisms where individuals are either male or female but not both or neither. In technical usage there is this wide difference between the two terms, that sexual processes may involve cells (or nuclei) either of the same or of different gender. Later, I shall review the curious but widespread processes which are sexual without involving any differentiation of gender, but for the moment I shall assume that any confusion on this point is a thing of the past, and shall use the more familiar term 'sex' to mean 'gender' in contexts where no ambiguity will be created.

More seriously, there still lingers some confusion about the distinction between 'sex' and 'reproduction', so that, as Ghiselin (1974a) points out, we refer to 'sexual reproduction' as though sex were a subset of reproduction. In fact, sex and reproduction are quite distinct processes: sex is a change in the state of cells or individuals, whilst reproduction is a change in their number. There is then the formal possibility of a separation of sex from reproduction

(a)

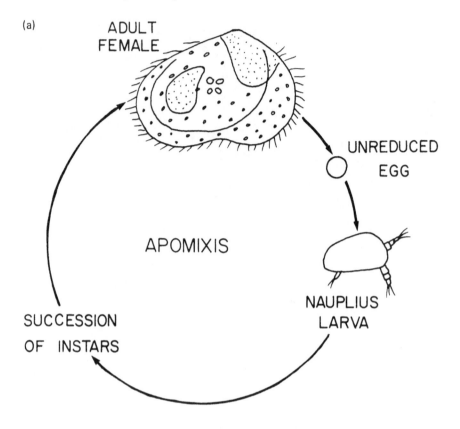

Figure 1.1D: The Life Cycles of *Cypridopsis vidua* (a) and *Philodina* (b). Sexuality of any sort is completely absent in these organisms (Sections 4.30 and 4.10).

which is realized in those unicellular protists where mixis can occur by an automictic process (roughly, self-fertilization) which reshuffles the genome without cell division, whilst reproduction occurs subsequently as the result of an equational (mitotic) nuclear division accompanied by cytokinesis. In multicellular organisms I presume that this process is not feasible, because it would result in a mosaic organism in which every cell would have a different chromosomal constitution. Instead, meiosis and syngamy are restricted to cells in the germ line, so that sex and reproduction go hand in hand. In this case it is legitimate to speak of sexual (or of asexual) reproduction, whilst bearing in mind that the sexual and the reproductive aspects of the process may have quite different consequences. This is nicely illustrated by a study of an unidentified acontiate sea-anemone by Smith and Lenhoff (1976). This anemone reproduced asexually by pedal laceration, the fragmentation of small pieces of tissue from the pedal disc; no sexual reproduction was found. The rate of laceration increased when food was

(b)

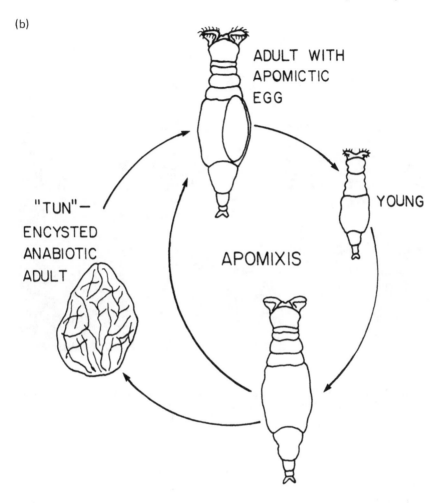

ADULT WITH
APOMICTIC
EGG

"TUN" −
ENCYSTED
ANABIOTIC
ADULT

APOMIXIS

YOUNG

withheld, so that a single large individual in effect propagated itself by cuttings in the same way that gardeners propagate plants, to produce a clone of genetically identical but very much smaller individuals. This was interpreted as being an appropriate response to low levels of food because a single large individual with a relatively small mouth and a relatively large ratio of weight to surface area had been replaced by a number of small individuals each with a relatively larger mouth and a smaller weight of tissue to support per unit area of oral surface. In effect, each anemone was responding to starvation by increasing the area of its mouth and thus increasing the efficiency of its feeding, accomplishing this by a process of asexual fragmentation. However, the lack of sexuality is trivial, since the same result would have followed if the anemone had produced genetically diverse progeny; the success of the behaviour depended not on the presence

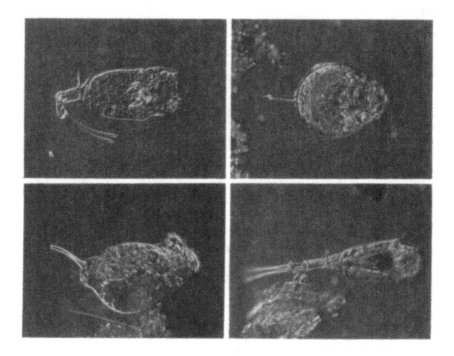

Figure 1.1E (Plate): The Four Monogonont Rotifers shown in these Photographs Pass Through Several or Many Generations of Apomictic Females Before Reproducing by Arrhenotoky. The four individuals shown here are all females (Section 4.10).

or absence of sexuality, but on reproduction. We must be careful, then, when evaluating the functional significance of sexual reproduction, to be sure that we are identifying the consequences of sexuality rather than those of reproduction.

Sex would be merely a curiosity, if it were not so widespread amongst animals and plants. It occurs in all major phyletic groups; it is by far the most important mode of reproduction in such large and diverse taxa as molluscs, arthropods, echinoderms and vertebrates. More than this, the nature of the sexual process differs only in its details between most of the members of these groups, and in particular the dance of the chromosomes during meiosis, despite its intricacy, is often almost identical in taxa which have only the most tenuous phylogenetic relationship. This combination of intricacy and uniformity makes it impossible to interpret sexuality other than as a highly adaptive character, precisely sculptured by selection to fulfill a function of central importance in the economy of a multitude of species. The nature of this function must be inferred from the consequences of sexuality; it must moreover be implicit in the most general features of sexual processes if it is to explain their very general dissemination.

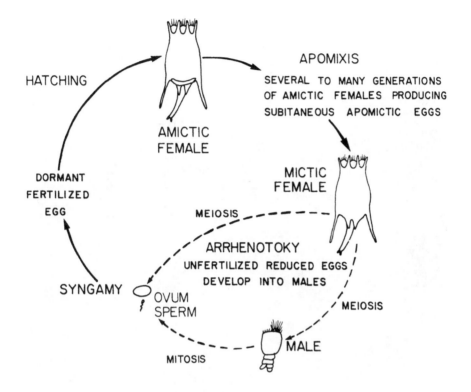

Figure 1.1F: The Life Cycle of a Monogonont Rotifer, *Brachionus*. This is a heterogonic cycle in which apomixis alternates with arrhenotoky, the males being haploid and morphologically reduced (Section 4.10).

The first step on the road towards a solution of the problem of sexuality is therefore to identify these consequences as precisely as possible.

1.2   Modes of Reproduction

We have set ourselves the task of inferring the functional significance of sexuality from its consequences, but it would be futile to attack on so broad a front without first spying out the land, and a necessary preliminary labour is to describe more precisely how sexuality modulates reproduction in different groups of organisms. Part of the reason for doing this is merely to define the technical terms that I shall be using later; I suspect that many people have steered clear of the study of life cycles and sexuality, despite its endless fascination, because of the almost impenetrable hedge of jargon with which it has

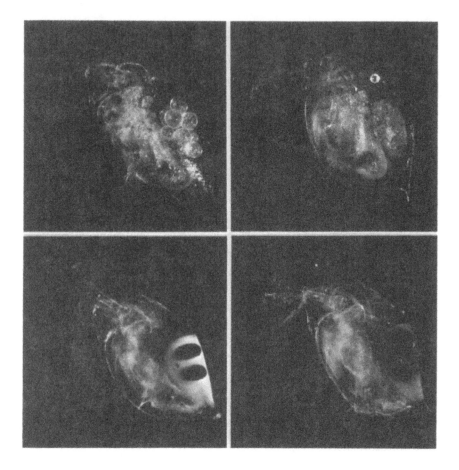

Figure 1.1G (Plate): Reproduction of *Daphnia pulex*. The animals in the upper photographs are producing subitaneous apomictic progeny. At upper left, the eggs have only recently been shed into the brood-pouch. The photograph at upper right shows a more advanced stage of embryogenesis. The animals in the lower photographs are producing an ephippium. At lower right the ephippium is in an early stage of development and the two eggs it contains are clearly visible. At lower right the wall of the brood-pouch has undergone further thickening and the ephippium is about to be released. The two embryos it contains will lie dormant in the sediment for a considerable period of time. In most cladocerans the ephippial eggs are produced by amphimixis, but this population has an entirely asexual life history, with both ephippial and subitaneous eggs produced by apomixis (Section 4.29).

been fenced round. The other part of the reason is more profound: by delineating a range of different sexual or asexual modes of reproduction, we greatly increase the power of the comparative tests that we can apply to competing theories of the functional significance of sexuality. The most important job of the evolutionary biologist is to interpret broad patterns of variation in nature. It is not enough just to think up likely stories. It is not even enough to think up stories in which

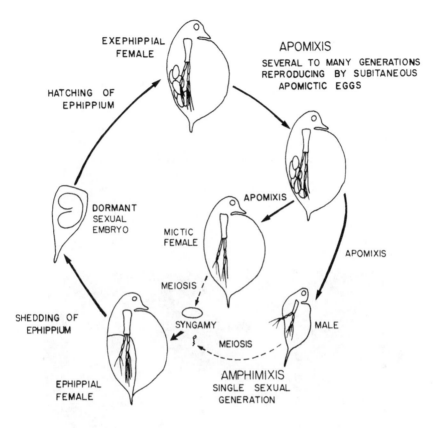

Figure 1.1H: The Usual Life-history of *Daphnia pulex*, a Heterogonic Cycle in which Apomixis Alternates with Amphimixis (Section 4.29).

an advantage for sexuality follows from unquestioned axioms with the inevitability of Greek tragedy. What we have to do is infinitely more difficult: it is to explain why a particular mode of reproduction has actually come to prevail in a given group of organisms, since we can be certain of what actually has occurred, but have only our untrammelled imagination to tell us of what might have occurred. Naturally, we must be sure that any hypotheses that we develop are consistent with what can be logically inferred about the action of natural selection in biological populations. But beyond this minimum requirement of plausibility, we shall prefer the most broadly applicable hypothesis, on the grounds that we wish to interpret as much as possible of the natural world as simply as we can. The first step is then to describe the range of phenomena that a successful theory will encompass.

Having advanced such an ambitious plan of campaign, I must now hurry to disavow it. An exhaustive enumeration and interpretation of modes of reproduction

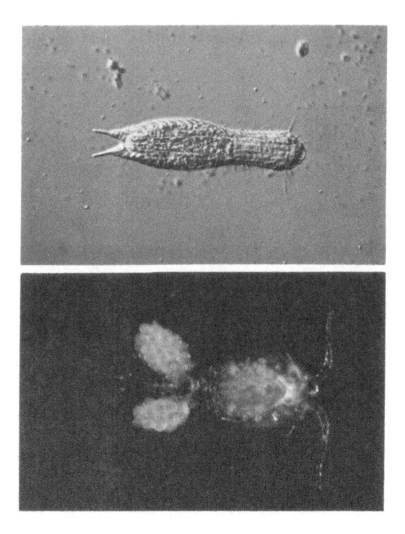

Figure 1.1I (Plate): Two Obligately Mictic Animals. The upper photograph shows the chaetonotoid gastrotrich *Lepidoderma* and the lower photograph the cyclopoid copepod *Cyclops*. Though both are mictic, they represent two fundamentally different types of mixis: the gastrotrich is automictic (Section 4.11), and the copepod amphimictic (Section 4.30). The next sequence of figures illustrates the cytology of automixis and amphimixis.

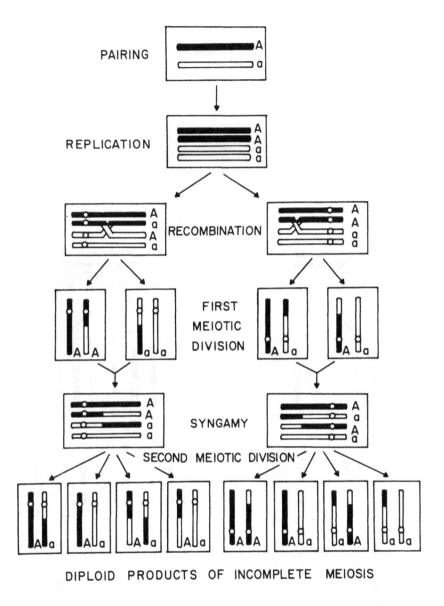

DIPLOID PRODUCTS OF INCOMPLETE MEIOSIS

Figure 1.1J: Automixis in Gastrotrichs may Involve either the Suppression of One of the Two Meiotic Divisions, or Autogamy (Section 4.11). This figure shows the results of suppressing the first meiotic division, by allowing its products to fuse before undergoing the second meiotic division. Note that the consequences of such an incomplete meiosis at any particular locus depend on whether crossing-over occurs between that locus and the centromere (postreduction, shown on the right-hand side of the diagram; the centromere is situated at the extreme left-hand end of the chromosomes) or not (prereduction, shown on the left-hand side of the diagram). Compare this figure with the description in Section 1.2.2.

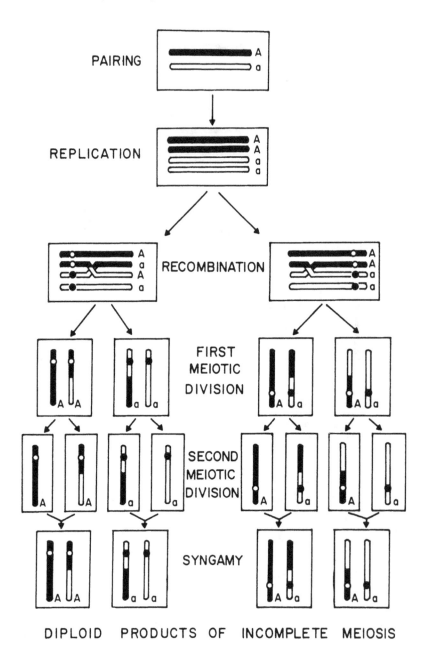

Figure 1.1K: The Results of an Automictic Suppression of the Second Meiotic Division, for prereduced and postreduced loci (Section 1.2.2).

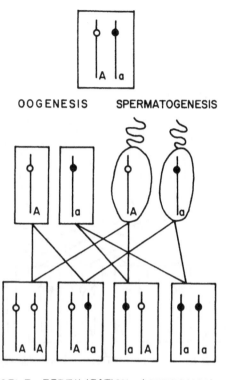

OOGENESIS   SPERMATOGENESIS

SELF - FERTILIZATION   (AUTOGAMY)

Figure 1.1L: Autogamy is an Automictic Process in which Syngamy Occurs Between Distinct Gametic Cells Rather than only Between Gametic Pronuclei (Section 1.2.2).

should be one of the central goals of evolutionary biology, but it is not yet attainable. The details are so complex and our knowledge so meagre that so Alexandrian a project must be set aside for the future. Rather, I shall adhere to the spirit of my opening section and confine myself to the most prominent features of the foreground, paying attention only to the commonest and most distinctive variants of the sexual theme. I shall also, at least for the most part, work only with comparative series amongst multicellular animals. The reason for this second restriction is quite simply that I do not know enough about protists or plants to be sure that I can review them dispassionately; a parallel treatment by a botanist would be of great interest. The scope of the problem has been narrowed, then, to an attempt to elucidate the functional significance of some of the commoner modes of reproduction amongst metazoans. These are exemplified in Figure 1.1 and summarized in Table 1.1, to which the rest of this section provides a gloss.

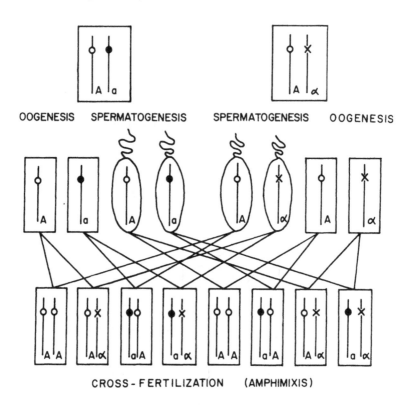

Figure 1.1M: Amphimixis: Syngamy Occurs Between Gametes Produced by Different Individuals, so that the Progeny Inherit the Characteristics of Both Parents (Section 1.2.2).

1.2.1 Reproduction without Sex

A fundamental distinction can be drawn between mixis and amixis. If the progeny are produced mitotically, without meiosis or syngamy, then the maternal genome is conserved and reproduction is said to be amictic. Within this category, which is erected on genetical grounds, I make a distinction on ecological grounds between colonial and clonal proliferation. In either case, the absence of any sexual process leads to the replication of identical progeny from a single ancestor: in colonial organisms these progeny are physically connected and necessarily live in close proximity to one another, whilst the progeny of clonal organisms are physically distinct and may become widely dispersed. Essentially the same distinction is made by botanists (see Gustaffson, 1946), who use the term 'agamospermy' for amictic reproduction which involves seed formation, and refer to propagation via suckers, runners or rhizomes as 'vegetative reproduction'. The distinction is not absolute: in plants, a shoot may be induced from a rhizome whose connection to the parental shoot is subsequently severed, and comparable

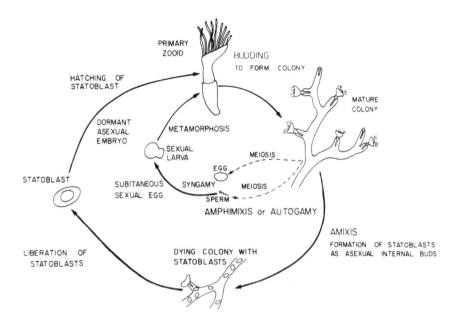

Figure 1.1N: Another Organism Found at the Same Site was the Bryozoan *Plumatella repens*, whose Life Cycle Displays Most of the Modes of Reproduction Illustrated Above (Section 4.17). A single primary zooid buds to form a colony of more or less autonomous zooids (see Section 1.2.1). These may reproduce by meiotically reduced gametes. If the gametes from one zooid fertilize those of another zooid in the same colony, this is equivalent to autogamy, since all zooids in the same colony have the same genotype; this may be usual among freshwater phylactolaemate bryozoans (see Section 4.17). On the other hand, if zooids from other colonies are fertilized, the mode of reproduction is amphimictic. When conditions become unfavourable, asexual internal buds called statoblasts are released and the colony degenerates; this mode of reproduction is functionally equivalent to apomixis. This animal thus displays both colonial and clonal amictic proliferation, autogamy and amphimixis in the same life cycle.

phenomena, such as the colony fragmentation of some bryozoans, are known from animals.

The least differentiated colonies are found amongst animals like corals and consist of no more than aggregations of genetically identical zooids with a continuous cytoplasmic connection. In more advanced colonial organisms such as scyphozoans there is a marked division of labour within the colony, different zooids undergoing the morphological differentiation appropriate for the performance of different functions. In undifferentiated colonies it is usual for all zooids to be capable of producing reproductive propagules, either mictically or amictically, whilst in more advanced forms this capacity is restricted to a special class of zooid. It is difficult to draw any sharp line between tissue-grade animals with a marked division of labour and the higher organ-grade metazoa. A jellyfish and a rabbit are equally the result of colonial proliferation, except that the grain of

Table 1.1: Modes of Reproduction in Animals.

Mode of Reproduction	Cytological Characteristics	Costliness	Diversity	Heterozygosity	Examples
1. AMIXIS	Nuclear divisions wholly equational; no syngamy.				
1:1. Colonial proliferation	Cell division adds to pre-existing colony or individual.				
- Growth and regeneration	Mitotically produced cell allocated to specific function in multicellular organism.				Universal in organ-grade metazoa.
- Vegetative proliferation in colony of differentiated zooids	Mitotically produced zooids allocated specifc function within highly coordinated colony with marked division of labour.				Scyphozoa
- Vegetative proliferation in colony of undifferentiated zooids.	Mitotically produced zooids function autonomously within colony of interconnected zooids.	Zero[1]	Identical	Conserved at very high level	Colonial Hydrozoa, Entoprocta Bryozoa, Pterobranchia.
1:2. Clonal Proliferation	Cell division creates new distinct individuals				
- Vegetative proliferation in non-colonial animals.	New individuals created by growth or fragmentation of body of parent;two major categories are architomy and paratomy.				Most 'lower' invertebrates also polychaetes, enteropneusts, echinoderms.
- Polyembryony	Embryo divides during early development to give rise to two or more identical individuals at term, e.g. by strobilation.				Scyphozoa, monogenean trematodes, loxosomatid entoprocts, cyclostomatous bryozoans.
- Apomixis (ameiotic thelytoky)	Unreduced eggs complete development without fertilization. May involve pseudogamy.				Digenean trematodes, bdelloid rotifers; some nematodes, tardigrades, coccids, cladocerans, aphids, cynipid wasps, teiid lizards.
2. PARTIAL MIXIS	Syngamy or meiosis restricted to one sex.				
- Hybridogenesis	Syngamy without meiosis in females: male of different species contributes a paternal genome which is expressed in the triploid offspring but excluded from their offspring by an unequal segregation during oocyte maturation.	Zero	Generated by paternal genome only	Very high	Teleost *Poeciliopsis*
- Arrhenotoky	Meiosis without syngamy in males: females produce reduced eggs which develop into diploid females if fertilized and into haploid males if not.	Varies[2]	Maximal in females	High in female progeny; males hemizygous.	Monogonont rotifers, most Hymenoptera; a few mites, thysanopterans, homopterans, beetles.
3. MIXIS	Nuclear divisions reductional; diploidy restored by syngamy.				
3:1. Automixis	Syngamy of nuclei derived from same zygote.				
3:1:1. Automictic (meiotic) thelytoky	Males and distinct gametic cells absent; reduction during oocyte maturation in females compensated for in some manner; syngamy involves nuclei of same gender; may involve pseudogamy.				
- Premeiotic	Endomitosis and subsequent nuclear fusion prior to meiosis 1.	Zero	Restricted insofar as syngamy is restricted	Reduced	Lumbriculids, planarians.
- Intrameiotic (type 1PB)	Meiosis 1 suppressed.	Zero		Reduced	Gastrotrichs, tardigrades,
- Intrameiotic (type 2PB)	Meiosis II suppressed.	Zero		Reduced	coccids.
- Postmeiotic	Meiosis normal; first two cleavage nuclei fuse.	Zero		Lost	Psocopterans, coccids.
3:1:2. Autogamy (Self-fertilization)	Syngamy of meiotically produced gametes of different gender in same simultaneously hermaphroditic individual.	Low	Restricted	Reduced	Rhabdocoel turbellarians, cestodes, hrvozoans, notostracans, basommatophoran pulmonates.
3:1:3. Minor systems in which both sexes occur	Males present; diploidy restored by automixis in at least one sex.				
- Deuterotoky	Males and females develop from reduced eggs in which diploidy is restored by fusion of first two cleavage nuclei.	Zero	Restricted	Lost	Coccids, mites.
- Diploid Arrhenotoky	Females develop from fertilized eggs, males from unfertilized eggs in which diploidy is restored by fusion of first two cleavage nuclei.	Maximal for ♀ progeny, zero for ♂.	Maximal in ♀	Lost in ♂	Coccids.

3:2. Amphimixis	Syngamy of gametes of different gender derived from different zygotes.

Amphimicts can be classified according to the nature of their gametes. There are four major criteria.

(1) THE SITE OF MEIOSIS

- Haplonts with a zygotic meiosis, the gametes produced by mitosis.
- Haplodiplonts with a sporic meiosis.
- Diplonts with a gametic meiosis, the gametes produced by meiosis.

(2) THE POLARITY OF GAMETE GENDER

- Bipolar with two genders, usually recognisable as male and female.
- Multipolar with several to many genders.

(3) THE DEGREE OF GAMETE DIMORPHISM

- Isogametic, all gametes being of the same size.
- Pseudoanisogametic, with a range of gamete sizes but no correlation between the size and the gender of a gamete.
- Anisogametic, with gametes of the same size having the same gender.
- Oogametic, with a massive immotile female macrogamete and a minute motile male microgamete.

(4) THE GENDER OF INDIVIDUAL ORGANISMS

- Gonochoric, with distinct male and female individuals.
- Hermaphroditic, with male and female function united in a single individual.

Thus, *Homo sapiens* would be classified as a diplontic, bipolar, oogametic, gonochoric amphimict.

---

1. For a simpler classification, see Williams (1975); for plants, see Gustaffson (1946) or Stebbins (1950).
2. There may be a cost of sex to the parent in the case of polyembryony; see text.
3. For the cost of arrhenotoky, see Section 1.5.

development, so to speak, is different: in the jellyfish the unit of functional organization is a group of zooids, each of which retains some trace of autonomy, whereas in the rabbit it is an aggregate of cells, each of which is wholly dependent on the others. Ordinary somatic growth must then be allowed as the limiting case of colonial proliferation, and there is a sense in which all life histories have an asexual phase. If this idea seems strange, it is only because of the ease with which sex and reproduction are confounded; certainly, no-one would normally speak of growth as being a process of reproduction. The corollary of this argument is that the proliferation of zooids in undifferentiated colonies, whilst certainly amictic, is only dubiously reproductive. At best, it may be very misleading to compare the production of zooids by a coral with the production of eggs by a fish; there are not only genetic but also ecological differences between the two processes, and any naive comparison runs the risk of failing to distinguish between two quite different sets of consequences.

Unequivocally independent progeny are produced amictically by what I have referred to as 'clonal' proliferation, whilst conscious that the word 'clone' is often used to describe a set of genetically uniform descendents whether or not these remain in physical connection. Two clonally produced rotifers are just as alike genetically as two colonially proliferated coral zooids: the difference between the two lies not in genetics but in ecology. The two rotifers may drift wide apart and grow up in very different circumstances, and their fates are to that extent largely independent, whilst the two coral zooids are indissolubly bound to one another, experience very much the same conditions of life, and can expect to meet similar ends. The distinction is crucial because our eventual goal is to use the comparative method in order to distinguish between conflicting interpretations of sexuality. Sexual organisms almost invariably produce offspring which live for the greater part of their lives independently of their parents, but if we wish to contrast them with asexual organisms we must be careful to compare like with like. The appropriate comparison lies between sexual and clonal alternatives. Naturally, organisms are not confined to any single sexual or reproductive habit: colonial organisms also produce mictic or amictic propagules, and at this stage in their life history can legitimately be compared with the sexual organisms which are themselves highly integrated colonies of mitotically generated cells. My caution is merely that, if we are to use colonial organisms as a source of comparative material, we must be careful to dissect the consequences of sexuality from those of reproduction.

The simplest mode of clonal proliferation is the fragmentation of the body into two or more equal or unequal parts, each of which develops into a new individual. The regeneration of new individuals from accidentally detached fragments is a casual method of reproduction in most groups of invertebrates (the main exceptions being those with a hard external covering, like arthropods and molluscs), but many species regularly reproduce in this fashion, possessing special devices for the dissection of the parental body. For example, the budding of solitary hydroids is essentially the same process as that involved in the production of new zooids in colonial forms, except that the connection between the parent and its budded offspring is eventually severed. A similar process occurs in annelids, where the bud may develop into a morphologically complete new worm before the connection with the parent is severed. This type of fission is called *paratomy*. In other animals, such as flatworms and starfish, both (or all) of the fragments resulting from vegetative fission become separated before regeneration occurs, so that the individual products of fission are incomplete and must pass through some sort of developmental sequence before regaining the final adult state. This type of fission is called *architomy*.

A special category of polyembryony must be erected for the case in which the individual undergoing vegetative fission is an embryo. Straightforward vegetative fission leads to the production of a number of complete animals or regenerating fragments, of which one can be identified (perhaps arbitrarily) as the parent and the rest as offspring; the offspring are genetically identical to one another and to their parent. Polyembryony, on the other hand, involves the fission of an embryo which may itself have been sexually produced; thus, whilst all the products of fission are identical to one another, they may not be identical with their parent.

Vegetative fission may be equal or unequal. If equal, it results in two regenerating fragments of the same size, neither of which can be unequivocally identified as being parental. If unequal, one fragment becomes smaller and smaller relative to the other and its status as an offspring becomes less and less debatable, until at the limit the offspring is only a single cell produced mitotically by a multicellular parent. This category has been called apomixis ('amictogametic' reproduction of Williams, 1975): apomictic mothers produce daughters who inherit a complete unrecombined maternal genome, and who go through a complete development from egg to adult. It is a category which is crucial for comparative studies, since it supplies a range of material which can be compared directly with sexual organisms. Sexual individuals which develop from fertilized mictic eggs and apomictic individuals which develop from unfertilized apomictic eggs have their method of reproduction in common, and any general differences between the two categories can be ascribed confidently to the difference in sexuality.

Haploid apomixis, the development of haploid embryos without fertilization, is known both from higher plants and from higher animals, but only as a laboratory curiosity (Gustaffson, 1946; Stebbins, 1950).

1.2.2 Sexual Reproduction

Mictic and amictic modes of reproduction are in principle perfectly separable; in practice there is a debatable ground between them in which both meiosis and syngamy occur, but only in one of the two sexes. Although systems of this sort are undoubtedly mictic, I find it convenient to put them into a separate category of 'partial mixis'. There are two systems of this sort; one is exceedingly rare whilst the other is very common.

Hybridogenesis has been reported only from a single genus of freshwater fish, *Poeciliopsis* (Schultz, 1969) and from the anuran *Rana esculenta* (Uzzell and Berger, 1975). Females do not undergo meiosis, but mate with males of another species, incorporating a haploid sperm genome into each of their unreduced diploid ova. Their daughters thus inherit both a haploid paternal genome and an unreduced, unrecombined maternal genome. The paternal genome is excluded from the germ line of the daughters by an unequal segregation during gametogenesis. Hybridogenetic fish are thus sexual parasites which utilize a paternal genome without transmitting it. This mechanism is reminiscent of the much more widespread practice of gynogenesis, or pseudogamy. Gynogenetic females produce unreduced eggs which will not develop unless the egg membrane is penetrated by a sperm; however, the sperm then degenerates and its genome is not expressed in the progeny. Hybridogenetic and gynogenetic systems are thoroughly amictic in that the maternal genome is strictly conserved; however, the requirement for fertilization, or pseudofertilization, by males of a related sexual species represents a rather severe ecological constraint.

Arrhenotoky (strictly, haploid arrhenotoky; 'haplodiploidy') is by contrast a very common mode of reproduction, especially amongst rotifers and hymenopterans. Females undergo meiosis and produce haploid eggs which may then develop either with or without syngamy. If the egg is fertilized it develops into a female, and if not it develops into a male. Males are thus hemizygous at all loci, and there is no recombination during spermatogenesis.

The remaining possibilities which are realized in nature are characterized by a mixis which includes both meiósis and syngamy in both sexes. The classification of mictic systems hinges on the ancestry of the reduced gametes or gametic pronuclei which fuse to form the zygote. If the two haploid nuclei involved were generated meiotically from the same diploid genome then the process is automictic; if this is not the case, the process is amphimictic.

Amongst automictic systems, I shall make a distinction between those in which distinct gametes are formed, and those in which they are not. In the latter case nuclear division occurs without cytokinesis, and the subsequent syngamy takes place between the gametic pronuclei rather than between separate gametic cells. I shall refer to this process simply as *automixis*. In simultaneous hermaphrodites distinct gametes are formed, and in such organisms self-fertilization may occur; I shall refer to this process as *autogamy*, or *selfing*. There seems to be a need to coin a simple term for automictic processes which are not autogamous, but I shall avoid creating neologisms and instead rely on context to differentiate between the broader and narrower senses in which I shall employ the term 'automixis'.

In the narrower sense of automixis, then, there are neither males nor microgametes, and there is therefore no differentiation of gender either in the haploid or in the diploid generation. The reconstitution of the diploid state from the meiotically reduced oocytes is instead achieved by the syngamy of nuclei of the same gender; or, in plainer language, once an automictic female has produced haploid ova, she must arrive at some internal solution for doubling the chromosome number, without the intervention of males. There is a bewildering variety of chromosome-doubling devices, which by combining the more detailed accounts of other authors (especially Beatty, 1967; White, 1973) I shall classify under three heads and illustrate by reference to a locus at which the maternal genome bears the alleles $A1$ and $A2$. In the first place, the most straightforward – although by no means the most common – solution is to double the chromosome number before meiosis by means of an endomitosis. The result is a tetraploid nucleus, containing two identical diploid chromosome complements, which can then be reduced to diploidy by a normal meiosis. If only sister chromatids pair there will be no segregation, and the progeny will be identical to themselves and to their mother; this is the most likely outcome, and the process is functionally equivalent to apomixis. On the other hand, if quadrivalents are formed during meiotic prophase, segregation can occur, and the $A1A1$, $A1A2$ and $A2A2$ genotypes will be represented amongst the progeny in some complex ratio. The second solution is to interfere with the reduction division itself, by suppressing either the first or the second meiotic divisions. These two possibilities are illustrated in Figures 1.1J and 1.1K. Asher (1970) has shown that suppression of meiosis I is equivalent to centric fusion of the reduced nuclei, whilst suppression of meiosis II is equivalent to terminal fusion. If the first division (and thus the first polar body) is suppressed, all the strands of the tetrad remain in the ovum. These strands have the genotypes $A1A1$ and $A2A2$ and will be reduced to $A1$ and $A2$ by the second meiotic division. If they are not separated by crossing-over, all offspring will be

A1A2; a single chiasma occurring between them will give a mixture of *A1A1*, *A1A2* and *A2A2* genotypes in the offspring, in the ratio 1:2:1. Suppressing the second meiotic division is a little more complicated. The first division of a normal meiosis is reductional for loci which lie in the region between the centromere and the most proximal chiasma, but equational for loci between the first and second chiasmata: loci which separate reductionally during the first division separate equationally during the second (prereduction) and vice versa (post-reduction). If the locus in question is prereduced, the secondary oocytes will be either *A1A1* or *A2A2*, and if the second polar body is suppressed the offspring will comprise equal numbers of *A1A1* and *A2A2* homozygotes. On the other hand, if the locus is postreduced, all the progeny will be *A1A2* heterozygotes. Disregarding the details, the essential point is that any intrameiotic device for doubling the chromosome number involves a substantial probability that some of the progeny of heterozygous mothers will be homozygous. Finally, a third possibility is that the oocytes may go through a normal two-step meiosis to end up as haploid ova; the only avenue then open to them is to divide equationally to produce two identical haploid nuclei which then fuse before or during the first cleavage division. The immediate consequence of this postmeiotic doubling is that all progeny are homozygous at all loci. Naturally, the progeny will not necessarily be identical since, if a locus is heterozygous in the maternal genome, we expect half her progeny to be homozygous for one allele and half for the other; the progeny have already been diversified meiotically even though they are not further diversified by syngamy. The homozygous lines formed in this way will thereafter conserve the maternal genome in each generation, and so behave like homozygous apomictic clones. In short, automixis increases homozygosity, and in extreme cases may result very quickly in the construction of entirely homozygous genomes. This is only a sketch of possible automictic mechanisms, however, and I shall later describe some ingenious devices by which even wholly automictic organisms manage to maintain a high level of heterozygosity.

Self-fertilized organisms can be distinguished from automicts proper on the grounds that they produce both eggs and sperm, so that there is a differentiation of gender at the gametic level (Figure 1.1L). Self-fertilization can be regarded as another automictic chromosome-doubling device, with the same expected increase in homozygosity, except that it retains more flexibility than strictly automictic processes which do not involve the formation of distinct gametes. The Knight-Darwin rule (Knight, 1799; Darwin, 1858) states that no organism self-fertilizes in perpetuity, and whilst exceptions may occur it remains a very powerful generalization: self-fertilization permits occasional outcrossing whereas automixis proper does not.

Both automixis and self-fertilization are rather common and taxonomically widespread modes of reproduction. There remain a few rare and aberrant processes which are most conveniently classified as automictic, and I shall give two examples which differ from the more common processes in that the organisms involved (a few species of mites and coccids) are gonochoric, individuals being either male or female. Deuterotoky involves the development of both males and females from reduced eggs, diploidy being restored postmeiotically by the fusion of the first two cleavage nuclei; in diploid arrhenotoky only the males are produced in this way, the females developing in the usual manner from fertilized eggs. The cytological details have been worked out by Nur (1972).

The term 'parthenogenesis' (coined by von Siebold, 1871) has some potential for causing confusion, since it has on occasion been used as a synonym for asexual reproduction. Its literal meaning is 'virgin reproduction', that is, reproduction in the absence of males. However, the presence of distinctly male individuals is not a prerequisite for mixis, since genomes can be reshuffled automictically in organisms where there is no differentiation of gender even at the level of the gametes. I shall use the term to include all those breeding systems in which the eggs produced by a given individual do not require fertilization by gametes produced by another individual: all kinds of apomictic reproduction, together with arrhenotoky, automixis proper (meiotic thelytoky) and obligate self-fertilization. The two major categories of parthenogenesis are arrhenotoky and thelytoky; in turn, the two major categories of thelytoky are ameiotic thelytoky (apomixis) and meiotic thelytoky (automixis).

A final piece of terminology: the word 'gamete' comes from a Greek root meaning 'marriage', and a gamete is an entity (usually a cell) which undergoes syngamy. The unreduced eggs produced by parthenogenetic females are not gametes. However, if the parthenogenesis is automictic the behaviour of the nuclei will have been gametic at some point, and for this reason they are usually spoken of as gametic pronuclei.

Automixis is one major category of mixis; the other is *amphimixis*. Amphimixis involves a syngamy between gametes produced by different parents; roughly speaking, amphimictic organisms are cross-fertilized, whereas automicts are self-fertilized (Figures 1.1L and 1.1M). I do not think that an hierarchical classification of amphimixis is practicable, but we can characterize amphimictic systems according to the nature and derivation of the entities involved in fertilization. Since any sexual process involves a reduction of ploidy through meiosis and its restitution through syngamy, perhaps the most fundamental dichotomy we can recognize refers to the ploidy of the gamete-producing individual. If it is haploid, the gametes are produced by mitosis, and meiosis occurs at some time during or after the germination of the zygote, to give rise to gametophytes; if it is diploid, the gametes are produced directly by meiosis. To put this in another way, we can classify amphimictic systems according to the timing of meiosis during the life cycle: it may occur at zygote germination and give rise immediately to haploid gametophytes, or it may occur after the zygote has developed into a multicellular diploid vegetative individual, in which case its products are either spores which germinate into gametophytes, or gametes. Amphimictic systems, whether haplontic, haplodiplontic or diplontic, may also differ with respect to the nature of the gametes: there may be only two gamete genders, or there may be several; there may be only one size-class of gametes, or there may be two, or there may be a range of gamete sizes; and there may, or may not, be a correlation between the size of a gamete and its gender. The site of meiosis and the gender and morphology of the gametes are useful in classifying the sexuality of protists and thallophytes, and their functional significance is discussed in Chapter 5. Among multicellular animals, however, all amphimicts are bipolar oogametic diplonts – that is, all are forms in which the only two gametic types are a massive immotile macrogamete and a much smaller, typically motile microgamete; in which syngamy occurs only between and never within these two types, which are therefore gender categories as well as morphological categories; and in which the gametes are produced directly from a diploid individual by meiosis. If we wish, we can extend the classification by considering the nature of the gamete-bearing individuals as well as that of the gametes. For instance, the gametes themselves invariably undergo fusion, and in organisms which release their gametes directly into the external medium are the only entities which undergo a sexual fusion of any kind; in other organisms, however, gamete fusion is preceded by some other variety of sexual fusion whose effect is to regulate and restrict the possibilities of syngamy, such as gamontic or gametangial fusion, or somatic copulation. We can also recognise a difference between those organisms in which each individual produces gametes of both genders, either simultaneously or at different times during its life, and those in which each individual is committed to the production of gametes of a single gender. These extensions of the classication, however, lead to topics in mating behaviour and gender allocation that I shall not follow up in this essay.

Williams (1975, pp. 114-17) proposes a much simpler classification of reproductive modes, which may be preferred by some, but which I have not been able to accept in full. He draws a fundamental distinction between recombinational and nonrecombinational modes of reproduction. This seems to obscure the essential role of syngamy in diversifying diploid genomes, which can occur without any recombination, and I have therefore followed Ghiselin (1974a) in drawing the fundamental distinction between mictic and amictic propagation. For the same reason, I have rejected the suggestion of Mahendra and Sharma (1955) that reproduction is fundamentally either meiotic or ameiotic. Williams proceeds to make more or less the same distinction between colonial and clonal proliferation that I have done, although without stating the reasons for doing so. He also considers it important to distinguish between 'primitive' and 'derived' conditions, presumably because of the historical nature of many of his arguments, but I have rejected this approach on the grounds that it introduces an unnecessary phylogenetic speculation. The same objection can be levelled against the scheme proposed by Winkler (1920), in which amixis is the primitive and apomixis the derived asexual state. Finally, Williams terms automixis 'degenerate sexuality', but this seems to prejudge its evolutionary significance. Indeed, I think that the most serious criticism of Williams' classification is that it seems to reflect not so much the nature of the processes involved as the author's opinions about the way in which they have evolved; his classification thus hinges on a particular interpretation of sexual processes, and cannot very well be used to discuss other interpretations.

The prevalence of amphimixis among multicellular animals is the major problem that I shall attempt to solve. The competence of a theory to account for the occurrence of amphimixis will be assessed by comparing the taxonomic and ecological distribution of amphimicts with the distribution that would be expected if the theory were true. As I have already stressed, the power of the comparative method depends on the appropriateness with which the categories being compared have been chosen; the most appropriate comparison lies between categories which are equivalent except with respect to the character under consideration. Since we wish to find out why progeny should be diversified genetically, we shall prefer to contrast amphimixis with apomixis. With this overall strategy, however, tactical considerations often make it desirable to broaden the range of taxa available for comparison. Lumping other forms of clonal reproduction in with apomixis involves, I think, only a slight relaxation of rigour, but much more caution must be exercised if we are to compare the vegetative proliferation of zooids within colonies to any sexual process. There is also a temptation, to which I shall frequently succumb, to compare parthenogenesis with nonparthenogenetic modes of reproduction. This is not strictly a comparison of asexuality with mixis; at best, it compares incompletely mictic systems in which genomes are reshuffled, if at all, without the addition of novel genetic material with completely mictic systems in which nonmaternal alleles are introduced into progeny genomes through outcrossing. This less direct comparison is forced on us in the many instances in which the occurrence of a meiosis during oogenesis is questionable or has not been investigated.

Finally, I should again emphasize that animals are not necessarily restricted to any single mode of reproduction. Cladocerans, for example, are well-known for their habit of sandwiching a round of amphimixis between sequences of apomictic reproduction; in plankton rotifers, apomixis alternates with arrhenotoky; and most of the other possible combinations of the commoner modes of reproduction are known from one group or another. This 'cyclical parthenogenesis', as it has been called ('intermittent mixis' would be a more precise term), is of great interest because it allows us to compare the biology of asexual with that of sexual reproduction within a single line of descent. Indeed, Williams (1975) opens his book on sex with the statement that, since mixis in organisms with cyclical parthenogenesis occurs as a response to changed conditions, the functional significance of mixis is not in doubt, and the only role of theory is to harness this observation to some plausible Mendelian framework. I am inclined to believe Bacon's aphorism: 'If a man will begin with certainties, he shall end in doubts; but if he is content to begin with doubts, he shall end in certainties.'

1.3 Population Consequences of Sexuality

1.3.1 Individuals and Populations

Darwinism and Mendelism have made such a successful marriage that we forget what uncomfortable bedfellows they once were. Until Fisher's classic paper of 1918, it seemed to many people, and especially to the biometrical geneticists of Pearson's school, that the particulate, uncontaminated genes of Mendel's theory were an insuperable obstacle to a belief in adaptive change through natural selection. After all, if the genes themselves did not change during transmission, whilst genotypes were put into Hardy-Weinberg proportions by a single generation of random mating, how could any population undergo permanent, heritable change? The breach between the biometricians and the Mendelians (which made the first half dozen or so volumes of *Biometrika* such lively reading) was healed by the 1920s, and by the end of the 1930s the Darwinian stem had been grafted onto the Mendelian rootstock to form a single synthetic theory of evolution. But deep wounds leave lingering scars; and these old quarrels are not of merely historical interest, since their influence can still be felt today.

Nowhere has this influence been stronger than in the study of breeding systems. Any characteristic of individuals will, of course, have an effect on the aggregate of individuals that we call the population since, when this character is selected, the composition and thus the properties of the population will change. Breeding systems occupy a unique position, however, in that any change in the mode of reproduction practised by individuals will effect a qualitative change in the nature of the population. Asexual organisms can be grouped into conspecific populations only on the basis of an ultimately arbitrary degree of morphological resemblance; each clone represents a separate, isolated line of descent, and genotype frequencies reflect only the past history of the population. Populations of amphimictic organisms are quite different: because individuals combine their genes through syngamy with the genes of other individuals, the species can be defined objectively by the criterion of interfertility. Each individual stands on many lines of descent and may be the ancestor to as many more; and genotype frequencies move towards equilibria whose positions are governed by Mendelian rules rather than by history. Sexual reproduction thus knits together the population into a single entity, to the point where we can speak in terms of abstractions such as 'gene pools' and 'stores of variability'. The fate of a mutant gene which arises in an asexual clone is bound up with the fate of the clone itself, since in the absence of mixis it cannot be transmitted to other clones. It might be that the effect of the mutant allele on fitness depends on the presence or absence of alleles at other loci, so that whether or not it is favourably selected depends to a large extent on the nature of the genome in which it arises. If it should originally find itself in uncongenial company, it has no recourse so long as reproduction is wholly amictic. But the effect of mixis, and especially of amphimixis, is to create a new setting for the mutant allele in every genome in which it occurs, and some of these settings may happen to show it off to great advantage. Amphimixis liberates a newly arisen allele from a single line of descent and disperses it amongst the innumerable different genomes which arise in every generation through meiosis and syngamy. Some of the genomes in which it is included may also happen to include newly arisen mutations at other loci, so that combinations of different rare alleles which could be built up only very slowly by recurrent mutation in a clonal organism appear very quickly in amphimictic populations as haploid genomes generated by meiosis in unrelated individuals are brought into combination by syngamy. In a word, genetic variation in sexual populations is protean, whereas in asexual populations it is frozen into the pre-existing structure of the clones.

This effect of sexuality on population structure has fascinated geneticists almost to the exclusion of any interest in the consequences of sexual reproduction for the fitness of the individual. The result was that in the century after Darwin an immensely sophisticated Mendelian theory of sexuality was developed, whilst the Darwinian consequences of sexuality remained utterly obscure. The eightieth anniversary of the publication of the *Origin of Species* saw the publication of a book by C. D. Darlington (1939) called *The Evolution of Genetic Systems*: as erudite as it was influential, Darlington's text nevertheless makes no important theoretical advance on Weismann's essay of 1886, so far as the relationship between sexuality and natural selection is concerned. The Mendelian sowing for which Weismann prepared the ground and which Darlington harvested proved so fruitful that the necessity for a parallel Darwinian treatment was forgotten; it was almost as though the theory of natural selection yet remained to be discovered. It is only in the last decade or so that any sustained attempts have been made to repair this omission, by a group of biologists amongst whom J. Maynard Smith, G. C. Williams and M. T. Ghiselin have been especially prominent.

The idea that a character may be selected because it is beneficial to the population or to the species, even though it is harmful to the individuals which bear it, can be traced back to antiquity. The modern history of the idea began in 1962, when V. C. Wynne-Edwards published his big book on social behaviour, in which he made a case for interpreting many aspects of the breeding biology of vertebrates in terms of their significance to the species rather than to the individual. Once the theory of 'group selection' as it came to be known, was made explicit, it attracted a vocal opposition (for example, the exchanges in *Nature*, 14 March, 1964), which culminated in the publication of Williams' (1966) book on *Adaptation and Natural Selection*. Williams attacked not only the explicit statement of group selection made by Wynne-Edwards, but also the implicitly non-Darwinian positions taken by many other authors. The attack was a devastating one, because the positions against which it was launched turned out to be indefensible. Imagine a set of populations, in each of which there are certain individuals who behave 'altruistically' — using the word in its widest sense to include any behaviour which benefits the population whilst harming the individual — and others who behave selfishly. If the altruists are to be favourably selected, we require that the selective elimination of whole populations should be so rapid as to overpower the selective elimination of individuals within populations. This is implausible, primarily because the variance of the mean values of a character between populations is usually much less than the variance of the character within populations, and because populations are usually much longer-lived than individuals. It turns out to be so difficult to conceive situations in which group selection is a more powerful influence on gene frequencies than individual selection that since 1966 there have been few explicit defenses of the position that group selection is primarily responsible for any widespread features of biological organization – although the implicit, and presumably ignorant, assumption that breeding systems are designed for the good of the species flourishes almost unchecked. (Readers wishing to follow the later stages of the debate, which has concerned a largely unsuccessful search for genetically plausible mechanisms of group selection, should consult Levins, 1970; Eshel, 1972; Boorman and Levitt, 1973; Charnov and Krebs, 1975; Gadgil, 1975; Gilpin, 1975; Levin and Kilmer, 1975; Wilson, 1975, 1977; Maynard Smith, 1976a; and Bell, 1978a. Williams, 1968, has edited a useful selection of the earlier papers, and the book by Ghiselin, 1974a, is in large part an attack on the use of group selection in theories of sexual phenomena. A book defending group selection has been published very recently by D. S. Wilson, 1980.) It was the very success of this attack which led population biologists to realize how embarrassing sex is. Most supposedly altruistic behaviours were quickly found either to have concealed advantages for the individual, or else to be directed towards the welfare of closely related individuals. But sex appeared to fit into neither of these categories: if it permits the rapid mobilization of genetic variation, then this may be a matter of vital concern for the population, but does not in itself concern the individual. Evolutionary biologists thus found themselves on the horns of a dilemma: either the apparently unsatisfactory hypothesis of group selection was indeed an adequate explanation for the maintenance of sexuality, or else a quite different hypothesis framed in terms of natural selection must be sought.

1.3.2 Genotypic Diversity

Sex is a complicated process, and it entails a complicated series of consequences. The almost endless variety of sexual mechanisms and the way in which each is tailored to genetic and ecological circumstances is a rich and still almost unexplored field for speculation and experiment, but for the present it does not concern us. The details of the process can be interpreted only when the primary functional significance of sexuality has been identified, and this must reside in the most general consequences of the sexual habit. There are three such consequences, and only three.

The most obvious of these is the constant creation of new genotypes in a sexual population. Since we know that sexual populations are exceedingly diverse, we expect parthenogenetic populations to be much more uniform. Indeed, if a constant, uniform environment were to be seeded with a variety of clones, it is to be expected that one of these clones would be superior to all the rest, and therefore that selection would eventually eliminate all its competitors. On the other hand, if the environment is neither constant or uniform this conclusion does not necessarily follow, and it is impossible to predict how many clones we would expect to find in a given geographical area which is diverse both in space and in time. The question must therefore be settled by observation.

Morphological Variation. Work on weevils and lizards has established that morphological variation is reduced in taxa which discard sexuality, although it is not lost entirely. Suomaleinen (1961) was able to demonstrate significant differences in certain morphometric ratios between polyploid parthenogenetic weevils collected from different localities, and suggested that this variation was adaptive. This implies that evolution is still proceeding in these apomictic populations, although perhaps very slowly, and therefore that each population possesses a significant quantity of genetic variation. A number of authors (reviewed by Maslin, 1971) have worked on functionally apomictic teiid lizards in the genus *Cnemidophorus*, obtaining results similar to those of Suomaleinen. Zweifel (1965) found significant differences in scutal characters between samples of *C. tesselatus* from different localities, and indeed claimed that the variation shown by the whole assemblage of clones referred to *C. tesselatus* was comparable to that of the amphimictic taxa *C. tesselatus marmoratus* and *C. gracilis*. However, variation within samples was much smaller for the apomicts, the coefficients of variation being roughly half those calculated for the amphimicts. A detailed study of morphological variation in *C. tesselatus* has been published recently by Parker (1979a). Parthenogenesis is also known in European lizards of the *Lacerta saxicola* complex, although in this case it is probably automictic. Here again, samples of the parthenogenetic taxa have much less morphological variation than is found in closely related amphimictic populations (Darevsky, 1966).

Weevils and lizards provide the best comparative evidence, since it is possible to contrast parthenogenetic taxa with the related amphimicts from which they are presumed to have evolved. Other groups provide weaker but broadly corroborative evidence. For example, it has been suggested that the low levels of morphological variation found in many freshwater snails is due to their habit of self-fertilization (Hubendick, 1951); but since the frequency of selfing has been questioned (Hunter, 1961, 1964) it seems wisest to suspend judgement. Interestingly, the facultatively apomictic snail *Potamopyrgus antipodarium* is much more variable with respect to colour pattern than the two other New Zealand species, which are amphimictic (Winterbourn, 1970). The obligately apomictic *P. jenkinsi* also has a number of morphologically different strains (Warwick, 1952).

Karyotypic Variation. Several karyotypes have been found in the automictic grasshopper *Warramaba virgo*, but within any given colony all individuals have the same karyotype (White *et al.*, 1963). *W. virgo* is monomorphic with respect to body colour, whereas amphimictic morabines are usually polymorphic. In contrast to this uniformity, the parthenogenetic Diptera which have been investigated cytologically are mostly polymorphic structural heterozygotes, whilst their nearest amphimictic relatives are monomorphic structural homozygotes. *Limnophyes virgo* is heterozygous for a single inversion (Scholl, 1956), *Drosophila mangabeiri* for three (Murdy and Carson, 1959). *Lonchoptera dubia* has four automictic cytotypes; two, three or all four may occur together at a single locality (Stalker, 1956b). *Pseudosmittia* cf. *arenaria* has a complex series of chromosomal rearrangements (Basrur and Rothfels, 1959). In *Phytomyza crassiseta*, Block (1969) found structural rearrangements, mostly inversions. Finally, in *Cnephia mutata* as many as 70 thelytokous cytotypes may occupy the same locality, with as many as 700 occurring in the whole range of the species (Basrur and Rothfels, 1959).

White (1973) has pointed out that in these structurally heterozygous automicts the inversions all seem to be paracentric; pericentric inversions, which would lower the fecundity of an amphimict but would not affect that of an apomict, or of an automict with premeiotic restitution, do not occur. He concludes that these inversions did not arise within the parthenogenetic clones, but were present in the amphimictic stock from which the parthenogenetic clones descended. The difficulty with this argument lies in the homozygosity of the sexual diploids most nearly related to the thelytokous forms, especially that of diploid amphimictic *Cnephia mutata*. In *Warramaba virgo* White allows that chromosomal rearrangements seem to have taken place after the evolution of thelytoky.

Genic Variation. The comparison of parthenogenetic taxa with amphimicts makes it certain that parthenogenesis is associated with a loss of variation, and it is likely that this loss is caused by genetic pauperization. However, definitive evidence on the crucial issue of the number of clones present in a sample of parthenogenetic animals was not obtained until the techniques of gel electrophoresis made it possible to scrutinize enzyme phenotypes which can be related directly to the underlying genotype.

The most extensive surveys of genic variation in apomicts have been carried out by Suomaleinen's group, who have measured genetic diversity in five parthenogenetic European insects. The details of these investigations are too complex to present here, and the reader should consult the original papers of Suomaleinen and Saura (1973), Lokki *et al.* (1975, 1976a, b) and Saura *et al.* (1976a, b). I have tried to extract the essence of their results in Table 1.2, which summarizes the distribution of genotypes in large samples of four apomictic beetles and an automictic moth. It is clear that, in most cases, individuals collected from the same locality do not all belong to a single clone; even with rather modest sample sizes, several different genotypes can usually be detected when a reasonably large number of enzyme loci are measured. However, there are exceptions to this rule; for example, of the 56 samples of the chrysomelid beetle *Adoxus obscurus* collected from 52 localities (which have been pooled in Table 1.2)

Table 1.2: Genotypic Diversity in Samples of Parthenogenetic Insects. Only samples of 20 or more individuals have been included; but data for *O. salicis* are pooled from seven localities, and those for *A. obscurus* from 52 localities. Column headings: *N1*, number of individuals in the sample; *N2*, number of different genotypes observed in the sample; *N3*, number of variable loci in the whole sample of the species, including individuals from all localities; *N4*, the number of alleles in the whole sample. The fourth column gives the numbers of individuals in a sample which have the same genotype at the loci measured; thus, in the last entry, for *S. triquetrella*, 151 individuals all had one genotype, five individuals had another different genotype and a third different genotype was represented in the sample by a single individual.

Species	Parthenogenesis	Ploidy	Number of individuals with same genotype	*N1*	*N2*	*N3*	*N4*
Otiorrhynchus scaber (Curculionidae)	Apomictic	3n	12,6,5,5,5,5,4,4,1	47	9	11	28
			87,18,4,3,2	114	5	11	28
		4n	29,5,2,2,2,1,1,1	43	8	11	36
			7,7,2,2,1,1,1	21	7	11	36
			20,3	23	2	11	36
O. salicis (Curc.)	Apomictic	3n	30,10,9,5,3,2, (17 × 1)	76	23	9	21
Polydrosus mollis (Curc.)	Apomictic	2n	35,3	28	2	8	21
Adoxus obscurus (Chrysomelidae)	Apomictic	3n	261,34,13,11,4,3,2	328	7	5	14
Solenobia triquetrella (Psychidae)	Automictic	2n:XY	33,7	40	2	8	29
		2n:XO	22,20,3,1,1	47	5	9	27
		4n	28,6,2	36	3	13	43
			52,2,1	55	3	13	43
			151,5,1	157	3	13	43

Source: Compiled from the series of papers by Suomaleinen and coworkers discussed in the text.

only ten contained two or more different genotypes (another three comprised only a single beetle). Moreover, the number of distinct genotypes is in all cases much fewer than the number of individuals, whereas in amphimictic populations we would expect (given the number of variable loci used in these surveys) almost every individual to have a different genotype. In most cases a single genotype predominates, comprising up to 90 per cent or more of the whole sample. These insects are, then, genetically depauperate: even though a locality is not usually stocked exclusively with a single clone, the paucity of different genomes is striking.

In other invertebrates a similar picture has emerged. Shick (1976) and Shick and Lamb (1977) worked on populations of the anemone *Haliplanella luciae* on the Atlantic coast of North America. These animals appear to propagate themselves exclusively by vegetative fission, since certain enzyme loci were uniformly heterozygous – if any sexual reproduction occurred, of course, one would expect recombination to generate substantial numbers of homozygotes. As in the previous case, individuals were assigned to the same clone if they had the same enzyme phenotype at all of the (four) variable systems studied. On this basis, two localities were occupied exclusively by members of a single clone, one locality by three clones and a fourth locality by five clones. Jaenike *et al.* (1980) found eight clones among some 2000 individuals of the oligochaete *Octolasion tyrtaeum* collected from New York, North Carolina and Tennessee. However, two clones were far more abundant than any others, both occurring across the whole spectrum of habitat types. These two clones differed at about 40 per cent of their loci. Clones of the apomictic snail *Potamopyrgus jenkinsi* also differed at about half their loci (Selander and Jones, cited by Selander and Hudson, 1976). In another apomictic snail, the diploid *Campeloma decisa*, only two clones were found in five localities by Selander *et al.* (1977), and the population at each locality was monoclonal. Mitter *et al.* (1979) found 18 clones of the gynogenetic (and probably apomictic) moth *Alsophila pometaria* in central Long Island, and another eleven near Princeton; four clones were found at both localities. Parker *et al.* (1978) identified ten distinct clones of the apomictic cockroach *Pycnoscelus surinamensis*, but with few exceptions these came from very distant localities.

Natural populations of gynogenetic and hybridogenetic fishes and amphibians usually comprise several, perhaps many, clones. (For hybridogenetic populations the term 'clone' is inappropriate since only half the genome of each individual is conserved; as an alternative, Kallman has proposed 'hemiclone', which I shall adopt.) Kallman (1962) found that 47 individuals of *Poecilia formosa* from a Texas drainage ditch represented at least four distinct clones, as judged by histocompatibility criteria, and Darnell *et al.* (1967) were able to recognize at least two clones in a sample of 13 *P. formosa* from a stream pool in northeastern Mexico; on the other hand, in the material examined by Kallman and Harrington (1964) at least two individuals derived from different parents found at different localities belonged to the same clone. Gynogenetic *Poeciliopsis* and hybridogenetic *Rana esculenta* are also polyclonal (Uzzell and Berger, 1975; Moore, 1977, 1978). Using electrophoresis, Vrijenhoek *et al.* (1977, 1978) found that most populations of hybridogenetic *Poeciliopsis* comprise several hemiclones; moreover, tissue-grafting studies have shown that electrophoresis tends to underestimate the number of clones present in a sample, perhaps even by a factor of two. However, two samples from rivers near the northern limit of the range of the biotype seemed to be genuinely monoclonal and, if not identical, the clones from the two localities were very similar to one another. Genetic uniformity over large geographical areas is even more marked in *Cnemidophorus*. Maslin (1967) found that four morphotypes of *C. tesselatus* represented only two clones, judged by histocompatibility criteria, and grafts were successfully exchanged between individuals collected in localities as much as 350 km distant from one another. Likewise, Cuellar (1976) found that virtually all grafts exchanged between individuals of *C. uniparens* collected at the same locality were accepted, and his extensive experiments leave no room for doubt that his samples were monoclonal. This extreme genetic uniformity was confirmed by the electrophoretic work of Neaves (1969), but somewhat modified by the more thorough study of Parker and Selander (1976). These latter authors found a total of twelve clones among the diploid thelytokous biotype of *C. tesselatus*, with on average 1.67 clones at each locality (range 1-4). However, one of these clones occurred in 22 of 27 samples, whilst the remainder are much less common, most being restricted to a single locality.

In general, clonal diversity seems to be much greater in sperm-dependent parthenogenetic fish than it is in thelytokous *Cnemidophorus*, and it is natural to ascribe this to a gradual 'leakage' of genetic material into the parthenogenetic populations through the rare incorporation of sperm genomes. In *Cnemidophorus* there is some slight evidence for recombination, which could occur if homologues, rather than identical sister chromatids, were occasionally paired during meiotic prophase, but the attempt by Parker and Selander (1976) to apportion clonal origins between mutation, recombination and hybridization must be regarded as highly speculative.

Cnemidophorus is automictic, but since restitution is premeiotic the maternal genome is conserved. Very little work has been done on automictic animals whose restitution is intrameiotic, or on obligate selfers. The intertidal oligochaete *Lumbricillus lineatus* has a peculiar restitution mechanism, but is basically a triploid automict with an asynaptic meiosis. On the basis of isozyme phenotypes at two variable loci, Christiansen *et al.* (1976) identified between three and seven clones in each of their study populations. *Rumina decollata* is an autogamous land snail which appears to comprise a single clone throughout its distribution in North America (Selander and Kaufman, 1973). Populations are also usually monoclonal, or have very low clonal diversity, in its native range in France and North Africa; in southern France, Selander and Hudson (1976) found only two clones, which differed from one another at about 50 per cent of their loci.

In organisms which alternate between amphimictic and parthenogenetic reproduction there should be an intermediate level of genotypic diversity, and this expectation has been confirmed in several cases. *Metridium senile* is an anemone which reproduces sexually and by pedal laceration (architomy), and is comparable to the strictly amictic *Haliplanella luciae*. Clones recognized on the basis of colour and isozyme phenotype were mapped by Hoffmann (1976) and more extensively by Shick *et al.* (1979). Shick *et al.* found at least 33 distinct clones in an intertidal cove with an area of about 80 m^2. These showed some tendency to spatial aggregation, but some clones seemed to have been quite widely dispersed by currents.

Natural populations of the cladoceran *Simocephalus serrulatus* are polyclonal, but Smith and Fraser (1976) do not specify the number of clones they found.

A series of excellent studies of electrophoretic variation in natural populations of another cladoceran, *Daphnia magna*, has been published by Hebert (Hebert,

1974a, b, c, 1975, 1976; Hebert and Ward, 1972; see also Young, 1975; short review in Hebert, 1978). Genic variation is less than in most strictly amphimictic organisms, but exceeds that of obligately asexual populations of *D. cephalata*. Moreover, genotypic diversity – Young identified 30 or more clones in one population – can be maintained even in the absence of sexuality for periods of a year or more, with no tendency towards the elimination of all but a single clone; this strongly suggests the operation of frequency-dependent selection in an ecologically complex environment, although there was good evidence for the large-scale spatial separation of genotypes within a pond at only one locality. Rather surprisingly, there is pronounced local genetic differentiation, with populations occupying ponds only a few metres apart showing large differences in allele frequency; populations 100 metres or more apart may even be fixed for different alleles. Hebert does not think that this pattern is created by selection, since nearby populations tend to be genetically similar even when they inhabit quite different sorts of ponds; instead, he infers that most populations are founded by one or a very few ephippia (resistant sexual eggs) derived from neighbouring ponds. This spatial heterogeneity may be accompanied by dramatic temporal change. The populations of permanent ponds, whose sexual periods are typically weak or infrequent, are usually out of equilibrium both within and between loci, often with an excess of multiple heterozygotes, and display large fluctuations of genotype frequency over comparatively short periods of time. These fluctuations are associated with, and perhaps caused by, fluctuations in the rate of production of parthenogenetic eggs by different genotypes, which Hebert attributes to differences in the ability of different genotypes to acquire energy at a given population density. On the other hand, the populations of transient ponds, whose sexual periods are typically intense and frequent, were found to be close both to Hardy-Weinberg and to linkage equilibrium. This fact, together with the sudden appearance of large numbers of homozygotes in a population previously consisting almost exclusively of heterozygotes, is at variance with the interpretation of the local genetic variation, and shows that populations are often founded or augmented by the hatching of large numbers of ephippia. Ephippial females appeared in both the permanent and the transient ponds at the time of the population maximum in May–June, but they never dominated the population to the exclusion of parthenogenetic females, and there is some evidence that the ephippial females were not a genetically random sample of the population. In particular, a few ephippial females persisted in permanent populations when fairly high densities were maintained for some time after the maximum, and both Hebert and Young make the remarkable observation that genotypes which were declining in frequency in the population as a whole were disproportionately common in this minority of ephippial females.

King (1972, 1977) has studied the seasonal succession of genotypes in the monogonont rotifer *Asplanchna*, whose populations are highly polyclonal.

Two simple generalizations can be culled from these results. The first is that parthenogenetic animals possess very little genotype diversity, when compared with related amphimicts. In organisms as different as *Haliplanella*, *Adoxus*, *Rumina* and *Cnemidophorus* it is not unusual to find the limiting case of populations made up exclusively of a single clone. But the second generalization must be that this limiting case is not the rule: in most obligately parthenogenetic animals genotypic diversity, though strongly reduced, is not entirely eliminated, and populations usually consist of several distinct clones. Thus, when we compare an amphimictic with an obligately parthenogenetic population we are usually comparing a highly diverse stock with one which, although genotypically depauperate, is not quite uniform.

1.3.3 Heterozygosity

A second general consequence of mixis which follows from an elementary description of meiosis and syngamy is that levels of heterozygosity will be depressed, since heterozygous parents will produce a large fraction of homozygous progeny through Mendelian segregation. Apomictic animals, on the other hand, should have relatively high levels of heterozygosity, since heterozygotes will continually be created by mutation whilst not being lost through segregation. Some new mutant alleles may produce functional enzymes, and thereby increase fitness by heterosis; most will produce defective enzymes and will thus be deleterious or lethal in the homozygous state. Although 'functional heterozygosity' will initially increase under mutation pressure, it must eventually decrease as nonfunctional alleles are acquired by the clone. These nonfunctional alleles may eventually terminate the existence of the clone, either because they are not perfectly recessive or because in time mutation alone will make some loci homozygous for a lethal gene. Any degree of polyploidy postpones this evil day by multiplying copies of functional alleles, and thereby increases the fitness and the lifespan of a parthenogenetic lineage. These ideas have been worked out formally by Lokki (1976a, b); they suggest that heterozygosity should be higher in apomicts than in amphimicts, and higher in apomicts with greater ploidy than in those with lesser ploidy.

These theoretical results may be sensitive to very low rates of autosegregation (Gustaffson, 1942; Suomaleinen, 1961; Asher, 1970). If any segregation occurs, apomixis begins to take on an automictic hue: it will lead to the same consequences as an automixis which normally conserves the maternal genome but permits the creation of some homozygous progeny, as for instance in the case of premeiotic chromosome doubling and the subsequent rare formation of quadrivalents. Apomixis with autosegregation will lead to complete homozygosity, although only very slowly if segregation is very infrequent.

For purposes of comparison, the most complete listing of electrophoretic data for amphimicts is that published by Nevo (1978), who reviews all large-scale surveys done up to 1976. The mean heterozygosity per locus per individual for 93 species of invertebrates is 0.1123 (standard deviation 0.0720; strongly biassed by the numerous studies of *Drosophila* species) and for 135 species of vertebrates 0.0494 (standard deviation 0.0365) (Nevo, 1978, Table IIIB).

Among curculionid weevils, apomictic forms are more highly heterozygous than amphimicts, and polyploid cytotypes are more highly heterozygous than comparable diploids. I have summarized the available data in Figure 1.2.

Figure 1.2: Mean Heterozygosity Per Locus Per Individual Among Insects with Different Breeding Systems and Ploidies. *Poeciliopsis* and *Cnemidophorus* are included for comparison.

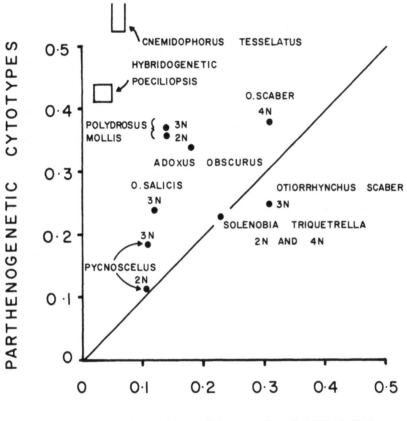

As might be expected, hybridogenetic *Poeciliopsis* are highly heterozygous. The average heterozygosity found by Vrijenhoek *et al.* (1977) was 0.426 (range 0.414-0.440), whilst sexual *P. monacha* and *P. occidentalis* had mean values of only 0.047 and 0.018, respectively. Vrijenhoek *et al.* (1978) found average heterozygosities of between 0.36 and 0.50 even among the offspring of hybridogenetic females mated to males from inbred laboratory stocks of *P. lucida*. More surprisingly, Parker and Selander (1976) have discovered an even more extreme situation in *Cnemidophorus*, which is an automict with premeiotic restriction.

All 67 triploid individuals of *C. tesselatus* that they examined were heterozygous at the same 15 loci from a total of 21 examined, giving an average heterozygosity of 0.714, whilst 339 diploids were heterozygous at eleven or twelve loci, giving an average heterozygosity of 0.524 or 0.571; by contrast, sexual individuals of *C. tigris*, *C. septemvittatus* and *C. sexlineatus* collected in the same general area had values of 0.050, 0.058 and 0.070, respectively. Parthenogenetic biotypes of the *Lacerta saxicola* complex are also highly heterozygous (Uzzell and Darevsky, 1975).

Much lower values of heterozygosity have been reported for amictic *Haliplanella luciae*, where average individual heterozygosity for two monoclonal populations was 0.10–0.20 (Shick and Lamb, 1977; it is not possible to calculate heterozygosities for their polyclonal populations, but they would probably be higher); and for *Campeloma decisa*, where Selander *et al.* (1977) found average heterozygosities of 0.095 and 0.333 in their two clones. Apomictic *Pycnoscelus surinamensis* appear to have about the same amount of heterozygosity as their amphimictic ancestor *P. indicus* (Parker *et al.*, 1978).

All automictic mechanisms of restitution, except premeiotic doubling, should lead eventually to complete homozygosity throughout the genome. This is nicely exemplified by *Rumina decollata*, where Selander and Kaufman (1973) found that about 750 individuals from 33 localities in four states of the USA were all homozygous for the same alleles at 25 isozyme loci. Of the nine populations sampled in the native range of the species in Europe, seven were completely homozygous.

Rivulus marmoratus, an almost obligately self-fertilized hermaphrodite, is also almost entirely homozygous (Harrington and Kallman, 1968). Kallman (1964) claimed that *Poecilia formosa* was also largely homozygous, since grafts between *formosa* and *formosa* X *sphenops* hybrids were usually accepted; since these hybrids have turned out to be triploid, however, his evidence can no longer be accepted.

However, just as apomixis may not invariably lead to very high levels of heterozygosity, so automixis may not create complete homozygosity. Asher (1970) has shown that if meiosis I is suppressed a residual amount of heterozygosity may still be maintained by mutation at loci close to the centromere. More importantly, Asher also points out that if there is powerful selection for heterozygotes some loci may continue to segregate indefinitely, and the heterozygosity of the triploid gynogenetic salamanders of the *Ambystoma jeffersonianum* complex reported by Uzzell and Goldblatt (1967) has been interpreted in this way.

Both diploid and tetraploid parthenogenetic races of the psychid moth *Solenobia triquetrella* appear to have the same level of heterozygosity as the diploid amphimictic race. This may seem curious, since the parthenogenesis is automictic with a true meiosis and the subsequent restitution of diploidy by the fusion of cleavage nuclei, a mechanism which usually results immediately in complete homozygosity. However, in *S. triquetrella* it was established (Seiler and Schaffer, 1960; Seiler, 1963) that one of the two haploid fusion nuclei contains one set of homologues from the premeiotic cell, whilst the other fusion nucleus contains the other set. Syngamy, therefore, precisely restores the original chromosomal constitution, and conserves heterozygosity. Parthenogenetic individuals are therefore chromosomally female, since the female is the hetero-gametic sex in Lepidoptera; and there is no recombination, since chiasmata are not formed during the lepidopteran oogenesis. The usual tendency of automixis to increase homozygosity is thus forestalled, and this particular method of automixis has the same effect on diversity and heterozygosity as would apomixis, or an automixis with premeiotic restitution. It would be reassuring to have direct electrophoretic evidence that more conventionally automictic animals are indeed extensively homozygous, but none seems to exist. In Figure 1.3 I have compared heterozygosity at two isozyme loci between triploid automictic and diploid amphimictic cytotypes of *Lumbricillus lineatus*, from the data provided by Christiansen *et al.* (1976). The automicts have somewhat less heterozygosity at one locus, but somewhat more at the other. Again, the aberrant mechanism of restitution explains the failure to obtain the expected result.

It has also been argued that arrhenotoky will lead to a loss of heterozygosity. The argument is not cytogenetic but evolutionary: if genetic variation is main-tained largely through heterosis or through selection which varies in time it will be lost in haploids. This idea received strong support from Snyder (1974), who found no variable loci at all in a survey of three social bees. Metcalf *et al.* (1975), however, measured mean heterozygosities of between 0.038 and 0.078 (mean 0.061) in seven species of solitary hymenopterans. These values are lower than those characteristic of *Drosophila* and will probably turn out to be low for insects in general, but they are considerably greater than unpublished estimates for diploid amphimictic orthopterans cited by Nevo (1978).

The relationships between diversity, heterozygosity and mixis are fairly clear in principle, but the empirical work that has been done so far is not sufficient to describe them in detail. Figure 1.4 is my guess at what they should look like, together with a very rough indication of where the animals discussed in the text might lie. Parker (1979b) has provided a more qualitative analysis for some of these animals, but unfortunately it is not possible from published data to draw a quantified version of Figure 1.4.

1.4 Individual Consequences of Sexuality

So far I have discussed sexuality as though it were a property of groups of organisms, with consequences at the level of the group but having no significance for individual members of the group. However, this point of view is obviously incomplete: meiosis and syngamy are properties of individuals, and the diversity and heterozygosity of a population is only the sum of the diversity and hetero-zygosity of the offspring produced by the members of the population. There are then at least two consequences of sexuality at the level of the individual.

Figure 1.3: Frequency of Heterozygotes in Diploid Amphimictic and Triploid Automictic *Lumbricillus lineatus*. Each point is a population; the two loci plotted are phosphoglucomutase (PGM) and phosphoglucose isomerase (PGI). If the automicts were less heterozygous than the amphimicts, the plotted points would fall below the solid line and the regressions (broken lines) would have slopes of less than one.

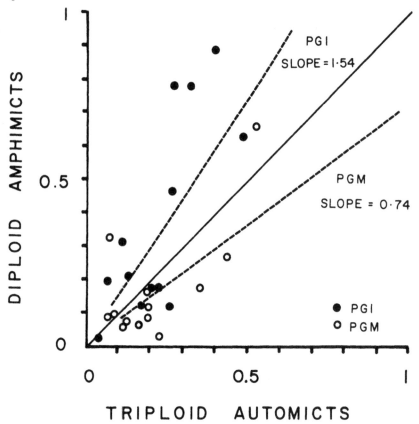

Source: Data from Christiansen *et al.* (1976).

The first of these is the generation of genotypic diversity in the progeny by assortment, recombination and syngamy, which I have already described as being the leading characteristic of sexuality. One can imagine an organism with only a single pair of chromosomes, no crossing-over and a strictly automictic syngamy; the first meiotic division then acts only to separate homologous chromosomes, without any exchange of genetic material, and syngamy acts immediately afterwards to stick these two chromosomes back together again. This would be a process which involved the meiotic formation of two haploid gametic pronuclei and the subsequent formation of a diploid zygotic nucleus by syngamy, and

Figure 1.4: A Speculative Mapping of Genetic Systems According to Their Effects on Genotypic Diversity and Heterozygosity.

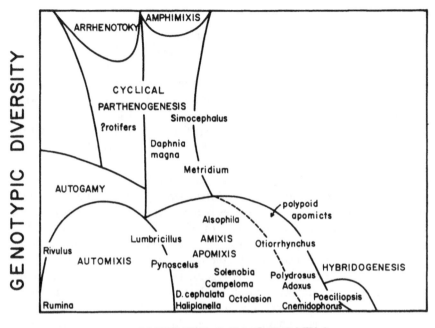

HETEROZYGOSITY

which would therefore have to be described as sexual, without involving the generation of any genotypic diversity whatever. The genome which enters meiosis is conserved, the same genes in the same order reappearing in the zygote. The important point, however, is not that such a process is imaginable, but rather that it can be imagined only by postulating the complete suppression of segregation and recombination together with the restriction of syngamy to the immediate products of the first meiotic division. Sex will conserve genotypes only if it is tied into a straitjacket with these three laces, and if any of them is loosened diversity will immediately follow. If there is more than one pair of chromosomes, random segregation will create different combinations of chromosomes in the haploid products of meiosis; if crossing-over is allowed, new sorts of chromosomes are created by recombination during meiosis; and if syngamy involves haploid genomes derived from different diploid nuclei, or even if a normal two-step meiosis takes place, the diploid genome will be diversified even if the haploid genome has not been. Not one but all three of these laces are unloosened in the great majority of sexual organisms, and the consequence is not merely diversity, but the creation of virtually unlimited diversity amongst their progeny. In this respect, sexuality and asexuality are polar opposites: discounting mutation, all the progeny of a sexual individual will have different genotypes, whilst all the progeny of an asexual parent will inherit the same genotype.

The second consequence of sex for the individual is an effect on the average heterozygosity of the progeny. Broadly speaking, since genomes are conserved when reproduction is asexual, there will be no tendency for heterozygosity to be lost from an asexual population; indeed, the number of heterozygous loci is expected to increase in time as recessive and often deleterious mutations accumulate. If such a highly heterozygous asexual population were suddenly to switch over to exclusively sexual reproduction, many of these deleterious alleles would be expressed in homozygotes and eliminated from the population, so that the average heterozygosity per genome would decline. Geneticists have called this effect the *segregational genetic load*.

The magnitude of the segregational load should depend on the nature of mixis. In particular, automixis should differ sharply from apomixis and amphimixis in that an excess of homozygotes is expected among the progeny, causing an inbreeding depression of fitness. This effect should be strongest in those organisms which occasionally produce a few parthenogenetic eggs whose ploidy has been restored postmeiotically by the fusion of haploid cleavage nuclei and which are therefore completely homozygous. This has been described in tychoparthenogenetic grasshoppers and phasmids, where the viability of automictic eggs and embryos is often very low, and Mockford (1971) has reported that in psocids, which also seem to have a postmeiotic doubling, eggs from virgin females have a much lower hatchability than fertilized eggs. In obligately thelytokous species of *Carausius* and *Bacillus*, however, over 80 per cent of eggs are viable, despite their postmeiotic restitution (Cappe de Baillon *et al.*, 1937b). This suggests that selection can increase viability from very low levels to near the average for an amphimictic species. Bergerard (1958) found an appreciable quantity of heritable variation in the viability of unfertilized eggs of *Clitumnus extradentatus*, and succeeded (Bergerard, 1962) in selecting for increased rates of eclosion. Cappe de Baillon *et al.* (1938) found some increase in the proportion of viable eggs between the first and fourth generations in an automictic lineage of *Menexenus semiarmatus*. However, Baud (1973) could not reduce embryonic mortality much below 25 per cent in the obligately parthenogenetic *Lonchoptera furcata*, and egg mortality may reach 40 per cent in *Drosophila mangabeiri*, though survival from larva to adult is a healthy 80 per cent (Carson *et al.*, 1957). White (1973) tentatively attributes these high rates of early mortality to crossing-over during automixis between mutually inverted chromosome segments, which would bring dicentric and acentric fragments into the zygote nucleus following centric fusion.

Considerable early mortality in automicts presents no general problem, if we posit the segregation of recessive lethals. These have been detected in automictic dipterans which have an intrameiotic restitution (Stalker, 1956b). It is more surprising to learn that the average hatching rate of *Pycnoscelus surinamensis* is little more than half that of its amphimictic relative *P. indicus* (Roth, 1974), since *P. surinamensis* is certainly apomictic.

Conversely, Schultz (1961, 1971) has found that gynogenetic *Poecilia* and *Poeciliopsis* surpass their sexual hosts, at least in aquaria, in vigour, size, disease resistance and generally in the ability to survive, and he attributes this superiority to heterosis.

There is some evidence, then, that automixis, at least in the early stages of its evolution, may lower the viability of progeny. However, the more fundamental issue here concerns the ecological significance of heterozygosity. If genomes in natural populations of organisms are entirely homozygous, with the exception of a few loci at which almost invariably harmful recessive genes have arisen through mutation, then automixis creates an inbreeding depression by bringing these deleterious recessives into homozygous combination. On the other hand, if genomes are usually highly heterozygous because some general virtue of hetero-zygosity leads to its maintenance by stabilizing natural selection, inbreeding depression arises because automictically produced progeny are less likely to possess the superior heterozygous genome at any given locus. The recent dis-covery (Harris, 1966; Lewontin and Hubby, 1966) that natural populations are indeed highly heterozygous has not really resolved these conflicting hypotheses, since it is always possible to argue that alternative alleles at a locus almost invariably have indistinguishable average effects on fitness (see the discussion in Lewontin, 1974). If heterozygous genomes are commonly favoured by selection because, by coding for two versions of the same enzyme, they permit a more flexible response to the environment, then breeding systems which depress heterozygosity, such as automixis, will persist only in environments where a flexible response is unnecessary. On the other hand, if inbreeding depression is caused by the segregation of a small number of recessive lethals, the loss of fitness experienced by automicts will handicap them equally in all environments. Only the first interpretation allows comparative predictions to be made, and I shall therefore use it as the basis for comparative work; it should be borne in mind, however, that, if comparison appears to falsify an hypothesis about the taxonomic distribution of automixis, it may really have falsified an assumption about the way in which heterozygosity is maintained in populations.

If this assumption is granted, apomicts and amphimicts might be thought of as packaging genetic variation in different ways. Apomicts produce uniform but highly heterozygous progeny, each of which is capable of dealing with a wide range of environments, even if in any given environment it may not perform as well as the best-adapted homozygote. Amphimicts forego a certain amount of heterozygosity in order to produce diverse progeny, each of which may be extremely fit in a particular environment. Apomicts, if you like, produce gener-alist progeny whilst amphimicts produce specialists; automicts compromise, since automixis depresses both heterozygosity and diversity below their maximum levels. However, it would be inadvisable to place too much reliance on this argument. Certainly, the heterozygote is sometimes fitter than either homozygote, although the ubiquity of this phenomenon remains a matter for debate. But it has not been established that heterozygote advantage follows naturally from some very general consequence of the heterozygous condition, and the idea that heterozygotes are ecologically more versatile than any given homozygote must be held tentatively, if at all. The topic has been discussed recently by Fincham (1972) and Berger (1976).

The problem of homozygosity in automictic animals is mirrored by the problem of hemizygosity in the males of arrhenotokous animals. The occasional parthenogenetic eggs of acridid grasshoppers are initially haploid, being later made diploid by endomitosis. Haploids or haploid/diploid mosaics which do not become fully diploid are inviable or develop abnormally. But this seems to be a property of haploidy itself, since it recurs with full force in the second parthenogenetic generation, by which time any recessive lethals should have been eliminated (King and Slifer, 1934). Tissues in mosaic embryos which develop abnormally have a higher than average proportion of haploid cells (Bergerard and Seugé, 1959).

An equally puzzling phenomenon occurs in cynipid wasps, which alternate between arrhenotokous and thelytokous reproduction. A few males often occur in the thelytokous generation, but they are 'weak', lack the instinct to mate and are haploid (Patterson, 1928). And yet if haploidy is the cause of their weakness, how is its effect cancelled in the following generation, when the sexual females are fertilized by vigorous haploid males?

In monogonont rotifers, the extreme precocity and short lifespan of the haploid males has led many authors to suggest that natural populations may be rather highly inbred, since there is little time for dispersal after the release of offspring from the brood chamber. If this were so, the lower hatchability of mictic eggs and the lower fecundity of mictic females might reflect inbreeding depression. However, this conclusion is not supported by the experimental work of Birky (1967), who studied the effect of selfing within clones of *Asplanchna*. He observed an immediate inbreeding depression of the viability of mictic eggs, but could recover almost complete viability by selecting among the selfed lines. These results strongly suggest the continued segregation of recessive lethals, which is inconsistent with the hypothesis of habitual selfing in natural populations. It also seems inconsistent with the occurrence of male haploidy; but Birky argues that males can tolerate a loss of function at many loci because they are initially brooded by the female, and even after release lack many of the functions of normal free-swimming animals.

1.5 The Cost of Sex

There is a third general consequence of sex which I have not yet mentioned. It is a reduction in the efficiency with which genes are transmitted from generation to generation: the 'cost of sex'. Like many simple concepts this seems to be difficult to grasp, and its validity has even been denied by some recent authors. At the same time it is crucial to an understanding of how sex evolves, and so I shall discuss it in some detail.

All higher organisms whose sexuality involves the formation of distinct gametic cells are oogamous, the female gametes (macrogametes, which, revealing my animal bias, I shall call ova) being large and immotile, whilst the male gametes (microgametes, sperm) are minute and highly motile. Let us concentrate on the females, and make the reasonable assumption that they produce the same number of ova; but those produced by the asexual females will contain a full diploid complement of chromosomes, whilst the ova produced by the sexual females are haploid and require fertilization. Disregarding complications such as sex chromosomes, we can see that a sexual female puts only half her genome into each ovum; or, equivalently, that a copy of a given gene is certain to be present in any asexual egg, but has only a 50 per cent chance of occurring in any given sexual ovum. The sexual female therefore propagates her genome, or any given element of her genome, only half as efficiently as the asexual female.

This is a very simple idea, and a part of the confusion which surrounds it is merely terminological. Maynard Smith (1971b), who appears to have been the first to state it in a rigorous fashion, spoke of a 'cost of sex', meaning that, in populations whose sex ratio at equilibrium is equal, the reproductive rate of females was halved by the production of male offspring. The 'cost of sex' (measured as the rate of replication of haploid genomes achieved by a sexual female relative to that achieved by an asexual female) is then 50 per cent. However, Williams (1975) prefers to speak of a 'cost of meiosis', by which he means the lower rate of replication of haploid genomes implied by the reduction of gametic ploidy following meiosis; from the argument given in the previous paragraph, the 'cost of meiosis' is also 50 per cent. It is important to realize that the 'cost of sex' and the 'cost of meiosis' are not different concepts. The cost of meiosis is the halving of ploidy in the gametes. If the diploid condition can be restored by syngamy with another haploid nucleus derived from the same zygote (automixis, self-fertilization), the cost of meiosis has been paid by syngamy, since both haploid chromosome complements descend from the same zygote. (The situation is not precisely the same as in asexual reproduction, since with random segregation there may be none, one or two copies of a given gene transmitted to the zygote; in asexual systems the number of copies transmitted per zygote is exactly one, whilst in automictic sexual systems the average number transmitted is one.) But if diploidy is restored by syngamy with a gamete which descends from a different zygote, the cost of meiosis will be 50 per cent. The usual halving of the rate of replication of copies of a given gene follows if, but only if, meiosis in an oogamous organism is followed by syngamy with a gamete descended from a different zygote or, in a word, if meiosis is followed by cross-fertilization. The 'cost of meiosis' and the 'cost of sex' are thus the same concept, the one term focusing on the reduction of ploidy at meiosis and the other on the necessity of its restitution at syngamy. I shall prefer to use the 'cost of sex', as having the broader meaning.

Two recent papers have denied the existence of a cost of sex. Barash (1976) argues that if genes coding for sexuality are fixed in a sexual population then each zygote will receive two copies of 'sexual' genes at the appropriate locus, just as an asexual egg will receive two copies of 'asexual' genes at that locus. However, as Barash himself acknowledges, this does not gainsay that a sexual female will put only one gene coding for sexuality into a given ovum, and must therefore produce two ova for the same genetic result that an asexual female achieves with a single egg. Barash's attempts to show that sexual females can produce two eggs as cheaply as asexual females can produce one are unconvincing (Maynard Smith and Williams, 1976). A more elaborate fallacy of the same general sort has been advanced by Treisman and Dawkins (1976). They imagine two haploid populations, one fixed for an allele which codes for sexuality and the other for an alternative allele coding for asexuality. Since both alleles are fully expressed, the two populations are reproductively isolated. If the alleles exert no average effect except on the mode of reproduction, so that they are neutral with respect to other aspects of viability and fecundity, then their proportions will not change in either of the two populations in time. Therefore their overall frequencies in the two populations taken together will not change in time, and we conclude that sexuality is in itself a selectively neutral trait which incurs no extraordinary 'cost'. This conclusion is true but trivial. It is not difficult to imagine that the fate of a gene for sexuality in hedgehogs will be indifferent to the evolution of asexuality in sharks, and this is the sort of situation which is modelled by Treisman and Dawkins. The critical assumption of their model is that the sexual population and the clone are not only genetically isolated but are also ecologically disjunct, so that the numbers of individuals are regulated separately in the two populations. But consider the more interesting situation where sexual and asexual variants have similar ecologies, which they certainly should have if they share the greater part of their genomes. Instead of a fixed number of sexual and a fixed number of asexual individuals, we have a fixed total number of individuals, and the frequency of sexual individuals in this total may vary. Since each asexual female has the same output of eggs as any two sexual parents, the asexual part of the combined population will reproduce twice as fast as the sexual part, and will quickly come to replace it entirely.

Elsewhere, however, Dawkins (1976) has made the valuable point that the cost of sex will be levied only on the differential elements of the genome – on those loci at which genic variance causes variance in the breeding system. To say that the state of these loci affects the rate of transmission of alleles at other loci is true but irrelevant, since there is no cost implied here; the dynamics of alleles at the differential loci depend wholly on the efficiency with which they are transmitted, and not at all on the efficiency with which the rest of the genome is transmitted. Williams' cost of meiosis is then misconceived; fortunately, this misconception has not affected the development of theories of sex, in which it is always assumed, implicitly or explicitly, that we are dealing with alleles at a differential locus.

Formal Model. The rather tortuous course of these verbal arguments make it desirable to construct a simple formal model in which the existence and magnitude of a cost of sex can be rigorously demonstrated. (Surprisingly, this has not been done before, with the exception of an extremely simple calculation by Maynard Smith, 1971b, 1978).

It is required to be demonstrated that an allele which causes the production of a greater proportion of unreduced eggs, but which is otherwise neutral, will tend to spread at the expense of any alternative allele which causes the production of a lesser proportion of unreduced eggs. An appropriate model to analyze is one of a diploid hermaphroditic organism in which each of the three genotypes at a diallelic locus is characterized by the investment of a certain fraction of total resources in female function and by the production of a certain proportion of unreduced eggs. The model can be summarized as in Table 1.3.

Table 1.3: Model of a Diploid Hermaphroditic Organism.

Genotype	Frequency	Investment		
		Unreduced Eggs	Reduced Eggs	Sperm
A1A1	F_{11}	$E_{11}S_{11}$	$E_{11}(1-S_{11})$	$(1-E_{11})$
A1A2	F_{12}	$E_{12}S_{12}$	$E_{12}(1-S_{12})$	$(1-E_{12})$
A2A2	F_{22}	$E_{22}S_{22}$	$E_{22}(1-S_{22})$	$(1-E_{22})$

Let us also assume that fertilization is external, with random gamete fusion occurring in the external medium, and that all reduced eggs are fertilized. The model is now sufficiently well-defined for allele frequencies to be followed for any desired number of generations. However, although a general analytical solution is feasible it is very tiresome, and instead I shall solve the model for the case of a population fixed for the *A1* allele, in which the *A2* allele appears as a rare mutant. It is required to be proven that the frequency of the *A2* allele will tend to increase if the allele causes the production of a greater proportion of unreduced eggs.

A2 alleles will be transmitted both sexually and asexually; sexual transmission is achieved both through reduced eggs and through sperm.

The reduced eggs produced by *A1A2* individuals make up a fraction

$$\frac{F_{12}E_{12}(1-S_{12})}{F_{11}E_{11}(1-S_{11})+F_{12}E_{12}(1-S_{12})}$$

of the total number of reduced eggs produced by the whole population. Because of the rarity of the $A2$ allele; virtually all these eggs will be fertilized by $A1$ sperm, so that half will give rise to $A1A2$ zygotes. Similarly, the sperm produced by $A1A2$ individuals makes up a fraction

$$\frac{F_{12}(1-E_{12})}{F_{11}(1-E_{11})+F_{12}(1-E_{11})}$$

of the total quantity of the sperm produced by the population. Half the sperm will bear an $A2$ allele and will, if successful, fertilize $A1$ eggs to give rise to $A1A2$ zygotes. To obtain the total number of sexual zygotes bearing an $A2$ allele derived from $A1A2$ parents, these fractions are multiplied by the total number of reduced eggs produced by the whole population. Finally, the total number of unreduced eggs produced by $A1A2$ individuals is

$$F_{12}E_{12}S_{12},$$

all of which are $A1A2$.

The total frequency of $A2$ alleles present after reproduction, F_2', can now be obtained by summing these three contributions, and dividing by the total number of genes transmitted, which will be equal to twice the total number of eggs produced by the population. This yields

$$F_2' = \frac{(1/2)\left[\dfrac{F_{12}E_{12}(1-S_{12})}{\Sigma F_{1i}E_{1i}(1-S_{1i})}+\dfrac{F_{12}(1-E_{12})}{\Sigma F_{1i}(1-E_{1i})}\right]\Sigma F_{1i}E_{1i}(1-S_{1i})+F_{12}E_{12}S_{12}}{2\Sigma F_{1i}E_{1i}},$$

where each summation is taken over $i = 1, 2$. Since the initial frequency of the $A2$ allele is

$$F_2 = \frac{F_{12}}{2}$$

and this is taken to be so small that $F_{11} = 1$ and $\Sigma F_{1i}E_{1i} = E_{11}$ very nearly, it is readily shown that $F_2' > F_2$ if

$$S_{12} > S_{11}\frac{E_{11}(1-E_{12})}{E_{12}(1-E_{11})} - \frac{E_{11}(1-E_{12})+(1-E_{11})(E_{12}-2E_{11})}{E_{12}(1-E_{11})}.$$

If the effect of the $A2$ allele is to alter the proportion of unreduced eggs without affecting the total investment in female function ($E_{12} = E_{11}$), this condition becomes simply $S_{12} > S_{11}$. Our theorem is therefore proven: any allele which directs the production of a greater proportion of parthenogenetic eggs, without any other effect on the phenotype, will tend to increase in frequency.

Other special cases are easily deduced; two are of some interest. First, if $A1A2$ individuals are exclusively female ($E_{12} = 1$) the condition becomes $S_{12} > 2E_{11} - 1$. If at equilibrium the original $A1A1$ population spent equally on male and female function ($E_{11} = \frac{1}{2}$), this is simply equivalent to $S_{12} > 0$ – any degree of parthenogenesis is favourably selected. More generally, selection for $A2$ is influenced by the investment in female function by the $A1A1$ genotype, and will be retarded when this exceeds one-half.

Secondly, by taking $E_{12} = 1$ we can rewrite the recursion equation for gene frequency as

$$F_2' = F_2 \left(\frac{1 + S_{12}}{2E_{11}} \right)$$

The numerator of the expression in parentheses represents the gain in fitness from producing unreduced eggs: it reflects the cost of meiosis. The denominator represents the reciprocal of the degree of female function in the competing genotype: it reflects the cost of producing males. The expression as a whole compounds these two effects into an overall cost of sex. If we take $E_{11} = \frac{1}{2}$ and $S_{12} = 1$ then $F_2' = 2F_2$: there is a full twofold cost of sex and the $A2$ allele doubles in frequency in every generation when it is rare.

As the $A2$ allele becomes more frequent it is necessary to take account of the formation of $A2A2$ zygotes, and the algebra becomes cumbersome. However, it can easily to be shown by iterating the general recursion equation given above on a computer that any allele which increases the production of parthenogenetic eggs is favoured, and the only population which is in equilibrium when fixed for a single allele is one which consists exclusively of parthenogenetic females.

Although these arguments have been developed for the case of a diploid hermaphrodite, it is easy to show that they apply equally to organisms which are haploid or gonochoric or both. Moreover, amphimixis may involve costs over and above the automatic genetic penalties of sex. In finding a mate, courting and copulating an amphimict will often run risks which will further reduce its fitness relative to a parthenogenetic female.

Automixis and Inbreeding. Automictic sexuality is cheap because when the same individual contributes both gametes or gametic pronuclei to the zygote the cost of meiosis will be paid by syngamy. In self-fertilizing hermaphrodites there will be a cost of some sort involved in the elaboration of separate male and female gametes, but it seems safe to assume that this will be very small when compared with the twofold cost of amphimixis.

If the cost of sex is nearly abolished by self-fertilization, it seems reasonable to argue that it will in general be reduced in proportion to the degree of inbreeding. Williams (1979) has argued that this is the case, and a simple formal argument proves the point.

Suppose that a single $A1A2$ heterozygote arises by mutation in an $A1A1$

population. The effect of the $A2$ allele is to cause a certain fraction of matings to occur with relatives of degree r; this fraction is K_1 for females and K_2 for males. The allele persists in the population for a certain period of time, at the end of which it is still sufficiently rare for the frequency of $A2A2$ homozygotes to be neglected, but any given $A1A2$ individual has an appreciable number of $A1A2$ relatives with which to mate.

The easiest case to analyze is when the $A2$ allele causes a certain fraction of matings to occur between full sibs. Consider the $A1A2$ progeny of a mating between $A1A1$ and $A1A2$ parents. These progeny will themselves reproduce, and the grandchildren of the $A1A2$ parent may be derived either through the male or through the female line. The number of $A2$ genes transmitted to these grandchildren is an appropriate measure of the fitness of the $A2$ allele relative to that of the $A1$ allele.

With probability $(1 - K_1)$ each daughter will be outcrossed to unrelated individuals, virtually all of which will be $A1A1$; half the progeny from such matings will be $A1A2$. With probability K_1 she will mate with a male sib, whose relatedness to her is $r = \frac{1}{2}$. Her mate will be $A1A1$ with probability $(1 - r)$, in which case half her progeny will be $A1A2$; he will be $A1A2$ with probability r, in which case half her progeny will be $A1A2$ and a quarter $A2A2$. Adding these contributions together, the expected number of $A2$ genes transmitted to each of her offspring by each daughter is

$$\frac{(1 - K_1)}{2} + \frac{(1 - r)K_1}{2} + \frac{rK_1}{2} + \frac{2rK_1}{4} = \frac{1 + rK_1}{2}.$$

The total number of $A2$ alleles transmitted to grandchildren through daughters is thus

$$\frac{fB(1 + rK_1)}{2},$$

where f is the number of daughters and B is female fecundity.

Sons will also make a contribution. Their contribution through inbreeding has already been included in the reproduction of daughters, and cannot be counted twice. Since they must otherwise mate with $A1A1$ females, the number of $A2$ alleles which they transmit through outbreeding is

$$m\left(\frac{F}{M}\right)B\frac{(1 - K_2)}{2},$$

where m is the number of sons, F the total number of females and M the total number of males in the population.

But we must constrain total expenditure on sons and daughters to be fixed, so let $f + m = 1$. Likewise set $B = 1$, so that the calculation now refers to the expected number of $A2$ alleles transmitted per zygote. Finally assume that when the population is at equilibrium males and females are equally frequent. The normalized rate at which $A2$ alleles are transmitted is then

$$R = \frac{[1 + frK_1 - (1-f)K_2]}{2}.$$

The comparable rate for the $A1$ allele is, of course, $R = \frac{1}{2}$.

Though this argument is developed for the particular case of sib mating, it can be applied to any regular pattern of inbreeding. Moreover, it can be applied to hermaphrodites as well as to gonochores, since f can be interpreted as investment in female function, rather than as investment in daughters. At one extreme, an apomictic female has $f = r = K_1 = 1$ and thus attains $R = 1$, whilst at the other extreme a random-mating amphimictic female has $K_1 = K_2 = 0$ and thus $R = \frac{1}{2}$. The $A2$ allele is transmitted twice as efficiently by the apomict, a reflection of the twofold cost of sex in this comparison.

An obligately self-fertilized hermaphrodite has $K_1 = K_2 = r = 1$ and hence $R = f$: the cost of sex decreases as investment in female function increases, vanishing as more and more extreme economies in sperm production are achieved, as the verbal argument had suggested. In the more interesting case of a mutation which causes some degree of inbreeding but does not affect proportional allocation to female function we have

$$R = \frac{1}{2} + \frac{(K_1 - K_2)}{4}.$$

Thus the cost of sex is reduced only if $K_1 > K_2$, that is, only if there is some gain through male function by outbreeding. This is Fisher's theory of the automatic increase of alleles causing higher rates of selfing, which is discussed in Chapter 5.

For the case of sib mating around which the original argument was constructed we have

$$R = \frac{1}{2} + \frac{(K_1 - 2K_2)}{8},$$

so that the cost of sex will be reduced only if $K_1 > 2K_2$ or, in other words, only if the commitment of male function to inbreeding is less than half the female commitment to inbreeding. More generally, for any regular pattern of inbreeding and for any degree of female function the cost of sex is reduced only if $K_1 > K_2(1-f)/rf$. Thus if distant inbreeding is to reduce the cost of sex it is necessary that only a very small quantity of sperm or pollen be sufficient to fertilize female relatives of given degree, the rest being used for outcrossing.

Finally, note that in general

$$\frac{dR}{df} = \tfrac{1}{2}(rK_1 + K_2),$$

so that with any degree of inbreeding there will be selection for increased female function. Given the requisite genetic variation we shall therefore expect alleles which increase the degree of inbreeding and which increase the degree of female function to be selected in concert, with the cost of sex becoming very small at equilibrium.

I should emphasize that these calculations concern only the efficiency with which alternative alleles are transmitted, without regard to the viability of the progeny into whose genomes the alleles are incorporated. The homozygosity associated with automixis may well create a 'cost of automixis' which is more than sufficient to balance the cost of amphimictic sex. Indeed, Lamb and Willey (1979) have argued that the cost of sex is in general balanced by a pro-portionate reduction of fertility among parthenogenetic insects. This seems to me too broad a statement (though the low fertility of apomictic *Pycnoscelus surinamensis*, which I mentioned in the previous section, certainly appears to support it), but there is little doubt that facultative automixis may usually encounter so severe an inbreeding depression that it will fail to spread, despite an increase in the efficiency of genetic transmission. The same reservation must be made for arrhenotoky, which is the subject of the next few paragraphs.

Arrhenotoky. Male haploidy presents a rather more difficult problem, because we have to answer two questions. In the first place, is there a cost of amphimixis relative to arrhenotoky? And secondly, is there a cost of arrhenotoky relative to apomixis? In the case of autogamy and inbreeding these two questions can be run into one, because the relationships are transitive: the fact that autogamy is cheaper than amphimixis implies that it will be less expensive relative to apomixis than is amphimixis. In the case of arrhenotoky this is not the case.

To solve the first problem, imagine an allele $A2$ whose effect in heterozygotes is to cause the production of a fraction k of fertilizable eggs which develop into diploid females and a fraction $(1 - k)$ of unfertilizable eggs which develop into haploid males and which arises in a population of amphimicts all of which are homozygous for the alternative $A1$ allele. As with inbreeding, we count the number of $A2$ alleles transmitted to grandchildren by an $A1A2$ individual. Half the kB daughters produced are $A1A2$, and these in turn produce $kB/2$ fertilizable eggs, which develop into $A1A2$ females, and $(1 - k)B/2$ unfertilizable eggs, which develop into $A2$ males. Half the $(1 - k)B$ sons produced from unfertilizable eggs are $A2$; these produce solely $A1A2$ female offspring by each fertilizing $B(F/M)$ eggs from $A1A1$ females. Adding together these contributions, the total number of $A2$ genes transmitted to grandchildren is

$$\frac{kB}{2}\left[\frac{(1-k)}{2}+\frac{k}{2}B\right]+\left[\frac{(1-k)}{2}\right]B\left(\frac{F}{M}\right)B = \frac{B^2}{4}\left[k+2(1-k)\left(\frac{F}{M}\right)\right],$$

whereas a comparable amphimict will transmit $B^2/4$ copies of any given $A1$ allele to its grandchildren. The $A2$ allele therefore spreads if

$$\left[k+2(1-k)\left(\frac{F}{M}\right)\right] > 1,$$

which if $F = M$ is equivalent to $k < 1$. A mutant allele will invade an amphimictic population with an equal sex ratio if it causes the production of any fraction of haploid eggs which develop without fertilization into males. We can calculate a cost of amphimixis relative to arrhenotoky per generation by taking the square root of the expression on the left-hand side of the equation above: if $F = M$ this is equal to $(2-k)^{\frac{1}{2}}$, so that the maximum possible cost is not two, as in the case of a comparison with apomixis, but $\sqrt{2}$.

The economy which is made through arrhenotoky is quite easy to understand intuitively. Female progeny are diploid and each element of the maternal genome suffers a halving of ploidy through meiosis in every generation, as in amphimixis. Male progeny are haploid and thus again receive only a haploid genome from their mother, but this haploid genome is then passed on unreduced and unrecombined to all their progeny. In this way genes are transmitted more efficiently through the male than through the female line, since each grandchild will inherit a haploid maternal genome through the male but only half, will do so through the female line.

However, it is far from established that the most economical form of arrhenotoky, in which virtually all the eggs are unfertilizable, could represent an equilibrium state, since while a population was being invaded by such a mutant its sex ratio would change, with the number of females and thus the number of fertilizable eggs declining. Suppose that the $A2$ allele passes to fixation but there then arises an $A3$ allele which in heterozygotes causes the production of a different fraction k' of fertilizable eggs. By an argument similar to that laid out above, the number of $A3$ alleles transmitted to grandchildren by an $A2A3$ individual is

$$\left(\frac{B^2}{4}\right)\left[k'+2(1-k')k\left(\frac{F}{M}\right)\right],$$

since each $A2A2$ female produces only kB eggs available for fertilization by $A3$ males. The $A3$ allele spreads if

$$\left[k'+2(1-k')k\left(\frac{F}{M}\right)\right] > \left[k+2(1-k)k\left(\frac{F}{M}\right)\right],$$

or in other words if

(1) either $k' > k$ given $2k(F/M) < 1$,

(2) or $\quad k' < k$ given $2k(F/M) > 1$.

But in a population fixed for $A2$, $(F/M) = k/(1-k)$. The conditions for the spread of $A3$ are thus

(1) either $k' > k$ given $k < \frac{1}{2}$,

(2) or $\quad k' < k$ given $k > \frac{1}{2}$.

Therefore a population fixed for $A3A3$ homozygotes which have $k' = \frac{1}{2}$ cannot be invaded by any allele causing a different fraction of fertilizable eggs. Such an equilibrium population will have an equal sex ratio, since $F/M = k'/(1-k') = 1$.

But what if such a population should be invaded by an apomictic mutant? Suppose that in an $A3A3$ population there arises an allele $A4$ such that $A3A4$ individuals produce a certain fraction s of unreduced eggs which develop without fertilization into $A3A4$ daughters, whilst a fraction $(1-s)k$ are meiotically reduced eggs which, when fertilized, develop into $A3A3$ or $A3A4$ daughters, and the remaining fraction $(1-s)(1-k)$ are reduced eggs which develop without fertilization into $A3$ or $A4$ sons. Adding up all contributions, the number of copies of the $A4$ allele which a rare $A3A4$ individual transmits to its grandchildren is

$$\left(\frac{B^2}{4}\right)\left\{s^2(2-k) + 2s\left[1 - k(1-k)\left(\frac{F}{M}\right)\right] + k\left[1 + 2(1-k)\left(\frac{F}{M}\right)\right]\right\}.$$

The factor by which $A4$ will be transmitted faster than $A3$ per generation is thus

$$\left\{1 + \frac{s^2(2-k) + 2s[1 - k(1-k)(F/M)]}{k[1 + 2(1-k)(F/M)]}\right\}^{\frac{1}{2}}.$$

If $A3A4$ is an obligate apomict ($s = 1$) this factor takes the value

$$\frac{2}{[k + 2k(1-k)(F/M)]^{\frac{1}{2}}}$$

It follows, as it should, that a gene causing obligate apomixis spreads twice as fast when rare as an alternative allele causing obligate amphimixis ($k = 1$, $F/M = 1$). But if the $A3A3$ individuals are obligately arrhenotokous and the population is in evolutionary equilibrium we have $k = \frac{1}{2}$ and $F/M = k/(1-k) = 1$ and the factor again takes a value of two. Arrhenotoky therefore incurs a two-fold cost relative to apomixis because, despite the development of males from unfertilized eggs, the cost of producing these males is not paid for by syngamy as it is in the case of automixis. We have arrived, then, at the curious conclusion that, whilst arrhenotoky is an efficient genetic system when it competes with amphimixis, it is just as inefficient as amphimixis when either competes with apomixis.

Cyclical and Facultative Parthenogenesis. When sexual progeny are produced only occasionally, or when they constitute only a fraction of each brood, the overall cost of sex is an average weighted by the frequency of the sexual progeny. In the case of cyclical parthenogenesis, a geometric mean is appropriate; in the case of facultative parthenogenesis, an arithmetic mean.

Polyembryony. All modes of amixis are equally cheap; vegetative proliferation and apomixis differ in their ecological but not in their genetic consequences. The argument is not quite as straightforward, however, for species with parental care in which the individual undergoing fission is a brooded, sexually produced embryo. Clearly the embryo itself bears no cost; however, the parent which subsidizes the division of the embryo has already borne the cost of sex. From the parent's point of view, polyembryony results in offspring which are less diverse but no less expensive.

Hermaphroditism. I have shown that a simultaneous hermaphrodite which mates at random in a large population will bear the full twofold cost of sex. This result fails if there is local competition for mates, since selection then favours an increase in female function which will retard the progress of an apomictic genotype.

Suppose there is a local mating group of $(N + 1)$ self-incompatible simultaneous hermaphrodites, each of which can devote a quantity B of resources to gametogenesis. Then each individual contributes

$$Z_e = B(1 - m)/G_e$$

macrogametes to zygotes, assuming that all macrogametes are fertilized, where m is allocation to male function and G_e is the quantity of resources needed to make a single macrogamete. Each individual also contributes

$$Z_s = \frac{NB(1 - m)}{G_e} \frac{mB/G_e}{(mB/G_s) + [(N - 1)\bar{m}B/G_s]}$$

microgametes to zygotes, where G_s is the quantity of resources needed to make a single microgamete and $\bar{m}$ is the average degree of male function amongst other members of the group. The factor $(N - 1)$ appears in the denominator because the microgametes produced by any given individual do not compete with those of any other individual either for its own macrogametes (since it may not self-fertilize) or for those of the second individual (which may not self-fertilize either). Thus the total number of zygotes receiving a haploid genome from an individual with allocation to male function m is

$$Z = Z_e + Z_s$$
$$= \left(\frac{B}{G_e}\right)\left[(1 - m) + \frac{mN(1 - \bar{m})}{m + (N - 1)\bar{m}}\right].$$

Setting the derivative of Z with respect to m equal to zero and solving for the equilibrium degree of male function, $\hat{m}$, by equating m with $\bar{m}$, yields

$$\hat{m} = \frac{(N - 1)}{(2N - 1)}.$$

This result has been obtained independently by E. L. Charnov (unpublished), who has also shown that the quantity maximized during this optimization is $E\,S^{(N - 1)/N}$, where E is the quantity of eggs produced and S the quantity of sperm. If the organism were self-compatible, the optimal allocation to male function would be

$$\hat{m} = \frac{N}{2(N + 1)}.$$

In large populations the optimal allocation to male function approaches 1/2, as we expect, but in small populations it may be much less. This is not difficult to understand. Suppose that each local mating group comprises just two self-incompatible hermaphrodites. Since neither can fertilize itself, it should produce only sufficient pollen or sperm to ensure the fertilization of its neighbour, and through this economy is free to make more ovules or eggs. A population with a structure of this sort, leading to the evolution of a high degree of female function, will resist a parthenogenetic mutant much more effectively than will a normal outbred amphimict.

This conclusion supports, and indeed is formally identical with, the contention of Maynard Smith (1978) that the cost of sex will be low in hermaphrodites which mate only once during their lives.

Male Parental Investment. If the participation of the male in parental care makes it possible for the female to rear more progeny, the cost of sex is reduced. If this cooperation results in twice as many offspring being reared as could be reared by a female on her own, the cost of sex is abolished.

This argument holds for all forms of male parental investment: it holds when the male helps to guard or feed the young, but it also holds for species with no parental care in which the male contributes cytoplasm to the zygote.

All the discussion so far has taken for granted that the organisms concerned are oogametic, so that paternal investment in a given zygote is negligibly small.

However, there are many protists and thallophytes where male and female gametes are not extremely different in size, and may even be the same size, a condition called isogamy (for a short review, see Bell, 1978b). In this case the concepts of maleness and femaleness lose their meaning, but for convenience I shall refer to paternal and maternal gametes, which make equal contributions to the zygote. Maynard Smith (1974) and Manning (1975) have pointed out that sex is cheap in isogamous organisms, and recent advances in the understanding of isogamy allow us to elaborate this idea (see Section 5.4). To introduce the topic, I should mention that the evolution of anisogamy (gametes of unequal size) from isogamy is thought to hinge on the relationship between zygote size and zygote survival. When zygote size is critically important to the viability of the zygote, then anisogamy is likely to evolve. Suppose that we symbolize size by m and survival by s, and that the relationship between the two can be expressed as $m = As^B$, where A and B are constants. Then if B is large a small increase in zygote size m will evoke a large increase in zygote size s, and it can be shown that anisogamy is likely to represent a stable evolutionary equilibrium; conversely, if B is small, and especially if it is less than one, the population is likely to remain isogamous. The relevant theory has been worked out in detail by Parker *et al.* (1972), Bell (1978b) and Charlesworth (1978) and is summarized in Chapter 5. Now, imagine an isogamous haploid organism in which syngamy is followed almost immediately by a zygotic meiosis. It seems clear that there will be no cost of sex if the zygote is so transient that its survival is not appreciably affected by its size: the reduction division leads to the production of four gametophytic cells of which two bear the 'maternal' allele and whose combined mass is equal to that of the 'maternal' isogamete. However, if the meiosis is gametic and the zygote has an appreciable lifespan it is no longer legitimate to neglect the effect of size on zygote survival. If the two variables are related by an equation of the sort given above in which $B = 1$, the halving of the birth rate caused by meiosis is just balanced by the halving of the 'maternal' isogamete; in this way syngamy pays the cost of meiosis and the overall cost of sex is zero. But if $B < 1$ the decrease in the death rate of the zygote is not sufficient to compensate for the reduced ploidy of the gametes, and the cost of sex is inversely proportional to B, rising towards 50 per cent as B falls to zero. Manning (1976) points out that the converse is also true: if $B > 1$ then sex in isogamous organisms will represent an immediate advantage under natural selection because of the disproportionate increase in the rate of survival of the sexually produced zygote. However, Bell (1978b) and Charlesworth (1978) have shown that isogamy is unlikely to be stable when zygote survival increases disproportionately with zygote size, so that if B did exceed unity we would often, but not always, expect anisogamy to evolve, and with it the reappearance of a cost of sex.

Meiotic Drive. An allele which is able to distort the normal meiotic segregation ratio, so that it becomes incorporated into a disproportionate number of gametes at the expense of an alternative allele, is said to undergo meiotic drive. Any allele with this property will be displaced less readily than an undriven allele by an allele for parthenogenesis. This effect will disappear, however, when the driven allele becomes fixed in the population.

(As this book was going to press, two important papers on the cost of sex were published by Charlesworth (1980a, b). Whilst the conclusions I have reached above seem sound, Charlesworth's approach is the more rigorous. He has also pointed out to me an article by Bull (1979), in which my argument regarding the economy of arrhenotokous reproduction is anticipated and elaborated.)

Autoselection. I have couched this section in terms of the effect of different phenomena of mixis on the cost of sex. The significance of these arguments is as follows. Suppose that a population is segregating for the two alleles *A1* and *A2*, where *A1* determines the production of mictic and *A2* the production of some proportion of amictic eggs. The *A1* allele will suffer the cost of sex and, if *A2*-bearing individuals produce only amictic eggs, the *A2* allele will be transmitted twice as efficiently as the *A1* allele. However, in some circumstances — for example, if *A1A1* individuals are autogamous or isogamous — the cost of sex is reduced and may even be abolished. Thus, in a population with these characteristics, an allele for parthenogenesis will spread more slowly than it would in an ordinary amphimictic population, and in the limiting cases it will not tend to spread at all.

I shall pause here to point out that this is only one facet of a much more fundamental proposition. The notion that an allele may be selected because of its proficiency in being transmitted from generation to generation is not a new one. It was fully grasped by Fisher, and has become familiar through the phenomenon of meiotic drive. What does not seem to be widely appreciated is that this represents a distinct category of selection. Darwin recognised artificial selection, natural selection and sexual selection; to these we must add a category of processes that I shall call *autoselection*. Autoselection is the process whereby a genetic element tends to increase in frequency by virtue of the nature of its transmission, even though it has no effect on the viability, fecundity or fertility of the individual which bears it. The discovery that a phenomenon, such as parthenogenesis or meiotic drive, is autoselected constitutes a theory of the phenomenon, in the same sense that natural selection constitutes a theory of crypsis or sexual selection of sexual dimorphism. It cannot be a complete theory, since otherwise parthenogenesis would everywhere have supplanted amphimixis. Rather, just as theories of sexual dimorphism are attempts to explain how sexual selection can be balanced by natural selection, so theories of parthenogenesis are attempts to explain how autoselection can be balanced by natural selection.

1.6 Paradox of Sexuality

The three major methods of making reproductive propagules, to which all other methods may be said to be tributary, are apomixis, automixis and amphimixis. The choice of any one of these three methods entails three sorts of consequences, which are effects on the diversity, heterozygosity and costliness of the progeny. This simple statement of the situation can be reduced to a table (Table 1.4).

Table 1.4: Consequences of Apomixis, Automixis and Amphimixis.

		Progeny Genomes:	
	Diversity	Heterozygosity	Costliness
Apomixis	Minimal	Maximal	Minimal
Automixis	Intermediate	Low	Low
Amphimixis	Maximal	High	Maximal

Switching from one method of reproduction to another involves trading, as it were, one set of putative advantages and disadvantages for another. Apomixis is an efficient way of transmitting genomes, and conserves their heterozygosity; but it does so at the expense of genotypic diversity. Automixis is a cheap method of generating at least some diversity amongst progeny; but through its tendency to bring together alleles in homozygous combination it may encounter a severe inbreeding depression. Amphimixis is much less prone to generate inbreeding depression, and realizes the greatest possible diversity of genotypes amongst the progeny; but in achieving these ends the rate at which haploid genomes are transmitted is halved. In short, the consequences entailed by different modes of reproduction are to a large extent mutually exclusive: none of the major breeding systems can realize advantages without at the same time incurring potential disadvantages. The balance of advantage and disadvantage in particular circumstances must determine the success or failure of apomixis, automixis and amphimixis in different taxa.

But how is this balance struck? Perhaps the great bulk of evolutionary change resides in rather small shifts in phenotype involving selection coefficients of at the most a few percent. Certainly Darwin thought so, and the demonstration by mathematical population geneticists that even very slight selection differentials will drive almost deterministic changes of gene frequencies in large Mendelian populations has given a good deal of credence to his opinion. An allele which is at a consistent slight disadvantage relative to some alternative allele will be lost from natural populations within a span of time which is quite short on the evolutionary scale. An allele which confers some large disadvantage on the individuals which bear it will be lost very quickly indeed. But what of alleles which code for sexuality? Sex is no minor readjustment having an inconsiderable effect on fitness. Sex tears apart every genome in every generation, and builds them up anew, every one different; and in doing so it does not merely reduce fitness, but halves it. If a reduction in fitness of a fraction of one per cent can cripple a genotype, what will be the consequence of a reduction of 50 per cent? There can be only one answer: sex will be powerfully selected against and rapidly eliminated wherever it appears. And yet this has not happened.

On the face of it, this might appear to be no more than a paradoxical gloss on what is fundamentally not a very difficult problem. To be sure, one consequence of sexuality is its costliness, and this requires an explanation. But we know that the other invariable concomitant of sexuality is the diversification of genomes amongst the progeny. It is elementary to reason that genetic diversity is sufficiently valuable to pay the cost of sex. The real difficulties begin to appear only when one attempts to discover where this value resides. It is not, of course, difficult to imagine benefits that might be gained by having diverse rather than uniform progeny, but it turns out to be very difficult indeed to conceive of any benefit in diversity, when uniformity appears to permit one to reproduce twice as rapidly. I shall devote the rest of this book to an attempt to uncover the evolutionary and ecological significance of diversity.

2 THEORIES OF SEX

2.1 A Scientific Method. (1) Hypotheses

There is something to be said for suppressing all discussion of the philosophy of biology. Too much of it is perpetrated by philosophers who know no biology, or by biologists with no knowledge of philosophy. However, philosophy cannot be completely excluded from any book about evolution, since the logical status of the theory of natural selection is still a live issue, at least among people who are not evolutionary biologists. But whilst the intrinsic merit of philosophical disputations by biologists may be doubted, they do at least serve to tell us how the author conducts his research, and in particular how he constructs hypotheses and on what grounds he accepts or rejects them. The following paragraphs should be read in this spirit. They are not intended to define the one true scientific method, but only to describe the method I have myself used in dealing with evolutionary problems.

It is impossible to argue, except from axioms which are not themselves in dispute. Until (in Section 5.1) forced to retract, I shall base all my reasoning on an axiom of perfection, which, in the present context, states that adaptation is extremely precise. I shall take for granted not merely that selection tends to favour more fit at the expense of less fit phenotypes, but that the phenotypes which are actually present are more fit than any alternative phenotypes; or, in a phrase, that selection alone is a sufficient explanation of organic diversity. Perhaps this is wrong, and the world is full of queer accidents that cannot be explained in the absence of a perfect knowledge of the past; in which case, biology is the study of miracles. It may be so; I do not insist on the truth of the axiom; but I do insist that it is absolutely necessary to assume its truth if we are to make any progress with explanation. If its truth is denied, then whenever we are confronted with a problem whose solution is not self-evident we shall be tempted to refer it to the operation of unique causes in an unknowable past. We shall become mythologists.

The force of this axiom is to compel the belief that genetic systems are highly precise adaptations. Otherwise we must believe that they are makeshifts, that several of them may perform the same function equally well and that a particular system has been adopted only by chance. We may even be driven to the belief that a particular system has been adopted even though it is demonstrably inferior to a number of alternatives. This is indefensible, since it means that any explanation we propose cannot be refuted. If our observations contradict our explanations, this merely points to the prevalence of miracles: our axiom permits any residue of unexplained or contradictory observations, however great. Since this is manifestly unscientific, we must instead argue from an axiom of perfection. It follows that an explanation of any particular genetic system can be judged to be satisfactory only if it refers quite specifically to that system, and not to any more general category of systems. For example, it has often been suggested that simultaneous hermaphroditism is adaptive when population density is low, because it permits individuals to reproduce through self-fertilization when there is very little chance of encountering any other individual with which to mate. This argument may very well be true, but it is nevertheless not an acceptable explanation of hermaphroditism, since either apomixis or a nongametic automixis would serve the turn as well as autogamy. Granted the truth of the relationship between population density and the genetic system, it still remains to discover why, in any given case, hermaphroditism and autogamy have evolved, rather than apomixis or automixis. If the reply is that hermaphroditism is only one of a number of possible solutions to the problem of being unable to find a mate, and whether or not it happens to be adopted in any particular case is an historical accident, then we shall have a theory of nongonochoric genetic systems, but not a theory of hermaphroditism. To obtain such a theory, we must infer the unique adaptive properties of the characteristics which are peculiar to hermaphrodites.

I must stress again that I do not advocate the truth of this axiom. Indeed, its truth could never be established, unless through an unimaginably minute and exhaustive survey of genetic systems. I do maintain that it is literally indispensible to any attempt at explanation, and that if it is denied no reputable theoretical work is practicable, or even conceivable. Of course, it is exceedingly likely that even the very best theories we are capable of imagining will leave some residue of unexplained variation. But we know we have advanced when this residue is reduced. It would be futile, indeed, to preface our investigation by assuming the existence of an arbitrarily large quantity of inexplicable variation.

Some of this can be translated into the technical language of population genetics. My position is the strictly Darwinian one that variation occurs indefinitely and in all directions, and that natural selection is universal. Like all approaches which assume that an optimal phenotype will evolve, this assumes the existence of adequate genetic variance for the character under selection. Although this might appear to be a straightforwardly empirical question, it is difficult to reduce it to empirical terms because the axiom requires only the small amount of variance in each generation necessary to fuel long-term evolution, and small amounts of genetic variance for characters such as genetic systems are difficult to detect. Moreover, even if they were detected in some particular cases it does not follow that the existence of adequate genetic variance is a general rule. It might also be argued that, when we attempt to predict which of a number of alternative phenotypes will respond most rapidly to selection, we assume not only that all have appreciable heritability but that all have approximately equal heritability. Obviously, a phenotype might be rapidly selected because of its high heritability even though it was less fit than some alternative phenotype with much lower heritability. I shall therefore adopt the usual subterfuge of assuming throughout that we are dealing with systems in evolutionary equilibrium, whose history has been effaced by continued selection. Finally, I shall assume that different genetic systems have different consequences and therefore respond to different selection pressures, and that the proper function of theory is to identify these selection pressures as precisely as possible. I shall take it that these selection pressures represent the ultimate significance of a genetic system, and that their identification constitutes an 'explanation' of the system.

Explanation itself is emphatically an art; it could not be otherwise, since if it were a science it would rest on axioms which themselves required explanation, and of which there would be an infinite regress. The criteria by which we judge explanations are therefore aesthetic and even poetic. I shall use three such criteria: aptness, generality and simplicity. An explanation can be apt only if it refers specifically to the process being investigated, and not to any more general category of processes. I have already covered this ground. When faced with a choice of several rival theories, all of them apt, we must make a decision between them on the grounds of generality and simplicity. Since the object of theorizing is to explain as much of the world as we can, the more important of these two criteria must be generality; we shall always prefer the theory which has the wider application. A theory may pass every test that we put to it, but if it can be applied only within narrow limits our interest in it will be small. The ideal hypothesis, then, will be highly specific, in that it refers only to the particular system requiring explanation, and very general, in that it provides an explanation for all cases of the system.

The human intellect is inclined to prefer the simple to the complex, and we are exhorted to use economy in the construction of our hypotheses. More precisely, we are enjoined not to multiply entities without necessity. The force of this injunction is considerably weakened by the practical difficulty of obeying it: like many edicts, Occam's Razor is easier to legislate than to enforce. If we wish to interpret the allelic diversity of natural populations, does drift or selection offer the more economical hypothesis? There is no straightforward answer, since there is no general agreement about the nature of an 'entity'. Instead of preferring economical hypotheses I shall prefer simple ones – or even, if you like, homely ones. Solving routine problems in population genetics often requires great subtlety and technical skill, but the big problem will yield only to a blunt instrument. Natural selection itself is the classical example, being so blunt an instrument that many philosophers have claimed it has no point at all. Perhaps simplicity is no easier to define than economy, but it is easier to understand. A simple explanation is one which invokes causes whose nature is immediately apparent to an untrained observer. Natural selection is a simple theory because it can be understood by everybody; to misunderstand it requires special training. A theory is unlikely to be of general interest unless it is possible to form a clear and direct mental image of how the proposed causes produce their effects. This is unlikely to be possible unless the theory can be stated in plain language, without resort either to jargon or to mathematics. Particularly I distrust the elaborate mathematical fantasies which are sometimes advanced as explanations of familiar natural phenomena. (Of course, whilst everyone is agreed that no more than a certain maximal mathematical sophistication is necessary in order to solve biological problems of any importance, everyone places that maximum at the point where their own mathematical training ended.) Since I see hackles being raised, I stress that extreme simplicity will be possible only for explanations of the most familiar and important processes; the further a process lies from common experience the more we must resort to a specialized vocabulary and to mathematical reasoning.

An explanation which is apt, general and simple is certain to be interesting, and will command our attention until it is refuted. Nothing is known about how such theories arise, as is proven by the notorious fact that they cannot be produced to order. Until we have a theory our mental processes are governed by poetic principles, and are unpredictable; once the beast has come into view we can act on scientific principles, in which predictability plays a leading part. What we have is a premise or a set of premises whose consequence is inferred to be the process or phenomenon we wish to explain. The premises and the manner of inference together constitute an hypothesis. (A theory is a well-confirmed hypothesis, or a set of related hypotheses; a speculation is an hypothesis which has not yet been clearly or completely stated.) Unfortunately, an hypothesis may meet all the criteria for being interesting, and yet be false. The business of the scientist is not only to formulate hypotheses which are as interesting as possible, but also to test them as searchingly as possible. The crucial tests of hypotheses are empirical; we strive to falsify hypotheses by testing the premises on which they rest or the predictions which follow from them. These tests are discussed in the first section of Chapter 4. But there is another sort of test which is not empirical at all. Between premise and prediction lies a verbal argument concerning causes and mechanisms which must be shown to be logically valid if the hypothesis is to be acceptable. For evolutionary arguments, this boils down to demonstrating that a selection pressure applied to a population which obeys the laws of inheritance will produce the end-result demanded by the hypothesis. When this cannot be done, the chain of deductive reasoning which forms the core of the hypothesis cannot be held to be valid; when it has been done, but the results run counter to expectation, the existence of a contradiction or inconsistency has been demonstrated. In order to achieve a rigorous demonstration that an hypothesis contains no internal contradiction we must build a population model; and since models in biology have been the target of much bad language over the last two decades, I shall describe how I mean to use them.

A model is a simple formal statement of an hypothesis. The purpose of a model is to provide a rigorous proof that the consequences envisaged by an hypothesis follow from the stated premises by a valid process of deductive reasoning. The need for rigour means that models are usually expressed mathematically, and this has led to a strenuous resistance to their use from naturalists who believe that the essence of their subject is too subtle to be captured by the stark formalism of algebra and geometry. It is also said that many models are incomprehensible to anyone who lacks a training in higher mathematics, and that the models which can be understood usually turn out to be merely tautologous. The second of these criticisms must be allowed; much of theoretical population genetics seems too abstruse ever to have any direct biological application, and many aspects of the subject are now beyond the reach of practising naturalists. Models are intended to assist in the development of scientific hypotheses, and when they fail to provide assistance they must be either abandoned or else bequeathed to pure mathematicians as purely aesthetic exercises. However, models should not be condemned because they are too simple to provide a faithful representation of reality, or because they are tautologous. A model is to an hypothesis as a diagram is to a picture; it is a representation so simple that the relationship of its parts can be established beyond dispute. In verbal argument it is often difficult to be sure of one's footing, even for quite simple problems in genetics. A perfect intuitive understanding of the operation of selection in Mendelian populations is given only to a few, especially when it is the genetic system itself which is being selected. A model aims to supply this certainty, but it can do so only through an extreme simplification. The behaviour of even rather simple models is often baffling; that of complicated models is usually unspeakable. If a model appears to be a mere caricature of nature, it is as well to remember that caricatures attempt to portray the irreducible essentials of their subject, even at the cost of ignoring all else – and a good model attempts to do the same thing. A model makes no comment on the truth or falsity of its premises, being concerned only that they are the simplest that will serve to represent the hypothesis; a model takes it premises for granted, and seeks only to validate the deductions which the hypothesis has suggested. A model, in short, adds nothing to its premises, but merely restates them in a different way. A model is therefore either tautologous or wrong. If any model appears to generate conclusions which are not implicit in its premises, this is due either to demonstrably incorrect reasoning or to a sleight of hand in which new premises are insinuated at some intermediate stage in the reasoning. It is, then, no severe criticism of a model to claim that it is tautologous; all good models are. A model can neither hold a mirror to nature nor generate new knowledge; nor is it intended to. A model is merely a mathematical crutch to supply the deficiencies of our intuition.

From what I have said, it will be obvious that I intend to develop strategic rather than tactical models. A tactical model is designed to produce precise predictions in particular situations, and it may as well be confessed at once that we do not understand enough about sex to build tactical models. A strategic model is designed solely to validate verbal arguments and analogies, and may be permitted, if necessary, an extreme elasticity in assigning values to ecological and genetic variables. Empirical tests will come later; for the moment I am concerned only with logical rigour, and will make almost any sacrifice of realism to attain that end. Naturally, the liberties we must take with our preconceptions of nature in order to make a strategic model work may tell us a good deal about the hypothesis under test; in particular they may suggest that its premises reduce to values of input parameters which seem absurd, or that it is excessively sensitive to slight variations in these input parameters. These will be legitimate matters for concern when the time comes to investigate the premises empirically.

2.2   The Role of History in the Evolution of Sex

Sex does not fossilize well, and its early history will perhaps always be a matter for speculation. It seems certain, however, that mechanisms of genetic exchange and recombination are such ancient features of life that to gain an adequate perspective of their history we must reach back to the origin of life itself.

Origin of Sex and Meiosis. In some of the oldest known unmetamorphosed sedimentary rocks there is a discontinuity in carbon-isotope ratios which has been held to indicate the appearance of autotrophic organisms about 3.2×10^9 years ago (Oehler *et al.*, 1972). The same deposits contain numerous organic spheroids some $20\,\mu m$ in diameter which may represent the earliest known cells (see the discussion in Schopf, 1970). The great variance in size of these structures has led Schopf (1976) to question their biological nature, in which case the oldest unquestionable remains of living organisms are the coccoid and filamentous forms found in the Transvaal Dolomite (about 2.3×10^9 years old; MacGregor *et al.*, 1974) and the Gunflint cherts (about 1.9×10^9 years old; Barghoorn, 1971). More recently, however, Knoll and Barghoorn (1977) have described organic spheroids from cherts some 3.4×10^9 years old which have a respectably low variance in size and which include figures strikingly reminiscent of binary cell division. We are left with something approaching certainty that primitive cellular organisms were alive $(2-2.5) \times 10^9$ years ago, and with a suspicion – based on the equivocal status of supposed microfossils, the discovery of very old structures interpreted as algal stromatolites and the isolation of chemically organic compounds from ancient sedimentary rocks – that cellular evolution may have begun as early as 3.5×10^9 years ago, soon after the first sediments were deposited.

Whatever the date of their origin, these earliest fossilizable organisms were prokaryotes, the remote ancestors of the modern bacteria and blue-green algae: they are externally similar in size and appearance to modern prokaryotes, and they contain no discernible intracellular organelles. During the middle of the Precambrian these prokaryotes diversified, until by about 10^9 years ago the range of growth forms – unicells, colonies and filaments – was nearly comparable to that of the modern flora. At some point during this diversification the first truly eukaryotic algae evolved from blue-green ancestors. They are certainly present in the Bitter Springs Formation, laid down about 0.9×10^9 years ago (Schopf, 1970), but cells judged to be eukaryotic, because of growth form or the presence of apparently membrane-bound inclusions, have been reported from a number of deposits of between 1.8 and 1.1 × 10⁹ years old (Cloud *et al.*, 1969; Schopf and Fairchild, 1973; Diver, 1974). It must have been about this time that mitosis and meiosis first arose. Tetrads of algal cells are known both from Bitter Springs and from a much older (1.5 × 10⁹ years old) Australian dolomite (Schopf and Oehler, 1976), and if they have been interpreted correctly may mark our first glimpse of eukaryotic cell division. At the base of the Cambrian, in rocks some 0.6 × 10⁹ years old, a diverse assemblage of multicellular organisms appears, and the familiar patterns of syngamy and meiosis must already have been ancient by this time. The phenomena we shall seek to explain, then, had their obscure beginning some thousand million or more years ago, and it is worth giving at least passing attention to the problem (even though it be eventually insoluble) of how they first came about.

The first living organisms, had anyone been there to describe them, would have been recognized because of their capacity for self-replication. No doubt the earliest processes of replication were inexact, with the progeny departing more or less widely from the condition of their parent; but this process would be in no way sexual, being more nearly analogous to mutation than to recombination, and there must have been ineluctable selection towards more exact and more reliable methods of replication. Obviously, genes which cause more exact replication of the genome will themselves be more exactly replicated, and will be transmitted more efficiently than alternative genes which are liable to lose their identity during reproduction. But modern bacteria have a sexual process of sorts, and for want of any direct evidence we must assume that some mechanism of genetic recombination arose very early in the prokaryotic line, and probably predates the evolution of eukaryotes. Can we derive eukaryotic from prokaryotic sexuality?

Baker and Parker (1973) argue that we can, and that in consequence the evolution of sex is thoroughly monophyletic. Their two main lines of evidence are that both chlorophyte algae and the higher plants are usually sexual, so their common ancestor must have been; and that the cytology of meiosis is very similar in most eukaryotes. Moreover, the biochemical similarity of modern prokaryotes and eukaryotes suggests a common ancestry, in which case the common ancestor of both must have been sexual. Now, they use the term 'sex' in a very general sense to indicate any process by which DNA transfer from one cell to another is followed by recombination and then by cell division, without necessarily implying the occurrence of true syngamy or meiosis. This implies that, in geological time, prokaryote sexuality has gradually evolved into eukaryote sexuality, a supposition which I find improbable.

In the first place, the two processes are very different: in prokaryotes some variable length of DNA is inserted into a circular chromosome which is not localized in a nucleus, whilst in eukaryotes the complete transfer of an entire genome by syngamy is followed by the regular segregation of linear chromosomes within a nucleus. Granted that both lead to some degree of genetic diversification, the two processes are about as different as they could possibly be. Secondly, everyone seems to be agreed that the immediate ancestors of the eukaryotes were not bacteria, in which a sexual process of sorts is well-documented, but blue-green algae, in which recombination has been only very recently described (Bazin, 1968), and is certainly very rare. Finally, their comparative arguments seem fallacious. The fact that both green algae and higher plants are generally sexual does not by any means imply that all their common ancestors, however remote, were sexual. It is a brute fact that, of the major taxa which are nowadays wholly or largely asexual, almost all are protists; and if this has any value at all as comparative evidence, it must indicate that asexuality is the primitive and sexuality the derived state of eukaryotes (Boyden, 1953). The converse view is defended by Dougherty (1955). It is true that meiosis is strikingly similar in almost all eukaryotes but, even taken at face value, and ignoring the possibility of convergence, this means only that meiosis evolved only once, perhaps after a long period of time during which all the ancestral eukaryotes were asexual. I conclude that, although contrary views are not excluded, it makes perfectly good sense to assume that the ancestral eukaryote was a haploid asexual unicellular alga.

All current ideas about the origin of sex lay great stress on fusion as the primary attribute of sexuality. This is usually supposed to have arisen through phagocytosis (e.g., Dougherty, 1955; Cavalier-Smith, 1975), an idea which eventually descends from archaic notions that sex originates in a 'hunger' for acquiring more cytoplasm (reviewed and criticized by Weismann, 1889). In more sophisticated versions the primal fusion is something akin to phage infection (e.g., Dougherty, 1955; Baker and Parker, 1973). To avoid an elaborate exposition of these schemes, I have reproduced Dougherty's ideas in Figure 2.1. A more sophisticated version has been advanced recently by Maynard Smith (1978), who envisages the initial formation of a dikaryon by the fusion of two haploid cells. This might be advantageous because any function that is lacking or inferior in one nucleus might be supplied by the other nucleus, and this function would be reinforced by the evolution of true diploidy and a regular mitosis. Recombination appeared first as a mitotic phenomenon which generated genotypic diversity at the expense of increasing homozygosity, and finally meiosis and syngamy evolved in order to maintain heterozygosity. This scheme is shown in Figure 2.2. It is an ingenious attempt to assign a function to each stage in the early evolution of sex; unfortunately, as Maynard Smith himself points out, each stage involves a reversal of the selective pressure which led to the evolution of the previous stage, so that diversity is first sacrificed for genetic complementation in a dikaryon, then regained at the expense of heterozygosity in a diploid, and finally suppressed in order to restore heterozygosity.

It is possible that this emphasis on fusion is an error, and that a plausible phylogeny can be constructed in which cell fusion is a comparatively late acquisition. Let us begin by assuming, as we must, that the immediate ancestor of

Figure 2.1: Two Schemes for the Origin of Sex. (a) Upper: a cell lyses and releases a self-replicating particle which is capable of infecting intact cells, behaving
like a bacteriophage. (b) Lower: self-replicating particles are transferred during
transient cell fusion. Note that there is no essential difference between these
two schemes, both of which assert that the transfer of part of a genome from
one cell to another is the primitive sexual process.

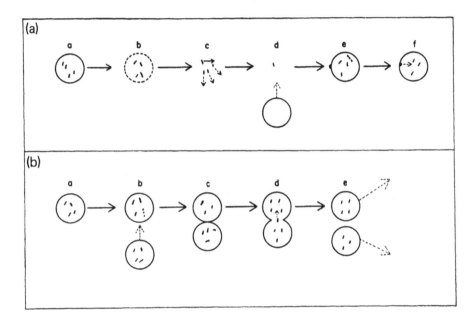

Source: Dougherty (1955), p. 160, Figures 2 and 3.

eukaryotes was a blue-green alga. Since all modern blue-green algae are thoroughly
asexual, we have no warrant for making any assumption other than that our
putative ancestor was also asexual. It follows that the first eukaryote was a
haploid asexual alga, a conclusion previously reached by Boyden (1954). I shall
endow it with a nucleus, a set of linear chromosomes and an efficient mitosis.
Suppose that a failure of cytokinesis during mitosis gives rise to an autodiploid
homozygous amictic line. Further mitoses will perpetuate the diploid condition,
but we can imagine that a variant form arises in which the homologous chromosomes separate, without cell division, and then reunite, an automictic process
functionally equivalent to amixis. If we now allow cell division to occur after
the separation of the homologues, but still require that the cells formed in this
way fuse immediately afterwards, we still have an automixis which precisely
restores the maternal genome, but one which involves the formation of distinct
gametic cells. If the organism concerned has more than one pair of chromosomes,
or if it forms more than two gametes, the progeny will be diversified to some

Figure 2.2: Maynard Smith's Scheme for the Origin of Sex. For a discussion, see text.

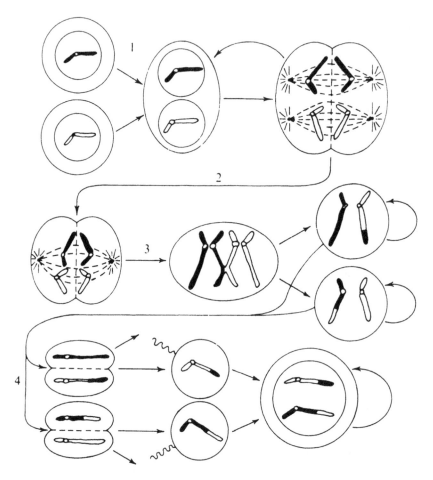

Source: Maynard Smith (1978), p. 8, Figure 1.

extent because the homologous and originally identical chromosomes will have accumulated somewhat different complements of mutations. Finally, when the gametes can escape from the parental soma and fuse with gametes of different parentage we shall have arrived at amphimixis, with much greater potential for diversification. Recombination will generate further genotypic diversity from mutationally created allelic diversity whether it first arises in the automict or in the amphimict. The merit of this scheme is that by laying stress on division rather than on fusion it proceeds by gradual steps, all of which are known to occur in living protists. Its two principle demerits are that the initial steps in the sequence are neutral or nearly so, and that it leads to the curious conclusion that haploidy is primitive in asexual but derived in sexual protists. However, I do not wish to defend the scheme, and describe it only in order to point out that the usual assumption that the primal sexual process was something akin to gamete fusion should not be made too lightly.

The Historical Hypothesis. Whatever may be the case in protists, meiosis is a remarkably uniform phenomenon throughout the higher eukaryotes, and I shall not dispute the conventional inference that the most recent common ancestor of these higher taxa was sexual, and perhaps even amphimictic. This means that the major historical problem in the higher eukaryotes is to account, not for the origin of sexuality, but rather for the repeated origin of parthenogenesis. But it also means something more to our main purpose, since if sex is primitive we have only to explain how its loss is prevented, rather than how its acquisition is accomplished.

This inference has recently been used by Williams (1975) as the basis of a distinct theory of sex, which I shall call the historical hypothesis. Williams devotes much of his book to an attempt to explain how sex can be favoured by individual natural selection. I shall describe these attempts below; but he is eventually forced to the conclusion that sex can be favourably selected only in organisms with high fecundity which continually experience intense selection with respect to characters determined at loci other than those concerned with the determination of sex itself. He is therefore driven to invoke evolutionary inertia: sex is consistently maladaptive in low-fecundity organisms, such as most vertebrates and insects, which nevertheless lack the preadaptations necessary to rid themselves of sexuality and adopt parthenogenesis. Naturally, one must agree that a character cannot evolve if it does not vary, or if it varies but is not heritable; and Williams' hypothesis, restated, is that there is no genetic variation for the expression of sexuality, at least in low-fecundity organisms. This is, in the first place, indicative of an important contradiction in Williams' account of sexuality. To explain the evolution of sex under natural selection, he adopts the extreme position that progeny of very high fitness for almost any conceivable environment will always be generated by sexual reproduction, so that whatever character or combination of characters is 'required' by the environment will immediately be available for selection. But when this explanation breaks down, he makes one exception to the rule, one character which entirely lacks genetic variation; and this is precisely the character of interest, the breeding system, which in the absence of variation can then be conserved no matter how intense is the selection which acts upon it.

This may seem gratuitous, but it is not indefensible, and Williams uses three main lines of defence. The first is that if haploid parthenogenesis (or certain types of automixis) arose in a diploid amphimictic stock, the first-generation progeny would suffer severe inbreeding depression. This argument is valid if most genetic variation is neutral, or if most balancing selection is due to heterosis; it will be of much less account if allelic variation is often maintained by frequency-dependent selection. If it is valid, and a good many deleterious alleles are suddenly exposed in the hemizygous or homozygous state, this will be a serious handicap only for the first generation after the change in breeding system. Secondly, diploid ameiotic parthenogenesis can evolve only if several genetic changes occur simultaneously, since the partial suppression of meiosis would be likely to produce unbalanced eggs. This argument is not watertight, since the partial suppression of meiosis (for example, the suppression of the first but not the second meiotic division) could lead to a meiotic partheno- genesis, which could later evolve to an entirely amictic system. However, this is an empirical problem which can be resolved only by a knowledge of the genetic basis of parthenogenesis, and of the immediate ancestors of diploid amictic taxa. Finally, it is argued that, even if meiosis is suppressed and viable unreduced eggs formed, all will have been in vain unless these eggs are capable of resisting fertilization; since if they are fertilized they will give rise to odd- ploid individuals which are likely to be sterile or inviable. This objection is nullified if it can be shown that parthenogenesis normally arises in association with polyploidy or some other device which acts as a primary isolating mechan- ism between sexual and asexual lines, or if reproductive isolation through modifications of behaviour or anatomy can evolve rapidly after the acquisition of parthenogenesis. These empirical questions are deferred until Chapter 4.

A more fundamental reason for objecting to the historical hypothesis is that it violates the axiom of perfection. It is well-known that theoretical population geneticists prefer equilibrium theories and tend to discount the role of history; there is a good reason for this. The equilibrium position of a population — roughly, the evolutionary stable strategy, in Maynard Smith's terminology — can be predicted in terms of parameters which are at least in principle measurable. If our results accord with reality, even by the optimistic standards that evol- utionary biologists are accustomed to use, then we have grounds for claiming that we have explained a phenomenon, that we understand it. Once we are driven to invoke the unknown effect of history, and especially if the history itself is not beyond dispute, then we have made, not an explanation, but a confession of ignorance. The only prediction generated by the historical hypoth- esis is that the correlates of parthenogenesis should be taxonomic rather than ecological; if we restrict ourselves to the measurement of ecological variables the only prediction we can make and test is that no prediction is possible; the historical hypothesis stands only so long as the predictions made by rival hy- potheses fail; and even so they stand precariously, for we can never be sure that no successful nonhistorical hypothesis can be imagined. Williams may be right: sex may be a genetic hangover from our distant past. But until alternative hypotheses have been tried and have failed, I shall continue to believe that history is bunk.

2.3 The Vicar of Bray

The idea that sex is selected because of its effect on the genetic structure of the populations can be traced back at least to Weismann, who wrote that: 'sexual reproduction will readily afford such combinations of required characters, for by its means the most diverse features are continually united in the same individual' (Weismann, 1889, p. 281). In the next sentence he makes it clear that the significance of sex lies in its effect on the population, or the species, rather than on the individual: 'I do not know what meaning can be attributed to sexual reproduction other than the creation of hereditary individual characters to form the material upon which natural selection may work'. This view became widely diffused, and fifteen years later we find Guenther (1906, pp. 256–68) writing that:

> The chief significance of amphimixis is that before a new individual is formed the characteristics of two animals must be blended . . . Thus the new individual has a selection of paternal and maternal traits and of the ancestors on either side . . . The great value of amphimixis, then, is that it adds new and different paternal qualities to those that the new organism receives from the mother . . . the constant re-combination of characters in the offspring is of the greatest value; it gives a wider field of operation to natural selection. It enormously increases the adaptive capacity of the animals, and the variations from which one animal would arise here, and another there, are united in one individual. It is due to amphimixis, therefore, that co-adaptations do not need to be selected slowly and successively, but may appear simultaneously.

Guenther's book, originally published in 1904, is essentially a pre-Mendelian work; a quarter of a century later the same ideas about the function of sex were harnessed to Mendelian population genetics by Fisher (1930) and Muller (1932). Fisher (1930, p. 137), with his usual terseness, wrote that:

> The comparative rates of progress of sexual and asexual groups occupying the same place in nature, and at the moment equally adapted to that place, are therefore dependent upon the number of different loci in the sexual species, the genes at which are freely interchangeable in the course of descent . . . even a sexual organism with only two genes would apparently possess a manifest advantage over its asexual competitor . . . from an approximate doubling of the rate at which it would respond to natural selection.

Muller (1932, pp. 120, 121, 123) was expressing the same idea when he wrote that:

> The essence of sexuality, then, is Mendelian recombination . . . the major value of recombination is the production, among many misfits, of some

combinations that are of permanent value to the species . . . the advantage
of sexual over asexual organisms in the evolutionary race is enormous.

Because of these contributions, this theory of sex has been referred to as the
Fisher-Muller theory. However, quite apart from a distaste for naming theories
after people as though they were diseases, this is a misattribution: the same
idea is clearly expressed by Weismann and Guenther, and doubtless by others —
it is even foreshadowed in some passages of Erasmus Darwin. At the same time,
we need names for theories, and preferably striking ones, if we are to avoid an
enervating circumlocution every time we want to refer to an idea. I have therefore named this theory for a quite different historical personage, an English
cleric noted for an ability to change his religion whenever a new monarch
ascended the throne. The 'Vicar of Bray' teaches that there may be great advantages of easily and gracefully adapting to changed circumstances.

The idea is a very simple and intuitively appealing one. Suppose that a haploid
population is fixed for the wild-type alleles $A1$ and $B1$ at two unlinked loci.
Conditions change so that the mutant alleles $A2$ and $B2$ become more fit than
their wild-type alternatives; the population will therefore become fixed for the
$A2B2$ genotype at equilibrium. However, each mutant allele will almost certainly
arise in a genome which bears the wild-type allele at the second locus, so that
$A1B2$ and $A2B1$ individuals will arise at mutation frequencies but the double
mutant $A2B2$ will arise only at a rate proportional to the product of the two
mutation frequencies, an exceedingly small number. In asexual lines of descent,
therefore, $A2B2$ genomes can be created in reasonable numbers only by the
mutation of $A1$ to $A2$ when the $B2$ allele has become common, or by the
mutation of $B1$ to $B2$ when the $A2$ allele has become common. But in a sexual
line of descent, $A2B2$ individuals will be created by recombination whenever the
$A1B2$ and $A2B1$ single mutants mate together. In this way recombination raises
the initial frequency of $A2B2$ individuals in the population, and thereby greatly
hastens their eventual fixation. To put it more bluntly: sex accelerates evolution.

However appealing, this argument rests on four assumptions, none of which
is entirely uncontroversial. In the first place, the process of evolution is pictured
as the replacement of wild-type alleles by newly arisen favourable mutations,
rather than as a shifting balance in the frequencies of alleles at polymorphic loci.
The Vicar of Bray must therefore be referred to the classical theory of genetic
variation, and stands or falls with the classical theory. Secondly, the advantage
invoked is an advantage that accrues to populations or species, rather than to
individuals, and the evolutionary process involved is therefore one of group
selection. Thirdly, it is taken to be self-evident that a greater rate of evolution
really does constitute an advantage, since populations which can evolve more
rapidly are less likely to become extinct when conditions change. Finally, the
verbal argument is not itself wholly conclusive in the absence of a mathematical
demonstration that recombination is capable of altering the rate of evolution in
genetically defined populations. Until quite recently, attention has centred on

this final assumption, and theorists have concentrated on obtaining a mathematical description of the effect of recombination on the substitution of mutant alleles at two or more loci simultaneously.

The first attempts to construct algebraic models were made by Muller (1958, 1964) and by Crow and Kimura (1965). The principles which lie behind the analysis are illustrated in Figure 2.3, which descends from the diagram originally published by Muller in 1932 and often republished since. In the asexual population, mutations can be incorporated only in series, so to speak, since two favourable mutations can be fixed only if the second occurs in the same line of descent as the first, whilst in a sexual population the two mutations can be incorporated in parallel, having been brought together by recombination. In a small population this difference may be trivial, since favourable mutations arise so infrequently that each has been fixed before the next arises. In a large population, however, mutant individuals carrying a novel favourable gene are likely to arise in every generation, and the fixation of the compound mutant genotype may be greatly accelerated by recombination. Suppose that the population size is N and the rate of mutation to favourable alleles per individual per generation is u, whilst on average a mutant gene which will eventually become fixed in the population arises every g generations, where g is a complicated function of N, u and the selection coefficient s. Crow and Kimura argue that the difference between asexual and sexual populations is that during each period of g generations one favourable mutation will be fixed in the asexual population whilst all the Nug mutations which have arisen during this period will eventually be fixed in the sexual population. The sexual population thus evolves Nug times faster than the asexual population, so that the importance of the effect hinges on the magnitude of the quantity Nug. If a small population experiences powerful selection then both N and g will be small, and the effect of recombination can be shown to be negligible; conversely, a large population which experiences weak selection has large values of N and g, and recombination may greatly hasten the fixation of mutant alleles. These conclusions are obvious enough from a scrutiny of Figure 2.3; what Crow and Kimura have added is a calculation of the numerical value of the factor by which sex accelerates evolution. For powerful selection ($s \gg u$) in small populations ($N < 10^5$ or so) this factor is not greatly different from unity, sexual and asexual populations evolving at nearly the same rate; but when selection is weak ($s \sim u$) and the population is large ($N > 10^8$ or so) the factor becomes nearly equal to the population size, and sexual populations fix favourable alleles in the blink of an eye whilst asexual populations dodder along for aeons without appreciable genetic change.

The case of neutral alleles was investigated by Karlin (1973), who was able to prove analytically that recombination hastened the appearance of double mutants in small populations as the inevitable consequence of its effect in decreasing the correlation between loci. By continuity, this result also holds for very weak selection. He also found that the fixation of the double mutant was hampered rather than hastened by recombination, especially in very small

Figure 2.3: The Vicar of Bray. An asexual haploid population (upper figure) is fixed for the three-locus genotype *abc* before a change in the environment which makes the alleles *A*, *B* and *C* more fit than their lower-case counterparts, the triple mutant *ABC* being fitter than any other genotype. The fittest single mutant type is *Abc*, which therefore becomes fixed, the two other single mutants *aBc* and *abC* failing to spread. Any double-mutant type is fitter than any single-mutant type, but since mutation is infrequent a double mutant will arise only when the single mutant *Abc* has become fixed. *ABc*, being fitter than the alternative *AbC*, is then fixed; and then one further mutation produces the optimal genotype *ABC*. In a sexual population (lower figure), on the other hand, the triple mutant *ABC* is soon produced through recombination and is thus fixed much more quickly.

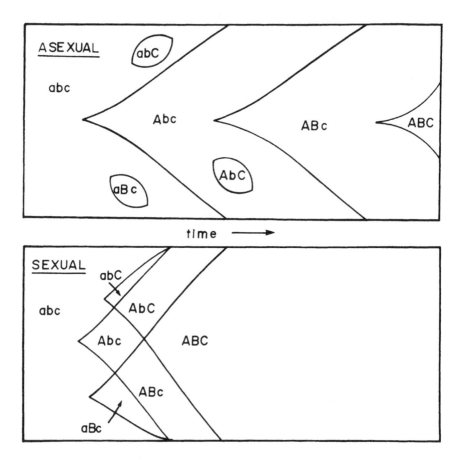

Source: Redrawn from Muller (1932).

populations, but this result follows from the strict neutrality of the alleles, and is of no consequence to the Vicar of Bray.

In this way, the early intuitions of Weismann, Fisher and Muller seemed to be vindicated: sex does accelerate evolution, and may do so to an enormous degree. If we are willing to accept two further premises – the validity of the classical theory of population genetics, and the potency of group selection – then we have a theory of sex: sex evolves because by increasing the rapidity with which genetic response to a changing environment can occur under selection it reduces the long-term risk of extinction. In 1965 this opinion seemed to be more firmly established than ever; but in retrospect we can see this as the high-water mark of the theory, and the following decade saw a gradual retreat into confusion and uncertainty.

The counterattack was begun by Maynard Smith (1968), who published a counterexample which appeared to falsify Crow and Kimura's logic. Imagine two loci in a haploid population at which the wild-type alleles are *A1* and *B1*, there being also two mutant alleles *A2* and *B2*. The effects on fitness of the mutant alleles are assumed to be small, negative and independent, so that they are only mildly deleterious and the fitness of the double mutant *A2B2* is obtained by multiplying the fitnesses of the two single mutants *A1B2* and *A2B1*. Under selection and recurrent mutation, genotype frequencies (denoted by the letter P with an appropriate subscript) will approach an equilibrium at which $P_{11}P_{22} = P_{12}P_{21}$, as a direct consequence of the assumption of independence of gene effect. We next imagine the environment to change so that the mutant alleles, mildly deleterious before, are now mildly beneficial. Genotype frequencies will of course change under selection, since the mutant types are now favoured and will increase in frequency. However, if the population reproduces asexually, the relationship $P_{11}P_{22} = P_{12}P_{21}$ will continue to be satisfied through the process of the substitution of *A2B2* for *A1B1*, since nothing has changed other than the signs of the selection coefficients. But this relationship is precisely that of linkage equilibrium, which will be established in a sexual population by the random segregation of alleles at unlinked loci. It follows that the rate at which *A2B2* replaces *A1B1* will be precisely the same in sexual and in asexual populations.

Maynard Smith concluded that the flaw in Crow and Kimura's argument was their assumption that favourable mutations are unique. It is this assumption which leads to the conclusion that *Nug* favourable mutations will be incorporated by a sexual population during the *g* generations which are required for the incorporation of a single favourable mutation into an asexual population. When this assumption is discarded, the enormously faster evolution of large sexual populations is no longer apparent.

However, Crow and Kimura (1969) were not slow to point out that Maynard Smith's counterexample was also flawed. The initial independence relationship between genotype frequencies on which Maynard Smith's argument hinges assumes that appreciable numbers of *A2B2* genotypes are created by recurrent mutation before the environmental change which favours the mutants. But if the balance of recurrent mutation with selection holds the frequencies of the *A2* and *B2* alleles down to, say, 10^{-5} before the environmental change, the double mutant *A2B2* will occur at a frequency of only about 10^{-10} and will therefore be absent most of the time even from very large populations. Naturally, *n*-tuple mutants, where *n* is greater than two, will be vanishingly rare even in the largest conceivable populations. Spurred by this criticism, Maynard Smith (1971a) proceeded to develop a more general model in which both population number and the recurrence of favourable mutations were explicitly taken into account.

His conclusions are most easily understood in the form of a diagram, as in Figure 2.4, which is essentially a generalization of Figure 2.3. We imagine a population which has lived in a particular environment long enough for wild-type alleles to be fixed at all loci. The environment then changes so that at a certain number of loci, say L, an alternative allele is favoured over the wild-type whenever it arises through mutation. As time goes on, favourable mutations arise, one by one, and are eventually fixed in the population; we wish to know how quickly this occurs in sexual and in asexual populations. If we isolate the spread of any one of these favourable mutations, we can break down its passage through the population into two phases. During the first phase, a certain length of time is required for the mutation to occur, after which the frequency of the mutant allele increases slowly under selection. Since it is rare, its dynamics are largely stochastic: it is likely to suffer large fluctuations in frequency from one generation to the next as the result of sampling error. However, it will eventually struggle up to a frequency at which the systematic effect of selection greatly outweighs random drift, and will thereafter proceed quite smoothly to fixation. The first phase, during which the allele is rare and its dynamics are largely stochastic, can be called the *establishment* phase, and occupies a period of time T_E; the second phase, during which the allele is common and its dynamics largely deterministic, can be called the *spread* phase, and occupies a period of time T_S. The values of T_E and T_S vary with the population number N, the mutation rate u and the selection coefficient s; sex may or may not accelerate evolution, depending on the magnitudes of N, u and s. In the first place, suppose that $T_E \gg T_S$, so that it takes much longer for an allele to become established than it does for it subsequently to spread through the population, which is the situation pictured in Figure 2.4a. In this case each favourable mutation which arises and becomes established has already spread through the population before the next favourable allele appears. Each new favourable mutation is therefore almost certain to arise in a line which already possesses all the favourable mutations which have so far had time to appear. Since the best available combination of alleles is then guaranteed by mutation, recombination does not hasten the appearance of these genotypes, and sexual and asexual populations will evolve at the same rate. This is likely to happen if N or u are small (so that mutants arise infrequently) or if s is large (so that any established mutant spreads very rapidly). Sex is unlikely to accelerate evolution in small populations exposed to intense selection. On the other hand, suppose that $T_S \gg T_E$, so that

Figure 2.4: Maynard Smith's Model of the Vicar of Bray. Explanation in text.

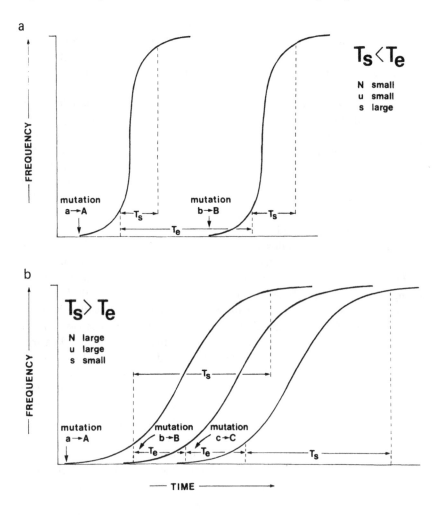

Source: Redrawn from Maynard Smith (1978), p. 18, Figure 3.

favourable mutations become established quite rapidly but afterwards spread only slowly, as in Figure 2.4b. When a new mutation arises, the previous mutation has not yet had sufficient time to become common, and so the two mutations are unlikely both to be borne by the same individual. In this case recombination will bring the two alleles together much more quickly than could be done by recurrent mutation alone, and a sexual population will fix the double-mutant type much more rapidly than an asexual population. This situation will arise if N or u are large (so that favourable mutants arise very frequently) or if s is small (so that established mutants spread only very slowly). Sex is then most likely to accelerate evolution in large populations which experience weak selection. This leads to a watered-down version of Crow and Kimura's original conclusion. They calculated the factor by which sex accelerates the process of gene substitution in large populations to be about $L(u/s)(N/\ln N)$, which may be very large if N is large; Maynard Smith shows that a more realistic estimate of the factor is about L or less.

As Williams (1975) has pointed out, the theory has by now developed an internal contradiction. Its motive force is group selection, and for very general reasons group selection requires a particular population structure in order to be effective: the population must be broken up into a great many small, semi-isolated groups (for discussion of a somewhat different point of view, see Wilson, 1975, 1980; Maynard Smith, 1976a; Bell, 1978a). But we have learnt that sex will accelerate evolution only in large populations. These two features of the Vicar of Bray are not easy to reconcile, and indicate that the whole basis of the theory may be seriously flawed.

In such a situation the reflex response of theoreticians is to fiddle with the fine tuning of a model, in an attempt to save it from becoming a heap of elegant junk. For instance, despite the fact that the primary effect of recombination is to change the relationship between alleles at different loci, the work that I have cited so far takes it for granted that there is no functional interaction between loci, so that the fitness of the double mutant in the two-locus case can be calculated by multiplying together the fitnesses of the single mutants. In a very interesting paper, Eshel and Feldman (1970) worked out the two-locus case in which the effects of alleles at different loci are not independent. Either both mutations are favourable when borne singly, but the fitness of the double mutant exceeds the product of the fitnesses of the single mutants, or else either mutation is mildly deleterious when it occurs alone but mildly beneficial when it occurs together with the other mutation. In either case the mutations are 'cooperative', in the sense that the double mutant is more fit than would be expected if the mutant alleles at the two loci had independent effects on fitness. It seems that, when mutations are cooperative, recombination actually slows down the rate at which the double mutant spreads through the population. For example, suppose that we begin with a population which is fixed for a wild-type allele at both loci, and then allow mutation and selection to occur. The population is assumed to be extremely large, so that an appreciable number of double mutants will appear whether the population is sexual or not. Powerful selection favouring these double mutants will create a positive linkage disequilibrium, which will be maintained in an asexual population as the double-mutant type increases to fixation. But in a sexual population the double mutants will be broken up by recombination, which will act in every generation so as to reduce linkage disequilibrium towards zero. Eshel and Feldman conclude that, when the assumption of multiplicative fitnesses is discarded, double mutants will often be more frequent in asexual than in sexual populations, because of the destructive effect of meiosis.

This conclusion was in part anticipated by Crow and Kimura (1965), who envisaged a haploid population in which both single-mutant types are less fit but the double mutant is more fit than the wild type. Since any increase of the single-mutant types will be opposed by selection until a sufficient number of doubly mutant zygotes have been formed, there will exist a point of unstable equilibrium which the frequency of the double mutant must exceed if either mutation is to spread. This unstable point does not exist in an asexual population, since the double mutants are not constantly being broken up by recombination to form the deleterious single-mutant genotypes. As Crow and Kimura point out, this idea can be traced back even further, to the adaptive topography of Sewall Wright (e.g. Wright, 1931).

These results have changed the whole tenor of the debate about the significance of recombination in evolution. It has long been clear that widespread heterosis would imply a serious disadvantage for the sexual habit in diploids, since meiotic segregation makes it impossible for heterozygotes to breed true. What Eshel and Feldman have proven is that the advantage for sexuality postulated by the Vicar of Bray also breaks down if fitness depends to any extent on positive epistasis. Sex will accelerate evolution only if the favourable interactions of genes, either within or between loci, can be ignored.

Theoreticians have in this way been led to the fork of an interesting dilemma. The classical view of genetic variation is that almost every locus in diploid organisms is occupied by two alleles which are identical in state or, at any rate, which have virtually indistinguishable effects on fitness. The major process of genetic change is the elimination of newly arisen deleterious recessive mutations; evolutionary progress results from the rare appearance of favourable alleles at mutation frequency and their subsequent fixation under directional natural selection. Heterosis is very rare, and the interaction of two or more mutant alleles can be ignored. In these circumstances sex may accelerate evolution, since sexual lines of descent can acquire favourable mutations at different loci in parallel through meiosis and syngamy. This opinion is not necessarily wrong, but it does necessarily involve a belief in the correctness of the classical theory, which includes, among other things, a belief in the selective neutrality of most of the variable enzyme systems detected by electrophoresis. Many geneticists, especially those who work in the field with natural populations, have not accepted the neutralist view of genetic variation, and they must therefore also reject the Vicar of Bray interpretation of sex. But there is a difficulty here, since the faster evolutionary response of sexual populations was long thought to represent the primary function of sex. It follows that if the adaptive significance of sex resides in its effect on groups of organisms rather than on individual reproductive success, and if the immense genetic variation of natural populations is maintained by selection, then sex must represent an advantage because it *slows down* evolution. Indeed, this view is also, but less strongly, urged on classical geneticists, by the contradiction between the population structure required for group selection and that required to fuel the faster genetic response of sexual populations to changed conditions. To save the situation, then, we must perform a complete *volte-face*: just as it was self-evident to Weismann, Fisher and Muller that a faster rate of evolution would benefit a population, so we must now contrive to believe in the self-evident desirability of evolving slowly.

This manouevre was accomplished by Williams (1975), through the device of erecting an alternative theory of extinction. Embedded in the Vicar of Bray argument is the notion that species become extinct because they are incapable of evolving fast enough to survive indefinitely in a changing environment. The rate of evolution is imagined to be limited by internal constraints such as the frequency of mutation or recombination. Williams gives a very different picture of extinction. He proposes that species are almost always capable of evolving fast enough to cope with changing selective demands. They become extinct not when their habitat changes, but when their habitat disappears. An asexual population can become fixed for a single genotype which is extremely well-adapted to a particular habitat; but if that habitat disappears, the clone, however superior to its competitors it may have been, disappears with it. A sexual population evolves more slowly and cannot attain the same precision of adaptation, but for this very reason is less likely to be wiped out by the destruction of its habitat, since many of its members will still be able to make a living elsewhere. Asexual taxa, in other words, tend to adapt themselves out of existence by perfecting a way of life which is eventually doomed to disappear. An essentially similar point of view has been put forward by Stanley (1976), who emphasized the greater frequency of speciation, rather than the lesser probability of extinction, in sexual lines of descent.

Now, sex is hardly likely to prevent extinction simply because it blunts the precision of adaptation. A taxon which is imprecisely adapted to a particular habitat (or niche, or biotope, or whatever), but which is nevertheless restricted to it, will become extinct when that habitat disappears just as readily as a taxon which has achieved precise adaptation. If sexual taxa are less sensitive to the disappearance of habitats, this can only be because they simultaneously occupy several of them, so that a disaster in any one of them does not threaten the continued existence of the taxon as a whole. The significance of the greater variance of fitness presumed to exist in sexual taxa is, therefore, that some part of this variance reflects the simultaneous occupancy of several habitats, each of which creates rather different selection pressures. An asexual population cannot diversify to the same extent, and instead consists of a single clone occupying a single habitat, or at best of a very few clones established in a very few habitats. Each clone may have a greater fitness than the average fitness of the sexual population with which it coexists, but the ecological narrowness of the asexual population implies a far greater sensitivity to the destruction of habitats. But by this point we have made another fundamental change to the original theory, by substituting for a uniform environment characterized by a single selection coefficient for any given genotype a complex environment whose spatial hetero-geneity will cause the fitness of an individual to vary according to the conditions in which it grows up.

In short, the original hypothesis seems almost as malleable as the populations which it claims to represent and the cleric for whom it is named. I shall pursue it no further, but close by emphasizing that in its final manifestation it has adopted a complex rather than a uniform model of environment. This is a theme to which I shall return.

2.4 The Ratchet

The Vicar of Bray, in its original form, suggests that sex is adaptive in a uniform but changing environment because it facilitates the rapid fixation of favourable mutations. A rival hypothesis of group selection asserts that sex may be adaptive in uniform environments which do not change in time because it facilitates the elimination of unfavourable mutations. This idea is due to Muller (1964), who referred to it as a 'ratchet' mechanism. Like the Vicar of Bray, the ratchet descends from classical genetic theory, but emphasizes the continual elimination of deleterious mutations rather than the rare fixation of favourable mutations. Unlike the Vicar of Bray it has undergone little theoretical development, but after a decade of silence some results have been published by Felsenstein (1974), Manning (1976) and Maynard Smith (1978, citing unpublished work by J. Haigh).

Imagine an infinite population of asexual haploid organisms. New mutations are constantly arising, and it is assumed that the great majority are slightly deleterious and have independent effects on fitness. The number of mutations borne by a representative member of the population will be determined by two opposed forces: mutation itself, which tends to increase the number of mutations, and selection, which tends to remove these mutations from the population. At some point the population will arrive at an equilibrium state where the contrary effects of mutation and selection are exactly balanced. At this point the population can be described by the frequency distribution of the number of mutations per individual, with a small proportion of individuals bearing very few mutations, an equally small proportion bearing very many, and with the majority bearing some intermediate number. Once attained, this distribution will be replicated indefinitely. It is defined mathematically by Maynard Smith (1978, pp. 34-6).

Now suppose that the population is finite, and perhaps small. Its dynamics are still governed by the deterministic equilibrium frequency distribution, but those classes which contain a very small proportion of the population now include only a small absolute number of individuals. For instance, the optimal class, comprising those individuals which bear the least number of mutations, lies on one tail of the curve, and so will include only a few individuals. Given any degree of stochasticity, this class will sooner or later fluctuate to extinction; that is, it will eventually disappear completely because of random sampling error in reproduction. The distribution has now shifted backwards one step, and the class which is now optimal bears one more deleterious mutation than the class which was previously optimal. The situation cannot be restored if the population

Figure 2.5: Muller's Ratchet.

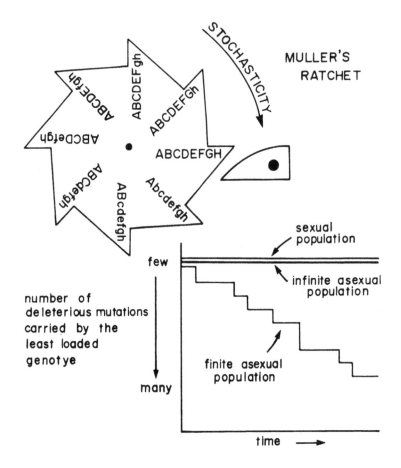

is small, since back-mutation is so infrequent that it will not have time enough to occur before the class which is now optimal has itself disappeared. As Muller originally put it: 'an asexual population incorporates a kind of ratchet mechanism, such that it can never get to contain, in any of its lines, a load of mutation smaller than that existing in its at present least-loaded lines.' Such a population will continue indefinitely to accumulate deleterious mutations until, its adaptation failing, it becomes extinct.

A sexual population is more fortunate. An infinite sexual population has the same frequency distribution of deleterious mutations as an asexual population, but in a finite sexual population the ratchet, though it still operates, will be rather ineffective. This is because, when two individuals mate together, they will both bear a certain number of mutations, but most of these mutations will have occurred at different loci. One effect of recombination will be to produce some individuals among the progeny which bear so many deleterious mutations that they are almost certain to be eliminated immediately by selection. But the complementary effect is to produce some offspring with recombined genomes which bear very few deleterious alleles. The significance of the greater fitness variance of sexual progeny is thus envisaged to be the production of a few individuals who have been almost completely purged of deleterious genes.

Sex thus acts as a new broom, which in every generation sweeps the population clean of newly arisen mutations. By contrast with the Vicar of Bray, the advantage of sex is seen to lie, not with the more rapid accumulation of favourable alleles, but with the more rapid elimination of unfavourable ones. This process does not represent a short-term advantage to the individual, since it clearly implies that the death rate among sexual broods will on average exceed that among asexual broods. Rather, sex is of long-term advantage in maintaining the adaptedness of the population, and the ratchet is thus a rival theory of group selection. The numerical experiments published by Felsenstein and Yokoyama (1976) support this interpretation. If an allele causing some degree of recombination is completely recessive to an allele which suppresses recombination, the population comprises two genetically isolated subpopulations; if there is any degree of penetrance, there will be gene flow between the two subpopulations. Felsenstein found that the ratchet is effective in the former case but not in the latter.

The operation of the ratchet is straightforward in a haploid population inhabiting a uniform environment, in which the effects of mutation at different loci are slight and independent, and in which a contrast is drawn between mixis and amixis. Discarding any of these assumptions weakens the theory.

First, any higher degree of ploidy will blunt the teeth of the ratchet, to the extent that newly arisen deleterious mutations are recessive. The mechanism will not stop, but it will operate more slowly. The ratchet will thus be made less effective by any correlation between parthenogenesis and polyploidy.

Secondly, the ratchet will be weakened by the occurrence of epistasis. This is because a tendency for compound mutants to be less fit than the product of the fitnesses of the single-mutant types will make selection for optimal genotypes more effective; the number of compound mutants will thus be made smaller, and the population will comprise a greater proportion of relatively lightly loaded lines.

Thirdly, the ratchet will vary in its effectiveness according to the mode of mixis. In a diploid population the most effective mode will be that which creates the greatest homozygosity; deleterious recessive alleles are then exposed more often to selection as homozygotes, and the continued adaptedness of the population is ensured by a plentiful supply of the favoured wild-type homozygotes. The ratchet therefore predicts that automictic populations will be the most successful in avoiding genetic deterioration; or, to put this another way, the advantage given by the ratchet mechanism to amphimicts relative to apomicts will be much greater than any advantage of amphimicts relative to automicts.

Finally, and most importantly, the ratchet operates in a uniform environment in which the simple concept of deleteriousness adequately encapsulates the competence and probable fate of a genotype. In complex environments, where a genotype may be optimal in one niche but virtually lethal in another, this concept, and with it the ratchet hypothesis, loses its force. I am not aware of any attempt to study the ratchet mechanism under these circumstances, but if my argument is valid it leads to the implied prediction that sex will be favoured in simple but not in complex environments.

2.5 The Best-man Hypothesis

In recent years, growing doubts about the validity of the Vicar of Bray have darkened the fair prospect of the 1930s, and the straight path cut by Weismann, Fisher and Muller has led to a quagmire. Even if we accept the dubious proposition that group selection is a potent evolutionary force, it is no longer clear whether sex would be an effective long-term adaptation, or even what it would represent an adaptation to. At first, it seemed as though this might be only a temporary setback, and that the situation might be saved merely by shifting the frame of reference of the theory. Williams (1966) pointed out that while the essence of the Vicar of Bray is the greater fitness variance expected in sexual populations, this inference could be applied as readily to progenies as to populations. It has been supposed that populations which comprise a great variety of genetically different individuals are preadapted to future changes in the environment; but we might equally well argue that if the environment is likely to change from generation to generation it will be advantageous for individual parents to produce a variety of progeny, since in this way they will ensure the survival of at least some of these progeny in the changed circumstances which the next generation will experience. To be sure, this point had not escaped Weismann, who wrote: 'As soon, however, as parthenogenesis becomes advantageous to the species ... it will not only be the case that colonies which produce the fewest males will gain advantage, but within the limits of the colony itself, those females will gain an advantage which produce eggs that can develop without fertilization' (Weismann, 1889, p.326). Despite this remarkable prescience, however, the credit must go chiefly to Williams for recognizing both the necessity for, and the absence of, a theory of sex based on individual selection.

The essence of the argument is shown by Figure 2.6, which has been adapted from Emlen (1973) and Williams (1975). On average, sexually produced progeny may be of somewhat lower fitness than the clonal progeny of an asexual female, because of the genetic loads created by recombination and segregation, but the much greater diversity of the sexually produced progeny implies that they will include a few individuals of extraordinarily high fitness. If only these individuals have any appreciable chance of surviving, then sexual parents will contribute a disproportionately large number of progeny to the next generation, and it may

Figure 2.6: The Best Man. Explanation in text.

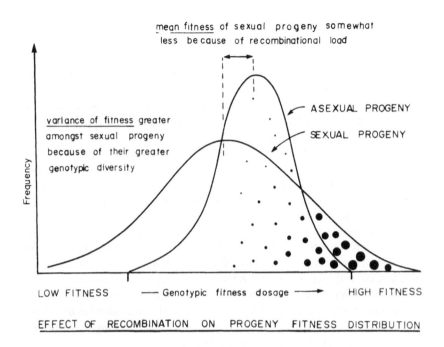

EFFECT OF RECOMBINATION ON PROGENY FITNESS DISTRIBUTION

be that this effect will be more than sufficient to offset the cost of sex. The important novelty in this line of argument is that sex is seen as an individual reproductive adaptation which maximizes the number of successful offspring in a changing environment, rather than as a device which, by increasing long-term evolutionary plasticity, prevents the extinction of the population or species. This is a perfectly respectable Darwinian proposition; since it hinges on the appearance through combination of a few extremely fit offspring in each generation, I shall refer to it as the 'best-man' hypothesis.

Capriciousness. This notion is an extremely attractive one, since it retains our preconception that sex must somehow represent an insurance against an uncertain future whilst evading our prejudice against group selection, and it nowadays dominates informed speculation about the evolutionary significance of sexuality. Nevertheless, some disturbing weaknesses were noticed by Maynard Smith (1971a), who reasoned as follows.

Imagine an environment which comprises a set of local habitats, or niches. Each niche is characterized by two variables, both of which may take either of two values. The state of any given niche may therefore be represented as AB, Ab, aB or ab. The frequencies of these states are P_{AB}, P_{Ab}, P_{aB} and P_{ab}; these vary with the overall frequencies of A and B, P_A and P_B, and with the correlation between them. The performance of an individual in any given niche depends on its genotype with respect to two diallelic loci. The organisms are haploid, so the four possible genotypes are $A1B1$, $A1B2$, $A2B1$ and $A2B2$; these have the frequencies P_{11}, P_{12}, P_{21} and P_{22}. In each different kind of niche a different genotype is optimal: $A1B1$ in AB, $A1B2$ in Ab, $A2B1$ in aB and $A2B2$ in ab. After selection the surviving asexual females produce dispersive propagules which settle randomly in niches. The surviving sexual individuals, which are imagined as isogametic self-incompatible hermaphrodites, first mate randomly, after which the zygotes are allocated randomly to niches. Should the dispersive propagules be produced sexually or asexually?

Since the environment is fully defined by the frequencies of the four niches, it will change in time only if these frequencies change. If they do not change, it is easy to see that sex will be neutral or disadvantageous. If they do change, then according to the best-man hypothesis sex will be favoured; but in fact the argument is a little more complicated.

First, suppose that the values A and B occur independently, so that $P_{AB} = P_A P_B$ and consequently $P_{AB} P_{ab} = P_{Ab} P_{aB}$. Even if P_A and P_B change in time, selection will act every generation to pull the population towards linkage equilibrium, and sex will only reinforce this tendency. To put this in another way: if an environment changes randomly in time the heritability of fitness will be zero, and the subsequent success of the progeny will be the same whether or not their genotypes differ from that of their mother. In general, ignoring the question of cost, sex will be neutral in random environments.

Secondly, suppose that the values A and B are associated in some way at all times. If the correlation between them is positive, then $P_{AB} P_{ab} > P_{Ab} P_{aB}$ at all times. The effect of a change in P_A or P_B will be to alter the ratios P_{AB}/P_{ab} and P_{Ab}/P_{aB} without upsetting this inequality. If AB is initially common but then becomes rare, the preadapted $A2B2$ genotype will be less common in the sexual than in the asexual population, since in the previous generation it will have been broken up by recombination in $A2B2/A1B1$ zygotes. Sex will therefore be disadvantageous.

Finally, suppose that the values A and B are always correlated, but that the sign of this correlation changes in time. Thus in one generation the correlation might be positive, with $P_{AB} P_{ab} > P_{Ab} P_{aB}$, and in the next generation negative, with $P_{Ab} P_{aB} > P_{AB} P_{ab}$. The best-adapted genotype for the following generation can then only be produced by recombination, and sex will be advantageous. To put this in another way: sex will be favoured only when the heritability of fitness is negative.

Environments in which the sign of the correlation between adaptively important features changes in time are neither merely changeable nor merely unpredictable: I shall call them *capricious*. The value of Maynard Smith's contribution is that, quite early in the development of the best-man hypothesis, he blocked off a number of avenues down which the simpler sorts of verbal argument might have led us. He asserted that sex will not create an individual reproductive advantage merely if the environment is changeable, nor even if it changes unpredictably. In order for sex to be favourably selected within a population the environment must change in this rather odd fashion that I have termed capricious. This conclusion was substantiated by Charlesworth (1976), who showed by numerical simulation that an otherwise neutral gene which caused a greater rate of recombination between two other loci will increase in frequency only if the sign of linkage disequilibrium between these two loci often changes in time. This is nothing more than a definition of capriciousness in terms of the genotype rather than the environment.

The results obtained by Maynard Smith and Charlesworth add up to a very cramping requirement since, although capricious environments may exist, it seems absurd to suggest that they are sufficiently common to explain the ubiquity of sexual reproduction. Even G. C. Williams, a most consistent and influential advocate of individualistic interpretations of sexual phenomena, was so impressed by Maynard Smith's arguments that he despaired of 'anyone ever finding a sufficiently powerful advantage in sexual reproduction with broadly applicable models that use only such general properties as mutation rates, population sizes, selection coefficients, etc.' (Williams, 1975, p. 14).

Truncation Selection. Faced with such apparently conclusive arguments, theorists did not, of course, abandon the best-man hypothesis. For one thing, there was no visible alternative. Rather, what happened was a burst of theorizing in which *ad hoc* assumptions were inserted into the hypothesis in the rather desperate hope that it would rise from the dead. What is common to all these attempts is that they postulate extreme forms of the hypothesis, in which either some genotypes are in every generation completely debarred from reproducing, or else only a single genotype is able to reproduce at all, in any given niche. In other words, all modern versions of the best man take for granted some form of truncation selection.

To illustrate how a truncate fitness distribution might resuscitate the hypothesis, let us take the population model discussed above, and add the feature that selection is so stringent that the fixed number of survivors from each niche all bear the same optimal genotype. The model is then sufficiently simple to be solved analytically; what follows is both a gloss on, and a reinterpretation of, the result obtained by Maynard Smith (1971a).

Consider any initial generation of colonists. Because only the optimal genotype survives in any given niche, the frequency of individuals of given genotype among the survivors is equal to the frequency in the environment of the niche in which that genotype is optimal. Thus, using the terminology explained above, $P_{11} = P_{AB}$ after selection, and similarly for all the other genotypes, whether the individuals concerned descend from sexual or from asexual parents. Subsequent reproduction by asexual females simply replicates the parental genome, so that the frequencies of genotypes among the progeny of asexual females newly settled into niches at the beginning of the next generation are

$$P_{11} = P_{AB};$$
$$P_{12} = P_{Ab};$$
$$P_{21} = P_{aB};$$
$$P_{22} = P_{ab}.$$

Random mating among the sexual individuals will give the corresponding frequencies

$$P_{11} = P_{AB} - D/2;$$
$$P_{12} = P_{Ab} + D/2;$$
$$P_{21} = P_{aB} + D/2;$$
$$P_{22} = P_{ab} - D/2;$$

where D is the coefficient of linkage disequilibrium

$$D = P_{11}P_{22} - P_{12}P_{21}$$
$$= P_{AB}P_{ab} - P_{Ab}P_{aB}.$$

These newly settled individuals now experience selection, such that immediately before reproduction the genotype frequencies among those which descended from asexual parents are

$$P'_{11} = P_{AB}P'_{AB}, \text{etc};$$

and those among the descendants of sexual parents are

$$P'_{11} = (P_{AB} - D/2)P'_{AB}, \text{etc};$$

where a prime indicates the value taken by a variable in this second generation. Assuming that the sexual and asexual individuals were equally frequent at the start of the first generation, we can simply sum these sets of genotype frequencies to calculate the fitnesses of sexual and asexual parents (W_{sex} and W_{asex}), measured as the relative production of grandchildren which survive to reproduce. The difference between these fitnesses is readily shown to be

$$W_d = W_{asex} - W_{sex}$$
$$= (P_{AB}P_{ab} - P_{Ab}P_{aB}) \; [(P'_{AB} + P'_{ab}) - (P'_{Ab} + P'_{aB})].$$

The quantity P_{AB} is the frequency of the niche AB. If the correlation between A and B is r, then we can write P_{AB} in terms of the overall frequencies of A and B

$$P_{AB} = P_A P_B + rQ,$$

where $Q = [P_A(1 - P_A)P_B(1 - P_B)]^{\frac{1}{2}}$; and similarly for the frequencies of the other three possible niches. Substituting these expressions into the equation for W_d we arrive, after a good deal of tedious algebra, at the conclusion that

$$W_d = rQ[4r'Q' + (1 - 2P'_A)(1 - 2P'_B)].$$

Sexual females have greater reproductive success if $W_d < 0$. Given the sign of r, and knowing that Q and Q' are positive, whether or not this condition is met depends on the sign and magnitude of r' and of the expression $(1 - 2P'_A)(1 - 2P'_B)$.

Suppose that $r > 0$, so that A and B are positively correlated during the first generation of selection. Then a necessary condition for $W_d < 0$ is that one of two possibilities should be realized; a sufficient condition is that both should be realized. The first is that $r' < 0$, in which case the sign of the correlation between A and B has changed between the first and second generations. This is the formal definition of a capricious environment. The second is that P'_A should be large and P'_B small, or vice versa. For instance, suppose that A is initially common and B rare. Since $r > 0$ most niches in the first generation will be Ab, with the rest being AB or ab. If the environment changes so that in the next generation A is rare and B common then the most frequent niche will then be aB, again with a small proportion of AB and ab, even though the correlation between A and B remains positive. Such an environment is not capricious, but merely mutable. Sex will be favoured because the fittest genotype in the second generation, $A2B1$, having been almost eliminated by selection in the first generation, can be recreated in moderate numbers only through recombination.

A complementary line of reasoning follows for the converse case, in which $r < 0$: sex may be favoured if the environment is capricious ($r' > 0$) or mutable (P'_A and P'_B both large or both small).

If we wish to distinguish between the effects of capriciousness and mutability, in the special senses in which I have used these terms above, then we can recognize three categories. The first includes environments which are neither capricious nor mutable. In such environments sex is never favoured, so the existence of either capriciousness or mutability is a necessary condition for the spread of sexuality. The second category includes environments which are both capricious and mutable. Sex is always favourably selected, so that the existence of both capriciousness and mutability is a sufficient condition for the spread of sexuality. The most interesting category is the third, which includes environments which are either capricious or mutable but not both. I have studied this category by solving the model numerically, for the case in which both correlation coefficients and gene frequencies are drawn randomly from rectangular probability distributions. The results suggest that sex is favoured in half of all instances, whether the environment is capricious but not mutable, or mutable but not capricious. I suspect, but have not proven, that the same result would follow from any set of symmetrical probability distributions. Capriciousness and mutability are thus equally effective in procuring selection for sexuality in a very simple model of truncate selection in a changeable environment.

I have described this model in some detail both because of its historical importance and because despite its simplicity it is about the most complex model for which an analytical solution is easily accessible. Contrary to Maynard Smith's original conclusion, it shows that, under truncation selection, capriciousness is neither a necessary nor a sufficient condition for the evolution of sex, and that mutability cannot be neglected. It also shows that in random environments sex is neutral: W_d will be negative in half of all instances and positive in the other half. Since in any realistic calculation any advantage of sex will be offset by the consequences of anisogamy and epistasis, a continuing advantage for sexuality will be generated only by environments with a rather strong tendency to be both capricious and mutable.

Lottery Models. However, it might be argued that this conclusion is unduly pessimistic, since the model includes only four possible niches and the four corresponding genotypes. Since the virtue of sex must lie in diversification, a more complex environment might provide a more favourable context for sexual reproduction. Since it is impracticable to obtain analytical solutions for more complex cases, they have instead been studied through numerical simulation, chiefly by Williams and Maynard Smith.

The most important of these models are linked by a common thread, the likening of truncate natural selection to the drawing of tickets in a lottery. If you enter a lottery by purchasing several tickets, you are more likely to win a prize if these tickets have different numbers, rather than all having the same number. Like most analogies, this is an attempt to state a difficult technical problem in familiar terms, in order to prime the pump – to facilitate making guesses about the solution of the problem, whose validity can later be investigated more rigorously. To retranslate the analogy, by increasing the phenotypic variance (range of numbers) of the progeny (tickets) we increase the probability that one or a few will be extraordinarily fit (prize-winning). If we can identify biological situations in which the fate of offspring is analogous to that of lottery tickets, we may be able to guess at the reason for the sexual diversification of offspring genotypes.

For example, imagine an organism which inhabits a checkerboard of small, discrete habitats. At the beginning of the growing season each habitat is stocked by a variety of individuals, which at first reproduce asexually, so that, as time goes on, the habitat begins to become saturated with a variety of clones, each descending from a single initial founder. This process of asexual reproduction is so rapid that, even if only a single individual were initially present, the clonal descendants of this individual would completely saturate the habitat by the end of the growing season. Moreover, there is such intense competition between the clones that, by the end of the growing season, selection has eliminated all but one, the one which happens to be best suited to that particular habitat in that particular season. The members of the surviving clone now produce propagules which are capable of some degree of dispersal; some remain in the same habitat whilst others are dispersed to nearby habitats; but any habitat which receives propagules from a given clone receives several, or many, rather than receiving only one. This is a sketch of what Williams has called the aphid-rotifer model (Williams and Mitton, 1973; Williams, 1975), from its rudimentary similarity to the life cycles and population structures of two heterogonic taxa, aphids living on herbaceous plants and monogonont rotifers inhabiting small ponds. The question which the model is designed to explicate is this: granted that several generations of asexual reproduction occur within the habitats, should the dispersing propagules also be formed asexually, or should they be diversified sexually?

Now, this model has been tailored so that reproduction approaches as closely as possible to the condition of a lottery: each entrant (clone surviving at the end of the previous season) buys several tickets (propagules) in a given lottery (habitat in present season), and in each lottery there is only a single winning ticket (clone surviving at end of present season). If there is any virtue in the lottery analogy, there must be some circumstances in which it will be better for a clone to produce a variety of propagules by sexual means. The advantage of diversity lies in the wastefulness of stocking a habitat with several identical copies of the same genotype, since only a single clone will survive the process of selection, and this may as readily descend from a single founder as from several. The greatest advantage of diversity will be realized, therefore, when uniformity is most wasteful. It can be foreseen that the pressure for sexuality will increase as the number of propagules from a single clone which stock a given habitat increases, and as the number of asexual generations intervening between the periods of propagule formation increases; and hence that sex will become the favoured means of making propagules beyond some combination of a large number of initial colonists per clone per habitat and a large number of asexual generations per growing season. These intuitions have been confirmed quantitatively by Williams, whose table (Williams and Mitton, 1973; Williams, 1975, p. 20) should be consulted for the details of a numerical example.

Yet even in the case of a model which approaches as closely as possible to an ideal lottery, there are reservations which must be made. Sex will diversify progeny effectively only if mating occurs between clones, so the condition that only a single clone survives in each habitat implies that the organisms must leave the habitat before reproduction; this is not the case for taxa such as rotifers and cladocerans. Each propagule genotype is allocated a fitness independently of the fitness of its parent or parents, and any heritability of fitness between habitats or between growing seasons would reduce the advantage of sex. At the same time, the differences in fitness between genotypes within a habitat during a single growing season must be great enough to permit the total elimination of all of its competitors by the one best-adapted clone, and this will be possible only if the rate of increase through asexual reproduction (Williams' zygote-to-zygote increase, or ZZI) is very great. The crucial parameters of the aphid-rotifer model are thus the grain of the environment (the number of colonists from a given sibship stocking a given habitat), the rate of increase of the colonists (the number of zygotes in the next sexually produced generation descending from a single zygote in the present generation) and the heritability of fitness (zero in a randomly mutable environment such as the one described by the model, negative in a capricious environment). By varying these parameters we can study the evolution of sex in a number of very different life histories.

If the environment is fine-grained, with a very large number of very small habitats available for colonization, each habitat would receive either one propagule or none at all, and the situation would be akin to a series of independent lotteries, in each of which only one ticket could be purchased. Purely asexual reproduction would make it possible to enter more lotteries without reducing the chance of winning any given lottery, and the costliness of sex would seem inevitably to imply a net disadvantage. Nevertheless, even in such an unpromising situation it is not utterly impossible to imagine the retention of sexuality under short-term natural selection. If each habitat of the aphid-rotifer model is reduced to a point, the environment comprises a large number of exceedingly small patches, most of which will not be colonized at all to begin with, whilst those which are colonized will receive only a single propagule. At the beginning of a growing season, then, we have a great array of habitats, a few of which receive propagules which, on germinating, find themselves surrounded by unoccupied habitats. Having established themselves, these propagules proceed to reproduce, and may do so either vegetatively or sexually. Vegetative increase spreads the initial individual to adjacent habitats, and ceases when a colony of identical zooids has spread itself as far as possible in all directions, until it has met with unfavourable conditions or with a superior competitor. Whilst retaining a level of vegetative production necessary to compensate for the death of individual zooids the colony now turns to sexual reproduction, producing diversified propagules which are able to spread to different habitats; sex is the only way in which the elements of this particular genotype can further increase their representation in the next generation, since the colony, having encountered the limits of its tolerance in all directions, can spread no more by vegetative means. At equilibrium there is thus a balance between the sexual and vegetative modes of reproduction, the sexual mode not being eliminated entirely.

This is a sketch of Williams' strawberry-coral model, so-called because it is intended to caricature the population dynamics of coral zooids or strawberry runners. I find it most unconvincing. As I have described it so far, it is not at all a model of sexuality, but rather one of reproduction. The difference between the budding of a primary zooid to form a colony and the growth and differentiation of an egg to form an adult is slight: in both cases the organism is at each successive moment of time faced with the problem of how to partition its surplus production between growth, by the mitotic proliferation of cells of zooids, and reproduction. No doubt it will be advantageous for a colonial organism to create a certain optimal number of zooids before embarking on the manufacture of reproductive propagules, just as it will be advantageous for a noncolonial organism to achieve a certain amount of somatic growth before reaching reproductive maturity. This is a matter for the theory of reproduction, not for the theory of sexuality.

On the other hand, it might be argued that once the decision to make propagules has been taken, these propagules may or may not be the result of mixis. Williams argues that as time goes by each local habitat will come to be occupied by a clone which has an extremely high fitness in that habitat, and which is thereby enabled to repulse all its competitors. Since this fitness is extremely high only with respect to the precise conditions encountered in a particular habitat, or rather small group of habitats, it is unlikely to be reproducible elsewhere in the environment; any other habitat is likely to be occupied already by a locally superior clone. It will be futile to put resources into amictic propagules, since these will everywhere encounter well-established clones which happen to be highly adapted to local conditions; rather, it is better to produce mictic propagules, even though they are genetically expensive, in the hope of coming up with a few devastatingly superior recombinants. This is a rather odd model of the environment: the heritability of fitness is asserted to be positive over small distances (otherwise a clone would not on average be fitter than its competitors in nearby habitats) but negative over large distances (since, if it were positive or zero, sex could not be selected). It is the negative heritability of fitness, or capriciousness, associated with long-distance dispersal that alone can create an advantage for sex in a lottery model when all individuals grow up away from their sibs. Granted this assumption that high fitness locally necessarily implies low fitness elsewhere, Williams' model seems likely to favour the evolution of sex; but I see no pressing reason to grant anything of the sort. It is a *petitio principii*, assuming what it is required to prove; the genetic destruction through meiosis of locally fit genotypes will inevitably be favoured whenever they are less likely to succeed elsewhere than a genotype chosen at random, but what we require is an explanation of how this state of affairs can arise, and the strawberry-coral model does not furnish us with one.

The evolution of sex is made difficult in the strawberry-coral model by the reduction in size of local habitats to the point where they are colonized only by a single propagule. The converse model is one in which each local habitat is so large that it receives thousands of propagules from the same sibship. Returning to the aphid-rotifer model, selection may favour the mictic production of dispersive propagules above some critical number of asexual generations per growing season. The greater the number of asexual generations, the greater is the rate of zygote-to-zygote increase, and the more intense selection can be. When the ZZI has a value such that the selection of the best men will barely pay the cost of sex, the system is in evolutionary equilibrium. Depending on fecundity, the requisite value of ZZI may be achieved after a greater or lesser number of asexual generations; but in the limit, organisms with enormous fecundity may achieve the necessary ZZI by the exclusively sexual production of zygotes, without any asexual proliferation at all. In this case, which is Williams' elm-oyster model (Williams and Mitton, 1973; Williams, 1975), a purely sexual life history is an equilibrium state. Elm trees produce so many seeds that the soil for some distance around is usually stocked far in excess of its capacity to support adult elm trees. If – and only if – the local habitat is likely to change in time so that fitness has virtually zero heritability, it will be better to stock the soil with a diversity of seeds, some of which will as seedlings display extraordinarily high fitness, than to rely on a multitude of identical seeds, all except one or two of which must be wasted. Elsewhere, Williams (1975) argues that positive assortative mating will inflate phenotypic variance, and thereby the variance of fitness, thus increasing the frequency of extremely fit genotypes resulting from recombination. In the cod-starfish model this phenotypic assortative mating is generated by mating between the survivors within a particular local habitat, all of whom will have experienced the same selective regime and may therefore be expected to possess somewhat similar phenotypes. The elm-oyster model is then a limiting case of the aphid-rotifer model in which the local habitats, rather than being reduced to points which can be stocked only by a single colonist, as in the strawberry-coral model, are instead expanded to areas which may receive legions of propagules. If we expand it still further, so that the environment contains only a single local habitat, the two models are opposites: sex will be favoured in the strawberry-coral model only if fitness has negative heritability in space, and in the extreme version of the elm-oyster model only if fitness has negative heritability in time.

Both of these modifications of the aphid-rotifer model, then, lead us back to capriciousness as the directing principle in the evolution of sex, without offering any hint as to how capriciousness might arise. Indeed, it is very difficult to imagine that capricious environments can be other than a special case. For instance, Williams (1975) invites us to imagine a marine invertebrate whose larvae are dispersed by a slow oceanic gyre, so that the colonists of a given stretch of coastline are the distant descendants of previous colonists, their more immediate ancestors having lived elsewhere (the triton model). In such a scheme of cyclical selection, recombination offers the only quick way of recovering the optimal genotype for the habitat to which the population has returned. This, however, is the triumph of ingenuity over common sense.

Maynard Smith's Lottery. The models that I have described above were developed in response to Maynard Smith's opinion that sex will be favourably selected only in capricious environments. Indeed, they can all be viewed as attempts to evade this stricture by invoking life histories or population structures incorporating some special feature. In the case of the most thoughtful and influential of these evasions, the aphid-rotifer model and its immediate derivatives, the feature is a special sort of truncation selection which has since come to be called 'sib competition'. Several siblings colonize an unpredictable habitat which will eventually support the descendants of, at most, one of them. The word 'competition' is here used in a sense which will be unfamiliar to most ecologists. There is no connotation of a restraint exercised on the utilization of a resource by the presence of other individuals, unless indeed one's own genotype is regarded as a resource, whose value is depreciated by the presence of other individuals, any of whom, by the possession of a superior genotype, might prevent the transmission of one's own. In the same way, one can envisage a process of competition in a lottery where only those tickets which have been sold are eligible to be drawn. If you make the only purchase the prize is certain to be yours; others may reduce your chance of winning, not by altering the number of your ticket, nor by attempting to destroy it, but merely because their additional purchases make it less likely that your ticket will be the one to be drawn. Because I shall invoke a very different concept of competition in a later section I have suppressed the term 'sib competition' and preferred some term such as 'lottery selection'.

The models developed by Williams are essentially thought-experiments through which it can be conceived how a process of lottery selection might favour the evolution of sex, given only zero rather than negative heritability of fitness – in random rather than in capricious environments. For a more general and rigorously quantitative account of lotteries, I turn to an important paper published by Maynard Smith in 1976. His model is structually similar to the aphid-rotifer model, and can be viewed as a generalization of it. He envisages an environment which comprises L local habitats; the state of each habitat is described by five ecological variables, each of which may take either of two values. Since the variables are independent and the two values that each may take are equally probable, there are 2^5 equally probable types of habitat. Genotypes are constructed on the same pattern, with five unlinked diallelic haploid loci controlling adaptation to local conditions. To quote Maynard Smith (1976b, pp. 229-30):

Each patch receives N offspring from each of R parents. Only one of the RN offspring survives to breed. The one survivor is chosen randomly from individuals in the patch (i.e. habitat) which have the largest number of genetic adaptations to the features of the environment. During simulation, for each patch, a female parent is chosen randomly from the L survivors in the previous generation; during this choice an asexual parent is K times as likely to be chosen as a sexual one. If the selected parent is asexual, N offspring identical to the parent are produced. If the selected parent is sexual, a second parent is chosen randomly from the sexual survivors of the last generation, and N offspring produced from the parents according to the usual laws of genetics. This procedure is repeated R times for each patch.

We have here a precisely realized model of genetic events in a population experiencing an environment which varies both in space and in time. The random spatial variation guarantees the maintenance of genetic diversity, whether or not the population is sexual: the local population is regulated in a density-dependent fashion — albeit a peculiar one, with each habitat both starting and finishing with the same number of colonists — which automatically gives a selective advantage to rare genotypes. The random temporal variation annuls any average advantage either in perpetuating the parental genotype or in rearranging it. The success or failure of sexuality therefore lies solely in the power of lottery selection to overcome the cost of sex as measured by the parameter K.

Maynard Smith's results are presented in his 1976 article. Suppose that each asexual parent, or pair of sexual parents, produces only a single offspring ($N = 1$). This is as though one bought a single ticket in a lottery, and there can be no advantage in sexuality (cf. the strawberry-coral model). On the other hand, suppose that parents produce many offspring, but that each habitat receives only a single set of offspring ($R = 1$). This is like a lottery in which one buys all the tickets, and again there is no point in diversifying the progeny (cf. the elm-oyster model). The lottery analogy is valid only if there is more than one ticket and more than one purchaser in each lottery or, in terms of the present model, if parents produce more than a single offspring and if each habitat receives offspring from more than one sibship. These are necessary conditions; if they are not met, then in a random environment sex is a merely neutral character which will quickly be lost from the population if it is in any degree expensive ($K > 1$). Provided that both $N > 1$ and $R > 1$, sex will be favoured provided that it is not too expensive — provided, in other words, that K does not exceed some threshold value. This threshold is set by a minimum value of the total intensity of selection, such that the dividend of providing a disproportionate number of the best men of each generation barely pays the cost of sex. Since N measures the intensity of selection within families and R the intensity of selection between families, the total intensity of selection is measured by the product RN. The formal problem which is set by the model thus boils down to finding how large RN must be for a given value of K in order that sex may be favourably selected. Perhaps this is an optimistic view of the simplicity of the problem, since even if it is granted that $R > 1$ and $N > 1$ the answer will depend to some extent on the relative magnitudes of R and N as well as on the absolute magnitude of their product RN. Nevertheless, it seems that sex may usually be favoured when RN exceeds a value of about 40, even when the maximum twofold cost of sex is levied. In other words, the advantage of producing a disproportionate number of best men becomes overwhelming when the best genotype in a given habitat is on average some forty times more fit than the average genotype.

Although this result is encouraging, since it involves no manifest absurdity, it is only illustrative: it depends rather sensitively on the values given to certain parameters of the model, and in particular it depends on the assumption that each habitat is characterized by a combination of exactly five independent ecological variables. Maynard Smith repeated one of his simulations (with

$L = 400, N = 8, R = 6, K = 2$) for the case in which two of these five variables are contingent, so that only two combinations of these two variables are permitted. Put in one way, this is equivalent to assuming the existence of epistatic fitness effects; the genotypes of individuals which survive selection will be in linkage disequilibrium with respect to the loci controlling adaptation to the two correlated variables, and sex will tend to restore equilibrium and thus to erode adaptedness. Put in another way, it is equivalent to reducing the number of independently assigned ecological variables from five to four. In either case, the effect of the change is to reverse the direction of selection so as to permit the clones to exclude the Mendelian population. I find this result encouraging rather than disturbing. If the effect of reducing environmental heterogeneity is to reduce the efficacy of sexual reproduction, then one is entitled to infer that increasing the heterogeneity will favour sexuality. Five ecological variables is by no means an excessive number, and it may be that if it were feasible to run simulations with many more variables it would be difficult to discover any circumstances in which sex was not favoured.

However, there is one aspect of Maynard Smith's simulations that I do find puzzling. He points out that although all genotypes experience frequency-dependent selection, whether they are transmitted clonally or sexually, they are far more likely to suffer random extinction during sampling if they are locked up in clones. Even with a healthy dose of sampling error it is relatively easy to maintain a sexual population polymorphic for n diallelic loci under frequency-dependent selection, but almost impossible to maintain all 2^n clones in an asexual population. In the simulation to which I just referred, for example, twelve of the 32 original clones had disappeared by the tenth generation of selection, and the decline in the total number of asexual individuals (from 200 to 146) can be attributed almost entirely to the decline in the number of distinct clones (from 32 to 20). This loss of genetic variation will inevitably restrain the spread of the asexual population, since the sexual population now has the broader ecological competence. Indeed, Maynard Smith gives an explicit illustration of the much lower degree of success achieved by a monoclonal as opposed to a polyclonal asexual population when in competition with a polymorphic sexual population. This phenomenon will assume a much greater significance in a later section; meanwhile, I wonder to what extent the success of sexual genotypes in Maynard Smith's model depends on the random loss of genetic variation from the asexual population, rather than on the lottery principle itself.

(Whilst the final draft of this manuscript was being prepared, I read two papers which propose analytical solutions to Maynard Smith's model (Taylor, 1979; Bulmer, 1980). I lack the time to discuss them properly here; in general they confirm Maynard Smith's results, and remove any residual doubts about the adequacy of the parameter space he explored, but they seem curiously insensitive to the number of genotypes present.)

Runt Models. In lottery models, selection in each niche is truncate, such that only one genotype has nonzero fitness. The converse situation is a truncate-fitness scheme in which only one genotype has zero fitness. Imagine an assemblage of clones which competes with a sexual population. In each generation, all genotypes, whether borne by sexual or by asexual individuals, have unit fitness, except that one genotype has zero fitness. This genotype is permanently lost to the clonal population but among sexual individuals will be restored by recombination. As times goes on the sexual population will maintain genotypic diversity, but the clones will disappear one by one until all have been irrevocably lost. This notion seems to be rather widely known but only vaguely appreciated: the only explicit statement I can trace is a short note by Treisman (1976). For obvious reasons, I shall call this version of the best man the runt model.

It is easy to build a simple simulation model. We start with a closed population in which a number of sexual and asexual genotypes compete. In each generation one genotype, chosen randomly, is lethal, whilst all the rest have the same fitness. Three replicate experiments with a four-locus haploid model gave much the same result, which is shown in Figure 2.7. Until generation 25 or so, there was little change in the average frequency of sexual individuals, though the variance of this frequency increased as some populations drifted upwards and others downwards. The variance continued to increase through time, so that by generation 50 the frequency distribution began to look bimodal. Nevertheless, by generation 100, sex had become fixed in every case. Adding a cost of sex, however great, does not alter this result: the bimodality of the frequency distribution becomes more pronounced but, since an infinite population is assumed, the sexual population neither becomes extinct nor loses any genotype permanently, whereas the clonal population is certain sooner or later to disappear.

The essential feature of the model which produces this result is that every genotype eventually encounters a period of time during which its fitness is precisely zero. In an infinite population this condition can be met only by a perfectly sharp truncation. If the truncation is incomplete, so that the runt genotype has very low but not zero fitness, then sex is either neutrally conserved or lost, depending on whether or not it is costly. However, if the population is finite this result no longer holds; a genotype with a very low expected current fitness is likely in practice to have zero fitness, being eliminated completely by sampling error. Figure 2.8 shows that there may be a pronounced tendency for sex to increase in small populations even when the runt has a fitness half as great as that of the other genotypes.

All these simulations refer to a uniform environment, in which any given genotype has a unique fitness at a given moment in time. In lottery models, more complex environments are more favourable to sex, but it is obvious that the reverse will be the case for runt models. I have run a model with 16 genotypes and 16 niches, in which the runt genotype is chosen randomly for each niche in each generation, the offspring of the survivors being allocated randomly to niches at the beginning of each generation. In ten simulations with a population of 128 individuals and no cost of sex, the average frequency of sexual individuals

Figure 2.7: A Runt Model with Complete Truncation. Plotted points are means of ten runs for each of three replicates; vertical lines are ranges; short horizontal bars are standard deviations of transformed data, but note that data are not even approximately normal. All populations were fixed for sex by generation 100. There is no cost of sex in this experiment.

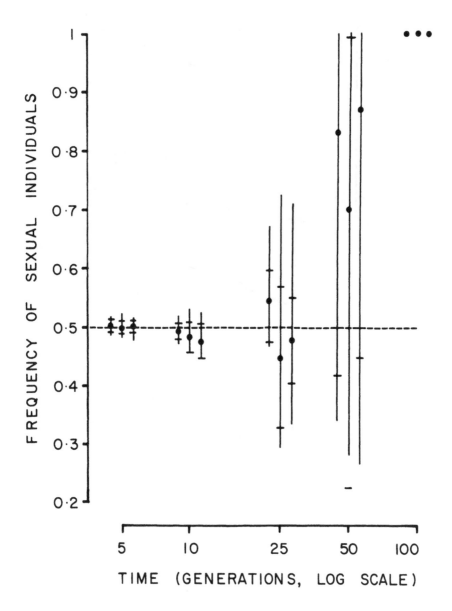

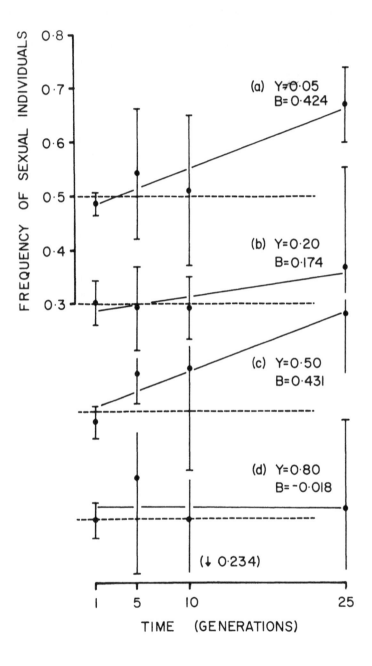

FREQUENCY OF SEXUAL INDIVIDUALS

(a) Y=0·05
B=0·424

(b) Y=0·20
B=0·174

(c) Y=0·50
B=0·431

(d) Y=0·80
B=⁻0·018

(↓ 0·234)

TIME (GENERATIONS)

declined from 0.479 at the end of the first generation to 0.334 at the end of generation 25, with one population being fixed for asexuality and another becoming fixed shortly afterwards. Sex fails to spread because the simultaneous existence of many independent niches blunts the edge of the truncate-fitness scheme; each genotype has more or less the same average fitness from generation to generation, and lethality in one or two niches is not disastrous.

Despite their different response to environmental complexity, the lottery and runt models are fundamentally similar. Both exploit the idea that, under truncation, selection or sampling error in a highly mutable environment the long-term geometric mean fitness of any genotype is zero. An asexual population is thereby reduced piecemeal by the successive elimination of clones, whereas a sexual population can maintain a full range of genotypes through recombination.

A much more sophisticated argument of this sort has been developed by Hamilton *et al.* (1981), who have kindly allowed me to see their work in manuscript. Consider an environment possessing a number of possible states, which succeed one another according to some rule, the state of the environment at any given time determining the fitness of any given genotype at that time. This variation in genotypic fitness through time is supposed to be caused by biotic interactions (see Section 2.8), but in practice is simply imposed on the population. We wish to compare the success of a sexual population with that of a clone or an assemblage of clones, and do so by comparing their geometric mean rates of increase over a long period of time, given the matrix of genotype fitnesses associated with each environmental state and the rule governing the occurrence of these states. Any given clone will vary in numbers through time, according to the vicissitudes of the environment. The corresponding genotype in a sexual population will also vary in numbers, but when it is forced to low frequency by selection it will tend to recover through reassortment or recombination, whilst if it becomes very common it will be pulled back by the same processes. The sexual genotype will therefore display less variance in number through time than its clonal counterpart, and as a consequence will have the greater geometric mean fitness. If this effect is sufficiently strong, a sexual population may have a greater geometric mean fitness than any clone, even when the full cost of sex is levied.

Figure 2.8: A Runt Model with Incomplete Truncation. The frequency axis is given for case (a) and is displaced 0.2 units for each subsequent case; the horizontal broken line is in all cases a frequency of 0.5. All genotypes have unit fitness in every generation, except that one randomly chosen genotype has a fitness of Y. Following selection, 128 individuals are chosen randomly to make up the next generation, the probability that a given individual will have a given genotype being set equal to the frequency of that genotype in the population after selection. The parameter B is the slope of the linear regression of the transformed frequency data. Plotted points are means for five replicates, with standard-error bars.

For instance, suppose that the environment may take either of two equiprobable states in any given generation: for the three genotypes at a diallelic locus in a sexual diplont the fitnesses corresponding to one state are r, $1/r$ and r, whilst those corresponding to the other state are $1/r$, r and $1/r$. The fitnesses of the same genotypes among individuals which reproduce asexually are the same quantities multiplied by s, which represents the cost of sex and takes a maximum value of two. Clearly, the geometric mean rate of increase of any clone will approach s over a sufficiently long period of time, and the sexual population will always exclude the clones only if its geometric mean fitness exceeds two for some values of r. But since the selection scheme is symmetrical the gene frequency at equilibrium will be $1/2$, and geometric mean fitness is thus

$$\left[\left(\frac{r}{4}+\frac{1}{2r}+\frac{r}{4}\right)\left(\frac{1}{4r}+\frac{r}{2}+\frac{1}{4r}\right)\right]^{\frac{1}{2}} = \frac{1}{2}\left(r+\frac{1}{r}\right).$$

The geometric mean fitness thus exceeds two if $r > (4 + \sqrt{12})/2$, or roughly if $r > 4$. Hamilton *et al.* interpret this to mean that sex can prevail if the fecundity exceeds eight; it seems to me more meaningful to say that sex can prevail only if there is at least a 16-fold difference in fitness between alternative environmental states.

Hamilton *et al.* go on to extend this sort of argument to a more general two-locus diploid model with no overdominance in which four possible environmental states, each favouring one of the four double homozygotes, occur in a random sequence. By analogy with the one-locus case, the geometric mean fitness of the sexual population will exceed that of any clone, provided that rates of increase (corresponding to r) are sufficiently great. The model has a certain artificiality (as they themselves acknowledge), since genetic variance is readily lost in a randomly fluctuating environment when there is no overdominance at any locus in any environmental state, and it is maintained in this model only because gene frequencies are permitted to become extremely small. Nevertheless, their preliminary results suggest that sex may prevail in a certain rather small region of parameter space, within which the effect of an occasional change of sign in linkage disequilibrium is sufficiently great to maintain a high arithmetic mean fitness for the sexual genotypes. I would interpret the results of their numerical experiments to mean that capriciousness depends both on a model of environment and on a model of fitness, the two being different expressions of the same concept, and that, given a stochastic model of environment, it is possible to specify the fitness model so that sufficient capriciousness is generated to favour meiotic recombination. If I am right, their work may prepare the ground for a deeper formal exploration of the concept of capriciousness.

2.6 The Hitch-hiker

The mathematical theory of selection at a single locus is a polite fiction, in which it is assumed that the locus in question is segregating independently of every other locus in the genome. In practice, a favourable allele at any stage of its spread through a finite population will be associated with a constellation of alleles at other variable loci, and this constellation will change through time because of sampling error, even in the absence of systematic effects such as epistasis. It follows that, since the rate at which an allele is propagated depends on the overall fitness of the individuals by which it is borne, this rate will vary through time, according to whether it finds itself on average in a congenial or in a hostile genetic context. The effect of such stochastically generated linkage disequilibria will be to increase the variance of fitness of any single-locus geno-type and thus to retard the progress of a favoured allele to fixation. Roughly speaking, there will be interference between the selection occurring at each of two linked loci, and the tighter the linkage the more serious this interference will be. The effect of recombination, however, will be to lessen interference by continually tending to reduce stochastic linkage disequilibria towards zero. An allele which causes or increases the rate of recombination may therefore be advantageous because it increases the rate at which a population can respond to an applied selection differential.

This argument is based on a description of the dynamics of two linked loci under selection published by Hill and Robertson (1966), and its significance for the evolution of recombination was pointed out by Felsenstein (1974). Felsen-stein remarks that the phenomenon seems identical with the Vicar of Bray, in that it invokes group selection favouring populations which can adapt more readily to changed circumstances. However, Strobeck *et al.* (1976) have since shown that in an important special case an allele causing a higher rate of recom-bination increases by a process of pure genic autoselection.

Their model involves three loci (Figure 2.9). At the A locus the two alleles $A1$ and $A2$ are maintained at a stable equilibrium by balancing selection; the nature of the balancing selection is not important. The second locus is fixed for the allele $B1$, as the result of prolonged directional selection at the B locus. The third locus is segregating for the alleles $C1$ and $C2$, whose only effect is on the rate of recombination between the two fitness loci A and B: the allele $C1$ is dominant, and completely or very nearly completely suppresses recombi-nation, whereas $C2$ is recessive and permits a substantial rate of recombination between the fitness loci in $C2/C2$ homozygotes. The A, B and C loci occur in the order ABC, fairly close together on the same chromosome. Now suppose that the environment changes in such a way as to reverse the direction of selec-tion at the B locus and favour the allele $B2$, which has hitherto existed only as a rare mutant. If the population is sufficiently small, we can assume that any $B1 \rightarrow B2$ mutation occurring soon after the change in the environment will be unique, and that in consequence the $B2$ allele will at first be associated exclus-ively with either the $C1$ or the $C2$ allele. Suppose that the original mutant chromosome has the genotype $A1B2C1$. This chromosome will increase in frequency because of the superiority of the $B2$ allele, and as it does so the

Figure 2.9: The Hitch-hiker.

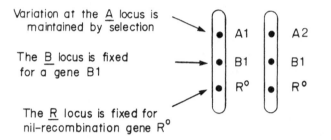

Variation at the A locus is maintained by selection → A1   A2

The B locus is fixed for a gene B1 → B1   B1

The R locus is fixed for nil-recombination gene R° → R°   R°

A mutation B1 → a beneficial allele B2 occurs at the B locus on an A1B1R° chromosome.

B2 replaces B1 on A1-bearing chromosomes but in the absence of recombination cannot spread to A2-bearing chromosomes.

A1   A2
B2   B1
R°   R°

A mutation R°→R⁺, an allele at the R locus which causes recombination between the A and B loci.

A1   A1   A2   A2
B2   B2   B2   B2
R°   R⁺   R°   R⁺

B2 alleles on A1B2R⁺ chromosomes can now spread to A2-bearing chromosomes and thus increase to fixation. In the process ABR⁺ chromosomes are favoured and R⁺ increases in frequency if it is linked to the B locus.

frequency of the *C1* allele must rise and thus that of the high-recombination *C2* allele will fall. However, the spread of the *B2* allele is eventually checked by its association with a dominant recombination-suppressing allele; having arisen in coupling with a particular allele at the *A* locus, it has reached the point at which any further spread is inhibited by the force of balancing selection at this locus. Eventually a cross-over between the recombination and the fitness loci will produce an *A1B2C2* chromosome, which when it occurs in a zygote together with an *A2B1C2* chromosome will generate *A2B2C2* gametes or gametophytes by the recombination permitted in *C2/C2* homozygotes. The *B2* allele now occurs in association with both *A1* and *A2*, and can resume its interrupted rise to fixation through selection of *A2B2C2* chromosomes. The selection of this chromosome, however, necessarily implies an increase in the frequency of the *C2* allele.

The dynamics of the *C2* allele during directional selection at the *B* locus thus falls into two stages. In the first, the allele decreases as the initial *A1B2C1* chromosome is selected; in the second it increases, once the *A2B2C2* chromosome has been created. It appears that the overall effect of selection at the *B* locus, given the operation of balancing selection at the *A* locus, is always to procure an increase in the frequency of *C2*, though it is not easy to see intuitively why this should be true. The converse case, in which the *B2* allele originally arises in coupling with the high-recombination *C2* allele, is much plainer, since the *C2* allele then increases during both stages of the passage of *B2* through the population. In the first stage, an *A1B2C2* chromosome arises and proceeds to spread, dragging up the frequency of *C2* until checked as before by balancing selection at the *A* locus. Its spread is resumed once an *A2B2C2* chromosome has been created by recombination in *C2/C2* homozygotes, resulting in a further increase of *C2*.

The scheme is, then, a simple realization of the sort of process envisaged by Hill and Robertson, with balancing selection at the *A* locus interfering with directional selection at the *B* locus and recombination annulling the effect of the initial stochastic linkage disequilibrium by transferring the favoured mutant allele from one genetic context to another. The increase of the recombination allele is an autoselected process in which (to use Maynard Smith's phrase) the allele hitches a ride with the high-fitness chromosome it has created; for this reason I shall call the hypothesis the hitch-hiker.

It is not entirely clear (to me) to what extent the success of the hitch-hiker depends on the particular genetic model employed. If we take an extreme case of Strobeck's model, it seems clear enough that a recessive recombination allele will be favoured when a favourable mutation arises in coupling with a recessive lethal (D. Charlesworth, cited in Strobeck *et al.* 1976). But what if we consider a large number of fitness loci, at any one of which, at any moment in time, a favourable mutation may not yet have appeared, may have just arisen or may be at any intermediate stage in its spread through the population? According to the simulations carried out by Felsenstein and Yokoyama (1976), a recessive allele causing free recombination at some locus linked to the fitness loci will become fixed more often than a dominant allele which suppresses recombination. This is an example of group selection, since the population as a whole is divided into two groups, one of which undergoes recombination between the fitness loci whilst the other does not, and these groups are genetically isolated. More surprisingly, he found an even stronger tendency for the recombination allele to be fixed when it is dominant. This result is very difficult to understand, but it suggests that the Hill-Robertson effect may be capable of procuring selection for high-recombination genes (relative, at least, to nil-recombination alleles) in a very large category of population models.

Returning to the more accessible machinery of the population model studied by Strobeck *et al.*, they assert that the $C2$ gene will tend to increase in frequency for any combination of input parameters. However, the conditions that this increase should be substantial are much more restrictive. The original paper should be consulted for numerical details, but, briefly, the $C2$ allele will receive a substantial lift only if $C1$ is dominant and almost completely suppresses recombination between the fitness loci. This is because the process which they describe will work only if the original mutant chromosome, say $A1B2C2$, can retain its integrity for an appreciable period of time; unless $C1$ were both dominant and also highly effective in suppressing crossing-over, $A2B2C$ chromosomes would readily be formed after syngamy with the common gametic type $A2B1C1$. The process also loses most of its power if the recombination locus is only loosely linked to the fitness loci, whilst the balancing selection at the A locus is weak; in this case the initial linkage disequilibrium will almost completely die away before the $C2$ allele hitches its second and decisive ride.

Two other criticisms might be advanced, though I have not investigated either of them numerically. The first is that the process appears to depend on small population size or a low mutation rate at the B locus, such that $Nu \ll 1$, together with intense selection such that $Ns \gg 1$; if this were not the case, several $B2$ alleles might arise at nearly the same time before $B2$ had increased substantially under selection (cf. the numerical results detailed by Felsenstein and Yokoyama, 1976). This would not be as serious if selection at the A locus were highly asymmetric, such that one allele was much more common than the other, but this would itself weaken the hitch-hiking effect, since it would prolong the first stage in the substitution of the $B2$ allele, during which $C2$ may either increase or decrease, depending on whether it is initially in coupling or in repulsion with $B2$. Secondly, the process depends on the complete substitution of the $B2$ allele; in other words, the environment must change from a state favouring one allele at the B locus to a state favouring the other with a period much greater than the time required to fix the newly favoured allele. If this were not true, the $C2$ allele would experience, during the second stage of the substitution process, at first an increase and then a decrease.

We are led to imagine, then, a highly mutable environment in which a locus controlling recombination is linked to a locus at which genic variance is maintained by balancing selection generated by some agent not affected by the mutability of the environment. At nearby loci, mutations arise, spread, become fixed and are in their turn superseded as the environment lurches from one state to another. Such a state of affairs will procure selection favouring recessive alleles that cause an appreciable rate of recombination between the fitness loci, and such alleles will gradually replace dominant alleles which suppress recombination. The situation envisaged by the hypothesis, therefore, is one in which a mutable environment maintains a polymorphism between dominant alleles which suppress, and recessive alleles which do not suppress, recombination in small nearby regions of the same chromosome. The hitch-hiker does not appear to generate any substantial advantage for an allele which causes an appreciable rate of recombination, relative to one which causes a lesser rate which is yet much greater than zero; nor for an allele whose effect is to raise rates of recombination throughout the genome. I would conclude that, although its status as a theory of recombination is unquestioned, if limited, its ability to explain the evolution of sex is far from clear. It is included here out of deference to the universal opinion that the correct theory of sex will also be the correct theory of recombination.

In the preceding five sections I have described the vicissitudes of Weismann's idea that sex facilitates adaptation to a changing environment. The same ground has been surveyed from a rather different point of vantage in a lucid and stimulating article by Thompson (1976). Despite its title, his article is concerned with recombination rather than with sex itself, and he stresses underlying theories of genetic organization much more than I have done. The article itself should be read as a complement to my own account, and I shall give only a very brief notice of it here. Thompson argues that there is little or no empirical evidence in support of the proposition that recombination increases the rate of response to directional selection, whilst simulation studies predict such an effect only when linkage is very tight. In general, recombination will accelerate evolution only if there is substantial negative linkage disequilibrium, since the effect of negative linkage disequilibrium is to reduce genotypic variance and thereby to reduce the rate of response to selection, which will be restored if the disequilibrium is broken down by recombination. However, there is little empirical evidence for widespread negative linkage disequilibrium in nature. Thompson concludes that the major function of recombination is not to hasten but rather to retard adaptation in changing environments, an hypothesis I have previously traced through a superficially different but really parallel chain of argument.

2.7 The Tangled Bank

Since the primary consequence of sex is the diversification of progeny, its advantage must relate to environmental heterogeneity. So far I have stressed the contribution of temporal heterogeneity, the fluctuation of genotypic fitness in time, since the idea that sex preadapts lines of descent to unknown future conditions has become entrenched in the literature at all levels, from introductory texts to technical monographs. Having encountered serious difficulties with this reasoning, I shall now turn to a much less thoroughly discussed alternative: that sex may be favoured in environments which are heterogeneous in space.

There can be two extreme opinions about environment. On the one hand, it may be thought of as being fine-grained and mutable. This implies that all individuals in a population experience more or less the same average of environmental conditions during their lifetimes, but that this average is liable to change from generation to generation. Holding this opinion, we are led to interpret sexuality through some version of the best-man hypothesis. On the other hand, the environment may be thought of as being coarse-grained and static. This implies that different individuals in a population experience different conditions of life, but that the range of conditions experienced by the population remains more or less the same from generation to generation. In this case the best-man hypothesis is inviable, and we must search for some radically different explanation of sex. Of course, neither of these opinions might be true. The environment might be uniform both in space and in time, as it is assumed to be in elementary population-genetics models; sexuality and recombination are then incomprehensible. More plausibly, the environment might vary both in space and in time, as we have been forced to assume in deriving the more advanced best-man models. In that case the analysis is difficult for even the simplest cases, and as a first step I shall discuss the consequences of purely spatial heterogeneity – the existence of a greater or lesser number of ecological niches in the same small geographical area – in an environment which does not change in time.

The only serious attempt to develop a theory of sex based on spatial heterogeneity was made by Ghiselin (1974a), whose ideas seem to have been universally misunderstood. He relied heavily on analogies drawn from economics and, because this is an excellent way of clarifying some of the principles involved, I shall begin in the same way. My example will concern a manufacturer of buttons, whose object is to realize the greatest possible rate of profit, without regard to any other consideration. A market for buttons exists, but happens to be completely untapped; there are no rival manufacturers. Two machines control the design of the buttons: one can be set to cut either round or square blanks, and the other to cut either two or four holes, so that up to four different types of buttons can be made. The advantage of producing a variety of buttons is that a larger market can thereby be exploited; the disadvantage is that resetting the machines is costly in terms of labour and time, and thereby diminishes the rate of production. How should the manufacturer proceed? At first the market is completely undeveloped, and every button that is produced can be sold immediately. The maximization of profit is achieved, therefore, by maximizing output, and this in turn is achieved by restricting the output to a single kind of button. Diversity would be merely wasteful. As time goes on, however, the market becomes more fully developed, and eventually approaches saturation, at which point buttons will be bought only to replace those which have been lost or broken. However, there may still be potential customers for other kinds of buttons, if these are imagined to serve slightly different purposes, and at some point the cost of diversification will be paid by increased sales in these under-exploited sectors of the market. At this point the manufacturer will choose to diversify, since profits will be maintained by increasing the variety of his product despite the inevitable fall in total output. This process of diversification will continue as successive sectors of the market become saturated, until the manufacturer can no longer produce novel kinds of button with the machinery he possesses. He will then install new machines, or resort to advertising to stimulate demand, but with these strategies we are not at present concerned. The moral that I wish to stress is that diversification can be used to maximize profits because different kinds of the same general product are in imperfect competition, in the sense that an increase in the sales of any given kind do not result in a one-to-one decrease in the sales of another kind: it may be possible to sell only a thousand round buttons with two holes, or only a thousand square buttons with four holes, when each is the only kind of button available, whilst it is possible to sell a total of 1500 buttons when the two kinds are marketed together. This interpretation may be contrasted with an analogy based on the principles of the best-man hypothesis – that diversification would be favoured in highly variable or cyclical economies, and especially in capricious economies where customers demanded round buttons with two holes or square buttons with four holes in one year but round buttons with four holes or square buttons with two holes in the following year.

We can translate the analogy of the button manufacturer into biological terms by imagining what will happen when an asexual clone is introduced into a favourable but unoccupied habitat. At first the clone will increase in numbers very quickly, but as time goes on population growth will be increasingly inhibited by the growing population density; the birth rate, initially high, will fall, or the death rate will rise, or both will happen together. Eventually reproduction will barely suffice to compensate for mortality, and at this point the population density will take some characteristic value, often called the 'carrying capacity'. Now imagine that a second clone is introduced. If the environment is uniform, then only one of two things can happen. The new clone may be able to increase at the carrying capacity of its rival; in this case it will tend to spread, and will eventually supplant its rival completely. On the other hand it may be unable to spread, in which case it will become extinct. In either case the population will be genetically monomorphic at equilibrium. This result holds whether or not the state of the environment varies in time. However, suppose that the environment varies in space, so that it comprises several patches or habitats or niches, and that the fitness of a clone varies between niches. Then the carrying capacity of the original clone in such a complex environment will depend on the frequency of different niches, to each of which it is more or less well-adapted. When the second clone is introduced it will be able to occupy those niches to which it is better-adapted than its rival, so that as long as this second clone is rare its fitness will be high, and it will tend to spread. Once it has occupied those niches to which it is better-adapted, however, its spread will be increasingly inhibited by competition with its rival, and at some point it will be able to spread no further. In this case the equilibrium population will be genetically diverse, the diversity being maintained by a frequency-dependent process of selection. Moreover, since the environment is now more fully utilized, the second clone being able to exploit niches unavailable to the first clone, the carrying capacity of the diverse population will inevitably exceed that of either single clone. We can repeat this argument for a third, a fourth and any subsequent number of clones, each tending to increase when rare, passing to some stable equilibrium frequency, and in the process raising the carrying capacity of the population. This will continue until there are as many clones as niches; the environment is then so fully exploited that no fresh opportunities exist, and any new clone arising subsequently can spread only by displacing one of the resident clones.

The logic of these verbal arguments is readily confirmed through simple algebraic models. The greater carrying capacity of a more diverse asexual population follows from most theories of competition, including the familiar Lotka-Volterra model and its generalizations; the maintenance of genetic diversity in complex environments has been studied by many authors since the seminal paper by Levene (1953), and Strobeck (1976) has recently given a rigorous proof that the maximum number of alleles that can be maintained at any given locus in a haploid sexual population is equal to the number of niches available. We are then free to speculate that populations living in a spatially heterogeneous environment will comprise a variety of genotypes, and that as genotypic diversity increases the carrying capacity will increase as well.

Now imagine a sexual population living in such a complex environment. When this population has reached its carrying capacity a single asexual clone is introduced at low frequency. If the environment were uniform, the clone would tend to increase through autoselection, and would eventually eliminate the sexual population. In a complex environment it will still tend to increase, supplanting sexual individuals in the niches to which it is best-adapted because they are handicapped by the cost of sex. However, although the sexual genotype corresponding to that of the clone will decline in frequency during any given generation by an amount proportional to the cost of sex, it will be replenished at the end of every generation by recombination between the remaining sexual genotypes. At the same time, the clone is unlikely to be able to eliminate these other sexual genotypes since, as they become less common, their fitness increases. The success of the clone is restrained by the narrowness of its ecological range. The result of the invasion will therefore be a stable equilibrium at which both sexual and clonal individuals persist.

Conversely, imagine a clonal population which is invaded by a few sexual individuals. In a uniform environment they will be eliminated by selection if sex is at all costly. But if the environment is complex then, by producing progeny which are genotypically diverse, the sexual population can exploit niches in which competition from the clone is ineffective, and will spread through the population until the same equilibrium point as in the previous case is attained.

In either case, a single clone is unlikely entirely to supplant a diverse sexual population. The process responsible is a sort of sib competition, but it is utterly different from the sib competition of lottery models. The diverse progeny of a sexual brood, occupying different niches, will compete imperfectly, depressing one another's fitness to a much lesser degree than would the uniform progeny of a clonal brood, all of which exploit the same niche. As a consequence, the sexually diversified brood may have a greater overall success than a uniform brood, not because it is more likely to include the single optimal genotype, but rather because, being genetically different, its members will have somewhat different ecological requirements. As an alternative to the concept of imperfect competition, one may prefer to think in terms of a contest between a clone which has the greater reproductive efficiency but a narrow ecological competence and an inefficient but broadly competent sexual population. Since the clone cannot displace the sexual population from the whole of its ecological range, whilst the sexual population cannot eliminate the clone from its favoured niches, the result is a stand-off, with clonal and sexual individuals coexisting at equilibrium.

I have called this notion the *tangled bank*, from a phrase in the concluding paragraph of the *Origin of Species* which expresses the complexity of natural habitats that must strike anyone who contemplates a clump of moss or a bed of weeds.

There are two aspects of this argument which deserve immediate attention. The first is that it rests wholly on a difference in diversity between the sexual and asexual populations. If the number of clones were as great as the number of sexual genotypes the sexual population would have no refuge and would diminish monotonically at a rate proportional to the cost of sex. In its simplest form, therefore, the theory explains only how a sexual population can resist invasion by a single clone, or a few clones.

Secondly, imagine the fate of a mutant allele, arising in a sexual population, which directs the production of a certain proportion of unreduced apomictic eggs. The fitness of the mutant allele will exceed that of the normal allele by an amount equal to the product of the cost of sex and the proportion of eggs which develop parthenogenetically. All these unreduced eggs carry the mutant allele and all, of course, have the same genotype as their parent. But in the following generation the mutant allele will also have become incorporated into a great diversity of genotypes, through the reduced eggs produced by the original mutant individual. Indeed, it is not difficult to appreciate that the mutant allele will eventually become incorporated into all the genotypes of the original sexual population. Its spread will therefore be unconstrained by the narrow ecological competence of any particular genotype, and it must eventually eliminate the allele which directs the production of exclusively sexual eggs. By a precisely similar process of reasoning it follows that any mutant allele which directs the production of a greater proportion of unreduced eggs will replace any alternative allele directing the production of a smaller proportion. By the successive substitution of alleles arising by mutation, the population will eventually become exclusively apomictic. It follows that the tangled bank has the hidden requirement that the competing sexual and asexual individuals should experience total reproductive isolation. In a sexual diploid, this is equivalent to the requirement that alleles which abolish recombination (by making meiosis achiasmate or asynaptic) should be completely dominant.

With this requirement in mind, whether one views the tangled bank as a theory of individual selection or of group selection seems to be a matter of taste. Any advantage generated through imperfect competition within sibships accrues to individual females in the short term; but on the other hand the requirement for reproductive isolation implies that sexual and asexual individuals constitute groups which compete against one another.

Formal Population Model. The verbal argument that I have developed above seems plausible, but its logical validity, and the strength of the effect that it invokes, can be firmly established only through numerical analysis. I have used a model in which the growth of a haploid clone in pure culture is represented by a logistic difference equation, assuming that the logarithms of birth and death rates vary linearly with population density. In mixed cultures where two or more clones coexist, the effect of habitat complexity is modelled crudely through a competition coefficient, c. For individuals of the same genotype, the strength of the competitive interaction (the depression of the birth-rate or the elevation of the death-rate) is defined as unity, while between individuals of different genotypes it is defined as c, where $0 \leqslant c \leqslant 1$. A value $c = 1$ represents the situation of complete niche overlap between genotypes; $c = 0$ implies complete niche separation; intermediate values of c indicate intermediate degrees of competition. The survival and fecundity of an individual of given genotype thus depend both on the density of the population and on its genetic composition. A sexually reproducing individual of given genotype has birth and death rates equal to those of an asexual individual with the same genotype, except that the logarithmic birth rate of the asexual individual is made to exceed that of the sexual individual by a quantity s, the cost of sex. A value $s = \log 2$ indicates that the full twofold cost of sex is levied; if $s = 0$ there is no cost of sex. The production of recombinant genotypes among the offspring of sexual individuals under random mating is modelled in one of two ways. The first is an approximate model in which all genotypes are put into linkage equilibrium at the beginning of every generation (LE model). The purpose of LE models is to simplify computation and to allow the number of loci involved to be altered at will through a single input variable. LE models exaggerate the effect of sexual reproduction in diversifying progeny. In the second type of model, expected zygotic frequencies are computed exactly from a mating table (MT models). These models are more realistic, and automatically take into account linkage disequilibrium.

Deterministic Models. Within the context of the general population model, the extreme situation most favouring sex is with $s = c = 0$. Starting with sexual and clonal individuals equally frequent, and with all genotypes within either population equally frequent, the outcome of competition is shown in Figure 2.10. When there are as many clones as sexual genotypes, sex is neutral, and the frequency of sexual individuals remains at 0.50. When the sexual population is the more diverse, the frequency of sexual individuals increases and, if only a very few different clones are present, sexual individuals greatly outnumber clonal individuals at equilibrium. In particular, when only a single clone is present, the final frequency of sexual individuals may exceed 0.99. Figure 2.11 shows how this final frequency increases as the number of loci segregating in the sexual population increases, becoming nearly unity when the number of loci is moderately large.

Figure 2.10: The Frequency of Sexual Individuals after 200 Generations in a Simple Deterministic Model of the Tangled Bank, as a Function of the Diversity of Competing Asexual Individuals. Discussion in text.

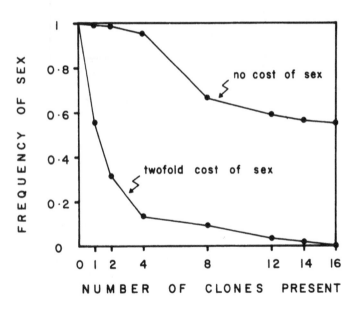

However, the success of the sexual population is sharply curtailed if sex is costly. Naturally, if there are as many clones as there are sexual genotypes, the final frequency of sex is zero if sex is to any degree costly. More generally, if the clonal population is moderately diverse, the final frequency of sex is low (but not close to zero) if the cost of sex is appreciable.

Increasing the competition coefficient has an effect similar to, but less pronounced than, raising the cost of sex in depressing the equilibrium frequency of

Figure 2.11: The Frequency of Sexual Individuals after 200 Generations in a Simple Deterministic Model of the Tangled Bank, as a Function of their Diversity when They are Competing with a Single Clone. Discussion in text. There is no cost of sex in this example.

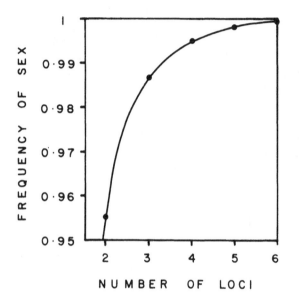

sexual individuals. In Figure 2.12, I have drawn a contour sketch of the final frequency of sexual individuals competing with a single clone for all combinations of cost and competition. Only if both cost and competition are small is the final frequency of sex close to unity. If either cost or competition are substantial, the final frequency still exceeds 0.50, but if both are substantial the final frequency is very low. If the cost exceeds log 1.2 and the competition coefficient 0.25, sex is virtually eliminated at equilibrium in a four-locus model.

These deterministic results confirm that the tangled-bank reasoning is correct, but also suggest that it may be fragile. Sex will be reduced to low frequency unless cost, competition and clonal diversity are all low. If the basic tangled-bank principle is correctly represented by these simple models, it will be effective only in checking the success of occasional inefficient asexual mutants. It will not prevent the mixed population from being dominated by an efficient asexual form with moderate genotypic diversity.

Elimination of Clones. If the sexual population is to any extent more diverse than the clones, the deterministic model inevitably leads to a stable equilibrium between sexual and asexual types because of the frequency dependence implicit in the scheme of competition. Under no conditions can asexual individuals be

Figure 2.12: A Contour Sketch of the Equilibrium Frequency of Sexual Individuals Competing with a Single Clone, when Both the Cost of Sex and the Competition Coefficient are Allowed to Vary in a Simple Deterministic Model of the Tangled Bank. Four loci are segregating in the sexual population. Discussion in text.

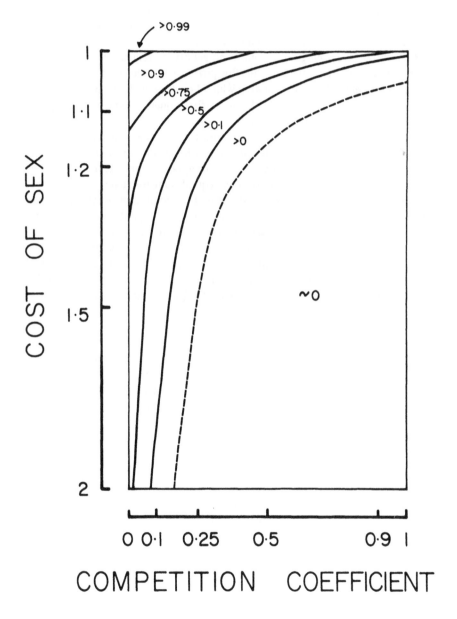

completely excluded at equilibrium. We can force this result, however, by altering the population model in one of two ways.

In the first place, we could admit sampling error. In a finite population there is always a nonzero probability that any given multilocus genotype will fail to be represented in the next generation of zygotes, merely through sampling error. In a sexual population this genotype is readily recreated by genetic recombination, but in the absence of mutation or migration it will be permanently lost from an apomictic population. Notwithstanding, a sexual population will lose genotypic diversity through sampling error even when genotype frequencies are protected by frequency-dependent selection; but this process will occur much more rapidly in an asexual population. The situation which I envisage is shown in Figure 2.13. If the sexual and clonal individuals are at first equally frequent and equally diverse, the sexual individuals will decline in frequency if sex is at all costly. In deterministic models this decline continues until the sexual individuals have been eliminated. But if the population is finite it may happen that a clone will become extinct before the sexual individuals have disappeared; the sexual individuals now being slightly the more diverse, their equilibrium frequency exceeds zero, and their frequency tends to move towards this equilibrium point rather than towards zero. When the next clone disappears, the equilibrium point shifts upwards and the frequency of sexual individuals tends to follow. Thus the frequency of sex will increase steadily as clone after clone goes extinct, and the process of successive clonal extinction will continue because the continued presence and increasing frequency of sexual individuals maintains the competitive pressure on each clonal genotype. With the disappearance of the final clone, sex has become fixed.

Alternatively, we might introduce some degree of temporal change in the environment. Imagine the environment to be divided into a number of niches, between which migration is limited. Within each niche, population density is regulated logistically, but in addition genotypes differ with respect to some component of fitness not related to population density. The total population is finite, with sampling error occurring during zygote formation and during the subsequent movement of a fraction of zygotes from the niche in which their parents lived to another niche. In every generation there is a certain probability that any given niche will not exist, in which case all the individuals present in that niche at the end of the previous generation are eliminated. In such a population, sexual individuals will tend to persist because of the tangled-bank principle of imperfect competition. As time goes on, each genotype will tend to accumulate in the niche to which it is best-adapted, but this accumulation will be more marked in clonal than in sexual lines of descent since the best-adapted sexual genotype will tend to be destroyed by recombination. When any given niche disappears for a generation, therefore, a very large fraction of the clonal individuals which possess that genotype best-adapted to the niche will be destroyed, and sampling error is likely to eliminate the few survivors which happen to be present in other niches. The genotype best able to exploit the niche when it

Figure 2.13: A Tangled-bank Model in which Asexual Individuals are Completely Eliminated by Imperfect Competition and Sampling Error in a Finite Population.

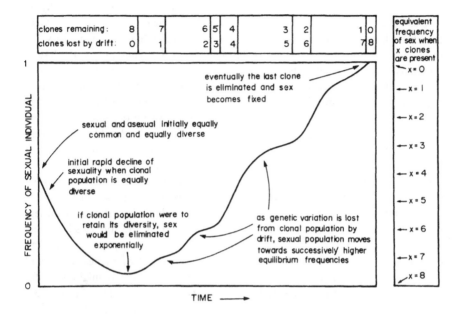

reappears in the following generation is then represented only in the sexual population. In time the clones will be completely eliminated.

This second model resembles Maynard Smith's (1976b) model of lottery selection, except that sexuality is maintained through imperfect competition rather than through truncation selection. It also brings to mind Williams' notion that sexual populations may be less liable to extinction because they cannot become highly specialized to a single ephemeral habitat type. To my mind, however, it is most reminiscent of a suggestion made by Maynard Smith in 1968 and later modelled by Slatkin (1975). Suppose that we have a series of environments, in each of which the optimal genotype has become fixed: $A1B1C1D1$ $\dots Z1$, $A2B2C2D2 \dots Z2$, $A3B3C3D3 \dots Z3$ and so forth. A new environment is now made available for colonization, in which the optimal genotype is, say, $A1B2C2D3 \dots Z1$. The most successful colonists will be recombinant offspring from sexual parents, the more numerous offspring of asexual parents lacking the necessary preadaptation. Applied to whole geographical areas, this hypothesis is perhaps extreme; applied to niches within a single diverse environment it becomes much more plausible, provided that some way of preventing the extinction of sexual individuals within niches can be found. However, I shall not develop this second model further, since it seems little more than a more concrete version of the first model; in either case clonal genotypes often fluctuate to extinction and thus irreparably reduce the ability of the clonal population to exploit the environment fully.

Arguments like these can be used to construct a theory of cyclical parthenogenesis on a tangled-bank foundation. Females should reproduce asexually as long as the environment remains undeveloped. If a niche becomes fully occupied, the clone has reached its maximum size and can perpetuate itself only by the production of dormant (or dispersive) propagules. When these propagules hatch, they may again find themselves in an undeveloped environment, for instance at the beginning of the growing season. If this environment is very simple, so that it can be fully exploited by a single clone, or if it is very transient or very large, so that there is insufficient time for all the niches to be occupied before the close of the growing season, then the dormant propagules should be produced asexually. But suppose the environment is complex, and persists for long enough for all the niches to be fully occupied by the end of the season. A clone which produced asexual dormant propagules would merely ensure that when they hatched they would already more or less fill those niches to which they were best-adapted, with little further room for expansion. The success of these propagules as a whole would thus be severely limited by the sort of sib competition envisaged by the tangled bank. On the other hand, propagules which had been diversified sexually would be spread over a variety of niches, and could expand into each. The combined success of these propagules might thus far outweigh that of a uniform sibship, despite their costliness. In a finite population with a large random mortality of propagules, sex might become the invariable way of producing propagules.

Stochastic Models. Models of the sort that I have described above are time-consuming and expensive to investigate numerically, and I have studied only the first and simplest, in which sampling error in a finite population is the only cause of clonal extinction. Because the numerical studies are experiments rather than calculations – their outcome cannot be accurately foreseen – I shall first describe how they are designed.

The model has no pretence to realism, and its function is purely strategic: it is designed to uncover any logical inconsistency in the hypothesis that, in finite populations inhabiting a stable but complex environment, sexual individuals will tend to increase in frequency through the operation of the tangled-bank principle of imperfect competition, and will eventually become fixed. The most appropriate initial state is thus an equal mixture of sexual and asexual individuals, the two competing populations having the same genotypic diversity. Since the only purpose of the experiment is to test the logical coherence of a general argument, the crucial case is that which the deterministic calculations have shown to be the most favourable to the hypothesis: no cost of sex, and no competitive overlap between genotypes. The result of an experiment is the frequency of sexual individuals through a series of generations, up to some specified final generation. To test the hypothesis, three predictions suggest themselves. First, if the simulations proceed for a sufficiently large number of generations, each population will eventually become fixed for sexual or for asexual individuals: it is predicted that sex will be fixed the more frequently. This prediction can be tested using statistics based on the expansion of the binomial distribution. Secondly, if we choose some generation far enough away from the initial generation for any systematic effect to have had time to manifest itself, the frequency of sexual individuals should by this time have increased. Using a number of replicate experiments, we can then compare the average frequency of sex at the end of the first generation (when sampling but not selection has occurred) with that at the end of the previously specified generation. The most powerful technique is to convert the observed frequencies to angles of equal information content through the arcsin transformation, to confirm the normality of the transformed data and then to calculate Student's t. Thirdly, if we choose a sequence of generations (say 1, 10, 25, 100 . . .) we predict that sex should tend to increase systematically with time. This approach is valid only if the frequency is calculated independently in each case; thus, it is necessary first to run a series of experiments each of which terminates in generation one, then a different series which terminates in generation ten and so forth, using only the terminal value from each experiment. The data can then be transformed into angles and normal parametric regression techniques employed.

The first series of experiments was run using an LE model for different numbers of loci. The frequency of sex increased sharply in time, departing significantly from its initial value of 0.50 from generation ten onwards, so that by generation 100 the clones were rare even if only two loci were segregating in the sexual population, and were extinct if the number of loci exceeded two. As the number of loci increases or, equivalently, as the environment becomes more complex, the slope of the linear regression of the frequency of sexual individuals on time becomes steeper (Figure 2.14). The relationship is not only highly significant, but also provides a good prediction (0.06) of the zero expected value of the slope when only one locus is present. Thus the rate at which sexual individuals spread increases linearly with the logarithm of environmental complexity.

Paradoxically, it seems that although a zero competition coefficient generates the highest equilibrium frequency of sexual individuals in deterministic models, it does not procure the fastest spread of sexual individuals in stochastic models. In three simulations with four loci, an initial population of 256 zygotes and $c = 0$, sex was in one case fixed in generation 75, was close to fixation at generation 100 in another case, and in the third case failed to increase much above its initial value in 100 generations. In five simulations with nonzero competition coefficients ($c = 0.1$–0.9) sex was fixed before generation 100 (in generations 33, 55, 56 and 96) in four cases and was close to fixation in the fifth. Moreover, when a single clonal genotype was initially present at a frequency (1/33) equal to that of each of the sexual genotypes it was eliminated most rapidly when $c = 0.25$, with fixation occurring more slowly both for smaller and for larger values of c (sex fixed in generation 81 for $c = 0$; 46 for $c = 0.1$; 31 for $c = 0.25$;

Figure 2.14: The Relationship Between Genotypic Diversity and the Strength of the Tangled-bank Effect. The plotted points are the slopes of the transformed frequencies of sexual individuals on time in stochastic LE models run for different numbers of loci. In all cases the sexual and asexual populations were initially equally diverse, and the population number was 256.

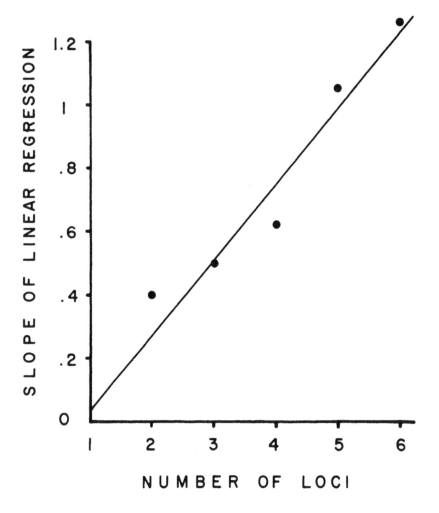

55 for $c = 0.5$). Naturally the clone should be eliminated in the majority of cases by sampling error even in the absence of any tangled-bank effect, but in a control simulation run with $c = 1$ the sexual population had not changed appreciably in frequency after 100 generations. What happens is that the number of distinct clones decreases more rapidly than the overall frequency of asexual individuals,

so that each surviving clone comprises a large number of individuals. These clones exert a selective pressure on genotypes in the sexual population, forcing some alleles to low frequency and thus reducing the ability of the sexual population to exploit the environment fully. If the competition coefficient is substantial, the surviving clones are restrained more effectively by sexual individuals with different genotypes, and their effect on the genetic variance of the sexual population is less severe.

This is encouraging, since it means that relaxing the condition that $c = 0$ does not impair but rather improves the efficacy of the tangled-bank process. However, relaxing the condition that there is no cost of sex has a drastic effect to the contrary. When the clones have a birth rate only 20 per cent greater than that of the sexual individuals, the spread of sex is halted or even reversed. In some cases sex persists at some quasistable frequency for long periods of time, but in most cases the sexual individuals are eliminated more or less rapidly. I am convinced that there is no combination of parameters that will prevent the elimination of sex when anything like a twofold cost is levied.

Moreover, the LE models exaggerate whatever advantage sex may entail by generating a greater genetic variance than would actually be the case. Table 2.1 shows that this effect is a large one. Two replicate experiments using four-locus MT models in which equally diverse sexual and clonal populations compete prove that sexual individuals tend to increase in frequency, but the linear rate of increase is an order of magnitude less than that observed in LE models. Only if the asexual population initially comprises a single clone is the sexual population able to prevail within 100 generations.

Conclusion. These admittedly rudimentary numerical studies have satisfied me that the principle of operation of the tangled bank is logically valid, and that it provides a mechanism by which sexual may replace asexual reproduction. However, they have also thrown grave doubt on the robustness of the hypothesis. It seems to require that sexual and asexual individuals should be completely (or very nearly completely) reproductively isolated; that the cost of sex should be very small; that the asexual population should comprise only a very few genotypes; that competition between genotypes should be very limited; and that the habitat should be very complex. The first and either the second or third of these conditions seem to be strictly required; the last two conditions are less critical. The field of operation of the hypothesis seems, therefore, to be very tightly circumscribed. Asexual individuals will tend to be eliminated in complex environments if they arise rather frequently but are reproductively inefficient or if, although reproductively efficient, they arise only very seldom. The tangled bank, therefore, might explain how isogametic sex arose from an asexual line; it might explain how contemporary sexual populations are protected against invasion by rather frequent parthenogenetic mutants whose eggs have low viability, or by extremely infrequent mutants whose eggs have high viability; but I doubt that it could explain the exclusion of asexual forms with even moderate levels of

Table 2.1: Results of an Experiment Using a Stochastic Four-Locus MT Model to Test the Validity of the Tangled-bank Hypothesis, with 16 (Two Replicates), Four or One Clones Initially. The number of independent observations in each experiment, N, is shown. The initial population size was in all cases 256, a compromise between a lower value at which any systematic effect would be obscured by noise and a higher value which would be prohibitively expensive. Any run in which the population rose above 500 or fell below 100 was pre-programmed to abort. All genotypes were in all cases equally frequent. Independent observations of gene frequency were made after one, five, ten, 25 and 100 generations; there were generally ten independent observations for each of these intervals. Random sampling of zygotes based on the FORTRAN pseudorandom-number-generating subroutine RSTAR. The three tests are as follows. (1) Elimination: number of cases in which all clones initially present were eliminated by generation 100, out of total number of cases observed. In no case was the sexual population eliminated. (2) Comparison: comparison of frequencies of sexual individuals (arcsin transformed) in generation one with that in generation 100, using appropriate t test. (3) Regression: slope b of regression of independently estimated frequencies of sexual individuals (arcsin transformed) on time, together with correlation coefficient r and probability that r differs from zero by chance. Since the hypothesis is directional, all tests of probability are one-tailed. None of the statistics are convincing when considered alone; taken together all are in the predicted direction; some taken alone seem unlikely to be attributable to sampling error; and as a whole they support the hypothesis that there is a weak tendency for sexual individuals to increase in frequency and for asexual individuals to be eliminated.

		16 clones				
		Replicate 1	Replicate 2	Pooled	Four clones	One clone
N		49	60	109	50	52
(1) Elimination test		0/9	0/10	0/19	2/10	8/10
(2) Comparison test	t	−0.65	−0.62	−0.93	−0.61	−1.69
	P	~0.25	~0.27	~0.18	~0.28	~0.07
(3) Regression test	b	+0.0445	+0.0267	+0.0356	+0.0265	+0.0752
	r	+0.188	+0.113	+0.151	+0.120	+0.322
	P	~0.10	~0.20	~0.05	~0.20	$0.01 > P > 0.005$

genotypic diversity and reproductive efficiency. The tangled bank must be rejected as a general theory of sex, unless patterns of sexuality in living organisms happen to mirror the shortcomings of the hypothesis.

2.8 The Red Queen

It will have become clear by now that the function we ascribe to sex depends crucially on the nature of environment. In the last two sections I have contrasted the consequences of interpreting environment as a primarily temporal variable with those of interpreting it as a primarily spatial variable. Both interpretations, however, take for granted that the important dimensions of the environment are physical factors such as temperature or moisture, or else that biotic factors can be treated in the same way as physical factors. In recent years, however, a number of authors have argued that biotic and physical factors cannot be equated, and that it is the biotic factors which are crucial to the evolution of sex. It is proposed that sex is favoured by interaction with other sexual species because the changing spectrum of genotypes among these other species creates a highly uncertain environment, and compels an adaptive genetic response which can be supplied only through recombination. An opinion of this sort was stated explicitly by Hamilton (1975) and Levin (1975), and was developed more fully by Glesener and Tilman (1978). At first I was not greatly impressed by the argument: granted that biotic factors will contribute greatly to environmental uncertainty, it seemed to miss the point that capriciousness, rather than mere uncertainty, is required to elicit the evolution of sex, unless some special scheme of selection is invoked, in which case the argument adds little to a theory based on physical factors. Late in 1978, however, I came to realize that species interactions may generate capriciousness as well as uncertainty.

To fix ideas, imagine two reproductively isolated populations of haploid sexual organisms, which for convenience I shall call 'predators' and 'prey' (Figure 2.15). At some given time the predator population happens to include a large proportion of individuals bearing the genotype $A1B1$, where $A1$ and $B1$ are alleles at two autosomal loci, which may bear the alternative alleles $A2$ and $B2$. This will create a selective stress on the prey population, favouring the prey genotype which is best able to avoid capture by $A1B1$ predators. Being favoured by selection, this genotype will increase in frequency in the prey population. In doing so it reduces the fitness of $A1B1$ predators and thereby creates an advantage for alternative genotypes in the predator population, so that in time the $A2B2$ genotype becomes common. But this in turn alters the relative fitness of genotypes in the prey population; and so on indefinitely. The fitness of any genotype in the predator or prey populations which is concerned with the interaction between them therefore varies in two ways. In the first place, fitness is frequency-dependent; when a genotype is common in one population, it evokes a genetic response in the other population which reduces its fitness. Secondly, this frequency dependence is lagged in time; the genetic response evoked in the interacting population takes a certain period of time in which to develop.

Because these inferences may be important, I shall pause to develop them more formally. The simplest model of the population genetics of the predator-prey

Figure 2.15: The Red Queen.

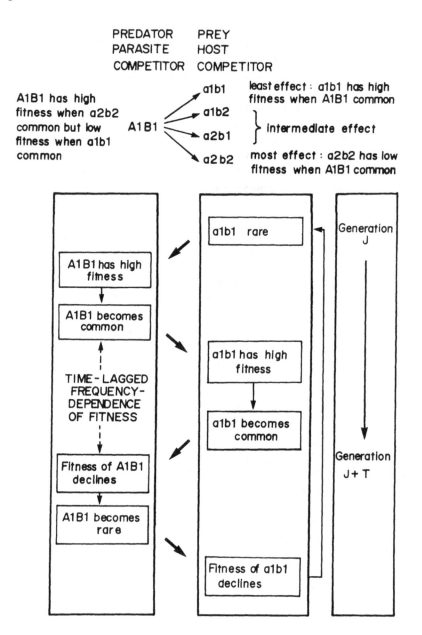

interaction that I can conceive concerns two haploid species, in both of which the strength of the interaction – the impact of a given predator genotype on a

Table 2.2: Simple Model of Predator-Prey Interaction, with Single-Locus Haploid Genetics and Constant Coefficients of Interaction whose Effect is Linear on Gene Frequency.

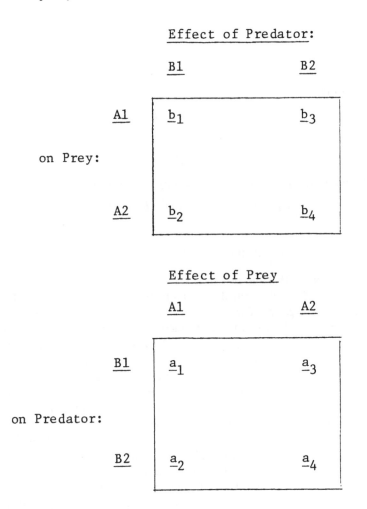

Effect of Predator:

	B1	B2
on Prey: A1	$\underline{b}_1$	$\underline{b}_3$
A2	$\underline{b}_2$	$\underline{b}_4$

Effect of Prey

	A1	A2
on Predator: B1	$\underline{a}_1$	$\underline{a}_3$
B2	$\underline{a}_2$	$\underline{a}_4$

given prey genotype, and vice versa — is mediated at a single diallelic locus. The interaction is therefore specified by the eight coefficients a_1-a_4 and b_1-b_4 of Table 2.2. (This model is identical with a two predator, two prey model in population dynamics where both predators and prey species have exponential population growth and the total numbers of predators and of prey are held constant.) The simplest version of this simple model is one in which the coefficients are symmetrical (i.e., $b_3 = b_2$, $b_4 = b_1$, $a_3 = a_2$, $a_4 = a_1$) and that their effect is linear on frequency. Given that the frequency of $A1$ among prey is p and the frequency of *B1* among predators is q, the fitnesses of the four genotypes are then as follows:

among prey, *A1* has

$$w_1 = 1 - b_1 q - b_2 (1 - q),$$

and *A2* has

$$w_2 = 1 - b_2 q - b_1 (1 - q);$$

among predators, *B1* has

$$v_1 = 1 - a_1 p - a_2 (1 - p),$$

and *B2* has

$$v_2 = 1 - a_2 p - a_1 (1 - p).$$

The evolution of the system is then governed by the inequalities:

$$w_1 > w_2 \text{ if } (1 - 2q)(b_2 - b_1) < 0;$$

and

$$v_1 > v_2 \text{ if } (1 - 2p)(a_2 - a_1) < 0.$$

For example, suppose that the system is set up with $b_1 > b_2$ and $a_2 > a_1$, and with the *A1* and *B1* alleles initially the more frequent, so that $(1 - 2q)$ and $(1 - 2p)$ are both negative. Then for the first few generations $w_2 > w_1$ and $v_1 > v_2$, so that *A1* decreases while *B1* increases. When *A1* becomes less frequent than *A2*, $(1 - 2p)$ becomes positive and makes $v_2 > v_1$, so that *B1* now begins to decrease in frequency. When *B1* becomes less frequent than *B2*, $(1 - 2q)$ becomes positive, so that $w_1 > w_2$ and the *A1* allele begins to increase. This increase will eventually make $(1 - 2p)$ negative and thus procure selection favouring the *B1* allele, whose subsequent increase will eventually make $(1 - 2q)$ negative and thus cause the decline of the *A1* allele; and so forth.

   The dynamics of even this simplest of schemes are quite complicated, but solutions which are not sensitive to small independent changes of the interaction coefficients seem all to be unstable. There is an unstable point if $a_1 < a_2$ and $b_1 < b_2$, and an oscillation of increasing amplitude otherwise. Only in special cases $(a_1 = b_1$ and $a_2 = b_2$ exactly, provided $a_1 > a_2$ and $b_1 > b_2)$ is there a stable point (at $p = q = 1/2$). The source of this instability, contrasting so strongly with the great stability of most simple frequency-dependent systems, is illustrated in Figure 2.16. When the fitness of *A1* individuals in the prey population is plotted against their current frequency (Figure 2.16a), the result is a spiral with a centre at $p = 0.50$, $w_1 = 0.85$ (for $a_1 = b_2 = 0.2$, $a_2 = b_1 = 0.1$). This spiral winds outwards in time, increasing the amplitude of the oscillations which it generates. Fitness thus varies with current frequency, but it is not uniquely determined by current frequency: at any given frequency there are two possible values of fitness, whose mean is (about) 0.85. These two values may be very different: at a frequency of 0.50 the fitness of the *A1* allele within the range of generations plotted in Figure 2.16a may be either 0.823 or 0.975, the smallest and the greatest values associated with any frequency. Which of the two values obtains depends on the recent history of the population. If we take any such pair of values and advance both 29 generations along the curve, they now both take the same value (29 is half the distance in time between the maximum and minimum points of the oscillation, so that $29 = P/4$, where P is the period of the oscillation). Thus, if we plot current fitness $w_1(t)$ against the frequency of the *A1* allele 29 generations previously, $p(t - P/4)$, each value of past gene frequency will be associated with a unique value of current fitness. The result is a linear regression with negative slope (Figure 2.16b; the slight scatter of points around the line is caused by the slow unwinding of the spiral of Figure 2.16a). Selection acting on the *A1* allele is thus both frequency-dependent and lagged in time.

Figure 2.16b suggests that the dynamics of gene frequency in the predator-prey model are controlled by two parameters: the lag T between a genetic change in one species and the development of the appropriate response in the other, and the slope $-a$ of the regression of current fitness on frequency T generations in the past. In the present example $T = 29$ and $a \sim 0.10$. Now, the effect of time lags in an analogous situation, the regulation of the number of individuals in a population through their density, is well-known and has been the subject of much recent work (for a review, see May, 1973). Roughly speaking, if a density-regulating factor operates with a lag T on a population whose exponential rate of increase at very low population density approaches r, the effect of the lag is to destabilize the equilibrium population number if the lag T exceeds the characteristic return time $1/r$. By analogy, I reasoned that the effect of a lag in the operation of a frequency-dependent selective factor would be to destabilize the equilibrium gene frequency if the lag T is large relative to a measure of the response time, $1/a$. In the present model the lag, although fully determined by the input parameters, cannot be conveniently manipulated. I therefore propose to write a single-species equation in which both the lag and the rate of response appear explicitly

$$w_1(t) = (1 - c) - a\,[p(t - T)].$$

The justification for using this single-species equation to study the consequences of a two-species interaction is that it constitutes an excellent description of Figure 2.16b. It differs from the explicitly two-species model in that the lag is fixed, and will not drift as it does in Figure 2.16a. This discrepancy is irritating, but I do not think that it is fatal. This drift does not occur, for instance, over a substantial volume of parameter space in the corresponding two-species diploid model, but the larger number of parameters, including dominance, that have to be specified for diploid models makes it impracticable to discuss them here. At this stage of the investigation the simplicity of the single-species analogue and the ease with which it can be investigated more than outweigh its inadequacies.

Figure 2.16: The Instability of a Simple Genetic Model of the Predator-Prey Interaction: (a) Response of Fitness to Current Frequency; (b) Relationship Between Fitness and Past Frequency. Details and discussion in text.

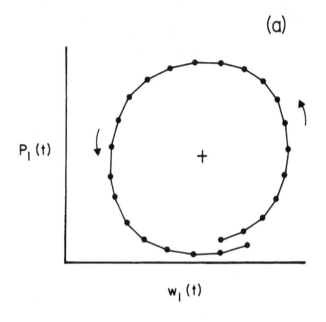

(a)

$P_1(t)$

$w_1(t)$

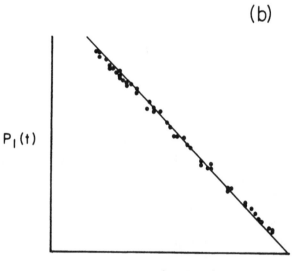

(b)

$P_1(t)$

$w_1(t+T/4)$

The results of iterating the equation for a range of values of T and a are shown in Figure 2.17. When the lag is short and the response slow, a stable point is approached asymptotically or via a series of damped oscillations. A very long lag and a steep response yield oscillations of increasing amplitude, so that one allele or the other is eventually fixed (or at least exceeds a frequency of 1-10^{-6}). Intermediate values for the lag and the response yield oscillations of constant and moderate period and amplitude, which, since they are generated from any initial gene frequency (except the neutral point at $p = 0.5$), represent stable limit cycles.

This is only a very brief sketch of a model which could be elaborated much further. (And which should be elaborated further: in view of the amount of attention which has been given to the effect of time lags in density-regulated systems by population ecologists, it is surprising that so little attention has been given to the phenomenon, presumably equally prevalent, of time lags in frequency-dependent systems by population geneticists.) I have avoided elaboration because the elaborations quickly become very elaborate indeed, but they are not in any case to the point. All that I am concerned to establish is that species interactions will generate time-lagged frequency-dependent selection; that the effect of a lag sufficiently long is to destabilize the equilibrium point and produce a series of oscillations; and that the nature of these oscillations depends on two parameters, the length of the lag and the magnitude of the restoring force. None of this is directly relevant to the evolution of sex; it becomes highly relevant when we inquire into systems where a species interaction is mediated by effects at two or more loci.

The simplest case is that of two diallelic loci in a haplontic organism. Retaining the model developed above, in which fitness is a decreasing linear function of past gene frequency, we can rewrite this appropriately as in Table 2.3. Again, any number of more complicated models could be built; I am concerned only with the simplest representation of basic principles. The complications that I have examined – diploidy; genetic variance of c; nonlinear models in which $P_{ij}(t - T)$ is raised to a power $b \neq 1$ – do not contradict the general conclusions I shall draw from the simple model.

Table 2.3: Fitness in Simplest-case Model.

Genotype	Frequency	Fitness
A1B1	$P_{11}(t)$	$(1-c) - a[P_{11}(t - T)]$
A1B2	$P_{12}(t)$	$(1-c) - a[P_{12}(t - T)]$
A2B1	$P_{21}(t)$	$(1-c) - a[P_{21}(t - T)]$
A2B2	$P_{22}(t)$	$(1-c) - a[P_{22}(t - T)]$

The state of the system at any given time t is fully defined by three parameters – the two gene frequencies $p(t) = P_{11}(t) + P_{12}(t)$ and $q(t) = P_{11}(t) + P_{21}(t)$, and the coefficient of linkage disequilibrium $D(t) = P_{11}(t)P_{22}(t) - P_{12}(t)P_{21}(t)$.

Figure 2.17: Sketch of the Dynamics of Gene Frequency Governed by the Equation

$$W_j = (I - C) - a[p_j(t - T)]$$

for a Between 0 and 0.4 and T Between 0 and 32, Given $c = 0.2$. Based on iterations of all the combinations of a and T marked on the inside of the axes. The double line is roughly the boundary between a region in which the point $P_j = 0.5$ is attracting and a region in which is is repelling. An attempt is made to separate the region of stable cycles into those with moderate and those with large amplitude (amp).

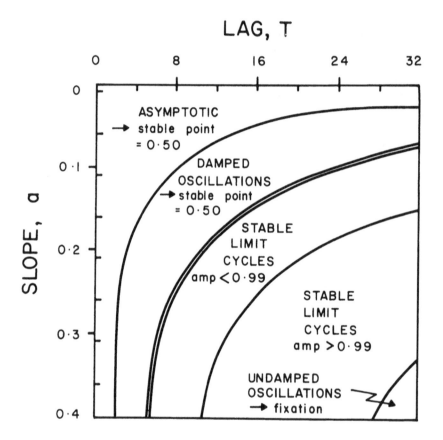

The behaviour of the system is illustrated in Figure 2.18. Setting $T = 0$ implies that the genotype frequency feeds back instantaneously on genotype fitness, and the model could be written and solved in the form of a set of ordinary differential equations. There is a single point of stable equilibrium defined by $p(t) = 1/2$; $q(t) = 1/2$; $D(t) = 0$. All initial states converge asymptotically on this joint equilibrium. The case of $T = 1$ implies that the fitness of a genotype is determined by its frequency in the preceding generation, and corresponds to an ordinary difference-equation model. There is no change in behaviour: all initial states converge on the joint equilibrium. For somewhat longer lags the joint equilibrium may remain globally stable, but be approached through a series of damped oscillations. But if both T and a are moderately large, gene frequencies continue to oscillate indefinitely. Moreover, not only the gene frequencies but also the coefficient of linkage disequilibrium oscillates, and as it oscillates it alternates between regions of positive and negative sign.

This last result is very curious. It implies that over a substantial range of conditions in a simple and unforced model the effect of a time lag in the operation of genotype-frequency-dependent selection is to destabilize the joint equilibrium in such a way that linkage disequilibrium repeatedly changes sign. This could be interpreted to mean that, whilst the physical environment is only very rarely capricious, the biotic environment often is. Interacting species do not merely add to environmental uncertainty, nor need they supply qualitatively new genetic challenges; rather, by making a selectively appropriate response they continually reverse the direction of selection in other members of the same community, thereby creating a permanently capricious environment for themselves. These are precisely the circumstances in which sex and recombination should evolve.

Whilst I arrived at this conclusion independently, I have since discovered that priority must be awarded to Jaenike. He wrote (Jaenike, 1978, p. 192):

> Consider an asexual prey species with two variable loci . . . With frequency dependent selection by enemies common genotypes are at a selective disadvantage. Under a wide variety of conditions, each asexual genotype and its enemy will undergo numerical fluctuations in a stable limit cycle . . . These numerical fluctuations of individuals of prey genotypes can be transformed to fluctuations in the frequencies of different genotypes within the prey species . . . these fluctuations are necessarily out of phase . . . Thus the selected, optimum sign and magnitude of the linkage disequilibrium varies temporally, and recombination is favoured.

Although Jaenike invokes the population dynamics rather than the population genetics of the interacting species, his logic seems valid, and his argument essentially the same as mine.

The effect of selection on recombination can be demonstrated directly by introducing a third locus, which is segregating for neutral alleles that determine the rate of recombination between the A and B ('fitness') loci. The alleles at this third locus are dominant or recessive according to their effect when heterozygous on the probability of recombination between the two fitness loci during a zygotic meiosis. The recombination locus is unlinked to the fitness loci.

Suppose that we take the system illustrated in Figure 2.18, and set the time lag T at eight generations. A dominant allele causing 50 per cent recombination introduced into a population which already contains a recessive allele causing zero recombination tends to approach an equilibrium frequency of about 10 per cent, irrespective of its initial frequency. This suggests that any mixture of a dominant allele for high recombination and a recessive allele for low recombination represents a stable equilibrium when the average rate of recombination (expectation that a randomly chosen zygote will undergo crossing-over between the two fitness loci) has a certain value which in this case is close to ten per cent. This in turn suggests that under any particular combination of genetic and environmental constraints there may be an equilibrium rate of recombination established and maintained by selection.

I have examined this idea more closely by means of the following numerical experiment. Dominant and recessive alleles causing different rates of recombination were introduced into the population at equal frequency, and allowed to compete for 250 generations. At the end of this period the frequency of the allele causing the higher rate of recombination was recorded. These numbers are given in Figure 2.19a, where each cell records the outcome of competition between a given pair of alleles at the recombination locus. In each column of the table there are two entries which are zero or nearly zero, indicating that

Figure 2.18: Phase Diagrams to Illustrate the Evolution of a Two-Locus Genetic System with Time-lagged Frequency-dependent Fitnesses. (a) For very short lags, the system moves asymptotically to linkage equilibrium. (b) For lags of between two and five generations, linkage equilibrium is approached through a series of damped oscillations. (c) Lags which exceed six generations destabilize the system, so that it oscillates indefinitely between a region of positive and a region of negative linkage disequilibrium. (d) The relationship between gene frequencies and linkage disequilibrium is this butterfly-shaped figure. On the left-hand edge of the left-hand 'wing' the $A1$ and $B1$ alleles are rare, but continue to decrease in frequency because of their abundance $T = 8$ generations previously. The population accumulates an increasing number of the favoured $A2$ and $B2$ alleles, and in doing so moves towards linkage equilibrium as they become concentrated into the coupling genotype. Once the frequency of the $A1$ and $B1$ alleles eight generations previously falls below 0.5 the direction of selection reverses, and they begin to increase in frequency. By this time the $A2B2$ genotype is present in excess, so that linkage disequilibrium has become positive; as selection reverses and repulsion chromosomes increase in the population, a negative disequilibrium evolves as the frequencies of $A1$ and $B1$ increase, passing down the right-hand edge of the left-hand wing. A similar sequence of events unfolds a few generations later as the population approaches the tip of the right-hand wing. The butterfly is symmetrical because of the symmetry of the fitness scheme, which constrains the oscillations of gene frequency above and below the equilibrium value of 0.5 to be of equal magnitude, so that $p(\text{max}) + p(\text{min}) = 1$.

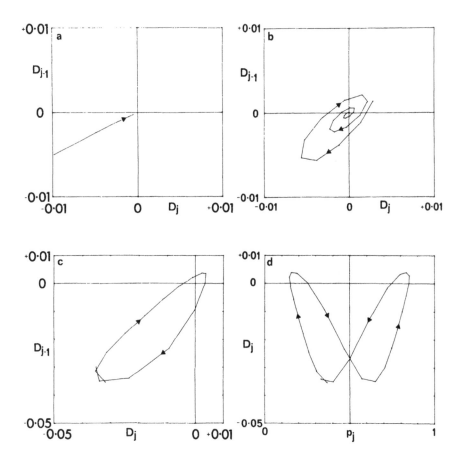

the competing alleles are equally or nearly equally fit. One of these cells represents pairs of alleles which cause equal or nearly equal rates of recombination, so that there is no basis for selection; these cells form a line which extends diagonally downwards from the uppermost left-hand cell. The other near-zero cell represents alleles which, when equally frequent, produce an average rate of recombination in the population as a whole which is close to the equilibrium rate; these cells form a line which extends diagonally upwards from the lowermost left-hand cell. These two diagonal lines divide the table into four sectors, in each of which a gene causing higher recombination is either more or less fit than one causing lower recombination, depending on its dominance and on the average rate of recombination in the population (Figure 2.19b). The two lines cross at an average value of about ten per cent recombination, which is therefore close to the equilibrium value under selection in this system.

For any given value of a, the periods of the oscillations in gene frequency and linkage disequilibrium increase with the length of the lag. The population spends

Rate of Recombination Determined by Dominant Allele:

A

	000	025	050	075	100	125	150	175	200
00	0'	+0447	+0266	+0228	+0091	+0038	-0009*	-0070	-0131
05	+1335	+0164'	0'	+0034	+0021	+0003	-0032	-0075	-0122
10	+1062	+0333-	+0089	+0014'	0'	-0005*	-0026	-0057	-0094
15	+0718	+0380	+0091	+0022	-0004*	-0011'	0'	-0022'	-0053
20	+0690	+0241	+0066	+0006*	-0032	-0045	-0041	-0025'	0'
25	+0429	+0143	+0040	-0030	-0075	-0092	-0093	-0082	-0062
30	+0290	+0100	-0003*	-0083	-0128	-0148	-0154	-0147	-0132
35	+0188	+0049	-0063	-0145	-0189	-0212	-0221	-0219	-0208
40	+0128	-0014*	-0134	-0211	-0256	-0282	-0294	-0295	-0288
45	+0060	-0090	-0208	-0281	-0327	-0355	-0370	-0376	-0373
50	-0015*	-0173	-0284	-0355	-0402	-0432	-0450	-0459	-0459

Rate of Recombination Determined by Recessive Allele:

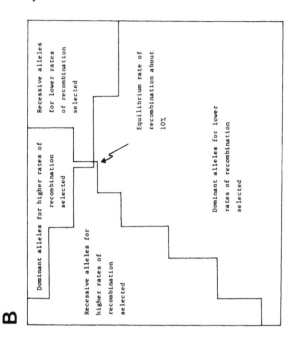

B

Dominant alleles for higher rates of recombination selected

Recessive alleles for higher rates of recombination selected

Recessive alleles for lower rates of recombination selected

Equilibrium rate of recombination about 10%

Dominant alleles for lower rates of recombination selected

longer and longer periods of time in the same state (positive or negative linkage disequilibrium), during which selection will favour lower rates of recombination. The equilibrium rate of recombination will therefore fall as the lag increases beyond the point at which a persistent oscillation in linkage disequilibrium is generated. The equilibrium rate of recombination can be estimated from the asymptotic frequency of a dominant allele causing 50 per cent recombination competing against a recessive allele causing zero recombination (this procedure is not entirely straightforward, since the oscillation in linkage disequilibrium between the two fitness loci induces a persistent oscillation in linkage disequilibrium between these loci and the recombination locus, which in turn induces an oscillation in the asymptotic frequency of alleles at the recombination locus). Figure 2.20 shows that there is a nearly linear relationship between the logarithm of the time lag and the logarithm of the equilibrium rate of recombination. For short lags ($T < 6$ generations in this case) the system eventually approaches linkage equilibrium, at which point all rates of recombination are selectively neutral. The maximal rate of recombination evolves under the shortest lag which generates a persistent oscillation in linkage disequilibrium, and average recombination falls steeply as the lag increases. Equilibrium recombination rates of one per cent or more were observed for lags of between about six and twelve generations.

These results are consistent with those reported by Charlesworth (1976) and already mentioned above but, in his model, fluctuations in linkage disequilibrium were externally imposed, rather than arising as a natural consequence of the properties of the system. Charlesworth also found that genes for higher rates of recombination were more likely to be selected when closely linked to the fitness loci, a result which I can confirm; the results given above are to this extent

Figure 2.19: The Outcome of Competition Between Alleles Coding for Different Rates of Recombination. Results of a numerical experiment. (a) For each pairwise combination of dominant and recessive alleles at the recombination locus, the two alleles were introduced at equal frequency and the simulation run for 250 generations. The number in each cell of the table is the change in frequency of the allele causing the higher rate of recombination; thus, when a dominant allele causing ten per cent recombination and a recessive allele causing 20 per cent recombination are equally frequent in the initial population, the frequency of the recessive allele after 250 generations of selection is 0.4968. Decimal points are omitted throughout. In each column, two cells are zero or nearly zero; those marked with a prime (') represent pairs of alleles which have similar effects and which are therefore not distinguished by selection, whilst those marked with an asterisk (*) represent pairs of alleles which, when equally frequent, produce an average rate of recombination close to the equilibrium value. (b) The two diagonal lines marked out by the primes and asterisks can be used to divide the table into four regions. They intersect near the point which represents an equilibrium value of recombination for the system; in this case, with a lag of eight generations, the equilibrium value is about ten per cent.

Figure 2.20: Relationship Between the Length of the Time Lag and the Average Rate of Recombination Between the Two Fitness Loci at Equilibrium Under Natural Selection. Plotted points are the maximal and minimal values observed between generations 9950 and 10,000; the minimal value for a lag of 20 generations was not observed in this interval. Further explanation in the text.

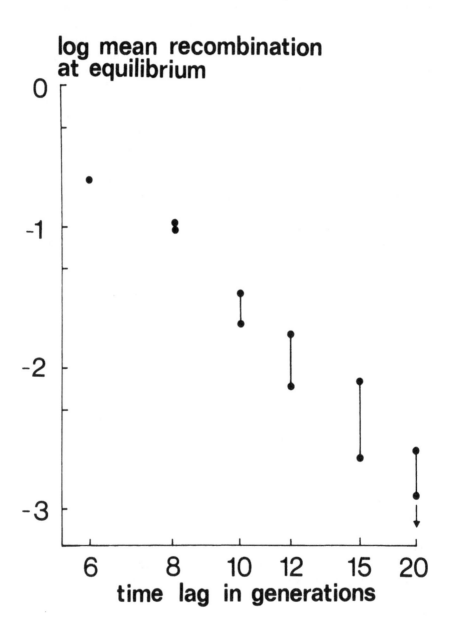

conservative estimates of the equilibrium rate of recombination, since they refer to a recombination locus which is unlinked to the fitness loci.

I shall summarize my findings by saying that when the genotype-frequency-dependent component of fitness is lagged in time, as it generally will be when it reflects the counteradaptation of interacting species, the average rate of recombination between the loci which mediate the interaction will often approach an equilibrium value greater than zero under natural selection. The fate of a newly arisen mutation which affects the rate of recombination between the fitness loci will depend on three factors: on the external environment (the length of the lag and the steepness of the adaptive response); on the state of the population (the average amount of recombination already occurring); and on genetics (the dominance of the new allele, and the linkage of the recombination locus to the fitness loci).

Since, according to this hypothesis, sex and recombination are selected because continual adaptation is necessary in order to stay in the same place, so to speak, I shall call it the Red Queen, after the character in *Alice*. The name was suggested by its generic similarity to L. van Valen's well-known explanation of extinction schedules, to which he gave the same title. It is the simplest, subtlest and most elegant theory of sex, and provides an ironically prompt retort to G. C. Williams' despair of ever finding a valid theory couched in terms of the familiar abstractions of elementary population genetics.

A much more complicated and explicit model incorporating similar principles has been published very recently by Glesener (1979). In this model a predator-prey interaction is mediated by up to ten diallelic loci in both predator and prey, and he compares the effects of the alternative states of amphimixis and apomixis, rather than those of continuous variation in the rate of recombination, in the prey. Glesener does not explicitly recognize the crucial role of time lags in destabilizing linkage equilibrium, but he finds that sex will be favoured when the prey are highly fecund and their population size is small, conditions which imply a short lag and a steep genetic response, and which are thus broadly comfortable with my results. He also finds that sex tends to be favoured if the genotypic diversity of the competing asexual form is low, as one would expect. His results show that the principle underlying the Red Queen hypothesis is not only capable of procuring the selection of nonzero rates of recombination in a system which models the predator-prey interaction at one remove, but may also favour expensive amphimictic sexuality in an explicitly two-species context.

2.9 Two Taxonomies

A theoretical population geneticist will perceive the most fundamental consequence of sex as being the capacity of meiotic recombination in amphimicts to digest, as it were, linkage disequilibrium between multilocus genotypes. In a very perceptive essay, Felsenstein and Yokoyama (1976) used this property as the basis for classifying theories of sex into two categories; I shall retain their two categories and add a third, as follows.

First, the digestion of stochastically created linkage disequilibria permits a faster response to an applied selection differential by lubricating the interference between selection at different loci, and thus permits a population (or sibship) to arrive at an optimal genotype more quickly under directional selection. In most such hypotheses, the primary function of recombination is to liberate a mutation arising in a finite population from its initial genetic context. The Vicar of Bray (in the usual Fisher–Muller version), the ratchet and the hitch-hiker belong together here. Felsenstein argues that lottery versions of the best man also operate on the same general principle, since they rely on the chance linkage disequilibrium between members of a sibship occupying the same niche.

Secondly, the unloosening of linkage disequilibria created deterministically by epistasis allows the population (or sibship) to respond more quickly if the direction of epistasis should change. Recombination will function (even in an infinite population) to untie the linkage disequilibria existing at the moment when the direction of selection reverses. The Red Queen falls into this category, along with any version of the best man requiring a capricious environment.

Thirdly, the dissolution of linkage disequilibria allows the population (or sibship) to exploit the environment more fully in space or in time. This consequence may arise through a reduction of competition in complex, stable environments, or through a reduction of risk in environments which are uniform and mutable. We can therefore classify together the tangled bank, Maynard-Smith's theory of successive invasion, Williams' theory of extinction and runt versions of the best man.

Obviously, some theories have found themselves strange bedfellows, and one is inclined to protest, as Maynard Smith does, at the inclusion of lottery models and the hitch-hiker in the same category of phenomena. However, this is not only one taxonomy, it is only one possible sort of taxonomy, which uses as its criterion a highly abstract description of how recombination greases the skids of selection. Its only purpose is to order certain formal notions about the consequences of sex, and it is not necessarily at all useful if we want a taxonomy to serve some other purpose. In particular, it is virtually useless if we wish to classify concepts in such a way that we can compare rival hypotheses according to the predictions they make about the occurrence of sex in nature. For this purpose, we require a taxonomy whose criteria refer to measurements that have been made (perhaps very crudely) for at least one large and diverse group of organisms. The taxonomy I have adopted for this purpose is as follows.

First, we can make the fundamental operational distinction between hypotheses which are capable of making falsifiable predictions about the occurrence of sex and those which are not. The latter group is to be rejected out of hand.

Secondly, we can divide the former group into functional and nonfunctional hypotheses. The nonfunctional hypotheses are those which invoke history as the main or sole factor in the evolution of sex, and which therefore predict that sex will be correlated with phylogeny rather than with ecology. Since I have already taken up a strictly functionalist position, I shall consider non-functional hypotheses only if the functional hypotheses can be shown to be inadequate.

Thirdly, the functional hypotheses can be divided between a group of balanced hypotheses, which require substantial contemporary genetic variation for the mode of reproduction and which operate through short-term individual selection, and a group of unbalanced hypotheses, which do not have this requirement and instead operate through long-term group selection. This is not a perfectly clean logical distinction; the Vicar of Bray and the ratchet fall clearly into the unbalanced category and the best man and the Red Queen into the second, but runt models and the tangled bank have an intermediate character. It is made for operational convenience, because I anticipate that it will be relatively easy to discover whether or not sex has substantial consequences at the level of groups of unrelated individuals.

Finally, when we have eliminated as many hypotheses as possible, we must decide between the remaining possibilities by comparing the different types of natural selection which they invoke in order to counter the autoselection of alleles coding for parthenogenesis. This boils down to comparing the different models of environment underlying the rival hypotheses: mutable versus constant, simple versus complex, physical versus biotic. Reducing these abstractions to operational terms is then the major remaining difficulty.

I shall amplify and implement this scheme in Chapter 4. We cannot proceed, however, without first establishing the facts which the successful theory is required to explain. These facts form the subject of the next chapter.

3 PARTHENOGENESIS AND VEGETATIVE REPRODUCTION IN MULTICELLULAR ANIMALS

Introduction

The scientist strives to make the internal mental world of theory and the external physical world of fact fit like ferrules. Each is incomplete and, indeed, functionless without the other; and when properly fitted there is no discrepancy or disagreement between the two, but only the sweet conformity of a well-made fishing-rod. So far we have made only the butt-end of the rod, and the female ferrule; or, rather, we have made a variety of butt-ends, some elegant and some crazy botched affairs, but all alike incomplete, and all awaiting the test of whether they will receive the male ferrule of fact. Some will no doubt prove too narrow to be united at all; others will be so broad that they could accommodate any conceivable partner, and the actually realized possibilities of the world will slop about inside them like a stick in a bucket. One – we can hope – will be just right.

The question with which we are faced concerns the utility of sex in the natural economy. Before we can proceed further, however, it will be necessary to decide more precisely what sort of question is involved. Some recent authors have expressed it very bluntly: what use is sex? Or, why reproduce sexually? Or, why does the genotype not congeal? But these blunt interrogatives are not perfectly apposite, since they can be given answers that I shall stigmatize as being illegitimate. For example, it is conceivable, as I have related, that sex and recombination are concerned with loosening the knots of stochastically created linkage disequilibria. However, I shall argue in the next chapter that this opinion cannot be used to explain the distribution of sex in nature; with great ingenuity, some slight and narrow basis for belief might be built, but scarcely a foundation capable of supporting the immense edifice of material facts about sexuality. This nuisance is created by attempting to explain sex in too abstract a fashion, as though it were an ideal process whose mere material manifestations could be neglected, so that we are led into speculations whose worth cannot be established. It can be abated only by framing a question which refers to sex as a phenomenon rather than as a concept. The most general question of this sort is: how can the occurrence and distribution of sex among living organisms be explained? It might be objected, though, that much remains unknown, and we can hardly explain what we do not know; so that the question might be better phrased as, how can what we know of the occurrence and distribution of sex among living organisms be explained? We shall then be able to make predictions about the sexuality of organisms which have not yet been studied, although, indeed, these predictions will amount to little more than extrapolations of previously established generalizations.

The fire in which our theories should be tested is then fuelled by observations of sexuality in living organisms. These should be extensive, but they should also be representative. They should be extensive, in the sense of embracing a wide variety of organisms, because of the value we place on generality. They should also be representative, since no test is fair if we are allowed to select instances which happen to fit any particular interpretation, whilst rejecting those which do not. But it is very difficult to choose dispassionately between material facts, any of which might profitably be cited; biases and prejudices insensibly pervert one's judgment. The only remedy, I believe, is to strive, not to be representative, but to be complete, and to omit nothing that seems to bear on the issue. The perfect fuel, then, would be an exhaustive catalogue of the modes of reproduction found among plants and animals, together with a minute account of their ecological and physiological correlates. This would be an immense undertaking, as much to read as to write. Instead, I have chosen to review the occurrence of parthenogenesis among multicellular animals. This is a large, diverse and relatively well-known group of organisms, and although my review is doubtless incomplete, I have not deliberately ignored or suppressed any pertinent fact. In this way I have tried to make the review sufficient for its purpose, which is to provide the basis of the comparative analysis in the following chapter, in which rival theories of sexuality are judged according to their ability to explain the facts set out below.

The review is divided into a number of sections, each corresponding to a major taxon or group of taxa, whose arrangement and nomenclature, with a few trifling exceptions, follow Blackwelder (1975). The text of each section is divided into three parts: the first part is a list of previous reviews, if any are available; the second delineates as concisely as possible the modes of parthenogenesis exhibited by members of the taxon; and the third is a more extended description of the ecology, genetics, physiology and geographical distribution of parthenogenetic species, comparing them wherever possible with their amphimictic relatives. Where material is sparse the first part may be omitted and the second and third parts combined.

The review is incomplete in two major respects. In the first place I have omitted or given only a very brief account of a number of topics which seem peripheral to the main issue Thus, I have made virtually no mention of artificial parthenogenesis; nor have I described the cytology of parthenogenesis in more detail than is necessary to understand its genetic consequences; nor have I attempted at this point to review chromosome number (other than degrees of ploidy) or crossover frequency (beyond noticing instances of achiasmatic meiosis). Secondly, I have read the original literature only when it has been published in English or French. I have cited German, Italian, Scandanavian, Russian and other authors only when their work has been translated, or when it has been summarized in English or French reviews whose treatment of other English and French articles I consider satisfactory. I hope that I have not thereby overlooked any large body of relevant facts; but at all events my account cannot be relied upon to attribute priority correctly.

I have no space for a detailed history of the study of parthenogenesis, but I must give some brief notice to my predecessors. The ancient writers hardly mention vegetative reproduction and parthenogenesis in animals, although, as is well known, Aristotle guessed at the development of honey bees from unfertilized eggs. The modern period opens with the eighteenth century, when Albrecht (1701; cited by Geddes and Thomson, 1901) noticed that a silkmoth, long isolated in a glass case, produced fertile eggs, and von Leeuwenhoek (1702) observed the budding of hydras. Somewhat later Bonnet (1745) described parthenogenesis in aphids, and Trembley (1744) published his classic studies of asexual reproduction in hydras. The discovery of a succession of sexual and asexual phases in the same life history is generally credited to the poet Chamisso (1819), but his results, together with those of Sars, Leuckart and von Siebold, were not appreciated until Steenstrup (1842) made them widely known. By this time the first reviews of parthenogenetic reproduction had started to appear: Owen (1849), Leuckart (1853), von Siebold (1871) and Karsten (1888). The closing decades of the nineteenth century were most remarkable for the tremendous contributions of Weismann to the study of genetics, through which the cytology of parthenogenesis, the nature of the polar bodies and the cyclical parthenogenesis of cladocerans and rotifers began to be understood (see Weismann, 1876, 1889). At this time, too, there was a great deal of interest in artificial parthenogenesis, the results of which were summarized by Delage and Goldsmith (1913). The turn of the century saw the publication of the widely read book by Geddes and Thomson (1901), who, despite their anti-Darwinian approach, summarized a good deal of earlier literature. The only other twentieth-century books which include substantial amounts of material on parthenogenesis are those by Winkler (1920) and Vandel (1931), the less systematic account given by Ghiselin (1974a), and the treatment of vegetative reproduction by Bounoure (1940). However, more or less extensive summaries have been published as articles by Ankel (1927, 1929), Suomaleinen (1950), Delavault (1958), Peacock and Weidman (1961), Narbel-Hofstetter (1964) and Cuellar (1977); those of Suomaleinen and Narbel-Hofstetter are especially useful. I located a short review by Clark (1973) too late to make use of it. Certain standard texts also include a good deal of relevant material, especially those by Hyman on invertebrates and White on cytology, together with Grassé's *Traité de Zoologie* and the series of volumes edited by Giese and Pearse on the reproduction of marine invertebrates. However, there has been no thorough review of animal parthenogenesis in any language for half a century, and it was the absolute indispensability of such a review for the testing of hypotheses about sexuality which led me to write one.

3.1 Porifera (Parazoa)

Reviews. Fell (1974b); Simpson and Fell (1974; gemmules).

Incidence of Parthenogenesis.

CALCAREA (*Calcispongiae*). Small calcareous sponges living in the littoral and intertidal. Gemmule formation is unknown but budding is common, and vegetative fragmentation may be a regular mode of reproduction in some species.

HEXACTINELLIDA (*Hyalospongiae*). Large siliceous deep-water sponges. Budding certainly occurs, but the production of gemmules is doubtful.

DEMOSPONGIAE. The largest class of sponges, and the one which is found in the widest range of habitats. Asexual reproduction by gemmules is widespread, being especially characteristic of the freshwater Spongillidae.

Discussion. Sponges are sessile aquatic filter-feeders. They are most common at moderate depths in the sea, but the glass sponges (Hexactinellida) inhabit the deep ocean, and a single family of Demospongiae, the Spongillidae, is found in fresh water. Most and perhaps all sponges can reproduce sexually; most species are hermaphrodite but there are a few gonochores. All are oogametic but, although the process of fertilization is extremely difficult to observe, it seems that the sperm is carried to the egg by a choanocyte, an arrangement without parallel in other multicellular animals. Fertilization is thus internal, although the animals are sessile and the male gametes are shed into the water.

Asexual reproduction by budding or fragmentation is probably nearly universal, and sponges are well known for their ability to regenerate completely from very small fragments, but many species can also produce the amictic propagules called 'gemmules', which are unique to sponges. Gemmules can be envisaged as internal buds of a sort. A number of amoebocytes come together, and divide until a little ball of cells has been formed; then the cells on the outside secrete a tough coat of spongin, into which spicules are incorporated. A thorough review of gemmule morphology and development has been contributed by Simpson and Fell (1974). The finished gemmule then has a core of cells packed with food reserves, which is protected by an almost impermeable covering of noncellular material. The gemmule is released from the sponge and carried away by water currents, eventually to fall to the substrate, where after a certain period of time it germinates, to develop directly into a young sponge. The gemmule is the first example we meet of what will become a very familiar class of structures: the resistant, dormant or overwintering stage. It is the means by which individuals survive when the environment changes for the worse, and will no longer support the growth or even the survival of adult sponges; it disperses the sponge population in time, permitting the continuous occupation of highly seasonal habitats. In very harsh environments gemmules may remain dormant for many years, and germination has been reported amongst gemmules stored in the laboratory for a quarter of a century (Harrison, in Harrison and Crowden, 1976, p. 34).

Gemmules and other fragments of sponge tissue may also disperse the population in space, colonizing new habitats to which they have been carried passively by water currents. The active dispersal stage is a ciliated parenchymella larva, which is usually the product of sexual reproduction. In some marine ceractino-morph sponges the gemmules are said to develop into parenchymella larvae, rather than directly into adult sponges. After the gemmules have been formed by cell aggregation and division, a two-layered embryo is produced by the migration of small cells towards the periphery of the gemmule, and further differentiation results in a ciliated larva which, when it hatches, can disperse actively. However, the amictic production of active ciliated larvae, identical in external form with the larvae which develop from fertilized eggs, was a subject of controversy eighty years ago (Wilson, 1891, 1894, 1902; Maas, 1896; Minchin, 1897) and is still controversial today (Bergquist *et al.*, 1970; discussion in Harrison and Crowden, 1976, pp. 20-3). The main objection to the claims of Wilson (1891), Sivaramakrishnan (1951) and Bergquist *et al.* (1970) seems to be senti-mental: there is a reluctance to believe that a group of diploid cells can undergo the extensive growth and differentation required to form a complex larva without meiosis and fertilization. Since precisely this process is known for certain to occur in other groups of animals, this objection cannot be sustained. More seriously, the accumulation of intracellular food reserves obscures cytologi-cal detail, and makes it difficult to distinguish between the amictic process of embryogenesis by cellular growth and aggregation, and the meiotic process of oogenesis.

Gemmules have been reported from many species in two orders of the Demo-spongiae; a complete list, with authorities, is given by Simpson and Fell (1974). The great majority of these are freshwater forms; among the few marine sponges for which gemmulation has been reported, the only known example of a species in which the gemmules are the only persistent phase for a part of the year is *Haliclona loosanoffi*, whose life history has been the subject of a number of reports (Hartman, 1958; Wells *et al.*, 1964; Fell, 1974a). Fell describes a popu-lation living in the Mystic Estuary, Connecticut. There, the sponge overwinters as gemmules, which germinate in late spring and develop directly into adult sponges. The adults, which are gonochoric, reproduce sexually all through the summer, but in July an abrupt decline in the production of sexual larvae coincides with an abrupt increase in the production of gemmules. The switch from almost exclusively sexual reproduction via larvae at the end of June to almost exclusively asexual reproduction via gemmules by the end of July is most striking. Wells *et al.*, working on a more southerly population at Hatteras Harbour, North Carolina, report two periods of gemmule formation, one in summer and the second in late fall; there are also two periods of sexual reproduction and hence two periods of larval settlement, one in June–July and the other in October–November. In North Carolina, gemmules represent the only persistent phase in the life cycle between July and September rather than during the winter, as is the case further north. In the related sponge *H. oculata* both gemmules and functional tissue are present through the year in New York waters (Fell, 1974a), and the continuous production of gemmules seems to be the rule in other marine sponges (Herlant-Meewis, 1948).

Most of the sponges living in Chilka Lake, India, which is subject to pro-nounced seasonal changes in salinity, reproduce by purely sexual means, but gemmules are formed by some individuals of *Laxosuberites lacustris* living in shallow water (Annandale, 1915).

The reproduction of intertidal marine Demospongiae has been reviewed by Bergquist *et al.* (1970). All of the sponges that they studied reproduce sexually, although some (e.g., the ceractinomorph *Halichondria moorei*) do so only very infrequently; likewise, all have some means of reproducing asexually, and indeed the asexual mode predominates over the sexual in most cases. Some of these sponges reproduce colonially, sending up buds from a network of attachment stolons; others practise a sort of clonal reproduction which results in the passive dispersal of detached buds or gemmules; and yet others are said to produce an active ciliated larva by ameiotic means, a claim that I shall accept, although it remains controversial. The frequency and variety of asexual reproduction is interesting in itself, but still more interesting is the relationship between the mode of reproduction and position on the shore. The eight species of ceractino-morph sponges that were studied all relied heavily on asexual propagation, pro-ducing ciliated larvae amictically; there is even a suggestion that mixis occurs in only one species of the eight, although the evidence is not conclusive. What does seem likely is that the predominantly sexual larvae of subtidal sponges become replaced by asexual larvae in the intertidal. The same pattern was found amongst the tetractinomorph sponges. Of the three common genera in the intertidal, one (*Plankina*) produced larvae by exclusively sexual means, but the other two (*Tethya* and *Aaptos*) achieved most of their reproduction asexually. *Tethya* produces detachable buds at all depths; *Aaptos* buds only in the intertidal. Sexual embryos are often found in specimens of *T. aurantium* and *A. aaptos* from the subtidal, but very rarely in specimens from the intertidal. Finally, it is only in the intertidal that these two species show extensive stolon budding, producing interconnected colonies of zooids. All in all, there is an unmistakeable tendency for the predominately sexual reproduction of the subtidal sponges to be replaced by asexual mechanisms in the intertidal. *Halichondria* appears to rely almost entirely on the ameiotic production of larvae, and occupies the highest position on the shore, at the margins of pools and rocks; moreover, the habitat of the adults seems to be determined by the behaviour of the larvae, which creep slowly upwards until they reach the water's edge, rather than by selective death follow-ing nonselective settling.

The life cycle of freshwater sponges has been the subject of a number of recent papers by Gilbert and Simpson, and their colleagues. Simpson and Gilbert (1973) worked on a population of *Spongilla lacustris* and *Tubella* (= *Trocho-spongilla*) *pennsylvanica* living in a bog pond in New Hampshire. The gemmules hatched during late April and early May, just before water temperature began to increase, and therefore presumably as a response to internal cues of some sort. Eggs were produced over a period of about six weeks during May and June, and exceptionally as early as the end of April; sperm production, on the other hand, was confined to a very short period in the first half of June. Both species appear to be gonochoric; eggs and sperm were never observed in the same individual. Larvae were found in the pond around the time of sperm production, and there seems no reason to doubt that all the larvae are of sexual origin. These larvae settle and grow into adult sponges during the remaining part of the summer. Gemmulation usually begins towards the end of October, and is completed by the middle of November: some large individuals of *Spongilla* produced a few gemmules in midsummer, but in the fall all surviving individuals gemmulated, irrespective of their size. The gemmules are the only phase of the life history present in the pond during the winter.

A good deal of interest has been aroused by the timing of gemmulation. After the early laboratory work of Rasmont (e.g., Rasmont, 1962) had appeared to implicate a wide variety of physicochemical factors, the field experiments conducted by Gilbert (Simpson and Gilbert, 1974; Gilbert, 1975) have established that the onset of gemmulation is triggered primarily by environmental cues associated with the onset of autumn. The evidence for this position is twofold. In the first place, the second generation of adult sponges, developing from larvae during the summer, undergoes gemmulation during October, at about the same time as the survivors of the first generation, which have developed from gemmules hatching in the spring. The second line of evidence is experimental. Branches of gemmulating sponges were collected from the pond in November, kept in cold storage over the winter, and then transplanted back into the pond at various dates between early May and late June. Although the sampling intervals were too coarse to permit a definitive conclusion, it seems that the later transplants gemmulate only slightly later than the earlier ones – certainly, the difference in the dates of gemmulation was much smaller than the difference in the dates of transplantation.

In later experiments, transplants were made over a much wider range of dates, between early May and mid-September, and differences in the date of gemmulation could now be clearly demonstrated; in the extreme case, the sponges developing from gemmules transplanted in mid-September had not themselves succeeded in producing any gemmules by the end of the growing season in mid-November. In general, it seems that sponges which hatch from gemmules put into the pond before the end of June all gemmulate more or less at the same time during October, whilst later transplants have difficulty in producing any gemmules at all. Since sponges transplanted as late as mid-September were by mid-November as large as those transplanted much earlier in the year, Gilbert concludes that the failure to gemmulate cannot be due to small size or inadequate nutrition, and instead prefers to implicate some undetermined effect of long storage or some undiscovered environmental cue necessary for later gemmulation and experienced in the pond before but not after the end of June. There may be a simpler answer: whilst these late transplants may be of normal size they may never have had any excess energy available for reproduction, having been forced to spend it all on growth, whilst the earlier transplants, having achieved the requisite size during the summer, could utilize all their surplus energy for gemmulation.

African freshwater sponges of the family Potamolepididae produce amictic dormant structures which Brien (1967) refers to as statoblasts. Simpson and Fell (1974) argue that this term is inappropriate, because of its extensive prior usage to denote the asexual dormant stages of bryozoans, and I agree that it should be suppressed.

The life cycle of gemmule-forming sponges, then, seems to be very similar in freshwater and in the sea. The gemmules usually hatch at the beginning of the growing season and develop into adult sponges which undergo one or more phases of sexual reproduction until, with the onset of unfavourable conditions at the close of the growing season, the surviving adults again produce gemmules. Naturally, many sponges do not produce gemmules at all. Most marine sponges do not, and a few freshwater sponges seem to have an exclusively sexual life history (for a review, see Penney and Racek, 1968). For example, an unidentified spongillid studied by Simpson and Gilbert (1974) was first observed to colonize an artificial substrate in June, but did not produce eggs or sperm until September, with the larvae appearing in late September and October; no gemmulation at all was found even in samples taken as late as early November, and there is a strong presumption that the life history lacks an asexual phase. Similarly, *Ochridaspongia rotunda*, a deepwater sponge living under conditions of constant low temperature, seems to be exclusively sexual (Gilbert and Hadzisce, unpublished observations cited by Simpson and Gilbert, 1974). Some interesting correlations of the presence of an asexual phase with the nature of the habitat have been observed by Poirrier (1969; cited by Simpson and Fell, 1974). In southern populations of *Eunapius gracilis* and *Anheteromeyenia ryderi*, gemmules are formed during summer in seasonal habitats, but in the fall in permanent ones; even more strikingly, *Heteromeyenia baileyi* has a midsummer gemmulation in seasonal bodies of water but does not form gemmules at all in permanent habitats.

Sponges which do not form gemmules either feed and grow throughout the year, as most marine species do, or else form reduction bodies, in which functional tissue is lacking and the vital processes are at a very low ebb (see, for example, Penney, 1933).

A final aspect of sponge biology which merits brief discussion is the fusion of adjacent sponge individuals. According to Rasmont (1970), when the gemmules of freshwater sponges are cultured together they will often fuse to give rise to a single adult individual, but they will only do so if they are of the same 'strain'. Strains appear to be determined genetically, rather than by the environments experienced by the parental sponges. This process does not seem to be analogous to the formation of heterokaryons by the fusion of fungal hyphae, since there is no suggestion that the presumably diploid nuclei from each gemmule ever come to coexist within the same cell; nor is it comparable with syngamy, both because the nuclei of the fusing gemmules are unreduced and no fertilization occurs and because fusion occurs between members of the same strain rather than between members of different genders. Unless the populations studied by Rasmont comprised only a few clones and each strain comprised all the members of a single clone – which is not impossible – the products of fusion will be a mosaic of cells with different diploid genomes, and I presume that this implies that the gemmules that they produce, although formed amictically, will reflect the genotypic diversity of the founding gemmules. There is no evidence that this process is sexual, since nuclear fusion and recombination have not been observed (or looked for, so far as I know), but perhaps it is not stretching a point too far to regard it as a distant cousin of sexuality.

3.2 Mesozoa (Agnotozoa)

Review. Stunkard (1954).

Incidence of Parthenogenesis. The Mesozoa are cellular endoparasites of uncertain affinities; they are often considered to be structurally degenerate Platyhelminthes, but Hyman (1940) makes a good case for considering their simple structure to be primitive. It is possible that the two subclasses usually recognized do not comprise a natural group. Their reproductive biology is interesting because it involves an alternation of sexual and asexual generations, and because they are unique amongst multicellular organisms in reproducing by means of agamonts.

DICYEMIDA (Rhombozoa). Endoparasites of cephalopods. A ciliated larva (larval stem nematogen) infests the kidneys of young hosts. The inner (axial) cells contain two nuclei, one vegetative and the other generative; the generative nucleus becomes surrounded by cytoplasm to form an agamete. The agametes divide to form a ball of agametes within the axial cell, which now constitutes an agamont. The larva is at this stage called an *adult stem nematogen*. The formation of the agamont resembles both the schizogony of some protists (e.g., Coccidia) and the polyembryony of digenetic trematodes. Each agamete divides several times to form a number of small somatic cells and a larger axial cell. By further division the somatic cells come to surround the axial cell, the individual formed in this way being called a *primary nematogen*. The primary nematogens escape into the host's kidney fluid and continue to proliferate asexually by the same process of agamogony. When the host reaches sexual maturity – most cephalopods are semelparous, dying soon after reproduction – the primary nematogens give rise to *rhombogens*, which are morphologically identical to nematogens but contain large reserves of food. Certain agametes within the axial cells of the rhombogens differentiate into a cluster of cells called *infusorigens*. The superficial cells of the infusorigen differentiate into ova and the axial cell agametes into sperm by a presumably meiotic process which involves the expulsion of a polar body. The fertilized eggs develop into a free-swimming infusoriform larva, which escapes from the host and is thought to infest an unknown intermediate host before a cephalopod is again infested to complete the cycle. This is a very brief account of a life cycle of which contradictory accounts have been given by Lameere (partial summary in Lameere, 1922), Nouvel (especially Nouvel, 1947), Gersch (1938) and McConnaughey (1951). According to Nouvel the oocytes extrude a single polar body, and the sperm becomes pycnotic and is eliminated, so that reproduction is gynogenetic. More plausibly, McConnaughey describes the expulsion of two polar bodies and the fusion of male and female pronuclei. It is possible that in some species, or populations, the oocytes develop partheno-genetically and in others after fertilization. At all events, any sexual process would seem to be autogamous, and the infusorigens are best interpreted as self-fertilizing hermaphrodites.

ORTHONECTIDA. Rare endoparasites of a variety of benthic marine invert-ebrates. Their life cycle was worked out by Caullery (see Caullery and Lavallée, 1912). The agamont is a large multinucleate plasmodium found in the gonads of the host. Each nucleus becomes surrounded by cytoplasm, and the plasmodium fragments into a large number of agametes. These differentiate into male and female sexual forms, elongate ciliated individuals which are released into the external medium where they copulate and produce fertilized eggs. These eggs develop into ciliated larvae which invade a host, where their outer somatic cells are lost and their inner generative cells disaggregate and become scattered in the host's gonads. They then develop into a multinucleate plasmodium by repeated nuclear division.

3.3 Monoblastozoa

Incidence of Parthenogenesis. The monoblastozoa are a phylum named for the single genus *Salinella*, a minute ciliated organism of the interstitial fauna. Asexual reproduction by transverse fission is normal; these are unconfirmed reports of a sexual process leading to the production of a ciliated larva (Hyman, 1940).

3.4 Coelenterata (Cnidaria)

Reviews. Campbell (1974); Berrill (1949; medusa budding); see also Russell (1953, 1970).

Incidence of Parthenogenesis.

HYDROZOA. A very large and diverse class in which asexual budding may or may not lead to colony formation. The coelenterates as a whole, and the hydro-zoans in particular, are characterized by a more or less regular succession of sexual (medusoid) and asexual (polypoid) phases in the life cycle. The sexual medusae are in most cases formed as asexual buds on the polyp; since most species are gonochoric, all the medusae produced by a single individual or colony are usually of the same sex. Medusae may also arise though asexual budding from other medusae (reviewed by Berrill, 1949), but the medusa characteristically reproduces sexually, producing either eggs or sperm. The fertilized egg, which may or may not be brooded by the female medusa, develops into a stereogastrula and thence by elongation into a ciliated free-swimming planula, which later becomes attached to the substrate and metamorphoses into a polyp. In the Hydrozoa the medusa is often suppressed, in which case the asexual buds (gonophores) of the polyp do not develop into free-swimming medusae but remain sessile during sexual maturation.

Hydroida (reviewed by Kanaev, 1952; see also Lenhoff and Loomis, 1961; Burnett, 1973). An almost exclusively marine group represented in freshwater by the solitary hydras and the colonial *Cordylophora*. Most hydroids are solitary polyps which undergo asexual reproduction by budding, or less often by fission. Asexual reproduction can also be accomplished by means of frustules, which are nonciliated planula-like bodies representing an asexual dispersive phase; they develop into polyps and are apparently produced by polyps. In stressful environments the polyp itself can become a reduced resistant phase. Marine genera such as *Obelia* and *Pennaria* have a free-swimming medusoid phase, and the polyp plays no part in sexual reproduction. Other marine genera such as *Sertularia* and *Plumularia* exhibit various stages in the suppression of the medusa, which remains attached to the polyp and eventually comes to consist of no more than a specialized structure for producing gametes (the 'sporosac'); finally, in the freshwater hydras the medusa cannot be distinguished, and gonads form directly in the polyp.

Milleporina (*Hydrocorallinae*). Colonial forms of shallow tropical seas, often on coral reefs. Colony growth and medusoid budding are asexual processes.

Stylasterina. Colonial forms of tropical and subtropical seas, extending to considerable depths. Colony growth is by asexual budding.

Trachylina. Oceanic gonochores in which the polyp is reduced or absent. Most develop from a planula into an actinula (a tentaculate polyploid lava) and thence into a medusa; but in some there is a form of polyembryony in which the egg is brooded and develops directly into an actinula which buds off other actinulae, all of which subsequently develop into medusae. The Oriental *Craspedacusta* and the African *Limnocnida* are exceptional freshwater medusae.

Siphonophora. Polymorphic oceanic colonies comprising modified medusoid and polypoid zooids; the gonophores do not develop into free-swimming medusae. There appears to be little asexual reproduction, other than budding to form the colony; however, the planula may bud in a manner similar to the actinula budding of trachylines.

SCYPHOZOA. Polymorphic oceanic medusae found in all seas, but most common in coastal waters. Most scyphozoans are free-swimming forms in which the polyp has been reduced to a transient scyphistoma larva; however, the Stauromedusae of polar seas are sessile animals whose attached medusa bears a superficial resemblance to a polyp. The adults are usually gonochoric, and some forms, including the familiar *Aurelia*, brood their eggs. The fertilized egg develops into a tentaculate polypoid larva, the scyphistoma, and subsequently into a free-swimming medusoid larva, the ephyra, before transforming into an adult medusa (their development is reviewed by Berrill, 1949). Asexual reproduction by polyp budding or fission is widespread. In Stauromedusae the planula puts out up to four stolons, which separate from one another and develop separately. The scyphistoma larva of Semeostomae and Rhizostomae may bud (like hydroid polyps) or send out stolons which bud off other scyphistomae. The scyphistoma may also produce ephyrae by transverse fission, a process known as strobilation (reviewed by Berrill, 1949; Chapman, 1966). Scyphistomae, ephyrae and adult medusae may all form reduction bodies when starved, but the characteristic resistant phase of scyphozoan life histories is the podocyst, a structure formed asexually by the scyphistoma (for a short review, see Chapman, 1966).

ANTHOZOA. Marine, solitary, or colonial polyps, without a medusoid phase; a majority of anthozoans are gonochores, but hermaphroditic species are not uncommon.

Alcyonaria. A diverse group of colonial marine coelenterates, including the sea pens, sea fans and pipe corals. Parthenogenetic eggs have been reported from *Alcyonium* (see below), but otherwise there seems to be no method of asexual reproduction, except colony formation by budding from a primary polyp.

Zoantharia. The sea anemones and typical corals, an exclusively marine group including both solitary and colonial forms. The Actinaria, or sea anemones, are solitary sessile polyps typically inhabiting rocky shores and most abundant in shallow tropical waters, but present in all seas and to considerable depths. Review of reproduction by Chia (1976). Some anemones regularly reproduce in an asexual manner by longitudinal fission, budding or pedal laceration: fission and budding are rare, but the former occurs regularly in *Haliplanella* (= *Sagartia*), and *Boloceroides* sheds tentacles which, either before or after their separation from the parental anemone, bud a complete new individual. Pedal laceration is the fragmentation of small pieces of tissue from the pedal disc of the anemone, all of which subsequently regenerate into new individuals; it is a regular mode of reproduction in several genera of sea anemones, but elsewhere occurs only amongst ctenophores. The Scleractinia (Madreporaria), or stony corals, include a few solitary forms but are mostly colonial. Like all the reef-building corals they are common only in the shallow littoral of tropical and subtropical seas. The colonies are formed by asexual budding from a single primary polyp, which is metamorphized from a sexually produced planula (but see below). Reproduction in the Fungiidae, a family of solitary scleractinian corals, is rather unusual (reviewed by Wells, 1966). The sexually produced planula settles and develops into a primary coral polyp, which grows upwards. When a certain size has been reached, it undergoes transverse fission: the broad discoidal oral end becomes detached, leaving a stump attached to the substrate. The stump often dies, but may grow back and pass through several more cycles of transverse fission. The adult polyps, lying free on the substrate, may also produce buds which themselves divide transversely to form new detached coral polyps. Transverse fission may thus give rise to clones which descend either from a sexual (planula) or from an asexual (polyp) ancestor, being in either case able to produce a new sexual generation of planulae. The remaining anthozoan orders are the colonial Zoanthidea (Zoanthiniaria) and Antipatharia, and the solitary Ceriantharia, all of which are exclusively marine and largely tropical in distribution. Other than the usual process of colonial proliferation by budding, I have found no mention of any regular means of asexual reproduction in these groups, although the Ceriantharia are known to have considerable powers of regeneration.

Discussion. Although the coelenterates are an almost exclusively marine group, the bulk of the literature on sexuality refers to the aberrant freshwater hydrozoan *Hydra*. Hydras are the familiar solitary polyps found attached to a variety of substrates in streams, lakes and ponds. They are an excellent example of the usual tendency for larval or dispersive phases to be lost from the life histories of freshwater organisms, the medusa having been entirely eliminated. They are also of historical interest, since the first unequivocal reports of asexual reproduction in animals describe the budding of *Hydra* (von Leeuwenhoek, 1702; Anon., 1704; Trembley, 1744). Since that time, dozens of articles have been devoted to the description of the sexual cycle in laboratory populations of hydras, and there has been the usual uproar of rival schools contending. Kanaev (1952) summarizes the earlier work; since then, collections of articles edited by Lenhoff and Loomis (1961) and by Burnett (1973) have appeared. I shall attempt only a brief sketch of the recent position.

Hydras may undergo longitudinal or transverse fission, but these processes have been observed only rarely, and budding is by far the most usual mode of asexual reproduction. Each polyp usually bears between one and three buds, but up to eight have been reported. The buds eventually detach and live independently of their parent. Whilst they are attached they are entirely dependent on the parent for nutriment until their mouths open, after which feeding is mutual, so that a rudimentary 'colony' is formed. Indeed, under optimal conditions buds which are still attached to their parent may themselves begin to bud, giving the appearance of a small colony of polyps. Buds usually detach after a minimum of about three or four days' growth.

Hydras also reproduce sexually, some species being hermaphroditic, e.g., the green hydra *Chlorohydra viridis*, and others gonochoric, e.g., *Hydra fusca*, *H. oligactis* and probably *H. littoralis*. The fertilized egg always passes through a period of dormancy. In some species there is a classical cyclical parthenogenesis, a period of budding being followed by sexual reproduction, after which the individual dies (Laurent, 1844, cited by Kanaev, 1952; Gross, 1925), but in other species (or, it seems, in other populations of the same species) a single polyp may undergo several generations of sexual reproduction, separated by periods of budding (Brien, 1949, cited by Kanaev, 1952). Park (1961) reports the case of a male *H. littoralis* which continued to produce spermaries for 103 days. Finally, in *H. littoralis* and *H. viridis* budding and sexual reproduction may occur simultaneously on the same polyp (Mrazek, 1907; Burnett and Diehl, 1964), although a partial inhibition of one mode of reproduction by the other is probably usual, if not inevitable (but see Park, 1961).

A great many workers have attempted to identify the factors responsible for eliciting sexuality, and have come up with a great many possibilities. Most, however, have plumped for an effect either of crowding or of temperature. Nussbaum (1892, cited by Kanaev, 1952) and Schultz (1906, cited by Kanaev, 1952) both reported the onset of sexuality in starved, or crowded hydras, and their early experiments were strikingly confirmed by the elegant experiments of Loomis and his collaborators (e.g., Loomis, 1957, 1959; Loomis and Lenhoff, 1956; and, especially, Loomis, 1961), who were able to show that elevated partial pressures of carbon dioxide are potent in eliciting sexuality. In his 1957 paper, Loomis showed that the rate of sexual differentiation is proportional to the initial CO_2 concentration in the culture. In his 1961 article he proved that, with free diffusion, the shell of water surrounding each hydra accumulates metabolically generated CO_2; in crowded cultures the 'haloes' of different hydras often overlap, producing very high local CO_2 concentrations, which elicit sexuality. If the culture is mechanically shaken the halves are dispersed, and no sexual differentiation occurs. In the 1957 experiments Loomis was unable to elicit sexuality when the hydras were cultured singly, even in very small volumes of water, but in 1961 he reported that by adding agar to the culture medium the CO_2 haloes were stabilized and sexuality followed. These experiments establish conclusively that budding hydras often switch to sexual reproduction in stagnant, crowded conditions.

There have also been repeated claims that the expression of sexuality varies with the physical environment, and especially with temperature. Gross (1925), echoing earlier authors such as Korotnev and Herting, found that *H. oligactis* will become sexual only at low temperatures; he also made the interesting observation that the response to temperature depends on the season, being greatest in winter and least in summer. Uspenskaya (1921, cited by Kanaev, 1952) found the converse situation in *H. grisea*, with sexuality being repressed by a transfer from high to low temperature. Burnett and Diehl (1964) found that *H. pseudooligactis, H. girardi, H. fusca* and *H. oligactis* rarely show spontaneous sexual differentiation when cultured at room temperature, but become sexual when transferred to culture at 8-10°C; by contrast, *H. littoralis* and *H. viridis* often became sexual under crowded conditions at room temperature, but were little affected by the cold treatment. Kanaev (1952) varied both temperature and food level in a series of experiments with green hydras: he found that a transfer from cool to warm cultures will elicit sexuality only if the animals are starved – the same treatment, but with a surplus of food supplied after the cold treatment, failed to elicit sexuality. Moreover, not only temperature itself but also change of temperature may contribute to eliciting sexuality (Nussbaum, 1909, cited by Kanaev, 1952); Goetsch (1927) even claimed that any of a wide variety of environmental shocks – change of diet or temperature, or even mechanical jolting – could lead to gonad differentiation. Hydra cultures may, however, go sexual even when maintained at constant temperature: Burnett and Diehl (1964) have suggested that natural photoperiod may be an additional factor, with decreasing day length inducing sexuality in *H. pseudooligactis*. Even more confusingly, Park and coworkers (Park, 1961; Park *et al.*, 1961) describe apparently endogeneous rhythms of sexual differentiation both in populations and in individual hydras maintained in the laboratory at nearly constant temperature, density, food level and photoperiod. There has been a suspicion (see, for example, Lenhoff and Loomis, 1961, p. 371) that the sexual periods of natural populations and laboratory cultures of hydras are linked in some subtle way – for example, by some agent in the laboratory water supply – but if this is so the agent responsible has yet to be identified.

Burnett and her students (e.g., Burnett, 1973; Burnett and Diehl, 1964) have proposed as a general hypothesis for the induction of sexuality that 'when growth in hydra ceases the animal enters the sexual state' (Burnett and Diehl, 1964, p. 246). It is claimed that the interstitial cells virtually all differentiate into gametes, at least in some species of *Hydra* and that, when all have been used up, further growth and budding is impossible. The evidence which is cited in support of this point of view is chiefly that growth ceases during the sexual phase; that reduced temperature, elevated CO_2 and other stimuli associated with sexuality may no more than reflect a basic depression of growth rate; and that budding and gonad differentiation do not occur simultaneously. Although the generality of this argument is attractive, some reservations must be made. There are many instances in which budding can occur during or even after gametogenesis, and these are not explained by the hypothesis. Reduced temperature and stagnation may slow down growth, but this does not constitute evidence of any direct effect of growth rate on sexuality. Finally, any general relationship between growth rate and sexuality is scarcely surprising, since sexual reproduction will inevitably reduce the quantity of energy available for somatic growth.

Hydra is not an ideal organism for the study of sexuality, since gametogenesis and budding are not equivalent modes of reproduction, in the sense that the production of mictic and amictic eggs is equivalent. A failure to distinguish between sex and reproduction has flawed many of the laboratory studies of hydras. For example, a number of authors (e.g., Uspenskaya, 1921, cited by Kanaev, 1952; Goetsch, 1927) have reported that fed *Hydra* produce more gonads, or have a greater proclivity to form gonads, than starved individuals, and Burnett and Diehl (1964) list 'overfeeding' as a stimulus for sexuality.

Likewise, budding rate is increased by liberal feeding or, equivalently, is depressed by crowding (e.g., Thorp and Bartholemew, 1975). However, it is not surprising to learn that *reproduction* is promoted by additional food; if extra energy is available, the organism may use at least some of it to achieve additional reproduction. Only those experiments which demonstrate a switch from budding to sexual reproduction, or vice versa, when food levels are varied are relevant to a discussion of the function of sexuality.

A second and even more damaging criticism of the laboratory studies is that almost all have been performed without any reference to natural populations, and indeed have often been conducted on clones of unknown provenance purchased from dealers. For this reason the significance of much of the experimentation cannot be assessed, since it is not known whether the factors which elicit sexuality in the laboratory are ultimate or proximal in nature. For example, we might find that stagnation is a reliable inducer of gonad differentiation in cultures derived from a certain population of some species of *Hydra*. Because stagnation results from crowding, sexuality might be characteristic of dense populations in nature; but on the other hand, the partial anaerobiosis caused by organic decomposition may signal the approach of fall, and thereby act as a cue to an abrupt change in physical conditions. Stagnation might then serve to indicate in nature either that conditions are crowded, or that the physical environment is likely soon to change. Laboratory experiments cannot provide wholly satisfactory answers to the questions which are of most interest to an evolutionary biologist, unless they are designed and interpreted in the light of observations made in the field.

Having made these reservations, my abiding impression of the laboratory work on *Hydra* is that it has demonstrated the crucial role of crowding in the elicitation of sexuality. The less extensive work on other coelenterates strengthens this impression. Sexual medusae are produced by the colonial marine hydroid *Podocoryne* when CO_2 concentrations are experimentally increased (Braverman, 1962), and the sexual zooids differentiate first in the crowded centre of the colony (Braverman, 1963). The scyphozoans *Aurelia* and *Cyanea* seem to strobilize, preparing for sexual reproduction by releasing ephyrae, only when they are neglected, in stagnant cultures.

There have been so few reliable studies of sexual periods in natural populations of hydras that the relevance of much of the very extensive laboratory work is questionable. Several authorities repeat the statement that green hydras undergo sexuality in spring and early summer, whilst brown hydras do so in fall and early winter, which seems consistent with the usual laboratory result that sexuality is induced in green hydras by transfer from high to low temperature and in *H. oligactis* by transfer from low to high temperature. However, it has not been adequately documented for natural populations. Kanaev (1952, pp. 262, 267) states that *H. attenuata* may have several sexual periods alternating with periods of asexual budding during a single growing season, with the asexual periods coinciding with temporary abundance of food. His basis for this claim appears to be Stolte (1928), but I have not been able to locate this work and the figure he reproduces is incomprehensible. The only quantitative data I have found are due to Miller (1936), who worked at Douglas Lake in Michigan. He found that budding agonadic hydra (*H. oligactis*) were present throughout the year, being most abundant in early summer and least abundant in midwinter. A rather large number of males, amounting to as much as a quarter of the total sample, were found in October and November; females were recorded a month or so later, but only in very small numbers. By contrast, in Lake Washington he found large numbers of females in November and December, but although eggs, presumably fertilized, were found later he never saw any males. Bryden (1952) found no sexually differentiated individuals at all during a large-scale systematic survey of hydra populations in Kirkpatrick's Lake in Tennessee, which is ice-free in most years.

The dynamics of natural populations of hydras have been studied by Boecker (1918), Welch and Loomis (1924), Miller (1936) and Bryden (1956), the last being by far the most satisfactory; the theses by Carrick (1956) and Griffing (1965) have not, so far as I know, been published. The review by Reisa (1973) is idiosyncratic and incomplete. It seems to be characteristic of hydra populations that their density and location within a lake are subject to abrupt change during the growing season. Bryden (1952) has contributed useful information about the rate of budding in natural conditions. He found that under the most favourable conditions most individuals carry a single bud, which detaches in two or three days, in agreement with Miller's (1936) opinion that summer populations of hydras can double in two days if a sufficient area of surface is available for settlement. Under less favourable conditions, ten days or more may be required for the development of the bud, and during this period other buds may have formed – presumably because, late in its development, a bud can contribute to the nutrition of its parental polyp. The presence of several buds per individual may therefore indicate unfavourable conditions and a low rate of reproduction, rather than the reverse. The average number of buds produced per week was maximal in late September and early October (about 4.5 buds/polyp/week) and minimal in February (about 0.33 buds/polyp/week). There is therefore a hint that when sexuality occurs (which it did not in Bryden's population) it may do so just after the period of maximal rate of asexual reproduction.

Protohydra is an atentaculate polyp found in shallow brackish water; no medusoid phase is known. It reproduces asexually by transverse fission, and occasionally by budding. In natural populations sexual reproduction occurs in late summer or fall; individuals captured in winter developed gonads when they were starved and exposed to fluctuating temperatures (Muus, 1966). The fertilized egg presumably develops directly into a polyp, but this has not yet been definitely ascertained.

The extraordinary hydrozoan *Polypodium* occurs as a budding stolon endoparasitic in the eggs of sturgeon. The budded polyps are released when the host fish spawns, each taking with it a small quantity of yolk, and then lives for a time free on the river bed. During this phase of the life history the polyps reproduce asexually by longitudinal fission. They are said sometimes to bear gonads, but how sexual reproduction occurs, and whether or not a medusa exists, remain unanswered questions.

Lytle (1961) has published an especially interesting study of *Craspedacusta*, one of the very few freshwater coelenterates which has retained a medusa. The polyp, formerly described as a separate genus (*Microhydra*) is minute and lacks tentacles. *Craspedacusta* is probably native to China, where it inhabits shallow seasonal pools on the floodplains of rivers (Kramp, 1950); where it has been introduced in Europe and North America, it seems to be a 'weedy' species usually found in small, often artificial, bodies of water, where it flourishes for a short time before disappearing. Russell (1953) gives a complete bibliography of *Craspedacusta*; more recently, its ecology has been discussed by Acker and Muscat (1975). The polyps (hydranths) reproduce asexually by budding, but produce three different kinds of bud: hydranth buds, which remain attached to the parental polyp and so create small colonies; planuloid buds (frustules), which detach and move away from the parent before differentiating into new polyps; and medusoid buds, which are released as free-swimming medusae. The medusae, of course, produce gametes and represent the sexual phase, so we can equate the induction of medusoid buds with the elicitation of sexuality. Previous authors (Reisinger, 1934, 1957; McClary, 1959) had found that medu-soid budding occurred when the temperature was raised to about 28°C. Lytle (1961) showed that the situation is more complicated. In cultures maintained at a variety of temperatures between 19 and 27°C there is first a spell of hydranth budding, which is gradually replaced by a slow rise in frustule production, and this, in turn, is replaced after about eight weeks by a rather abrupt increase in medusoid budding. His data make it quite clear that the medusae are produced only under crowded conditions, as the cultured populations approach their carrying capacities. Acker (1976) also found that medusae are produced at a maximal rate when food supply is low. If the food supply is varied, the emphasis shifts from one sort of budding to another. Under a starvation regime a few buds of each type are produced; with more liberal feeding about the same number of hydranth buds but many more frustules and medusoid buds appear. When an excess of food is supplied there is an enormous increase in frustule budding, a lesser increase in hydranth budding, but hardly any increase in medusoid budding. Thus, high food levels inhibit sexual reproduction but favour a dispersive mode of asexuality. At very low food levels medusoid budding is suppressed, but all forms of reproduction are shut down almost completely.

The promotion of asexual dispersal through planuloid budding by high food levels is a little surprising, since it differs from the situation in *Hydra*. Adult hydra polyps can secrete a bubble of gas beneath the pedal disc, which lifts them to the surface of the water, where they hang from the surface film and can be dispersed by wind and currents. Lomnicki and Slobodkin (1966) have shown that this behaviour is characteristic of starved animals. Surprisingly, however, floating does not seem to be a simple response to population density:

a density of eight individuals in 10 cc of medium was found to be most effective in eliciting floating, with less floating observed both in less crowded and in more crowded cultures. The dispersal and planktonic occurrence of *Hydra* is discussed at length by Reisa (1973).

Slobodkin (1964, for example) has also described the growth of *Hydra* populations in laboratory culture, and the effect of different levels of inoculum, food supply, predation and interspecific competition (see also Stiven, 1962a, b; Schroeder, 1969).

There is relatively little pertinent information on the sexuality of marine hydrozoans. The anthomedusan *Margelopsis haeckeli* appears to have an entirely asexual life cycle, despite the presence of both polyp and medusa. The medusae are formed during spring and summer in the usual way, by budding, from a solitary pelagic polyp. They produce two sorts of eggs: subitaneous eggs, which develop immediately into young polyps; and larger resting eggs, which sink to the bottom and do not develop into polyps until the following spring. According to Werner (1955, 1963), both subitaneous and resting eggs are produced parthenogenetically (although Russell, 1953, documents the existence of occasional male medusae of *M. haeckeli*). Whilst asexual reproduction by budding or fission is so common in coelenterates, *M. haeckeli* provides one of the few examples of a true apomixis, in which the propagules are unreduced eggs, rather than mitotically derived masses of tissue.

North Sea populations of the hydrozoan *Coryne* bud medusae in midwinter, but the medusae do not themselves reproduce until May or June (Kunne, 1962; cited by Werner, 1963). Sexuality is thus confined to the early part of the growing season. In the laboratory, *Coryne* reproduces vegetatively at 14°C and buds medusae at 2°C (Werner, 1963). Werner also describes the case of the anthomedusan *Rathkea octopunctata*. The polyps of this species produce medusae at the end of the growing season, in fall, and these medusae reproduce throughout the winter. Moreover, this winter reproduction is asexual, with medusa buds arising from the manubrium of the parental medusa, and gametogenesis does not begin until the following spring. In the laboratory, gametes are formed at temperatures higher than about 6°C.

Although hydrozoan medusae typically arise by budding from polyps, they may also arise directly by budding from other medusae. There is an excellent review of the subject by Berrill (1949), who emphasizes that asexual budding is usually characteristic of young medusae, and ceases when the gonads form. In a few species, such as *Eleutheria dichotoma* and *Hybocodon prolifer*, budding and sexual reproduction may occur simultaneously, but in the other examples cited by Berrill, from all the major hydrozoan taxa, budding is characteristic of earlier and sexuality of later life. An extreme example is provided by the anthomedusan *Niobia dendrotentaculata*, which has twelve tentacles, each of which has arisen at a different time in the past. Each tentacle bulb differentiates into a new medusa, the oldest tentacles first and the youngest last, until all twelve have been shed. What remains of the original medusa then develops gonads and reproduces sexually. I have already mentioned a similar tendency in hydras, where budding may continue indefinitely but sexual reproduction is often followed by exhaustion and death. However, the age of a medusa will often be correlated with the time of year; if the medusae are liberated in spring, for example, they will be young early in the year and old in fall. Berrill mentions that in *Hybocodon*, where medusa budding and sexual reproduction may occur at the same time, the medusae liberated during the period when budding and sexuality overlap are themselves sexually precocious, and proceed to reproduce sexually without any intervening period of budding, whereas medusae which are liberated earlier pass a certain period of time before developing gonads. This suggests to me that it is the time of year rather than the age of the medusa which is the critical factor in determining the onset of sexuality.

The medusae of *Cladonema radiatum* and *Gastroblasta raffaeli* reproduce asexually by a complete transverse fission, from which two new medusae result.

Polyembryony appears to occur in some hydrozoans which brood their progeny. In *Pegantha smaragdina* the elongate tentaculate larvae bud within the gastric cavity of their parent, and are subsequently liberated as medusae (Bigelow, 1909). Berrill (1949) reviews other examples of embryonic budding in narcomedusan hydrozoans, amongst which the extraordinary life history of *Cunina proboscidea* is worth describing. Large female medusae are said to produce eggs which can develop with or without fertilization. The fertilized eggs give rise to a gamma larva, which metamorphoses into a dwarf male medusa. The unfertilized eggs (which I presume are diploid, though there appears to be no cytological confirmation of this) develop into alpha larvae, which by budding generate beta larvae, and both these types of larvae develop into dwarf female medusae. Large male medusae develop from the eggs of beta female medusae which have been fertilized by gamma males; large female medusae develop from the unfertilized eggs of alpha medusae. Thus, adult female medusae appear to be reproduced by an entirely asexual process, involving both budding and apomixis, whereas the large male medusae are the product of a cycle which involves two episodes of meiosis and syngamy – the fertilization of the dwarf beta female medusae by the dwarf gamma male medusae.

The most pertinent studies of marine coelenterates concern sea anemones. *Haliplanella luciae* is a colonizing species, remarkable both for the ephemeral nature of its populations, which after flourishing for a time often disappear quite suddenly (Parker, 1919; Stephenson, 1935), and for its tolerance to extremes of temperature (Shick, 1976). This precarious way of life is associated with an exclusively asexual mode of reproduction – usually by longitudinal fission, but in Japanese populations by pedal laceration (Minasian, 1976) – and I have already mentioned the discovery (Shick, 1976; Shick and Lamb 1977) that local populations comprise only one or a few clones. Shick and Lamb present a most interesting comparison between *H. luciae* and a related anemone, *Diadumene leucolena*, with which it is not only sympatric over most of its range but with which it occurs in very similar habitats in the intertidal. Whilst

H. luciae was exclusively asexual in the habitats that Shick and Lamb sampled, *D. leucolena* was exclusively sexual. This difference in reproductive biology is parallelled by an ecological difference: *D. leucolena* is most common in sheltered habitats, is not usually found in exposed positions and is intolerant of dessic-cation. Nor have the frequent mass extinctions characteristic of *H. luciae* been reported for *D. leucolena*; Shick and Lamb comment that, whilst virtually all individuals of *H. luciae* will die together when certain tolerance limits are passed, the mortality of *D. leucolena* increases gradually as the environment becomes more severe, which demonstrates the greater variance of fitness in the sexual form. Shick and Lamb also draw attention to the fact that the most stressful environments sampled were each occupied by only a single clone of *H. luciae*, whilst the most variable population was found at a subtidal locality.

Metridium senile is intermediate between *H. luciae* and *D. leucolena*, in that it may reproduce either sexually or asexually. Sexual reproduction leads to the development of dispersive planktonic larvae from externally fertilized eggs, whilst asexual reproduction by pedal laceration gives rise to clones of identical individuals. Asexual individuals are found higher on the shore than sexually reproducing individuals (Hand, 1955). Hoffmann (1976) mapped the micro-distribution of the clones in a harbour on the east coast of North America, using similarity at an isozyme locus as a criterion of clonal membership. The pronounced aggregation of clonemates suggests a rather high heritability of local fitness, although there was no evidence of intense larva-to-adult selection at the locus in question.

Francis (1979) found that small individuals of *Anthopleura elegantissima*, reproducing by longitudinal fission, lived higher in the intertidal than larger, sexual individuals. These small individuals formed clonal aggregations which, even at the same level on the shore, occupied more exposed microhabitats than the sexual individuals. The clonal form extends further to the north than the sexual form.

Cooke (1976) studied the anemones *Zoanthus pacificus* and *Polythoa vestitus* in Hawaii. *Polythoa* lives higher in the intertidal than *Zoanthus* and reproduces by asexual means about six times as fast. It is also better able to colonize sand flats, where there is often massive mortality caused by extreme changes in salinity during the winter rains.

Hartnoll (1977) has described a comparable situation in a very different group of anthozoans, the alcyonarian corals. *Alcyonium digitatum* and *A. hibernicum* occur together off the coast of the Isle of Man. *A. digitatum* forms large gonochoric (occasionally hermaphroditic) colonies which, on reaching maturity in their second year or later, release their gametes during midwinter; fertilization is external and the larval stage is a pelagic planula. Although a close relative, *A. hibernicum* reproduces in a completely different manner: ova are produced parthenogenetically in the first year of a colony's life, and are brooded in the gastric cavity of the parental polyp until early fall, when the planulae are released and take up a benthic existence before metamorphosis. This remarkable reproductive divergence is not associated with any clear-cut ecological dichotomy, although Hartnoll mentions that *A. digitatum* extends to a greater depth and is very much more abundant than *A. hibernicum*, which was represented by a few colonies only at each locality sampled.

The occurrence of asexually reproducing individuals at a higher level on the shore than sexual individuals has also been reported from *Phymactis clematis* in Chile (Stotz, 1979).

Rossi (1975) has described an interesting pattern of variation in *Cereus pendunculatus*. Two populations living near the surface in heavily polluted waters of variable salinity were found to be parthenogenetic — whether or not a meiosis occurs during oogenesis was not reported — whilst another population living in shallow unpolluted water was sexual and hermaphroditic, and a fourth population living in the subtidal was sexual and gonochoric. The parthenogenetic and hermaphroditic populations were viviparous, the gonochoric population oviparous. The observation that hermaphroditic anemones live near the surface, whilst gonochoric species or populations inhabit deeper water, is apparently quite general (Schmidt, 1967). In the parthenogenetic populations studied by Rossi, the eggs and larvae develop within the gastrovascular cavity, and when the actinulae are released they grow up close to their parents, forming dense populations. Sexually produced planula larvae, on the other hand, may become widely dispersed, and Rossi implies that population densities are lower as a result.

The brooding of larvae by adult anemones was observed in *Haliplanella luciae* by Verrill as long ago as 1898, but is still something of an enigma. Chia and Rostron (1970) suggest that eggs or planulae of *Actinia equina* are shed into the plankton by their parents and later enter other adult anemones, by which they are brooded — a sort of intraspecific parasitism. Some scyphozoan planulae are 'brooded' externally by foster parents of a different species (Berrill, 1949), and in this case the parasitic nature of the relationship is clear. However, Cain (1974) observed that the brooded planulae of *Actinia* almost invariably develop into individuals of the same colour as their host, and suggested that the relationship could be explained by self-fertilization; the brooding behaviour, in other words, was not parasitism but parental care. Ottaway and Kirby (1975) have proven the genetic identity of parent and offspring using isozyme phenotypes in a related species, *A. tenebrosa*. Although Chia and Rostron reported only a single simultaneous hermaphrodite amongst a collection of more than eighty gonochores, Cain suggested that the species is really a sequential hermaphrodite which possesses functional male and female tissue when it is changing sex, and which can achieve a good deal of self-fertilization at this time. Since populations consist predominantly of males and females the change of sex must be quite rapid, and most progeny will be produced by cross-fertilization; Cain's hypothesis refers only to the brooded larvae, but the almost invariable resemblance of brooded offspring to their parents argues a high degree of homozygosity, if the hypothesis of self-fertilization is true, and this seems inconsistent with the inference that most gametes are cross-fertilized. I would suggest that an ameiotic parthenogenesis, of the sort that perhaps occurs in *Cereus*, provides a more economical explanation of the data.

There are a few laboratory studies of the regulation of asexual reproduction in sea anemones, but they often do not clearly differentiate between sexuality and reproduction. For example, Schmidt (1970) found that fission in *Anthopleura* is inhibited by high salinity or low temperature, whilst Minasian (1976) described a positive correlation between the frequency of fission and the frequency of feeding; but in neither case is any switch from sexual to asexual reproduction reported. Schmidt involved the excretion of specific growth-stimulating substances to explain some of his results; similar claims have been made for *Hydra* by Burnett (1966).

There may be intense competition for space amongst anemones. Clones in aggregating species such as *Anthopleura elegantissima* are kept separate by aggressive behaviour towards other anemones at the boundary of the clone, and there may also be cooperative feeding within the clone; if a mixed collection is put into an aquarium, it quickly sorts itself out into separate clonal aggregations (Francis, 1973, 1976). In such cases the clonal aggregations, although comprising physically distinct individuals, have an unmistakeably colonial quality. On the other hand, clones of *Metridium senile* may grade into one another, and there is no evidence of aggression towards individuals in neighbouring clones nor of cooperation between members of the same clone (Hoffmann, 1976).

The reproduction and population ecology of stony corals, including competition for space and light, has been reviewed by Connell (1973). Some species have hermaphroditic polyps, but the extent of self-fertilization, if any, is unknown. In other species, only female polyps have been found; the possibility that some of these are parthenogenetic does not seem to have been investigated. Some reef corals may spread vegetatively, with detached heads rolling downslope to colonize habitats that might not be suitable for larval settlement.

Hydras and anemones are the best-known of coelenterates, so far as reproduction is concerned; it is a pity that both are atypical in being groups of sessile, benthic animals in which the medusa is wholly suppressed. Other coelenterates often have both a polyp and a medusa in their life histories, the polyp being sessile and reproducing asexually, whilst the medusae is free-swimming and reproduces sexually. This succession of sexual and asexual phases in the life history is conventionally referred to as an alternation of generations, a mischievous usage which invites a quite invalid comparison with the alternation of haploid gametophyte with diploid sporophyte in plants. It has been memorably attacked by Hyman (1940). The correct word to describe the life history is 'heterogonic' (originally 'heterogenic': Weismann, 1889). A great deal of quite fruitless controversy has turned on the issue of whether the polyp or the medusa is 'primitive' (i.e., geologically antecedant). Since, when both are present, the medusa is always specialized for sexual and the polyp for asexual reproduction, and since asexuality is likely to be a derived state in all multicellular animals (an issue I have touched on in Chapter 2, and shall return to in Chapter 4), the medusa might seem to be the obvious choice, a conclusion arrived at by Brooks as long ago as 1886 and widely accepted since (although not for the reasons originally given by Brooks). However, this argument begs the more fundamental question of whether specialization for sexuality is necessarily connected with a medusoid way of life, or for asexuality with a polypoid existence, and both statements are manifestly incorrect since medusae are known which reproduce asexually and polyps which reproduce sexually. Even deeper than this lies the problem of whether the benthic and planktonic ways of life are generally associated with different modes of reproduction: but I have reserved this more general topic for the next chapter, having made the point that the genetic structure of life histories must eventually be related to their ecological concomitants.

The budding of hydrozoans and the fission of anthozoans are two of the three major modes of asexual reproduction in coelenterates: the third is the transverse division of the scyphistoma larvae of scyphozoans, a process known as *strobilation*. Scyphozoans are gonochores (except one genus, *Chrysaora*, of protandrous hermaphrodites) and are therefore cross-fertilized; the eggs develop internally before being released as free-swimming planulae. In some genera (e.g., *Aurelia* and *Haliclystus*), large planulae are sometimes produced which later bud off other planulae (Berrill, 1949). The planula eventually attaches aborally to the substrate and transforms into a tentaculate polyp, the scyphistoma. Scyphistomae may bud, and the budded polyps may either separate, swimming away by ciliary action, or remain attached to form a small colony; Gilchrist (1937) gave a detailed description of the budding of *Aurelia* scyphistomae, and the topic of scyphistoma budding was reviewed by Berrill (1949). *Chrysaora* sometimes constricts off tentacles, which have the appearance of planulae, and which, like planulae, settle, attach and develop into a polyp (Hérouard, 1913). Scyphistomae may also form resistant bodies called podocysts, from which polyps develop (see Chapman, 1968). However, their most characteristic mode of reproduction is strobilation, a transverse fission of the polyp to form free-swimming immature medusae, the ephyrae. A single scyphistoma may produce only a single ephyra (monodisc strobilation), or, by a series of constrictions, up to thirty ephyrae may be formed (polydisc strobilation); during its lifetime a single scyphistoma may give rise to hundreds of ephyrae. Scyphistomae may also produce polyps rather than ephyrae by strobilation, but this has been observed only under abnormal conditions in the laboratory.

Temperature, illuminations, food supply and thyroxine levels have all been implicated in the elicitation of strobilation in the laboratory culture (e.g., Lambert, 1936; Custance, 1964, 1966, 1967; Spangenberg, 1971; reviews by Berrill, 1949; Russell, 1970), but the only extensive field study is that of Thiel (1962, cited by Russell, 1970). He worked on a population of *Aurelia aurita* in Kiel Bay, which produced planulae in August. These attached and transformed into scyphistomae, which budded off further generations of scyphistomae between September and December. In mid-winter, however, the majority of buds remained attached to the parental polyp by stolons, so that small colonies were formed. Like most scyphozoans (Berrill, 1949), *Aurelia* scyphistomae strobilated during the winter, and by early spring only reduced polyps remained; later in the spring these revived and produced podocysts and more daughter polyps. The ephyrae produced by strobilation developed into adult medusae, which completed the cycle by reproducing sexually in mid-summer. (Other scyphozoans may reproduce sexually in winter or in summer; see authors reviewed by Berrill, 1949; Russell, 1970; Campbell, 1974). Some strobilation occurred during the summer, but much less than in winter. Thus, *Aurelia* has not only a seasonal cycle of sexual and asexual reproduction, but also a cycle of different modes of asexual reproduction. By putting artificial substrates with attached scyphistomae into the harbour, Thiel was able to demonstrate that polyps which were fed regularly were much more likely to strobilate than unfed polyps. He also noted a correlation between peaks of strobilation and the appearance of zooplankton blooms. It seems, therefore, that the asexual production of ephyrae is related to the temporary superabundance of food.

The scyphozoans of the open ocean may produce large eggs which develop directly into ephyrae (*Pelagia*; Berrill, 1949) or incubate a modified scyphistoma which is only released after it has developed into a medusa (*Stygiomedusa*; Russell and Rees, 1960).

The resistant phase of scyphozoan life cycles – other than a reduced polyp – is the asexually produced podocyst, which is analogous with the hydrozoan frustule. However, in *Cyanea capillata* many hundreds of sexually produced planulae may encyst together within a chitinous envelope.

One of the most difficult problems in coelenterate biology concerns the nature of individuality. In the same species we have well-marked individuals, clones, colonies and populations, and since we are concerned with the effects of individual natural selection it is important to know where to draw the line between, say, a clone of distinct individuals and a colony of interdependent zooids. I have previously recommended the criterion of physical separation: individuals are physically disjunct and may therefore experience different conditions of life, whereas colonial zooids are indissolubly bound together and share one another's fate. Although this is almost always the clearest and most appropriate criterion to use, there are extreme cases in which it seems to break down. For example, there may be anastomoses between the roots of forest trees belonging to different species: are we to infer that these two trees comprise a single individual? Clearly not: but then what criterion can we substitute for that of physical continuity? The problem is especially acute in coelenterates; we can recognize all grades of distinction between zooids from complete and unequivocal separation through a more or less tenuous cytoplasmic connection to complete interdependence. When a more precise definition of individuality is required, I suggest that it will depend on the particular aspect of individuality in which we are interested. For example, if distant zooids respond to the electrical stimulation of part of a physically interconnected clone, then the clone is a 'nervous individual' – it is an individual if we are interested in the conduction of the nervous impulse, and construct our criteria appropriately. Our purpose in defining individuality is to compare modes of reproduction between equivalent entities, and we are seeking, therefore, the definition of a 'reproductive individual'. I suggest that an individual comprises those units of organization (cells, organs, zooids) which draw upon a common pool of stored resources for the purpose of reproduction. A hydra and its attached bud are thus a single individual, whereas a hydra and its detached bud are different individuals; budding is strictly a process of growth, whilst detachment is a process of reproduction. We should have reached the same conclusion if we had used the criterion of physical separation. In animals such as corals, scyphozoans or colonial hydrozoans, the colonies are individuals to the extent that altering the nutritive status of a given zooid alters the reproduction of neighbouring zooids. Our definition of individuality is in this way quantitative, since the food reserves of a given zooid may be only imperfectly shared with other zooids. *Hydra* polyps or scyphozoan colonies create little difficulty, being clearly individuals; but in cases such as corals, or clones of physically separated anemones which nevertheless feed and repel competitors cooperatively, we may be forced to admit grades of partial individuality. Pearse and Muscatine (1971) have demonstrated the transfer of energy-rich material between adjacent parts of a stony coral, but apart from this work there seems to be no information on the response of colony reproduction to changes in the nutrition of single zooids, or of small groups of zooids, so that although the reproductive criterion can be operationally defined it has not yet been empirically assessed. This is bound to cause some uncertainty in comparative studies.

(This approach to the nature of individuality repeats the continuing controversy over the nature of the 'unit of selection'; I have merely restated its terms. Since we are concerned with the operation of individual natural selection, I have taken it that the 'unit of individual selection' is self-evidently the individual, and then attempted to define the individual. I use this subterfuge because it holds out the hope of being able to identify the units of selection as discrete entities in nature, rather than as conceptual abstractions; but the coelenterates illustrate the difficulty of doing so, and demonstrate that in some organisms there may be no qualitatively distinct units of selection.)

To make matters worse, the nature of many hydrozoan and anthozoan colonies has not been certainly decided. It is usually held that coelenterate colonies are formed from the budding of a single primary polyp which has metamorphosed from a sexually produced planula, and there is a good deal of evidence to support this notion: for example, individual coral polyps are either hermaphroditic or male or female, and in all the cases reviewed by Connell (1973) only a single type of polyp was found in each colony. However, Duerden (1902) and Williams (1976) argue that colonies may often result from the gregarious settlement behaviour of unrelated planulae. Under controlled conditions in the laboratory, the larvae of a number of colonial hydrozoans tend to settle near other larvae, or near established adults (Williams, 1976). The colonies which result from the subsequent proliferation of polyps by budding will be genetically diverse. Duerden (1902) noticed similar behaviour by the planulae of the reef coral *Siderastrea*, whilst Nishihira (1967, 1968) has described the preferential setting of hydroid planulae on certain algae; and, of course, the gregarious settlement of planktonic larvae is well-known in other marine invertebrates such as barnacles, molluscs and serpulid worms (e.g., Knight Jones, 1951, 1953; Hidu, 1969; Wilson, 1968). It seems that we must not assume too lightly that coelenterate colonies are always single clones, even though the weight of evidence suggests that in most cases they are (Connell, 1973).

Williams (1976) also described the interesting case of the hydrozoan *Plumularia alleni*, which releases large numbers of planulae embedded in a mass of mucus. After this mass has adhered to a surface the planulae develop synchronously to produce a colony of zooids which are related to one another as sibs.

Although coelenterates as a whole are characterized by an alternation between sexual and asexual modes of reproduction, this bald statement glosses over the enormous diversity of the group. It is not merely that some species are wholly sexual and others wholly asexual, but rather that asexual reproduction is both of paramount importance in the life history and of unparallelled complexity. There may be several different methods of asexual reproduction in the same life history; they are usually, but by no means invariably, associated with a particular morphological type, the polyp; and in many cases it may not be possible, even in principle, to draw any unequivocal distinction between asexual reproduction and growth. The dispersal phase may be asexually produced, as medusae are; or it may be sexually produced, as planulae are; or both planulae and medusae may be present, and both may contribute to dispersal. The resistant stage, which disperses the organism in time, may also be produced either asexually, like the scyphozoan podocyst, or sexually, like the encysted eggs of freshwater hydras. In the diversity and mutability of their reproductive habits, the coelenterates have a claim to be called the most protean of metazoans.

3.5 Ctenophora

Review. Pianka (1974).

Incidence of Parthenogenesis. Ctenophores are an exclusively marine group with a structural resemblance to coelenterates. Most are pelagic, but creeping benthic forms are not uncommon amongst platyctenid ctenophores. All known ctenophores are simultaneous hermaphrodites, with the exception of some apparently protandrous platyctenids. The only regular means of asexual reproduction involves the regeneration of small fragments of tissue cast off by adult ctenophores, a process comparable to the pedal laceration of sea anemones (Dawydoff, 1938). Depending on the species, sexual and asexual reproduction may occur simultaneously (for example, in *Coeloplana gonoetena*; Krempf, 1921) or at different times of the year (e.g., in *C. mitsukurii* and *C. echinicola*; Tanaka, 1932). In the two species studied by Tanaka, asexual reproduction occurred during summer, before the disappearance of both species from the littoral in August; on their reappearance in autumn they reproduced sexually. It seems that pedal laceration has been observed only in creeping benthic species; pelagic ctenophores are exclusively sexual.

Feeding appears to stimulate asexual reproduction in *Vallicula* (Freeman, 1967) and sexual reproduction in *Pleurobrachia pileus* and *Beroe gracilis* (Greve, 1970), but whether or not the level of feeding plays any part in the switch from sexual to asexual reproduction in species where both occur is not known. Pianka (1974) cites an unpublished observation by J. Mivota that gametogenesis in *Pleurobrachia bachei* is promoted by starvation, but this is a pelagic form which does not reproduce asexually.

Since most ctenophores are simultaneous hermaphrodites, there is obviously a potential for autogamy. Moreover, it seems certain that pelagic ctenophores, at least, are self-fertile (unpublished observations cited by Pianka, 1974). This circumstantial evidence inclines Pianka to believe that ctenophores are often self-fertilized, but in my opinion their spawning behaviour renders this conclusion very doubtful. During spawning the sperm are released first and the eggs only later. The sperm are emitted in a series of spurts: before each spurt the comb rows stop beating, but immediately afterwards resume rapid beating to disperse the gametes. Oocyte release follows a similar pattern. These facts suggest to me that gamete release is designed to facilitate cross-fertilization. The sperm, which will be much more rapidly dispersed than the vastly larger oocytes, are released first, and the renewed beating of the comb rows must then disperse the gametes. A system designed to facilitate self-fertilization would work in the opposite way: the oocytes would be released first, and the comb rows would cease beating for a considerable period of time following the later release of sperm. Nor is there any good evidence for internal fertilization, which would make selfing much easier, since early observations of fertilization within the gastrovascular cavity are convincingly attributed by Pianka to the consequences of working with damaged individuals. Moreover, ctenophores usually spawn simultaneously in the laboratory, and the swarms of ctenophores which are sometimes reported may be spawning aggregations (see, especially, Agassiz, 1874); these are not behaviours which would be expected in organisms which are predominantly self-fertilized. Finally, although ctenophores have a complicated reproductive anatomy, the volume occupied by testes does not seem (from the published diagrams and preserved specimens I have seen) to be markedly, if at all, less than that occupied by the ovaries, and so the evasion of the cost of sex we would expect to see in a regularly selfed organism has not occurred. I conclude that whilst self-fertilization may occur casually, it is of little importance in natural populations of ctenophores.

The aberrant platyctenid *Gastrodes* is endoparasitic in the tunicate *Salpa*; it has a planula larva, quite unlike the cydippid larva of most ctenophores. Certain large endodermal cells have been claimed to be oocytes, but sperm have never been found, and the genus may be parthenogenetic (Korotnev, 1891). There seems to be too little evidence to make a definite decision possible.

3.6 Platyhelminthes

Reviews. Henley (1974; marine turbellarians); Cable (1971; parasitic forms); Ax and Schulz (1959; turbellarian fission).

Incidence of Parthenogenesis.

   TURBELLARIA. A large group of predominantly free-living flatworms, common in the marine littoral and in freshwater, with some terrestrial members. Almost all are simultaneous hermaphrodites with internal fertilization.
Acoela. A marine group which is almost exclusively sexual. There have been reports of a paratomical fission from a few genera, including *Convoluta* (Wager, 1913; Marcus and Macrae, 1954), *Amphiscolops* (Hanson, 1960) and *Paratomella* (Dörjes, 1966).
Rhabdocoela. Common in lotic and littoral habitats; occasional in lentic, marine pelagic and damp terrestrial situations. Usually sexual and cross-fertilized. The Catenulida are a freshwater group which reproduce by paratomy, forming chains of zooids in different stages of differentiation; sexual reproduction is only rarely observed in nature, and is very difficult to induce in the laboratory. Marine catenulids have only recently been described from the littoral of Europe and North America. No asexual reproduction has been reported, and indeed the lack of paratomy is listed as a leading characteristic of the marine forms by Sterrer and Rieger (1974). The Macrostomida comprises two families, Macrostomidae and Microstomidae, both of which are found in marine and freshwater habitats. The Macrostomidae are exclusively sexual; the Microstomidae reproduce sexually and by transverse fission, forming chains of zooids similar to those of catenulids. The remaining rhabdocoel taxa consist predominantly of cross-fertilized sexual forms, though self-fertilization appears to be common amongst the freshwater and terrestrial Typhloplanidae.
Alloeocoela. Most abundant in benthic communities of the marine littoral, with a few freshwater and terrestrial genera. Reisinger (1940) has described parthenogenesis in *Bothrioplana semperi*.
Tricladida. The classification of triclad flatworms coincides with their habitat. The Maricola inhabit the littoral of temperate and subpolar seas, and are exclusively sexual, and self-fertilization occurs casually if at all. The Paludicola inhabit freshwater. If we follow Ball (1974) and other authorities in suppressing the Kenkiidae (Hyman, 1937), they can be arranged in two families, Planariidae and Dendrocoelidae. Asexual reproduction by architomical transverse binary fission is common in the Planariidae, and especially in the genera *Dugesia*, *Phagocata*

and *Polycelis*. Some populations of these forms exhibit alternating periods of sexual and asexual reproduction, and so may be said to have a cyclical parthenogenesis, whilst others are either exclusively sexual or exclusively asexual. In *Phagocata* asexual reproduction sometimes involves fragmentation followed by encystment, rather than binary fission followed more or less immediately by regeneration. Pseudogamy in triclads has been described by Benazzi (1950), Benazzi-Lentati (1961, 1962) and Benazzi-Lentati and Bertini (1961). Obligate self-fertilization has been reported for *Cura*. Both asexual reproduction and habitual selfing are unknown in the Dendrocoelidae. A polyembryonic fission of the early embryo may occur within cocoons of *Dendrocoelum lacteum* and other planarians (Koscielski, 1973). The Terricola are terrestrial triclads of tropical and subtropical forests; a few live in humid environments in the temperate zone. Certain genera, especially *Bipalium*, commonly reproduce by fragmentation.

Polycladida. A large and numerous group of benthic marine flatworms, with a few pelagic genera and a single freshwater species. They are exclusively sexual and are almost always cross-fertilized. Fission in turbellarians has been reviewed by Ax and Schulz (1959).

TREMATODA. The flukes, a class of parasitic helminths, almost all of which are simultaneous hermaphrodites.

Monogenea. Ectoparasites of fish, or occasionally of crustaceans. Most are cross-fertilized hermaphrodites. *Udonella* lacks an intromittent organ and may be obligately self-fertilized. The yolkless eggs of *Gyrodactylus* undergo polyembryony, with a second embryo developing inside the first, and a third inside the second, and a fourth inside the third; the second embryo does not develop until the first embryo has developed into an adult (Katheriner, 1904).

Aspidobothria. Endoparasites of molluscs and vertebrates. All are hermaphroditic and are probably usually cross-fertilized, but self-copulation is anatomically possible and may be quite common.

Digenea. Endoparasites with very diverse life cycles which involve a succession of sexual and asexual phases; there are many exceptions to the following summary. Capsules containing uncleaved eggs or embryos pass out of the host into the surrounding medium. Cleavage results in the formation of a smaller propagatory and a larger somatic cell; the propagatory cell continues to cleave, at each division forming a somatic cell and a new propagatory cell. The ball of somatic cells formed in this way differentiates into a larva, the miracidium. This is a small, actively swimming creature which seeks the first intermediate host, usually a gastropod. On entering the host it undergoes extensive alteration to form the sporocyst, which moves around in the host tissues absorbing nutriment. The germ balls of the sporocyst, derived from the propagatory cells at the miracidium, develop into rediae (or into daughter sporocysts, in which case rediae are missing from the life cycle); mature sporocysts are filled with rediae. A succession of one to several redial generations follows, before the germ balls of the rediae develop into an actively swimming larva, the cercaria, which escapes from the host into the medium. Each miracidium may produce hundreds or even many thousands of cercariae, all of which are derived from a single fertilized egg. These mitotically produced cercariae encyst either on plants or stones or in a second intermediate host, and are therefore an asexually produced resistant phase. The encysted larva (metacercaria) is eaten by the final (definitive) host, a vertebrate, where it develops into the adult fluke and produces sexual eggs. Adult digenean trematodes, although hermaphroditic, are usually cross-fertilized, but the occasional production of viable eggs from single-worm infections shows that self-fertilization can occur. There have been several reports of parthenogenesis from the dioecious blood flukes (Taylor *et al.*, 1969), and one of a parthenogenetic triploid strain of *Bunodera* (Cannon, 1970; cited by Cable, 1971). Cable (1971) gives details of several other arguable examples.

CESTODA. A class of habitually self-fertilized hermaphroditic endoparasites. The asexual reproduction characteristic of trematode miracidia is usually lacking in cestodes, where in most cases a single fertilized egg gives rise to a single adult. Some authorities interpret the growth of the strobila as a process of asexual budding of proglottids. In *Echinococcus*, scoleces proliferate asexually within a cyst formed by the larva in the intermediate host; when such an hydatid cyst is ingested by the definitive host it releases many (up to several thousand) scoleces, each of which develops into an adult. A few cestodes (*Dioecocestus*) are gonochoric, but most are hermaphroditic. Cross-fertilization can occur only when two or more adults infest the same definitive host, and most cestodes are usually self-fertilized. Selfing usually occurs by self-copulation between different proglottids, but ocassionally occurs within a single proglottid. Rogers and Ulmer (1962) have found some inbreeding depression of viability in selfed lines of *Hymenolepis nana*. The only confirmed example of nongametic parthenogenesis concerns a triploid caryophyllaeid tapeworm which infests carp (Jones and Mackiewicz, 1969; see also Coil, 1970).

Discussion. Almost all turbellarians are simultaneous hermaphrodites with internal fertilization following cross-copulation or, more rarely, hypodermic impregnation. Von Gelei (1924) found that isolated *Dendrocoelum lacteum* produced only infertile cocoons, and the same result was obtained by Goetsch (1925) with a species of *Dugesia*. Goetsch raised the possibility of some physiological mechanism of self-incompatibility, but this was disproven by Dahm (1958), and the elegant studies of Ullyot and Beauchamp (1931) made the hypothesis unnecessary by describing features of anatomical design which ensure cross-fertilization. For example, in *D. lacteum* the penis and the opening of the oviduct both lie within the inner part of the genital atrium, and it might seem inevitable that sperm would leak from the penis into the oviduct. However, the top of the penis is provided with a valve (the 'flagellum') which normally seals the penis and prevents any emission of sperm, but which is erected during copulation with another individual to permit the free passage of sperm. Hyman (1951a) lent the considerable weight of her authority to deny the occurrence of selfing amongst turbellarians, on the grounds that sperm did not leave the male system except during copulation.

There seems no doubt that this is too extreme a position. Some rhabdocoels (Sekera, 1906) and maricole triclads (Wilhelmi, 1909) are known to be capable of 'self-copulation' where, by folding the anterior part of the body, sperm can be introduced into the seminal bursa. Selfing in this way may well be only casual or occasional, but the existence of obligate selfing appears to have been proven in the freshwater triclad *Cura foremanii* (Anderson, 1952a, b; Anderson and Johann, 1958). In this species the copulatory bursa is represented only by a rudiment, which communicates with a medial branch of the intestine. Kenk (1935) suggested that the genito-intestinal canal formed in this way serves to carry off surplus sperm for digestion, but Anderson points out that it also provides an anatomical basis for self-fertilization, and has succeeded in maintaining entirely selfed lines for nine generations. In mature *C. foremanii*, masses of sperm are always present in the seminal receptacles, and in individuals which have been isolated since their emergence from the cocoon this sperm cannot have a foreign origin. The sperm appears in the seminal receptacles about 35 days after hatching and thus about 20 days before the first cocoon is laid, so the eggs must come into contact with the sperm before being laid. At this stage the oocytes have completed the first meiotic division, and undergo the second during or shortly after the deposition of the cocoon, when serial sections have shown a body within the oocyte which can be plausibly interpreted as a sperm nucleus. In mass cultures of the worm, cross-copulation was never observed and may well be anatomically impossible. It seems, therefore, that *C. foremanii* is an example, rare amongst free-living animals, of an obligately selfed simultaneous hermaphrodite.

The rhabdocoel Typhloplanidae are also said to be self-fertilized (Bresslau, 1903, cited by Hyman). In genera such as *Mesostoma* subitaneous eggs are produced by selfing early in the growing season and are brooded for some time, eventually being released when the body wall of the parent ruptures. Later in the season, thick-shelled dominant winter eggs are produced by cross-fertilization. These observations have been treated with reserve by some authorities but, if confirmed, they describe a cyclical parthenogenesis in which autogamy alternates with amphimixis.

I have found no definite reports of apomictic turbellarians, although the stenostomid *Rhynchoscolex* is said neither to reproduce by fission (but see Pennak, 1978) nor to possess a male system (Sterrer and Rieger, 1974). By contrast, asexual reproduction by transverse fission is a common mode of reproduction amongst freshwater catenulids, freshwater and marine microstomids and freshwater and terrestrial planarians. The rhabdocoels reproduce by paratomy, with differentiation occurring before detachment. *Catenula*, *Microstomum* and *Stenostomum* are common freshwater genera in which individuals almost always consist of chains of more or less well-developed zooids – a sort of linear colony. As one might expect, there is interference between fission planes, with new planes forming farthest away from the most completely differentiated zooids. Sexual specimens of these genera are observed only occasionally, and apparently usually in autumn.

By contrast, fission in freshwater and terrestrial triclads is by architomy, with differentiation occurring after detachment. Asexual reproduction by fission occurs only in species with a high regenerative capacity, but the converse is not true: many triclads have great powers of regeneration but do not regularly reproduce by fission. Moreover, asexual forms seem to be rather isolated taxonomically. Among freshwater triclads, fission occurs regularly in the Planariidae, but not in the Dendrocoelidae; among the Planariidae, most often in certain genera such as *Dugesia* and *Polycelis*; and even among these genera some species or species-groups reproduce asexually whilst closely related species do not. For example, asexual strains are common in the *Dugesia gonocephala* group, but in the *D. lugubris-polychroa* group have been recorded only once, in a population of *D. lugubris* near the southern limit of the species range, in Spain (Benazzi, 1974); in the genus *Polycelis*, *P. felina* is often asexual, whereas *P. nigra* and *P. tenuis* seem to be exclusively sexual, even though they have considerable powers of regeneration. This division into sexual and asexual groups continues even at the level of single species, which in *Dugesia*, *Polycelis*, *Fonticola* and *Coenobia* often consist of wholly sexual and wholly asexual populations, as well as populations which reproduce sexually at some times of the year and asexually at others. This phenomenon was documented as long ago as 1902 by Curtis for *Planaria maculata* (= *Dugesia tigrina*). When taken into the laboratory, asexual strains usually continue to multiply exclusively by fission (e.g., Kenk, 1937 for *D. tigrina*), and attempts to induce sexuality by manipulating culture conditions have not been very successful. Some species can be induced to form gonads by exposure to low temperature. The asexual strains are often polyploid or aneuploid; the subject has been reviewed by Dahm (1958). Benazzi-Lentati (1964) found that one subspecies of *D. etrusca* was diploid (with $2n = 16$) and remained wholly sexual in laboratory culture, whereas cultures of two other diploid subspecies developed high and presumably aneuploid chromosome numbers ($2n = 30$–40) after a period of fissioning. The genetics of fissioning has been reviewed by Benazzi (1974), who concludes that the trait is controlled by nuclear genes.

It would be interesting to know if wholly sexual and wholly asexual populations of triclads occupy ecologically different types of habitat, but I have not found any conclusive evidence in the literature. Castle (1927) says that he never found sexual individuals of *Planaria velata* (= *Phagocata velata*) in collections from temporary pools and ditches, and Pennak (1978) states that sexuality occurs in *Phagocata vernalis* and in *P. velata* only in permanent bodies of water. Reynoldson (1961) pointed out that there is a striking correlation between the genetic systems of the common European triclads and the productivity of their habitats; species in which asexual reproduction is common are characteristic of unproductive lotic habitats, and exclusively sexual species of productive lentic habitats. Kawakatsu and coworkers have made extensive observations on the geographical and altitudinal distribution of freshwater triclads in Japan (Kawakatsu, 1965, 1974; Kawakatsu *et al.*, 1967a, b). Their data (Kawakatsu *et al.*, 1967a) show no correlation between altitude and the incidence of sexuality. However, as in other parts of the world, the epigean species were found to be exclusively sexual (Kawakatsu *et al.*, 1967b).

Calow has investigated the energetics of triclad architomy, and has compared architomy with cocoon production (Calow and Woollhead, 1977; Calow *et al.*, 1980). He concludes that architomy is the more efficient mode of reproduction when the supply of food is always limiting, but his work does not show why cocoons should contain amphimictic rather than thelytokous eggs.

In populations which exhibit both sexual and asexual reproduction, gametogenesis usually occurs in winter or spring and fission in spring and summer. It is not always easy to interpret the timing of sexuality in relation to the environmental changes that might be expected to occur from year to year. For example, Curtis (1902) found that *Dugesia tigrina* was sexual in May and June, then ceased reproduction until a period of fission in August and September, and then again ceased reproduction until May; it is hard to decide which of these periods of reproduction is before and which after the transition from the conditions of one growing season to those of the next. These are, however, hints that the mode of reproduction may vary with population density. Castle (1927) observed that sexuality in *Phagocata velata* occurred when the worms were most common, and fission when they were most difficult to find. In three Japanese planarians studied by Kawakatsu (1974) there was a heavy mortality of adults following sexual reproduction, and the succeeding period of asexual reproduction must therefore have occurred when population density was low. By contrast, in *Dugesia akestii* there was a low rate of mortality during and after the breeding season, and no asexual reproduction was observed. Reynoldson (1961) observes that where sexual races or individuals occur in unproductive habitats the cocoons are laid in winter; in productive habitats sexual reproduction occurs in spring or early summer. Asexual reproduction is on the whole more common among stream-dwelling triclads than among the inhabitants of the more favourable evironment provided by the littoral zone of lakes.

The reproduction of turbellarians resembles that of coelenterates, in that sexuality is normally an irreversible state; individuals which can vary their mode of reproduction reproduce asexually as juveniles, and die following one or more periods of sexual reproduction as adults. Old individuals occasionally undergo an abnormal type of fission after having reproduced sexually, but there is no doubt that, in general, fission is restricted to younger and sexuality to older individuals.

The dormant overwintering stage of most triclads is a cocoon containing zygotes and yolk cells, produced by cross-fertilization. The thick-shelled resting eggs of rhabdocoels are also usually cross-fertilized. In *Phagocata velata*, however, the resting stage is asexually produced (Child, 1913, 1914; Castle, 1927, 1928). In spring the worms fragment into a number of small pieces, each of which encysts. The young worms develop within the cysts, and either hatch within a few weeks if the environment remains suitable, or else remain encysted until favourable conditions return. In temporary ponds which dry up in summer, there is often an unbroken succession of asexual generations whereas, in permanent bodies of water, the worms which emerge from the cysts in autumn reproduce sexually in the winter. In general, adult turbellarians, even terricole triclads, are unable to resist dessication, although they can survive quite prolonged periods of starvation by 'degrowing' − simply reducing their body size whilst remaining active.

3.7 Gnathostomulida

Reviews. Sterrer (1971, 1974).

Incidence of Parthenogenesis. The Gnathostomulida are a recently described phylum of minute marine worms. They are members of the interstitial fauna, and are especially characteristic of anoxic sands. All are hermaphroditic; some may be protandrous. Ovary and testes appear (from published diagrams) to be of comparable bulk, which may point to habitual cross-fertilization. No regular means of asexual reproduction has been described. Elongate genera may undergo fragmentation, but the fact that nearly all those found have been anterior fragments suggests mutilation rather than reproduction.

3.8 Rhynchocoela (Nemertinea)

Reviews. Coe (1930); Gibson (1972); Riser (1974).

Incidence of Parthenogenesis. This is a phylum of moderately sized acoelomate worms, most abundant in the littoral zone of colder seas, but also found in pelagic, deep-water and tropical habitats; a few species occur in freshwater and on land. Most nemerteans are gonochoric, but some genera, especially those of freshwater and the land, are simultaneous hermaphrodites, often with a tendency to protandry. The possibility that these hermaphroditic forms are self-fertilized is mentioned in the literature, but there is rarely any sound evidence to support the claim. Pennak (1978) states that individuals of the hermaphroditic freshwater nemertean *Prostoma* release both eggs and sperm into a mucous sheath, which certainly implies that this genus is commonly selfed. Asexual reproduction by the division of the entire body into a large number of fragments occurs regularly in some species of the elongate littoral nemertean *Lineus*; reproduction in other species of the same genus is strictly sexual (Coe, 1930; Gontcharoff, 1950,

1951). In some species fragmentation is followed by encystment of the pieces; in others the fragments regenerate directly. Fragmentation usually occurs during summer and is followed by sexual reproduction in winter, although at least one species reproduces both sexually and asexually in the winter (Gontcharoff, 1951). In the laboratory, fragmentation occurs spontaneously at temperatures near 20°C and is repressed at 5-10°C (Coe, 1930).

3.9 Acanthocephala

Incidence of Parthenogenesis. This is a small phylum of endoparasitic worms, whose intermediate host is an arthropod and whose final host a vertebrate. All are gonochoric. There is no record of any means of asexual reproduction at any stage of the life cycle.

3.10 Rotifera (Rotatoria)

Reviews. Thane (1974; marine rotifers); Wesenberg-Lund (1930), Hutchinson (1967), Birky and Gilbert (1971), Ruttner-Kolisko (1972) and Gilbert (1977; planktonic freshwater rotifers); Dobers (1915; bdelloids); Gilbert (1974; dormancy); Wesenberg-Lund (1923; male rotifers).

Incidence of Parthenogenesis.

SEISONIDEA. An exclusively marine group, parasitic or commensal on crustaceans. All known species are gonochoric, with the males always present, although sex ratios have not been reported (Hyman, 1951b, says that males are less abundant than females). Seisonids are believed to be exclusively amphimictic.

MONOGONONTA. Planktonic freshwater rotifers, with a few benthic, soil-dwelling and marine representatives. It is probable that all species are apomictic. In many species, mictic females produce meiotically reduced eggs at some time during the year. If unfertilized, these eggs develop directly into males, which are therefore presumed to be vegetatively haploid. If fertilized by a male, the egg develops into a dormant encysted embryo, which hatches, usually in the following growing season, into a diploid apomictic (amictic) female. Genetic recombination was demonstrated electrophoretically by King and Snell (1977). It has been assumed (for example, by Hyman, 1951b) that all monogonont life histories are arrhenotokous, but neither males nor resting eggs have been observed in many common and well-known species (Wesenberg-Lund, 1923, 1930), and it is quite possible that many species are purely amictic.

BDELLOIDEA. Benthic freshwater rotifers, found in every type of freshwater habitat from the littoral zone of large lakes to rain pools and the water film on mosses and lichens. They reproduce exclusively by the production of subitaneous apomictic eggs: males, meiosis and resting eggs have never been observed. The adults can withstand desiccation and freezing.

Discussion. Parthenogenetic reproduction in rotifers is usually taken to be purely apomictic. Hsu (1956a, b) found that the oocytes of two bdelloid rotifers had 13 chromosomes and underwent two mitotic maturation divisions. Maynard Smith (1978) has expressed doubts that bdelloids can be so thoroughly asexual as they seem to be, but other than a highly qualified remark by Wesenberg-Lund (1930, p. 186) there is no empirical basis for his reservations. The cytology of parthenogenesis in monogonont rotifers has been reviewed by Birky and Gilbert (1971). The eggs of amictic females have a single polar body, and it seems certain that they undergo a single maturation division. It is not entirely clear whether this process is thoroughly amictic, or whether it is automictic, with the second meiotic division being suppressed. Neither is it clear whether the males are vegetatively haploid, or whether the male eggs undergo an endomitotic doubling, to become diploid but homozygous, the sperm being produced by a subsequent gametic meiosis. There is direct evidence for vegetative male haploidy only in two species of *Asplanchna* (Robotti, 1975; Jones and Gilbert, 1976), a genus in which the males exhibit a high degree of morphological reduction. If the male germ line is haploid then the sperm are produced by mitosis; but in many monogononts the males produce both normal motile sperm and immotile rod-shaped bodies; these may assist in the penetration of the female cuticle during copulation (Hyman, 1951b), but Hutchinson (1967) suggests that they are the functionless products of abortive meioses. The balance of the evidence seems to favour the hypotheses of apomixis in parthenogenetic females and vegetative haploidy in males, but a more definitive answer would be welcome.

Rotifer eggs are enormous, relative to the size of the female, and usually only a single egg is matured at any one time. In *Asplanchna* the eggs which give rise to mictic females are smaller than those which hatch into amictic females, apparently because they contain less yolk (Tauson, 1925). Eggs which develop into males are always much smaller than eggs which develop into females, and many more of them can be produced: *Brachionus* may produce an average of 25 unfertilized male eggs, but when fertilized produces only about three resting eggs (Ruttner-Kolisko, 1972), and for *Asplanchna* the corresponding figures are 15 male eggs and six resting eggs (Buchner *et al.*, 1967). The quantity of cytoplasm devoted to egg production in male- and female-producing mictic females and in amictic females seems to be about equal (Wesenberg-Lund, 1923). Mictic and amictic females both live for three or four days, but the males live scarcely half as long (King, 1970; Ruttner-Kolisko, 1972; Snell, 1977). It is difficult to arrive at any precise estimate of the cost of sex in these circumstances, but it is presumably less than 50 per cent.

Hatching from smaller eggs, the males are reduced in size and may show an extreme reduction in morphological complexity. They do not feed. In the most extreme cases, such as *Polyarthra*, the males are indeed little more than sacs of sperm provided with a locomotor organ and a supply of food. The reduction of the male is most pronounced in planktonic and least pronounced in benthic monogonont rotifers. The males are also extremely precocious; active sperm can be seen even before hatching (Wesenberg-Lund, 1923), and the males may copulate during the first hour of life. The biology of male rotifers was monographed by Wesenberg-Lund (1923), who described all the males known at that time. Half a century later, the males of many common species have still never been seen, and it is difficult to maintain the assumption that these species are arrhenotokous; perhaps many monogonont rotifers are perennially apomictic with, at most, only very occasional sexuality.

In many species, mictic females retain the capacity to be fertilized for only a few hours after birth (Buchner *et al.*, 1967), and if they remain unfertilized during this period thereafter produce only unreduced subitaneous eggs.

Both bdelloid and monogonont rotifers are able to survive unfavourable conditions by becoming dormant; in bdelloids the dormant phase is the adult, whilst in monogononts it is an embryo. The subject has been admirably reviewed by Gilbert (1974). Adult bdelloid rotifers respond to desiccation by shrinking into a lemon-shaped structure, the 'tun'. They do not usually encyst, but remain permeable to water vapour and air. Bdelloid tuns have an extemely low rate of metabolism, and may remain viable for many years. They show an almost incredible resistance to extremes of temperature, withstanding temperatures of more than 100°C or of within a degree of absolute zero for an hour or so. After rewetting they will often recover full activity within an hour; I have found large populations of fully active *Philodina* in rock pools only half an hour after rain. Bdelloids can also withstand very low temperatures even when wet. They are one of the few metazoan taxa common in Antarctic freshwater habitats, where they are regularly exposed to rigorous cycles of freezing and thawing (Dougherty and Harris, 1963). However, even the moss-dwelling bdelloids of temperate regions will survive at temperatures far below those which they normally experience. The ability of the apomictic eggs of bdelloids to withstand freezing and drying has not been investigated systematically.

Adult monogonont rotifers cannot survive the abuse that bdelloids tolerate, although some soil species can withstand freezing. The characteristic dormant stage of monogononts is an embryo, which, developing from a fertilized egg, is protected by a thick envelope. After release from the female, these 'resting eggs' sink into the sediments or float to the shore. They may remain viable in the sediments for many years, and can withstand modest amounts of freezing and drying. The hatching of resting eggs seems to be extremely sporadic: a certain minimum period of time passes before any eggs will hatch, but thereafter hatching occurs at irregular intervals for a considerable time. In the laboratory, hatching is often induced by transfer to fresher, warmer or more dilute medium.

In monogonont rotifers it is usual for the fertilized egg to be a dormant and resistant stage, secreting a thick coat and undergoing a diapause, whilst unfertilized eggs are subitaneous and hatch almost immediately. In *Asplanchna sieboldi*, however, Tannreuther (1920) found that fertilized eggs could be either dormant or subitaneous; and in *Keratella hiemalis* unfertilized eggs may secrete a shell and undergo a diapause. The difficulty introduced by these phenomena is illustrated by the fact that dormant parthenogenetic eggs are usually called 'pseudosexual', a misleading term which implies that dormancy is necessarily an attribute of fertilized eggs. It is impossible to discover how widespread may be the occurrence of dormant unfertilized eggs, or of subitaneous fertilized eggs, because eggs with a thick shell are usually assumed to be fertilized and those without to be apomictic. The occurrence of males is the only reliable indicator of mixis in rotifers.

It is usual amongst monogonont rotifers for a female to produce either mictic or amictic eggs, but not both. In recent years, however, an increasingly large number of apparent exceptions to this rule have been reported. Females which bear both mictic and amictic eggs are said to be 'amphoteric', and have been recorded from *Asplanchna*, *Sinantherina* and *Conochiloides* (Champ and Pourriot, 1977; Ruttner-Kolisko, 1977; Snell and King, 1977; earlier literature reviewed by Gilbert, 1974). Amphoteric females seem to occur only in a few populations, and to be in a minority, usually a very small minority, when they do occur. Most of the observations which have been made are not quite unambiguous; they refer to individual females bearing both subitaneous eggs (or embryos) and resting eggs, but without direct proof that the resting eggs are mictic. However, Bogoslovsky (1960, cited by Ruttner-Kolisko, 1977) saw a female *Conochiloides coenobasis* carrying both male and female embryos, which certainly implies that a single female can produce both reduced and unreduced eggs, and King and Snell (1977) have obtained electrophoretic evidence that amphoteric females simultaneously produce eggs by meiosis and by mitosis.

The induction of sexuality in monogonont rotifers has been a favourite pastime of laboratory biologists for nearly a century. If any simple generalization has emerged from their efforts, it is that the most consistently effective stimulus for inducing sexuality in laboratory culture is crowding. Naturally, this generalization is subject to many exceptions and qualifications.

I shall not discuss the early work of Maupas, Nusbaum, Whitney, Shull and others; Birky and Gilbert (1971) present a short review, but point out that many of these studies were inadequately controlled. These early results are tabulated in the article by Halbach and Halbach-Keup (1972). Most of the more careful modern work refers to three genera — *Brachionus*, *Asplanchna* and *Notommata*. In *Brachionus* crowding is by far the most effective stimulus in eliciting sexuality; in *Asplanchna* and *Notommata* crowding plays a secondary role, modulating the effects of other environmental variables.

An effect of population density on the frequency of mictic females in *Brachionus calyciflorus* was reported by Buchner (1941) and later confirmed by Gilbert (1963). In short-term laboratory cultures mictic females appear when population density exceeds a certain value, whether the rotifers are cultured singly in small volumes or in larger numbers in greater volumes. The proximate mechanism involved is not known; crowding with individuals of related species is effective, but crowding with *Paramecium* is not, so any soluble substance which mediates the effect of crowding has some degree of taxonomic specificity (Gilbert, 1963). In long-term laboratory cultures the frequency of mixis tends to decline even though high densities are maintained.

Asplanchna reproduces asexually when fed on *Paramecium*, and mictic females are produced only when green algae are added to the culture (Mitchell, 1913; Birky, 1964; Buchner and Kiechle, 1965). The substance responsible has been identified as the vitamin E compound, tocopherol (Gilbert and Thompson, 1968; see also Gilbert, 1967, 1968). Most of the work has been done with *A. sieboldi*, in which a dramatic morphological change accompanies the switch to mixis: amictic females hatching from resting eggs are small and 'saccate', whilst mictic females develop prominent body-wall outgrowths, the so-called 'cruciform' morphotype. The proximate causation of mixis appears to be connected with the degree of development of the cruciform phenotype; the role of tocopherol is indirect, in that it tends to induce the development of cruciform rather than saccate females. Other plant lipids have no demonstrable effect. Crowding intensifies the development of the cruciform phenotype at a given concentration of tocopherol, so that population densities in culture qualify the otherwise highly specific effect of tocopherol.

Mictic females are produced in cultures of the periphytic rotifer *Notommata copeus* only when the daily photoperiod exceeds about 15 hours (Pourriot, 1963; Pourriot and Clément, 1975). This effect is also qualified by crowding, but in a rather complicated fashion (Clément and Pourriot, 1973a, 1975). When one or more females were cultured in varying volumes of medium the frequency of mictic offspring was smaller in lesser than in greater volumes. However, at any given density a group of females produced more mictic offspring than did an isolated female. Moreover, when a large number of females were cultured together in medium which was changed at intervals, the production of mictic offspring was greatest when the medium was changed most often. Thus, the effect of photoperiod was modulated by crowding – in a direction opposite to that observed in cultures of *Brachionus* and *Asplanchna* – and this effect of crowding was in turn qualified by a 'group effect', which depended on the absolute number of females rather than on their density.

It is difficult to draw out an unequivocal generalization from these experiments, unless it be that sexuality is triggered by different treatments in different rotifers. Crowding has always been implicated, however, either as the major proximate cause (*Brachionus*) or as modifying the action of biotic (tocopherol, and hence green algal biomass; *Asplanchna*) or abiotic (photoperiod; *Notommata*) factors. Halbach and Halbach-Keup (1972) carried out a very thorough investigation of the relationship between sexuality and various environmental factors in three species of *Brachionus* cultured under seminatural conditions without finding any consistent pattern of correlation at all. In all cases there was a positive relationship between the frequency of mictic females in a given species and the population densities and growth rates both of the same species and of the other two species, but in no case was this correlation formally significant.

There is an extensive but scattered literature on the seasonal incidence of sexuality in natural populations of monogonont rotifers. It has not been adequately collated since Wesenberg-Lund's (1930) article, and it would be the labour of many months to prepare a complete review of the last half-century. Hutchinson (1967) has reviewed much of the more recent work, including the important paper by Carlin (1943) but not that by Nauwerck (1963). At some risk of oversimplification, I have tried very briefly to summarize the observations of Wesenberg-Lund, Carlin and other authors in Table 3.1. In some genera and species (e.g., *Lecane, Macrochaetus, Polyarthra minor*) sexual periods appear never to have been observed, and are at least very infrequent. In other genera (e.g., *Trichocerca*) the sexual periods are well-known, but only seldom occur in pelagic populations. The sexual period usually occurs at the time of the population maximum, and characteristically occurs towards the end of the maximum, just before a return to much lower population densities. Species which are scarce throughout the year have sexual periods less often than species which usually exhibit a pronounced maximum at some time during the growing season (e.g., *Platyas; Gastropus minor*). Two or more maxima during the same growing season are often associated with two or more sexual periods (e.g., *Asplanchna priodonta, Polyarthra platyptera*), though this is not always true; in dicyclic populations of *Brachionus* and *Synchaeta* sexuality is usually associated with the spring but not with the summer or fall maximum. The association of the sexual period with the maximum population implies that, if the timing of the maximum changes, the sexual period will change too. This is usually true, but in *Cephalodella* there seems to be little correlation between the maximum and the sexual period, and in *Euchlanis dilatata* the sexual period occurs only after mid-August, often well past the period of greatest density. If a residual population of adult rotifers persists after the maximum, it almost invariably consists exclusively of amictic females.

3.11 Gastrotricha

Review. Hummon (1974).

Incidence of Parthenogenesis. Gastrotrichs are minute pseudocoelomate metazoans, forming part of the benthic and interstitial fauna; a very few may be planktonic (see Hutchinson, 1967).

CHAETONOTOIDEA. Five families of chaetonotoid gastotrichs live in freshwater and appear to be exclusively parthenogenetic, other than a single species reported by Remane (1927) to be hermaphroditic; two live in the marine littoral and are simultaneously hermaphroditic. The typically freshwater family Chaetonotidae includes two marine genera, both of which are hermaphroditic (Hummon, 1966); rudimentary male structures have been reported from several freshwater gastrotrichs. The mode of reproduction in parthenogenetic gastrotrichs has not been definitely determined. Hummon (1974) credits Sacks (1964) with the discovery that the second meiotic division is suppressed during oogenesis in *Lepidoderma squammata*; in fact, Sacks (1955) describes a single polar body being extruded before the first cleavage division. However, it would certainly be premature to conclude that all freshwater gastrotrichs are automictic. Weiss and Levy (1979) have recently provided the first unequivocal description of sperm from *Lepidoderma squammata*. Sperm were found in only a small proportion of individuals, and the frequency of these hermaphrodites in culture varied in time; Weiss and Levy suggest that there may be an alternation between autogamy and automictic thelytoky. *L. squammata* has many small chromosomes and may be polyploid. Two sorts of parthenogenetic eggs are produced by *L. squammata*: subitaneous ('tachyblastic') eggs which hatch after a few days, and resting ('opsiblastic') eggs, which have a much thicker shell and undergo a diapause (Brunson, 1949). The resting eggs are resistant to desiccation over calcium chloride for one or two weeks, to air-drying for one or two years and to freezing for three months when wet (Brunson, 1949). Crowded cultures which were allowed to stagnate elicited the production of resting eggs in *L. squammata*, though resting eggs were produced by *Chaetonotus tachyneusticus* when an individual carrying an immature ovum was isolated in a drop of fresh medium (Brunson, 1949). Resting eggs are generally the last produced by a female. Packard (1936), Brunson (1949) and Sacks (1964) give further life-history data; these are suggestions in the last paper of genetic differences between clones.

MACRODASYOIDEA. A strictly marine class, all of whose members are sequential or (usually) simultaneous hermaphrodites. Neither asexual reproduction nor resting eggs have been reported. There is a separate male gonopore, and cross-fertilization is assumed to be the rule; the male structures are comparable in bulk and complexity to the female. The sperm are among the longest, relative to the size of the adult, of any animal: they may measure 120–180 μm in an individual only 160–320 μm in total length (Hummon, 1974). As with all small metazoa the egg is enormous relative to the size of the adult, and in chaetonotoids causes a pronounced distension of the body wall.

3.12 Kinorhyncha

Review. Higgins (1974).

Incidence of Parthenogenesis. Kinorhynchs are minute pseudocoelomate metazoans found in shallow marine sediments. All known species are gonochoric,

Table 3.1: Reproduction in Freshwater Monogonont Rotifers. The second column gives the number of species listed by Ruttner-Kolisko (1972). In the third column, the abbreviation M denotes males; RE denotes resting egg; Red. denotes reduction of males, expressed as the ratio of male length to female length; the final score is given in eighths, so that 4.0 indicates a male half as long as the female. Different values of Red. represent different species. The fourth column gives a brief description of the habitat and population dynamics.

Genus	Spp.	Sexuality	Ecology	Sexual periods
BRACHIONIDAE				
1. *Rhinoglena* (= *Rhinops*)	2	M, RE Red. 7.0	Cold stenothermal; pelagic in open water of permanent ponds. Max April–May, followed by abrupt disappearance	Males and mictic females common late in max; small numbers of amictic females may be found throughout summer
2. *Epiphanes* (= *Hydatina*, *Notops*)	4	M, RE in all spp. Red. 4.0, 2.8	Benthic or semipelagic; common in small temporary ponds polluted by organic matter. Often enormous max after thaw but disappears by May	Males and resting eggs increase in frequency through period of abundance. Occasional females found in summer and fall are amictic
3. *Brachionus* (= *Schizocerca*)	29	M, RE in most spp. Red. 3.0, 3.7, 3.6, 4.3	Littoral, occasionally pelagic, benthic or ectocommensal; usually in small ponds. Generally thermophile but often common under ice. Max in spring (*B. calyciflorus*, *B. angularis*, *B. urceolaris*) or summer (*B. diversicornis*); sometimes dicyclic	Mictic females usually confined to a short period associated with max, usually in spring. In dicyclic populations, later summer or fall max usually comprises amictic females. Residual populations in summer, fall or winter wholly amictic. WL reports one locality where *B. calyciflorus* was sporadic, but sometimes abundant, with no sexual period in three years
4. *Platyas* (= *Noteus*)	1	None reported	Free-swimming littoral rotifer; thermophile. Local and seldom abundant. Summer max	

No.	Genus				
5.	*Keratella* (= *Anuraea*)	10	M, RE; also amictic RE. Red. 3.6, 3.6	Widespread in many different freshwater habitats. Typically perennial with a large spring max; often common throughout winter. A well-defined max may not occur every year and sometimes occurs in fall	Males and mictic females usually appear during spring max, and populations often comprise amictic females only between June and April. In localities and years with no pronounced max there is no sexual period. Mixis may occur during a fall max in pelagic populations of large lakes but late max often comprises amictic females, and mixis is generally rare in large lakes. In some localities mixis seems not to occur despite high population densities, for example in the *K. cochlearis* population studied by C; in *K. quadrata* he saw males only during a very intense max, and then only briefly; in smaller population of following year they were absent
6.	*Kellicottia* (= *Notholca*)	2	M, RE	Pelagic in large lakes. Often perennial with an early spring max; later max occur	Mixis during spring or later max. Perennial lake populations comprise amictic females for most of year. Mixis not observed during summer max of *K. longispina* (C)
7.	*Notholca*	20	M, RE for a few spp.	Cold stenotherm, inhabiting open water of small ponds, littoral of lakes, and streams. Euryhaline. Max in early spring; often common under ice	Mixis during early spring max, after which population disappears. Winter populations comprise amictic females
8.	*Anuraeopsis*	12	M, RE. Red. 6.7	Warm stenotherm of ponds. Max in summer	Mictic females during summer max (highest water temperature) but never seen in some localities

Table 3.1 (cont.)

Genus	Spp.	Sexuality	Ecology	Sexual periods
9. *Euchlanis*	8	M for most spp., RE for one sp. Red. 5.0+, 4.2, 3.2	Littoral; pelagic during blue-green blooms. Often no well pronounced max	Males and mictic females common during an enormous max of *E. triquetra* reported by WL. A pelagic species (*E. oropha*) was mictic during a fall max
10. *Mytilina*		M for several spp.	Creeping benthic genus; also found in soil	
11. *Trichotria* (= *Dinocharis*)		M (rare)	Littoral, often in acid pools. WL records it as common and widespread, but without distinct maxima	In July?
12. *Macrochaetus*		None reported	Thermophile; 'sparsely distributed among the macrophytes of the littoral' (RK)	
13. *Colurella* (= *Colurus*)		M in two spp. only	Widespread benthic rotifer, esp. common in small eutrophic ponds; known from salt water. WL reports max in spring or early summer	
14. *Lepadella* (= *Metopidia*)			Benthic in littoral of lakes, also in bogs etc. RK says perennial but never abundant; WL has seen large early spring max	WL mentions discovery of male in February
15. *Squatinella* (= *Metopidia*)			Benthic or epiphytic in littoral; usually found in small numbers in summer	
LECANIDAE				
16. *Lecane* (= *Cathypna*, *Manostyla*)			Benthic in littoral of lakes	WL reports *Lecane* to be acyclic

NOTOMMATIDAE

17. *Cephalodella* (= *Diaschiza*, *Notommata*)	?200	M in several spp. Red: 2.8, 4.9, 4.0, 6.6, 4.7	Benthic, epiphytic or ectocommensal in ponds and littoral of lakes; WL describes populations on algal mats. Typically perennial with slight max in spring and sometimes also in late summer or fall	Males rare; sexual periods do not always and may not generally coincide with max. 'In many years and in many localities sexuality may be totally suppressed' (WL)
18. *Monommata*			Benthic; typically in bogs and eutrophic ponds	
19. *Scaridium*	1	M (rare)	Among macrophytes in littoral. Summer max	In July?

TRICHOCERCIDAE

20. *Trichocerca* (= *Rattulus*, *Diurella*, *Mastigocerca*)	c. 100	M in many spp.; RE in a few spp. Red. 3.0, 4.2, 2.1	Typically among macrophytes in littoral, occasionally planktonic. Perennial with a summer max, sometimes found in small numbers in winter; max is not very pronounced	Sexual period usually towards end of summer max; residual populations comprise amictic females. Males of planktonic species seldom found; C did not record them for any of three species

GASTROPODIDAE

21. *Ascomorpha* (= *Sacculus*, *Anapus*, *Chromogaster*)	5	M, RE in all spp. except one. Red. 4.4	Variable; littoral or planktonic in large or small bodies of water. *A. agilis*, has max during spring and sometimes also in summer or even late fall. *A. ecaudis* has no pronounced max and can be found April–October. *A. ovalis* is an uncommon species	Males of *A. agilis* rarely reported despite intensive sampling, though thick-shelled eggs seen during later part of max. Mictic females of *A. ecaudis* seen throughout growing season. Males of *A. ovalis* reported, but not by WL. Resting eggs but not males described for *A. saltans*

Table 3.1 (*cont.*)

Genus	Spp.	Sexuality	Ecology	Sexual periods
22. *Gastropus* (= *Postclausa*)	3	M, RE in all three spp. Red. 1.8, 3.4	Pelagic cold-water genus. May be common in winter plankton. Max in spring or early summer	Spring max of *G. hyptopus* and *G. minor* followed by sexual period; weak max of *G. stylifer* in June usually followed by appearance of males. Where *G. minor* was rare no sexual period was observed
ASPLANCHNIDAE				
23. *Asplanchnopus*	3	M, RE in all spp. Red. 4.0	Littoral or marshy habitats; WL found it in very small bog-hole. Scarce	Amictic females only found by WL except, rarely, resting eggs (October) and males (July)
24. *Asplanchna*	7	M, RE in all spp. Red. 3.0, 1.5, 2.75, 2.3, 2.4, 2.8	Very variable genus. Typically pelagic but often found in small ponds. Max in spring and sometimes also in the late summer	Sexual period towards end of spring max; in dicyclic populations (e.g., of *A. priodonta*) sexuality may also occur during smaller summer max. Outside max population comprises amictic females
SYNCHAETIDAE				
25. *Synchaeta*	c.30	M, RE in most spp.; amictic RE reported. Red. 4.3, 4.9, 4.4, 4.6, 4.4	Most spp. brackish or marine. Freshwater species pelagic, especially in large lakes. Spring max, often short and intense	Where a sexual period has been observed in freshwater species it occurs late in the spring max. Occasional later max not associated with sex in *S. tremulla*. WL did not observe sexuality during an enormous spring max of *S. oblonga*, in which, RK states, males are very rare; but elsewhere they have been seen late in the max. C did not record males of this species. Males of *S. pectinata* are not seen in large lakes. C records males of

25. (cont.)				*S. lakowitziana* during a small winter max as well as during spring max. *S. kitina* was rare in the samples taken by C, and no males were observed
26. *Ploesoma* (= *Bipalpus*)	4	M, RE. Red. 1.4	Thermophile, pelagic in large ponds and lakes. Large max in July	Usually in late summer at time of max; not observed by WL
27. *Polyarthra*	6	M, RE in all spp. except one. Males strongly reduced	Most species are planktonic, with varying thermal preferences. Often perennial, and may have several max during year, but is often scarce for long periods of time	In *P. platyptera* WL records up to three sexual periods in a single season, which may occur at almost any time; however, they are usually associated with max. *P. minor* inhabits small, often temporary water bodies and even wet moss, and no sexual period is known. Males noted during May–June max of *P. dolichoptera* by C; in *P. vulgaris* males appeared in a fall max but not in a large spring max; in *P. remata* only a fall max appeared, with an associated sexual period. *P. major* was dicyclic in one year but had no sexual period in the following year
TESTUDINELLIDAE				
28. *Testudinella* (= *Pterodina*)	c.40	M. Red. 7.5	Littoral or benthic, usually in small ponds; occasional in plankton. Often no pronounced max; large max of *P. patina* in November seen by WL	Males rare; WL did not observe a sexual period
29. *Pompholyx*	1 or 2	M, RE. Red. 3.5	Thermophile genus of open water of lakes and ponds. Summer max, sometimes intense	Sexual period of *P. sulcata* associated with summer max, with only amictic females present from August onward, according to WL; RK says sexual periods occur 'mostly in spring and autumn'

Table 3.1 (*cont.*)

Genus	Spp.	Sexuality	Ecology	Sexual periods
30. *Trochosphaera*	2	M	Warm stenothermous in shallow eutrophic pools	
31. *Horaëlla*	2	RE	Warm stenothermous in plankton of small pools and slow streams	
32. *Filinia* (= *Triarthra*, *Tetramastix*, *Fadeewella*)	9	M, RE. Red. 3.8, 2.8, 4.0	Widespread, in most habitats except temporary or highly polluted ponds. Time of max differs between species: may be in spring, summer or winter	Sexual period in later part of max in three spp. studied by WL. In years when no max develops there may be no sexual period (*F. longiseta*). C. observed sexual period during slight June max of *F. terminalis*
HEXARTHRIDAE				
33. *Hexarthra* (= *Pedalia* = *Pedalion*)	7	M, RE in most spp. Red. 1.1	Usually thermophile. Pelagic with summer max, which in *H. mira* is not intense	Sexual period of *H. mira* in September, following max; later in year only amictic females present
CONOCHILIDAE				
34. *Conochilus* (= *Conochiloides*)	4	M, RE. Red. 0.6, 1.8	Pelagic rotifers, some species colonial. Usually early summer max, which may be intense	Sexual period late in max. *C. hippocrepis* common in most ponds sampled by WL, with max between mid-April and mid-May and sexual period in June. Amictic females present throughout year. *C. natans* usually rare, one enormous max being followed by a sexual period. In one locality amictic females found below ice; mictic

34. (*cont.*)

females appeared shortly after thaw, followed by disappearance of the population. No mixis observed during two years in *C. unicornis* despite high population density throughout much of the year

COLLOTHECIDAE

35. *Collotheca* (= *Floscularia*)	c.50	M, RE in many spp. Red. 0.5	Mostly sessile, some pelagic. Typically summer max	Resting eggs produced in fall, following summer max

Source: Mainly Wesenberg-Lund (1923, 1930; abbreviated WL), Carlin (1943; abbreviated C) and Ruttner-Kolisko (1972; abbreviated RK), although other authors have also been used.

with approximately equal numbers of females and morphologically unreduced males. The youngest females always contain sperm in the seminal receptacle. Asexual reproduction has never been reported, but the natural history of the phylum is so little known that its occurrence cannot be ruled out.

3.13 Priapuloidea (Priapulida)

Review. Van der Land (1974).

Incidence of Parthenogenesis. A phylum of benthic marine worms, most common in the littoral of colder seas. All known species are gonochores with external fertilization. Their powers of regeneration are limited, and vegetative fission or fragmentation is unknown. Collections of the recently discovered aberrant species *Maccabeus tentaculatus* consist entirely of females, suggesting that parthenogenesis may occur.

3.14 Nematoda (Nemata)

Reviews. Triantaphyllou and Hirschmann (1964; soil and plant-parasitic genera); Cable (1971; animal-parasitic genera); Hope (1974; marine free-living genera).

Incidence of Parthenogenesis. Nematodes are vermiform, bilaterally symmetrical, unsegmented pseudocoelomates. They are amongst the most cosmopolitan of metazoans; there are no planktonic nematodes, but otherwise they ·occupy almost every conceivable habitat, from sediments and soil to the interior of plants and animals. Nematodes reproduce exclusively by eggs: budding, fission, fragmentation and polyembryony are unknown. Most nematodes are gonochoric and amphimictic, but most other egg-based modes of reproduction are known. These are scattered through the phylum in a manner which prohibits any convenient prose summary, so I have instead summarized them in Table 3.2. Most hermaphroditic nematodes are habitually self-fertilizing; when cross-fertilization occurs it seems to be due to the presence of occasional pure males. In syngonic hermaphrodites the first gonial cells develop into spermatocytes, each of which gives rise to four spermatozoa; later gonial cells give rise to oocytes. In digonic hermaphrodites sperm are produced in a spermatogonium and oocytes in an ovary. The sperm are nonflagellate. Pseudogamy was first described in the nematode *Rhabditis aberrans* by Krüger (1913), but the proof was incomplete since some eggs developed without sperm penetration; it was first established conclusively for *Rh. pellio* by Hertwig (1922), and was subsequently discovered in other species of *Rhabditis* and *Mesorhabditis* (Belar, 1923; Nigon, 1949). Nongametic modes of reproduction certainly occur in nematodes, but their cytology has rarely been studied. Automixis has been proven to occur in species of *Rhabditis* and *Meloidogyne*, and apomixis in plant-parasitic nematodes in the family Heteroderidae. Parthenogenetic eggs lack the vitelline membrane characteristic of fertilized eggs, and may thus be subitaneous. The resistant phase of free-living nematodes is not, however, an egg but the anabiotic adult (review in Crowe and Madin, 1974). Some nematodes have a resistant larval phase (dauerlarva), which is induced by starvation.

Discussion. Hermaphroditic nematodes are protandrous, with a small quantity of sperm being produced first and then stored, later being used to fertilize the eggs. I have found no explicit reference to copulation between hermaphrodites, or between hermaphrodites and females. It seems, then, that all the sperm produced by an individual is used to fertilize its own eggs; but since the pioneer work of Maupas (1900) it has often been observed that not enough sperm is produced to ensure the fertilization of all the eggs (see Honda, 1925). In *Rhabditis vigueri* only about one-third of the egg complement can be self-fertilized, because of an insufficiency of sperm, and the remaining eggs can develop only if they are fertilized following copulation with a male. Males comprise only about five per cent of the population; there is a somewhat higher proportion of pure females, which can reproduce only through cross-fertilization. Even in species such as *Rhabditis duthiersi* and *Caenorhabditis elegans*, where males are rare and females unknown, the same extreme economy in sperm production is practised. Moreover, the rare males, although producing physiologically active sperm, often seem to lack the sexual instinct (Maupas, 1900; Potts, 1910). In other species, such as *Rh. gurneyi* and *C. dolichura*, sperm production may recommence once the initial stock has been exhausted, but ova may still go unfertilized if there is a temporary lack of sperm. This niggardliness is difficult to understand, and doubtless explains the lower fecundity of hermaphroditic species, relative to the nearest gonochoric forms. The hermaphroditic species, however, are often the more abundant in nature. Several authors (see Hirschmann, 1960) have noticed that hermaphroditic species often have extremely similar gonochoric relatives; for example, occasional hermaphrodites of the normally gonochoric *Diplogaster l'heritieri* are morphologically indistinguishable from normal individuals of the hermaphroditic species *D. maupasi*.

Hermaphroditic individuals and taxa seem, therefore, to represent a balance between autogamy and amphimixis. A similar balance seems to be struck between automixis proper and amphimixis in the gonochoric root-knot nematode *Meloidogyne hapla* (Triantaphyllou, 1966). In most of the populations which were studied, males were fairly abundant. The oocytes invariably underwent a first meiotic division, which if the female was inseminated was followed by a second meiotic division after sperm entry, and the fusion of male and female pronuclei – a perfectly normal amphimixis. In uninseminated females the second division was suppressed and diploidy thus restored automictically. Even when males were abundant, however, the supply of sperm in the spermatheca received during copulation was insufficient to fertilize all the eggs, so that individual females

Table 3.2: Parthenogenesis and Autogamy in Nematodes

Family	Ecology	Genus	Reproduction
Rhabditidae	Monogenetic parasites of invertebrates and amphibians; some saprophagous forms	Caenorhabditis	Mostly autogamous hermaphrodites with low but variable proportions of males. C. dolichura has an amphimictic race
		Diploscapter	D. coronata is parthenogenetic
		Mesorhabditis	M. belar is a pseudogamous hermaphrodite, apparently with some pure males
		Rhabditis	Mostly autogamous hermaphrodites; some spp. have amphimictic races, whilst others are wholly amphimictic
Rhabdiasidae	Heterogenetic parasites of reptiles and amphibians, with free-living saprophagous generation	Rhabdias	Parasitic phase is a protandrous hermaphrodite which stores sperm and may be autogamous
		Entomelas	Parasitic phase is a parthenogenetic female
Diplogasteridae	Free-living in soil; also associated with plants and insects	Diplogaster	D. robustus is an autogamous hermaphrodite; other spp. are amphimictic
Strongyloididae	Heterogenetic parasites of vertebrates with free-living saprophagous phase	Strongyloides	Parasitic adults are probably parthenogenetic females. In S. ratti an irregular male meiosis appears to be followed by pseudogamy, but this is not certain. In S. papillosus the eggs of parasitic females may extrude no, one or two polar bodies, and it is not clear whether there is pseudogamy, hybridogenesis or amphimixis; eggs of free-living females said to be apomictic
Cephalobidae	Saprophagous; often associated with plants or insects	Cephalobus	Two spp. lack both males and sperms and so may be parthenogenetic
Tylenchidae	Parasites of plants, especially roots	Helicotylenchus	Eight of 13 spp. lack males and are probably parthenogenetic (autogamous?)
		Hoplolaimus	Two spp. parthenogenetic, others amphimictic
		Pratylenchus	Eleven of 23 spp. probably parthenogenetic since males are rare or absent and females lack spermatheca

Family	Description	Genus	
Heteroderidae	Parasites of plant roots	*Rotylenchus*	Males unknown in *R. sheri*
		Scutellonema	Five of ten spp. have only females, without a spermatheca, and are probably parthenogenetic
		Tylenchorhynchus	Ten of 45 spp. lack males, and female gonad has neither spermatheca nor sperm; these are probably parthenogenetic
		Tylenchulus	Isolated, uninseminated females produce offspring, but males frequent
		Heterodera	*H. trifolii, H. lespedezae* and *H. galeopsidis* are apomicts in which males are rare or unknown; all are triploid or tetraploid. At least six other spp. are diploid and amphimictic
		Meloidodera	At least one species is a triploid apomict
		Meloidogyne	Males reported from all 14 spp. but may not be functional in some. Some populations of *M. javanica* and *M. arenaria* are high-ploid apomicts. *M. icognita* is a pentaploid apomict. Of 19 populations of *M. hapla*, 16 were facultative automicts and three were apomicts; all were high polyploids
Criconematidae	Ectoparasites of plant roots	*Criconema* *Criconemoides*	All 21 spp. probably parthenogenetic
		Hemicriconemoides *Hemicycliophora*	In all, 24 of 32 spp. believed to be parthenogenetic; remaining eight are hermaphroditic, with a spermatheca filled with sperm, and may be autogamous
			Males are rare in eight spp., which may be autogamous
		Paratylenchus	Altogether, 17 spp. probably parthenogenetic, since males are rare and sperm is not present in female gonad; a few spp. probably amphimictic
Allantonematidae	Homogenetic endoparasites of insects; occasionally heterogenetic with free-living phase in soil	*Heterotylenchus*	In 13 of 31 spp. males are rare or absent, and most individuals may be autogamous hermaphrodites. Four other spp. may be parthenogenetic
			Parthenogenetic phase occurs within host; sexual phase in both parasitic and free-living worms

Table 3.2 (*cont.*)

Family	Ecology	Genus	Reproduction
	Insect parasites	*Fergusobia*	Parthenogenetic phase occurs outside host; sexual phase in both parasitic and free-living worms
		Scatonema	Larvae may exceptionally be produced whilst their parent is still *in utero*, presumably by parthenogenesis
Aphelenchidae		*Aphelenchoides*	About half of 24 spp. are known to be amphimictic, rest may be parthenogenetic
		Seinura	At least some spp. are autogamous; others are known to be amphimictic
Monhysteridae	Free-living in marine sediments	*Monhystera*	Males uncommon (about 5%) in nature; isolated females can reproduce in culture
Chromadoridae	Free-living in marine sediments	*Chromadorina*	Males abundant in nature but disappear in culture; females continue to reproduce
Enoplidae	Free-living in marine sediments	*Phanoderma*	Males uncommon; no proof of parthenogenesis
Oncholaimidae	Free-living in marine sediments	*Viscosia*	Parthenogenesis inferred in culture; thought to be amphimictic in nature
Mononchidae	Free-living in soil and freshwater	*Mononchus*, etc.	Males rare or unknown in many species of this and other genera of soil and freshwater nematodes
Dorylaimidae	Free-living in soil and water	*Longidorus*	Eleven spp. lack males and lack sperm in female gonad, and are probably parthenogenetic; females of *L. tarjani* have spermatheca with sperm and may be autogamous
		Xiphinema	Most spp. may be parthenogenetic; at least one seems to be autogamous
Diphtherophoridae	Free-living in soil and water	*Trichodorus*	Parthenogenesis suggested in a minority of spp. by lack of males and by absence of spermatheca in females
Mermithidae	Heterogenetic parasites of arthropods; adult free-living in soil or water	*Allomermis*	Males of *A. myrmecophilia* very rare; females can produce viable eggs in absence of males
		Mermis	Females of *M. subnigrescens* can produce viable eggs in absence of males

Source: Cable (1971), Christie (1974), Hope (1974), Hyman (1951b), Maupas (1900), Mulvey (1960), Nicholas (1976), Nigon (1949), Triantaphyllou and Hirschmann (1964). Nomenclature and arrangement of families follows Chitwood and Chitwood (1974).

generally produced some offspring by amphimixis and others by automixis. In three other populations males were very scarce, and the eggs of all females, whether inseminated or not, developed apomictically.

Among free-living nematodes parthenogenesis appears to be much more common in freshwater and terrestrial than in marine forms: both sexes are known in 70 per cent of marine but in only 46 per cent of freshwater and 34 per cent of soil species; only 20 per cent of marine species are known from the female alone, but males are unrecorded in 53 per cent of freshwater and in 63 per cent of soil species (Micoletzky, 1922). Where males are known, they are usually as abundant as females in marine nematodes, but are often very scarce in freshwater and terrestrial species. Parthenogenesis is fairly common in plant-parasitic nematodes, especially amongst the Criconematidae, but is unusual amongst the parasites of animals, with the exception of *Strongyloides* and some mermithids. The great bulk of the obligate nematode parasites of animals which have a vertebrate as the final host are exclusively sexual and gonochoric.

The sex ratio of some parasitic nematodes varies with the state of the environment; the subject has been reviewed by Triantaphyllou (1973). In general, the proportion of males increases as the environment deteriorates; high infection density, low host nutrition, high host resistance and a number of other factors are effective in procuring an increase in the number of males. This need have no implications for sexuality: for example, high temperatures are associated with a high ratio of males to females in *Naccobus serendipiticus* (Prasad and Webster, 1967), but *Naccobus* is a wholly amphimictic genus and the effect of elevated temperature may have been merely to increase the death rate more in female than in male larvae. However, I am impressed by the fact that most reports concern genera which are known to be parthenogenetic, and I suspect that in most cases a shift in the sex ratio implies a shift in the mode of reproduction. Christie (1929) proved that the frequency of males in the parthenogenetic mermithid *Mermis subrigrescens* varied with the number of individuals harboured by the host, a grasshopper: light infestations of fewer than five eggs resulted in preponderantly female populations, whilst heavy infestations stimulated the development of many or most of the larvae as males. In natural populations the usual level of infestation is between one and five nematodes per host, though much heavier infestations are not uncommon. Most of the more recent work has been concerned with plant-parasitic forms. Tyler (1933) recovered only 0.7 per cent males from single-larva infestations of a parthenogenetic species of *Meloidogyne*, but 16.4 per cent from multiple-larva infestations and 56.5 per cent from plants reinfested by second-generation larvae. Later work on *Meloidogyne* reviewed by Triantaphyllou (1973) has in general supported the hypothesis that development into males is stimulated by unfavourable conditions, although there seem to be considerable differences in response between populations. The same preponderance of females at low infection densities has been noted in *Heterodera rostochiensis*, *H. schachtii* and *H. avenae* — see, for example, van den Ouden (1960) and Trudgill (1967), for *H. rostochiensis* — but here the situation is quite different: all three are obligately amphimictic species, and even at very low infection densities males and females are about equally frequent. In the triploid apomict *Meloidodera floridensis*, males appear in large numbers only when the larvae are starved for long periods of time. Male production in some but not all populations of *Aphelenchus avenae*, a parthenogenetic parasite of insects, is stimulated by high temperature and by high CO_2 concentration (Evans and Fisher, 1970; Hansen *et al.*, 1971). In short, it seems that when conditions become unfavourable, and in particular when, for whatever reason, food becomes difficult to obtain and growth ceases, development switches from female to male in obligately amphimictic species and from parthenogenesis to amphimixis in facultatively parthenogenetic species.

Plant-parasitic nematodes in the family Heteroderidae provide a good example of the tendency for parthenogenetic taxa to be polyploid (see Table 3.2).

3.15   Gordiacea (Nematomorpha)

Incidence of Parthenogenesis. Gordiacea are extremely elongate worms, perhaps allied to nematodes. Juveniles are parasitic in arthropods whilst the adults are free-living in fresh water; there is a single genus of pelagic marine gordiaceans. All known forms are gonochoric. No means of asexual reproduction, regular or casual, has been reported.

3.16   Endoprocta (Entoprocta, Calyssozoa, Kamptozoa)

Review. Mariscal (1975).

Incidence of Parthenogenesis. The endoprocts are minute sessile pseudocoelo-mates; all are suspension feeders, with a crown of ciliated tentacles, and almost all are marine. Asexual reproduction by budding is common in all three families. *Loxosomatidae.* Solitary zooids, ectocommensal on marine invertebrates, especially polychaetes. Buds form from the calyx, usually singly but sometimes six to twelve or more simultaneously; buds may be formed every five or six days in nature (Nielson, 1966b). Immature buds may themselves bud before separation from the parental zooid. Detached buds tend to settle close to the parental zooid, and in this way clonal aggregations may be built up. Most loxosomatids are gonochoric, and clusters of individuals are usually of the same sex; it is not certainly established, however, that budded offspring always have the same gender as their parent (see Atkins, 1932). The budded offspring of *Loxosomella kefersteinii* are capable of swimming freely away from their parent (Ryland and Austin, 1960). In some loxosomatids between one and four small adults develop as internal buds of a sexually produced larva, and are released by rupture of the larva (Jägersten, 1964; Nielson, 1966a). In one species the internal bud develops within the larva whilst the larva itself is developing within the ovary of its parent; the larva is subsequently released and lives for a few days in the plankton before rupturing. Indeed, up to four generations may coexist: the parental zooid, the sexual larva, the asexual internal bud of the larva and even secondary buds on the primary internal bud. Males are rare in some species of *Loxocalyx*, *Loxosomella* and *Loxosoma*, but there is no direct proof of parthenogenesis. *Pedicellinidae*. Marine colonial forms which bud from a stolon, or occasionally from the stalk, but never from the calyx. All members of a colony are attached by their bases to a common ramified stolon, and have arisen by budding from a single primary zooid. In most species of *Barentsia* and *Pedicellina* both colonies and individual zooids are gonochoric, but in *Myosoma* the zooids are usually gonochoric and the colonies hermaphroditic. Some species appear to be protandrous hermaphrodites (see Nielson, 1966a). At least casual self-fertilization seems to be likely in species with hermaphroditic colonies, but this is not definitely established.

Urnatellidae. A small family comprising two species of freshwater endoproct in the genus *Urnatella*. There is no well-developed stolon, but rather a non-ramifying basal plate from which several zooids arise by stalk budding to form a small colony (see Davenport, 1893). In all endoprocts unfavourable conditions cause the adult calyces to degenerate; the stalks represent a resistant phase and regenerate the calyces when favourable conditions return. The rate of budding varies with nutrition, but the balance between budding and sexual reproduction has not been studied. The reproductive periods are very imperfectly known; there is a summary table in Mariscal (1975).

3.17 Bryozoa (Ectoprocta, Polyzoa)

Reviews. Ryland (1970); Bushnell and Rao (1974; dormancy).

Incidence of Parthenogenesis. Bryozoans are minute sessile colonial coelomates. In all cases the colony increases by budding from a single ancestral zooid, which may itself have been produced sexually or asexually.

GYMNOLAEMATA. An almost exclusively marine group, most abundant in the littoral. The zooids are usually hermaphroditic, and bear ova and sperm simultaneously. The method of fertilization is still somewhat controversial, and is discussed below. The embryo is usually brooded before being released, and after a brief period of swimming settles and metamorphoses into an ancestrula. After a few days the ancestrula buds off a single zooid (occasionally two), from which the colony develops by repeated budding. Some species are perennial; in others the colonies degenerate with the approach of winter, and the overwintering phase is the ancestrula (produced by the metamorphosis of late larvae), a very small colony derived from the ancestrula or a dormant stolon. In the unusual freshwater gymnolaemates *Victorella* and *Paludicella* the overwintering phase is the hibernaculum, a dormant asexual winter bud; hibernacula may also be produced in response to unfavourable conditions at other times of year.

STENOLAEMATA. A second class of marine ectoprocts, nowadays represented only by the order Cyclostomata. The unique feature of cyclostome reproduction is a highly developed polyembryony (Harmer, 1890, 1893; Robertson, 1903; Borg 1926). After fertilization (which has never been observed) the zygote is brooded within the zooid, where it develops into a ball of cells. This buds off a number of secondary embryos, which may in turn bud off tertiary embryos, the process being subsidized by the disintegration of the zooid. In this way, as many as a hundred or more identical embryos may be formed asexually from a single fertilized egg. They are released as simple ciliated larvae. Robertson (1903), who worked on *Crisia*, thought that eggs might be formed parthenogenetically, but this suggestion has received no support from later workers.

PHYLACTOLAEMATA. A small group of exclusively freshwater bryozoans. The colonies are formed in the usual manner by budding (see Brien, 1936), and consist of hermaphroditic zooids. Gametogenesis begins in late spring and early summer and the sexually produced larvae can be found (in the Northern Hemisphere) between May and August. They attach, metamorphose and bud to form new colonies in the same growing season that they were produced. Some phylactolaemates reproduce asexually by fragmentation, a portion of the colony breaking off and drifting away. However, the characteristic asexual propagules are the statoblasts. These are internal buds which acquire a thick protective coat. Statoblasts formed early in the summer may germinate in the same growing season and thus give rise to a second generation of statoblasts in the fall. With the disintegration of the colonies in the fall, enormous numbers of statoblasts are released. These are the overwintering phase of all phylactolaemates, being resistant to freezing and drying, and germinate in the following spring.

Discussion. One of the most prominent features of the reproductive biology of bryozoans is the production of asexual resting structures. Their morphology and physiology have attracted a good deal of attention (see Bushnell and Rao, 1974), but their ecology is still little-known. Amongst the gymnolaemates the most specializing resting structures are the hibernacula produced by the brackish-water and freshwater genera *Victorella* and *Paludicella*. These are external buds containing cells and stored reserves which become invested by a heavily sclerotized wall and can resist freezing and desiccation. They are usually formed in autumn, but may be produced at other times of year in response to stress. In a brackish lake in Italy, hibernacula were produced in summer, apparently as a response to high salinity or pollution (Carrada and Sacchi, 1964). The brackish-water *Bulbella* survives unfavourable periods as a dormant stolon.

The statoblasts of phylactolaemates have been much more intensively studied, perhaps because their variety of form makes them unique amongst asexual propagules. The barbarity of statoblast terminology is perhaps also unique. 'Floatoblasts' possess an annulus of air-filled cells which provide enough buoyancy to ensure adequate passive dispersal within a lake. 'Sessoblasts' are non-buoyant and become cemented to the substrate side of the zooid wall, and germinate at the same site as the parental colony. Other statoblasts have spines or hooked processes which are thought to function in dispersal or attachment. Most genera produce only one sort of statoblast, but *Plumatella* produces both sessoblasts and up to three kinds of floatoblast (see Wood, 1973). There have been many elaborate investigations of variation in statoblast morphology (the most important are listed by Bushnell and Rao, 1974), but they have largely been confined to taxonomic and biometrical studies, and the functional significance of statoblast morphology remains obscure. One or two (in *Fredericella*) or up to 20 (in *Plumatella*) statoblasts are produced by individual zooids (Bushnell, 1966). Many authors have commented on the enormous abundance of statoblasts at some seasons; Brown (1933) estimated that *Plumatella repens* liberated 8×10^5 statoblasts per square metre in the littoral of a Michigan lake, and found drifts of statoblasts a yard wide extending along half a mile of lakeshore.

In temperate localities there is only one bout of sexual reproduction every year, usually in early summer. Statoblast formation occurs later in the season, following sexual reproduction and the metamorphosis of the sexual larvae. In some genera (e.g., *Pectinatella*) only one generation of statoblasts is produced every year, whilst in others (e.g., *Plumatella*) up to three generations of colonies derived from, and themselves producing, statoblasts may replace one another during the course of a single growing season. Gametogenesis and statoblast formation may occur simultaneously in the same colony (Bushnell, 1966). Sexuality appears to be suppressed entirely in the short growing seasons of the far north or of alpine lakes (Wesenberg-Lund, 1907). By contrast, tropical phylactolaemates may reproduce both sexually and asexually during most months of the year (Vorstmann, 1928), although in India statoblasts are produced most abundantly just before the hot season (Annandale, 1911).

The most detailed study of budding rates and other aspects of the ecology of natural populations of freshwater bryozoans is that due to Bushnell (1966). He states that 75 per cent or more of viable statoblasts which have found a suitable substrate may germinate within a few days in spring, and infers that sexually produced larvae may only rarely succeed in founding colonies. His populations of *Plumatella repens* usually exhibited two distinct population maxima, in spring and in summer: it is noteworthy that, whilst the spring maximum was associated with a period of sexual reproduction, the later maximum was not.

Fertilization has not yet been observed either in phylactolaemates or in stenolaemates, and is the subject of controversy in gymnolaemates. For more than a century after Huxley (1856) it was believed that bryozoans were exclusively self-fertilized. The reasons for this belief were: the simultaneous ripening of eggs and sperm in the same zooid; occasional direct observations of syngamy; the belief that fertilization must occur within the coelom; and the failure to observe emission of sperm. This belief was shaken by Silén (1966), who demonstrated sperm emission in *Electra*. The sperm migrate into the coelomic lumen of the lophophore and exit through the terminal pores of the two dorsomedial tentacles; they then drift away until sucked down into the tentacle crown of another zooid by its feeding current. About halfway down the tentacle they move to the outer, unciliated surface, where they cling until an ovum is emitted by the zooid; they then move swiftly down towards the ovum and are presumed to fertilize it, although sperm entry was not observed directly. These results seem to have turned opinion around, so that it now seems to be generally believed amongst bryozoologists that cross-fertilization is the rule.

Nevertheless, apart from the direct observations of autogamy reported in the earlier literature, two objections remain. The first is that most gymnolaemates brood the embryos; *Electra* is an exceptional form which sheds eggs directly into the water. Silén argues forcefully that this brooding is usually external to the body cavity, and thus involves at least a transient exposure of the ovum to the external environment, during which fertilization could occur; against this must be set a number of genera, admittedly a small minority, in which brooding occurs in the coelom. The second objection is that the fertilization of a zooid by another zooid in the same colony has on average precisely the same genetic result as self-fertilization within a single zooid; and if the colony is extensive and sperm movement restricted then the sperm are very likely to attach to tentacle crowns of other zooids in the same colony. Silén argues that water currents are sufficiently strong in nature to sweep the sperm away from the colony, and that mature sperm remain active for several hours in seawater. His arguments have received indirect support from the characterization of genotypes at two isozyme loci by Gooch and Schopf (1970) and Schopf (1974b). At both loci the genotypes were quite close to the expected Hardy-Weinberg proportions, and the simplest explanation of this (others are possible) is that syngamy occurs more or less at random between colonies in each local population. Nevertheless, whenever a substantial difference between the observed and expected proportions was found, it involved a deficiency of heterozygotes, and for one of the loci the average deficiency of hetrozygotes was statistically significant. On balance, I think that a compromise is called for. The shedding of male gametes by *Electra*, through the terminal pores of extended tentacles, seems designed to maximize gamete dispersal, and thus cross-fertilization, as far as possible, and this may apply generally to other gymnolaemates; nevertheless it seems likely that a substantial quantity of sperm will be acquired by other zooids in the same colony, especially if water movement is slow or the colony is large, and therefore that a substantial proportion of zygotes will be the result of self-fertilization.

In phylactolaemates the circumstantial case for self-fertilization is stronger; the egg is brooded in an invagination of the body wall and must be fertilized in the coelom or before it leaves the ovary.

The question of individuality is as vexed in bryozoans as it is in coelenterates. In phylactolaemates the coeloms of adjacent zooids are in direct connection. In cyclostomes the zooids are separated by a wall pierced by a number of rather small openings, whose size presumably restricts the degree of communication and exchange between zooids. Gymnolaemate zooids also communicate through pores in the zooid wall, but do so indirectly: the pores are plugged by soft tissue and traversed by a system of mesenchymatous strands. These strands, the funicular cords, certainly accumulate reserve materials and probably facilitate their transfer between zooids. In stoloniferous gymnolaemates the funicular system runs along the stolon and extends from the stolon into each zooid. It seems, therefore, that gymnolaemate colonies, possessing a special system for the redistribution of energy and nutrients, may display an unusually high degree of physiological integration; this in turn may help to explain the well-known polymorphism of gymnolaemate zooids. The problem of individuality in bryozoans has recently been the subject of interesting articles by Bobin (1974), Boardman and Cheetham (1974), Sandberg (1974) and Schopf (1974a).

3.18 Phoronida (Phoronidea)

Incidence of Parthenogenesis. Phoronids are sedentary tubiculous marine worms, most common in the upper littoral of tropical and temperate seas. Most species are hermaphroditic. In some species the gametes are shed directly into the sea; in others fertilization appears to take place in the coelom (Rattenbury, 1953). *Phoronis ovalis*, a small worm which bores into rocks or mussel shells, can reproduce by transverse fission (Harmer, 1917; Brattström, 1943) forming tangled clonal aggregations of tubes; it has been said that it can also reproduce by budding (Marcus, 1949, cited by Hyman, 1967), but this seems doubtful. The autotomy of tentacle crowns, perhaps as an antipredator behaviour, is common among phoronids; the posterior stump always regenerates, and in *Phoronis ovalis* (Silén, 1955) and *Phoronopsis albomaculata* (Gilchrist, 1919) the anterior fragment may do so too, autotomy in these species being equivalent to transverse fission.

3.19 Brachiopoda

Incidence of Parthenogenesis. Brachiopods are sessile bivalved lophophorates, most abundant at moderate depths in cooler seas. Most are hermaphroditic with external fertilization, and many species brood their eggs in the lophophore: the only gonochoric brachiopod, *Argyrotheca*, broods its eggs in an enlarged nephridium. So far as is known, all brachiopods are exclusively amphimictic.

3.20 Mollusca

Reviews. Coe (1943, 1944) and Fretter and Graham (1964) give general surveys of molluscan reproduction; Duncan (1975) and de Larembergue (1939) describe pulmonates, especially self-fertilization; see also Purchon (1968).

Incidence of Parthenogenesis.

APLACOPHORA. The solenogasters; hermaphroditic but apparently cross-fertilizing. No report of parthenogenesis.

POLYPLACOPHORA. The chitons are gonochoric. No report of parthenogenesis.

MONOPLACOPHORA. The single genus, *Neopilina*, is gonochoric. No report of parthenogenesis.

GASTROPODA: PROSOBRANCHIA. The prosobranchs are generally gonochoric, with a few protandrous and simultaneous hermaphrodites. Self-fertilization has been inferred in the minute intertidal *Omalogyra* (Fretter, 1948) and in the freshwater *Viviparus* (Alonte, 1930, cited by Hyman, 1967). Yonge (1953) found only females of *Hipponix antiquatus*, and suggested that sperm stored from an earlier male phase was used in self-fertilization. The few apomictic molluscs are all freshwater prosobranchs. The absence of males in *Potamopyrgus jenkinsi* was noticed by Taylor (1900) and Boycott (1917), and the occurrence of apomixis was confirmed by Boycott (1919) and Quick (1920). Winterbourn (1970) has described obligate and facultative apomixis in the New Zealand species *P. antipodarium*. Males are generally scarce in species of the North American viviparid *Campeloma*, and none were found by van Cleave and Altringer (1937) during extensive sampling of *Campeloma rufum*; apomixis was later confirmed cytologically by Mattox (1938). Most taxa of *Campeloma* from the northern United States and Canada seem to be thelytokous; their ecology has been discussed by Van der Schalie (1965), who also describes the reproductive anatomy of the southern sexual forms. The scarcity or absence of males in oriental melaniids was noted by Abbot (1948), and apomixis was subsequently established for three species of *Melanoides* by Jacob (1957). Reproduction in the thelytokous Malaysian *Melanoides tuberculata*, a very common freshwater snail which thrives especially in polluted areas, has been described by Berry and Kadri (1974). There is a single mitotic maturation division in *Potamopyrgus* and *Campeloma*, but two in *Melanoides*.

GASTROPODA: OPISTHOBRANCHIA. The opisthobranchs are typically hermaphrodites with cross-copulation and mutual insemination. There is no report of parthenogenesis.

GASTROPODA: PULMONATA. The pulmonates are simultaneous hermaphrodites, some of which undoubtedly have the capacity to fertilize themselves. A minimal list of the families in which self-fertilization is known to occur is: Lymnaeidae (Colton, 1918; Walton and Jones, 1926; Crabb, 1927; Boycott *et al.*, 1931; Colton and Pennypacker, 1934; DeWitt and Sloan, 1958); Planorbidae (Richards and Ferguson, 1962); Physidae (Crabb, 1927; Baker, 1933; Brown, 1937; DeWitt, 1954; Duncan, 1958; DeWitt and Sloan, 1954); Bulinidae (de Larembergue, 1939); Ancylidae (Colton, 1918; Basch, 1959); Arionidae (Williamson, 1959); Philomycidae (Ikeda, 1937); Limacidae (Oldham, 1942; Maury and Reygrobellet, 1963, cited by Achatinidae Duncan, 1975); (Ghose, 1959). A more complete list (which includes, however, some dubious cases) is given by de Larembergue (1939). The vexed question of self-fertilization in pulmonates is discussed more fully below.

BIVALVIA. Bivalve molluscs are generally gonochores with external fertilization; a few genera are sequential hermaphrodites, but simultaneous hermaphrodites are exceptional. There is circumstantial evidence for self-fertilization in *Anodonta cygnaea* (Bloomer, 1940, and references therein). *Xylophaga dorsalis* is a protandrous hermaphrodite which accumulates sperm in a seminal receptacle near the opening of the oviduct, and may be predominantly selfed (Purchon, 1968). Morris (1917, 1918) induced development in unfertilized eggs of *Cumingia tellinoides*.

SCAPHOPODA. The scaphopods are gonochores with external fertilization. There is no record of parthenogenesis.

CEPHALOPODA. The cephalopods are gonochores with copulation and internal fertilization. There is no record of parthenogenesis.

Discussion. Despite the enormous diversity of molluscs, what is known about their capacity for parthenogenesis can be summed up very briefly: vegetative reproduction of any kind is unknown; apomixis is known only from a few freshwater prosobranchs; and self-fertilization is rare except, perhaps, among the freshwater and terrestrial pulmonates.

Potamopyrgus jenkinsi is a recent colonist of freshwater. It was first recorded from brackish habitats in the Thames estuary in 1889, had spread to fresh water by 1893 and in the following years succeeded in establishing itself throughout the greater part of England and Wales. From there it spread to northwestern Europe, at first in brackish water and subsequently in freshwater. It often occurs in enormous numbers, but in some areas seems to disappear as rapidly as it first appeared. Fretter and Graham (1964), who should be consulted for a brief account of the species, note that it seems to be free of the parasitic trematodes with which related marine species of *Hydrobia* are often heavily infested. Winterbourn (1970) has given a detailed account of New Zealand species of *Potamopyrgus*, from which *P. jenkinsi* may have been derived. Two of these species inhabit brackish water and are probably gonochoric amphimicts, although only ten to 28 per cent of individuals in large samples of *P. pupoides* were male. The third species, *P. antipodarium*, inhabits freshwater and seems certainly to be apomictic, although functional males were found in nine of 24 populations; in *P. jenkinsi* only a single male has ever been collected (Patil, 1958). Apomixis in *P. antipodarium* may, therefore, be facultative in some populations: parthenogenetic females retain a bursa copulatrix and seminal receptacle.

P. jenkinsi has been studied cytologically by Rhein (1935) and Sanderson (1939, 1940). British snails have 36-44 chromosomes, whilst Rhein's European material – collected from only a single locality – had 20-2 chromosomes; it is possible that the British populations are tetraploid and the continental ones diploid. All three New Zealand species of *Potamopyrgus* appear to be diploid, with $2n = 24$. *Campeloma rufum* is also a diploid, with $2n = 12$. The apomictic species of *Melanoides* are generally high polyploids, but in *M. tuberculatus* there is both a widespread and abundant polyploid form with 90-4 chromosomes and a rare and local diploid with 32 chromosomes (Jacob, 1957). The single known male of *P. jenkinsi* had an apparently normal spermatogenesis (Patil, 1958), but in males of *Melanoides* a mixture of univalents, bivalents and quadrivalents occurs during first metaphase, leading to the degeneration of spermatids and consequently to sterility (Jacob, 1957).

It has been known for many years that pond snails are able to reproduce when isolated. Pelseneer (1906) believed that they reproduced by apomixis in the absence of males and by gynogenesis when males were present; and he claimed in support of this belief that isolated *Lymnaea* gave off only one polar body during oogenesis. This was contradicted by Colton (1912, 1918, 1922), who found that the eggs of isolated *L. columella* regularly emitted two polar bodies, and who also found sperm and ova within the same acinus of the ovotestis. The matter was settled by the careful observations of Crabb (1927), who established the existence of a normal meiosis, pointing out that the first polar body migrates back into the albumen before cleavage, and so is easily overlooked. Since then, self-fertilization has been reported for a large number of freshwater and terrestrial pulmonates. What remains at issue is the frequency of selfing in natural populations. It is, at least, certain that many species have the capacity for self-fertilization: individuals which are isolated as eggs are able themselves to produce viable eggs. Crabb (1927) reports fertilization within the ovotestis in *Lymnaea stagnalis*, and Lams (1907) even claims to have seen early cleavage within the ovotestis of *Arion*. Moreover, the use of genetic markers has permitted the identification of selfed offspring: Boycott *et al.* (1931) found that the first egg capsules laid after copulation in *Lymnaea pereger* were crossfertilized, whilst later capsules were selfed, and Williamson (1959) found mixed broods of selfed and crossed offspring in *Arion*. On the other hand, crossfertilization appears to be obligatory in some planorbids and limacids, in most ancylids and perhaps in all helicids. (There are some early reports of selfing in helicid snails, cited by de Larembergue, 1939, but it was not then appreciated that these snails are capable of storing foreign sperm for long periods of time.) More generally, it is possible to argue that self-fertilization is to a large extent a laboratory artefact, since in most natural populations copulation will occur with sufficient frequency to ensure an overwhelming preponderance of cross-fertilized offspring; thus, Hunter (1961, 1964) is able to argue that selfing is rare in nature. The evidence supporting this position is indirect. First, foreign sperm appears to be used preferentially in fertilization in three species of *Lymnaea* (Boycott *et al.*, 1931; Horstmann, 1955) and in *Biomphalaria* (Paraense, 1959). Secondly, it seems inconceivable (to me) that an elaborate system of male accessory structures would be maintained if it were functionless: self-fertilization in pulmonates very rarely requires self-copulation. The arguments in support of an hypothesis of frequent self-fertilization in nature are equally indirect. First, there is the enigma of the bursa copulatrix, a blind sac opening via a long narrow passage into the vaginal tract; its function remains questionable, but foreign sperm appear in the bursa after copulation, and at least some of them are autolysed here. It is possible to interpret the bursa as a trap whose function is to destroy foreign sperm. Secondly, Crabb (1927) observed no inbreeding depression, either of fecundity or of viability, in exclusively selfed *Lymnaea stagnalis*; unfortunately, however, he does not present his results quantitatively. Finally, Hubendick (1951) has attributed the low levels of morphological variability found in many freshwater snails to their habit of frequent self-fertilization.

It must be admitted, I think, that none of these arguments are conclusive, or even very compelling, and that the problem remains unresolved. There is a moral to the story: it is relatively easy to show that a given species does or does not possess a capacity for self-fertilization; it is extremely difficult to assess the frequency of selfing in natural populations. This applies to all hermaphroditic organisms, whether they are snails or flowering plants. If we are able to use genetic markers, and know the genotype of the parent, the genotypes of the progeny and the frequencies of the relevant genotypes in the local random-mating population, then we can in principle estimate the proportion of the progeny which was produced by self-fertilization; in practice, unfortunately, any such estimate will have a large variance. So far as I know, there is no other way in which the frequency of selfing can be objectively determined. For the vast majority of hermaphroditic organisms, therefore, our estimate of the natural frequency of selfing must be either zero or unity or somewhere in between; it will rarely be possible to be more precise.

There is little information on the ecological correlates of self-fertilization, beyond the fact that the ability to self is much more common amongst freshwater than amongst marine molluscs. In *Omalogyra* cross-fertilization has been inferred to occur in spring (although the manner of copulation remains speculative), with selfing in the summer during a period of rapid population increase

(Fretter, 1948). In populations of *Bulinus contortus* there are different frequencies of aphallic individuals, up to nearly 100 per cent, but these do not seem to vary in any obvious way with latitude (Table 3.3); aphallic individuals comprised half the total sample at one locality in Corsica, but over 90 per cent at another; in North Africa there are enormous differences between samples collected in Algeria and Tunisia (about 95 per cent aphallic), in Morocco (26 per cent aphallic) and in Egypt (less than one per cent aphallic). It may be that there are substantial local differences that would repay investigation.

Table 3.3: Proportion of Aphallic Individuals, Presumed to be Self-fertilizing, in Large Collections of *Bulinus contortus* from Different Regions.

Locality	Percentage of aphallic individuals	Sample size
Greece	84	88
Crete	100	35
Corsica: Ajaccio	43	79
Corsica: P. Vecchio	92	61
Algeria	100	35
Tunisia	93	1112
Morocco	26	246
Egypt	0.7	596
Senegal	98.7	839

Source: de Larembergue (1939), Table 6.

3.21 Sipunculoidea (Sipunculida, Sipuncula)

Review. Rice (1975).

Incidence of Parthenogenesis. Sipunculids are sedentary unsegmented coelomate worms, found over a wide range of depths in most seas. They are gonochores with external fertilization. In several species there is known to be a great preponderance of females among the adults (list of references in Rice, 1975), but this can scarcely be attributed to parthenogenesis, since unfertilized eggs are arrested in the first meiotic metaphase, and the completion of oocyte maturation and the initiation of cleavage require sperm entry and syngamy. Two sipunculans are known to reproduce vegetatively. Specimens of *Sipunculus robustus* kept in stale seawater developed lateral buds on the posterior half of the body which separated from the parent after a short period of growth (Rajulu and Krishnan, 1969); the same species can reproduce by transverse fission (Rajulu, 1975). Transverse fission appears to be a regular mode of reproduction in *Aspidosiphon brocki*, a small intertidal rock-boring species, where a constriction separates a smaller posterior from a larger anterior portion. Gonadal tissue has been identified in *A. brocki*, which is therefore presumed to be capable of reproducing sexually.

3.22 Echiuroidea (Echiura)

Incidence of Parthenogenesis. This is a gonochoric phylum of coelomate marine worms. The dwarf males and sex determination of *Bonellia* are the subject of a large literature (briefly reviewed by Gould-Somero, 1975). No means of asexual reproduction has been observed.

3.23 Myzostomida

Incidence of Parthenogenesis. The Myzostomida are a group of marine worms ectocommensal or parasitic on echinoderms, especially crinoids; often classified as polychaetes. They are hermaphroditic, with internal fertilization by means of spermatophores. Neither self-fertilization nor asexuality have been described.

3.24 Annelida

Reviews. Schroeder and Hermans (1975; polychaetes); Malaquin (1893), Potts (1911), Okada (1929; syllids); Stephenson (1930; oligochaetes); Lasserre (1975; marine oligochaetes); Dehorne (1916), Christiansen (1961), Muldal (1952), Omodeo (1952; oligochaete families); Gavrilov (1935; self-fertilization); Berrill (1952b), Herlant-Meewis (1958; vegetative fission).

Incidence of Parthenogenesis.

CHAETOPODA: POLYCHAETA. A group of marine worms which includes both active and sedentary, usually tubicolous, forms. The great majority are amphimictic and gonochoric. However, asexual reproduction by vegetative fission, though exceptional, is widespread amongst polychaetes: a list of known instances is given by Schroeder and Hermans (1975). Two modes of fission are conventionally recognized. Paratomy involves regeneration before fission, so that immediately prior to fission two nearly complete worms are separated by a short peduncle, which is subsequently ruptured by muscular contraction to yield two independent individuals. In architomy, fission precedes regeneration, so that the immediate products of reproduction are a greater or lesser number of fragments, each of which regenerates a complete worm. Although they remain convenient terms, all sorts of conditions intermediate between paratomy and architomy have been described. Paratomy has been reported from the

Spionidae and Ctenodrilidae (Scharff, 1887); it is the only type of fission known for the Serpulidae (Malaquin, 1895; Faulkner, 1930; Vannini, 1950; Cresp, 1964). Fission can be induced in the serpulid *Salmacina* by removing an individual from its tube and isolating it in a small volume of water. Three anterior segments are budded from the middle of the body of the parent, and the abdominal segments of the parent are then transformed into the thoracic segments of its budded offspring. The final connection between the two almost completely regenerated individuals is severed after about a week (Cresp, 1964). Architomy has been reported from the Spionidae (Rasmussen, 1953) and Ctenodrilidae (von Zeppelin, 1883), and is the only type of fission known from the Syllidae (Allen, 1921; Okada, 1929), Chaetopteridae (Potts, 1914), Amphinomidae, Cirratulidae (Martin, 1933; Berkeley and Berkeley, 1953) and Sabellidae (Caullery and Mesnil, 1920). The minute ctenodrilid *Zeppelina* is especially interesting because, despite intensive study (in the laboratory; Korschelt, 1942), it has never been known to reproduce sexually: no other polychaete relies exclusively on asexual reproduction. Architomy in *Zeppelina* may involve merely binary transverse fission, with the eventual production of two complete individuals from a single parent; however, the two halves may again fragment before regeneration, and by repeated fission very small fragments comprising only a few segments each may be formed, each of which is capable of regenerating a complete worm. More rarely, an individual may spontaneously divide into a large number of small pieces. Many syllids also spontaneously undergo fission at rather precisely predetermined points along the body, usually giving rise to fragments of two to four segments each, which subsequently regenerate anteriorly and posteriorly. Architomy reaches a limit ('schizometamery') in forms which spontaneously divide into isolated segments; this occasionally happens in *Zeppelina*, but is a regular mode of reproduction in the cirratulid *Dodecaceria caulleryi* (Dehorne, 1933). Each segment regenerates anteriorly and posteriorly to form a complete organism, but then immediately divides again: the anterior piece regenerates posteriorly and the posterior piece anteriorly to form two new worms, whilst the original segment, once more isolated, goes through the whole process of regeneration and fission again before dying. *Syllis ramosa* inhabits deepwater sponges, and exists as a profusely branched structure which ramifies by budding through the body of the sponge (McIntosh, 1879); Schroeder and Hermans (1975) correctly interpret the budding process involved as growth, with reproduction occurring only if branches become detached and subsequently regenerate anteriorly. In many polychaetes the vegetative individual ('atoke') gives rise to an individual specialized for sexual reproduction ('epitoke'). The epitoke may arise either by the metamorphosis of the whole pre-existing atoke ('epigamy'), or by a process involving the modification and liberation of the posterior end of the atoke ('schizogamy'). In schizogamous forms the posterior sexual section is called the 'stolon' and the remainder of the worm the 'stock'. Schizogamy is especially characteristic of the two syllid subfamilies Syllinae and Autolytinae, and has been reviewed by Potts (1911), Berrill (1952b), Herlant-Meewis (1958),

Durchon (1967) and Schroeder and Hermans (1975); Schroeder and Hermans also give a list, with authorities, of the other polychaete taxa from which schizo-gamy has been reported, including the palalo worm, a eunicid polychaete whose synchronous nocturnal swarming has often been described. The stolon may regenerate a head either before or after separation from the stock, parallelling the distinction between paratomical and architomical fission of vegetative individuals (see Malaquin, 1893). In some syllines, and especially in *Trypanosyllis*, a group of thirty or more stolons may arise simultaneously. In some autolyne syllids a chain of stolons is produced by repeated intercalary budding, so that in *Myrianida* and *Autolytus* as many as thirty individuals may be attached end to end ('gemmiparous' stolonization: see Malaquin, 1893; Potts, 1911; Okada, 1935). Mature stolons are packed with gametes. They may settle on the bottom and undergo further morphological change; some have a short-lived free-swimming existence. An analogy can be drawn between the schizogamous production of epitokes in polychaetes and the budding of medusae by hydrozoan polyps; in both cases an individual specialized for sexual reproduction and with a longer or shorter free-swimming phase is produced asexually by a vegetative individual. So far as is known, the gametes of polychaetes always go through a normal meiosis, although abnormal embryos have been produced from unfertilized eggs in the laboratory (see, for example, Scott, 1906, for *Amphitrite*); however, the unusual freshwater polychaete *Nereis limnicola* (= *Neanthes lighti*) appears to be an internally self-fertilized hermaphrodite (Smith, 1950, 1958; Baskin and Golding, 1970), and Durchon (1957) thought that embryos found within the coelom of *N. diversicolor* may have been produced parthenogenetically.

CHAETOPODA: OLIGOCHAETA. A cosmopolitan group of hermaphroditic worms. Oligochaetes reproduce in such a diversity of ways that I shall treat the subclass family by family.
Aelosomatidae and *Naididae*. Two families of freshwater oligochaetes which reproduce primarily by vegetative fission. Architomy has been reported from a few naidids (species of *Nais*, *Branchiodrilus* and *Aulophorus*; see Stephenson, 1930), but in most forms fission is paratomical. Dehorne (1916) recognized two modes of fission, which she called slow (*lente*) and rapid (*hâtive*). In the slow paratomy of *Dero*, *Aulophorus* and *Ophidonais*, fission separates two distinct individuals which have not yet themselves formed fission zones. In the rapid paratomy of most other naidids and aelosomatids new fission zones appear before the separation of the first two zooids, so that chains of zooids in various stages of regeneration are created. Dehorne further divides this latter class of paratomy into two categories. In 'naidian' paratomy (*Nais*, *Chaetogaster*, *Aelo-soma*) the new fission zone arises at the anterior limit of the previous fission zone, so that the new (third) zooid comprises entirely new segments; there is no necessary limit to the length of a chain of zooids constructed in this way, although in practice chains of (for example) *Aelosoma* rarely comprise more than eight zooids. Chains tend to be longest in spring and in fall. In 'stylarian'

paratomy (*Stylaria, Pristina, Macrochaetina*) the new fission zone forms one segment behind the anterior limit of the previous zone, so that the new zooid includes one segment from the original worm; the length of chains built up in this manner is thus limited by the finite length of the original worm. The rules which regulate the formulation of new fission zones have received a good deal of attention; they have been reviewed by Herlant-Meewis (1958). In the laboratory, naidids and aelosomatids can perpetuate themselves solely by vegetative fission for many generations, and perhaps indefinitely (see the early articles reviewed by Stephenson, 1930). In nature, sexually mature individuals are occasionally seen in late summer and early autumn; according to Pennak (1978) they may be quite common in *Chaetogaster*, but sexual development has never been observed in some genera of naidids, and is very infrequent in *Aelosoma*.

Tubificidae. Cernosvitov (1927) found that up to half the individuals in samples of *Tubifex tubifex* lacked spermathecae, and showed that isolated worms were capable of producing viable progeny; he inferred that the species commonly reproduced by self-fertilization. However, Christiansen (1961) refers to unpublished evidence that some tubificids have an automictic parthenogenesis which involves a premeiotic doubling of ploidy. Hrabe (1937, cited by Lasserre 1975) has described vegetative fission in *Aulodrilus*.

Enchytraeidae. Most enchytraeids are probably cross-fertilized amphimicts, although there is no anatomical barrier to self-fertilization. However, an unexpected wealth of cytological behaviour has been revealed by the brilliant work of Christiansen (1959, 1960, 1961; Christiansen and Jensen, 1964; Christiansen and O'Connor, 1958). Of the 88 cytotypes described in his 1961 article, 15 were found to reproduce parthenogenetically. In most cases the parthenogenesis is automictic, with suppression of the second meiotic division; but many of the parthenogenetic species produce some quantity of sperm, whose function remains unclear. In *Achaeta affinis* and *Fredericia maculata* spermathecae are absent; *Mesenchytraeus pelicensis* possesses spermathecae but they never contain sperm. Copulation is therefore infrequent or absent in these species, which are undoubtedly automictic, and yet sperm is always found in the egg cocoons. Perhaps each individual uses its sperm for a pseudogamous stimulation of its own eggs. In *Cognettia glandulosa* and *Fredericia ratzeli* sperm penetration seems to be responsible for suppressing the second meiotic division: in the absence of sperm, haploid embryos develop from the products of the first meiotic division, though it is not known whether they hatch. Some highly aberrant cytologies have been described. *Lumbricillus lineatus* has sympatric diploid and triploid cytotypes; the diploid individuals are amphimictic, whilst the triploid individuals produce triploid eggs by an unusual automictic process involving an asynaptic meiosis. The parthenogenetic eggs commence development, but hatch only if they are penetrated by sperm from diploid individuals, so that pseudogamy does not act in the usual way to elicit cleavage, but rather to procure the successful completion of embryogenesis. In *Enchytraeus lacteus* somatic nuclei contain 170 chromosomes, eight being larger than the others. The large chromosomes

form bivalents at meiosis but the small ones do not, so that the female pronucleus contains four large chromosomes, together with all 162 small chromosomes, the small chromosomes dividing equationally during the second meiotic
division. During spermatogenesis two types of nucleus are formed: a large
nucleus containing all the small chromosomes, which fails to develop, and two
small nuclei each with four large chromosomes, which develop into sperm. At
syngamy the somatic complement of 170 is restored by the fusion of a male
pronucleus containing four large chromosomes with a female pronucleus containing four large and 162 small chromosomes. Even amongst amphimictic
enchytraeids gametogenesis is often abnormal: in both amphimictic and parthenogenetic species of *Fredericia*, *Bucholzia* and *Hemifredericia*, and in some
species from four other genera, meiosis is achiasmatic in both sexes, a condition
unknown in the rest of the animal kingdom. Finally, vegetative fission, although
uncommon, has been recorded for a few species. In *Enchytraeus fragmentosus*
sexually mature worms have never been seen, and fission appears to be the only
mode of reproduction (Bell, 1959); in *Cognettia sphagnetorum* a few sexual
individuals were discovered by Christiansen (1959), but their eggs failed to hatch.
C. glandulosa and *Bucholzia appendiculata* can reproduce both by fragmentation
and by eggs; the sexually mature worms are present for a short period during
winter.

Lumbricidae, etc. The 'earthworms' comprise a group of families of terrestrial
oligochaetes, in some of which parthenogenesis has been inferred from the
reduction or absence of male structures (see Gates, 1959, 1974; Reynolds,
1974; and other authors cited by Edwards and Lofty, 1972, who give a list of
examples). The familiar Eurasian earthworms in the family Lumbricidae are
much the best known, and are the only ones whose cytology has been studied.
It has been known for some time that several genera are capable of producing
viable offspring when kept in isolation (Evans and Guild, 1948). This may
sometimes be the result of self-fertilization since, as with the enchytraeids, there
are few anatomical barriers to selfing in lumbricids. Gavrilov (1935) described
self-fertilization, both with and without self-copulation, in a number of earthworms, and should be consulted for a review of the early literature; André
(1962a, b, 1963; André and Davant 1972) reported self-fertilization, preceded
by self-copulation, in *Eisenia* and *Dendrobaena*. However, nongametic parthenogenesis was discovered in several species by Muldal (1952) and Omodeo (1952,
and other articles by the same author cited therein). Earthworms resemble
enchytraeids in that parthenogenesis is usually automictic, but differ in the type
of automixis: in most parthenogenetic earthworms there is a premeiotic doubling,
followed by the formation of chiasmatic bivalents and a regular meiosis with
the extrusion of two polar bodies. (Muldal, 1952, p. 71, mistakenly states that
obligatory parthenogenesis is usually apomictic. On the same page he refers to
a projected article on the cytology of parthenogenesis in earthworms which I
have not been able to locate.) Parthenogenesis is obligatory in most cases, but
facultative in *Dendrobaena subrubicunda*: parthenogenetic eggs are found in

experimentally isolated individuals, and are common in natural populations in winter. *Dendrobaena octaedra* appears to be genuinely apomictic, with a purely mitotic oogenesis and no formation of bivalents. Some earthworms (e.g., *D. rubida* including *D. subrubicunda*) comprise complex taxonomic units with a number of amphimictic and parthenogenetic cytotypes, some of them occurring sympatrically. Muldal states that the parthenogenetic forms are morphologically the less variable. The degree to which male function is reduced in parthenogenetic earthworms is very variable, with some species (e.g., *Allolobophora caliginosa*) producing apparently normal sperm; such species may be pseudogamous, but experimental proof is lacking. Vegetative fission is rare amongst earthworms, but a simple architomical binary fission has been reported from *Lumbriculus variegatus* (von Wagner, 1900), and *Lamprodilus mrazeki* is also said to reproduce vegetatively (see Brinkhurst and Jamieson, 1971). Stephenson (1922) discovered a process of fragmentation within the overwintering cysts of *Lumbriculus variegatus*.

HIRUDINEA. The leeches are carnivores and ectoparasites found in salt and fresh water; like the oligochaetes, all are simultaneous hermaphrodites. Their reproduction appears to be exclusively amphimictic.

ARCHIANNELIDA. A predominantly gonochoric group of marine annelids allied to polychaetes. Vegetative paratomy has been reported from *Dinophilus* (Schultz, 1902) and schizogamy from *Polygordius* (Dawydoff, 1949) and *Protodrilus* (Goodrich, 1931).

Discussion. There are two obvious ecological correlates of vegetative fission in annelids. The first is its association in marine polychaetes with a sedentary, often tubicolous, existence: cirratulids (such as *Dodecaceria*) and serpulids (such as *Salmacina*) live an almost colonial existence in dense clonal clusters. The correlation is not perfect, however; some motile polychaetes (syllids and ctenodrilids) reproduce vegetatively, whilst many sedentary forms (such as the familiar *Spirorbis*) do not. Secondly, vegetative fission is far more common in fresh water than in the sea, being the principle mode of reproduction among naidids and aelosomatids, though it is less prominent in the other freshwater and terrestrial taxa. In freshwater the correlation between vegetative reproduction and a semicolonial ecology breaks down; the nearest freshwater analogues of the marine sabellids and serpulids are the tubificids, which rarely reproduce by fission. Two other less distinct patterns can be recognized: the first is a tendency for motile vegetatively reproducing forms to be small, or even microscopic (*Zeppelina*, for example, or *Aelosoma*), and the second is the usual tendency for vegetative fission to be characteristic of younger individuals, preceding a sexual phase, if one should occur.

Nearly half of all the polychaetes which regularly reproduce by vegetative fission are species of the cirratulid *Dodecaceria*. Gibson and Clark (1976) have recently described the reproduction of *D. caulleryi* on the Northumberland coast, where it inhabits burrows in *Lithothamnion*. Vegetative fission, following the pattern I have already described, occurs between September and November, with the regenerating fragments occupying the parental burrow for several months afterwards. They disperse in spring, suffering heavy mortality. By the following summer regeneration is complete, all the worms having fully developed anterior ends, and it is no longer possible to distinguish the products of single segments from those of larger fragments. Epitokal individuals are rare, being found occasionally in spring and summer; sexual reproduction seemed not to contribute at all to the recruitment of the local population. Individual worms live for several years, reproducing solely by fission before becoming epitokous and, shortly afterwards, dying. *D. concharum*, which lives on the same shore, rarely develops epitokes, and males have never been seen.

The work of Muldal (1952), Omodeo (1952) and especially Christiansen (1961) has furnished unusually extensive data bearing on the relationship between ploidy and reproductive mode in lumbricids and enchytraeids. I have abstracted this information in Table 3.4. It enables us to confirm what has often been suspected in other groups, that there is a striking and highly significant association between polyploidy and parthenogenesis. We can imagine this correlation to consist of two parts. In the first place, polyploid cytotypes tend to reproduce parthenogenetically, although there are numerous exceptions to this rule. Secondly, parthenogenetic cytotypes are almost invariably polyploid; of the 82 diploid cytotypes identified in the two families, only two are believed to be parthenogenetic. One of these is the single exception to the rule amongst enchytraeids, *Mesenchytraeus glandulosus*, in which all individuals produce morphologically normal sperm, and about half bear sperm in their spermathecae. Of even greater interest is the correlation between ploidy and the potential for genetic recombination in strictly amphimictic species: the meiosis of polyploid amphimicts tends to be achiasmatic. Thus, although 27 of the 41 polyploid enchytraeids reproduce amphimictically, only seven of these 27 have a normal chiasmatic meiosis. Five of these seven are in the single genus *Enchytraeus*, and three are cytotypes of the single species *E. lacteus*. Most of the other amphimictic polyploids belong to *Fredericia*, all the members of which have an achiasmatic meiosis. Christiansen thinks that the polyploid cytotypes which have a normal chiasmatic meiosis may be alloploid; amphimicts which are thought to be auto-ploid avoid the formation of multivalents by virtue of their abnormal achiasmatic meiosis.

Polyploid cytotypes or species of lumbricids are often morphologically different from their nearest diploid relatives: they tend to have larger cells, larger gametes, greater size and more segments (Table 3.5). There is also an ecological difference: polyploid automicts tend to be more abundant and widespread than related diploid amphimicts. Muldal gives a list of examples: the triploids *Eisenia rosea* and *Bimastus tenuis* are the most common members of their genera; the most cosmopolitan species of *Octolasion* are two parthenogenetic forms,

Table 3.4: Relationship Between Ploidy and Reproductive Mode in Enchytraeids and Lumbricids. (A) Ploidy in amphimictic and parthenogenetic enchytraeids. (B) Ploidy in amphimictic enchytraeids, according to whether meiosis is chiasmatic or achiasmatic. There is no relationship between ploidy and the nature of the meiosis among the 14 automictic enchytraeids. (C) Ploidy in amphimictic and parthenogenetic lumbricids.

A. *Enchytraeidae*

	Amphimictic	Parthenogenetic	Total
Diploid	46	1	47
Polyploid	27	14	41
Total	73	15	88

B. *Enchytraeidae*

	Chiasmatic	Achiasmatic	Total
Diploid	40	6	46
Polyploid	7	20	27
Total	47	26	73

C. *Lumbricidae*

	Amphimictic	Parthenogenetic	Total
Diploid	34	1	35
Polyploid	8	16	24
Total	42	17	59

Source: Muldal (1952); Omodeo (1952); Christiansen (1961).

O. lacteum (a diploid) and *O. cyaneum* (a decaploid); and so forth. The exceptional diploid automict *O. lacteum* suggests that parthenogenesis, rather than polyploidy *per se*, is responsible for this success; but the evidence is inconclusive.

Different cytotypes of the same species may be found in the same region, but their distributions are usually disjunct on a smaller scale: Christiansen (1961) found more than one cytotype at the same locality in only four species.

3.25 Tardigrada

Review. Bertolani (1975).

Incidence of Parthenogenesis. Tardigrades are microscopic aquatic animals allied to arthropods. They inhabit streams, lakes and the sea, and are often prominent members of the interstitial fauna of sandy marine and lacustrine beaches, but their most characteristic habitat is the water film on mosses and other lower plants. Tardigrades are gonochoric, but males are often very scarce or entirely absent, and the suspicion is strengthening that parthenogenesis is common throughout the group. The presently known and suspected cases of parthenogenesis are

Table 3.5: Relationship Between Ploidy and Morphology in Lumbricids. Column headings: A, nuclear volume of oocyte (in μm^3); B, nuclear volume of cutaneous epithelium; C, nuclear volume of intestinal epithelium; D, length of sperm (in μm); E, volume of mature oocyte (in μm^3); F, body length (in mm); G, body width; H, number of segments.

Taxa	Ploidy	A	B	C	D	E	F	G	H
Bimastus eiseni/B. tenuis	2						47	2–5	93
	3						52	3	105
Eisenia venata/E. rosea	2	179	152	131	15.4	334	34	3–4	102
	3	317	210	112	29.1	509	55	3–4	135
	4	317	194	122		399			
	6	381	273	121		588			
	10	506	372	196	42.0				
Dendrobaena rubida	2	193	184	75	11.2	319	43	3.5	75
	4						60	4	85
Octolasion lacteum/O. cyaneum	2						97	4	126
	10						122	7.5	127
Allolobophora chlorotica	2	285	220	102	21.3	481			
	3	309	225	85		1048			
	4	278				1856			

Source: A–E, Omodeo (1952); F–H, Muldal (1952).

Table 3.6: Confirmed and Probable Instances of Parthenogenesis Among Tardigrades.

Taxon	Ploidy	Mode of reproduction
Echiniscus	? Diploid	Males unknown; perhaps apomictic
Hypsibius dujardini	Diploid	Automixis
H. oberhauseri	Diploid	Amphimixis
	Diploid	? Automixis
	Triploid	Apomixis
	Tetraploid	Apomixis
Isohypsibius granulifer	Diploid	Amphimixis
	Tetraploid	? Autogamy
Macrobiotus coronifer	Diploid	Males scarce or absent
M. hufelandi	Diploid	Amphimixis
	Triploid	Automixis (with achiasmatic meiosis)
	Triploid	Apomixis
	Tetraploid	Automixis
M. pseudohufelandi	? Triploid	Males scarce or absent
M. richtersi	Diploid	Amphimixis
	Triploid	Apomixis
Milnesium tardigradum	Diploid	Automixis
Oreella	–	Males unknown
Parechiniscus	–	Males unknown

Source: Bertolani (1975).

listed in Table 3.6. The first cytological results were announced by Ammermann (1962, 1967) for *Hypsibius dujardini*, which is automictic. The first meiotic division is reductional, with a haploid polar body being expelled. The second division is preceded by a redoubling of chromosome number and results in the expulsion of a diploid polar body and the formation of a diploid pronucleus. *Milnesium tardigradum* was also found to be a diploid automict (Baumann, 1964). However, a series of papers by Bertolani (summarized by Bertolani, 1975) have established the existence of both automictic and apomictic modes in *Hypsibius* and *Macrobiotus.* In one population of the automict *M. hufelandi*, meiosis is achiasmatic. Bertolani also describes a population of *Isohypsibius granulifer* in which sperm was found in the females despite an absence of males; he suggests that this form is a self-fertilizing hermaphrodite. Finally, there are several large genera, such as *Echiniscus*, in which males have never been reported, but which have not been investigated cytologically. There are probably many examples of parthenogenesis among tardigrades yet to be reported, and their cytology promises to be of considerable interest. Parthenogenesis is unknown in marine tardigrades, where reported sex ratios are usually near equality, and is most conspicuous in moss-dwelling forms. Vegetative reproduction does not occur in tardigrades.

It seems to be quite common for a species to include both parthenogenetic and amphimictic cytotypes. Moreover, the two types are often found in the same locality, resulting in mixed populations with a skewed sex ratio; Bertolani found apomictic and amphimictic *M. hufelandi* on the same rock.

Parthenogenetic tardigrades tend to be polyploid, despite the existence of a number of diploid automicts (see Table 3.6); the apomictic cytotypes are all polyploid usually triploid. *Echiniscus* has a diploid number of 14, whilst amphimictic types in other genera have ten or twelve. Bertolani says that body size increases with ploidy.

Many tardigrades produce both thin-shelled and thick-shelled eggs, and the latter may be resting eggs (Pennak, 1978); it is not known whether the cytology of the two types is different.

The marine tardigrade *Batillipes* produces thin-shelled pliant eggs in spring and thicker-shelled, or at least more rigid, eggs in fall (Pollack, 1970). The characteristic resistant stage of tardigrades is the 'tun', a reduced adult similar to the tuns formed by bdelloid rotifers (see Crowe, 1975; Crowe and Madin, 1974). Tuns are usually produced as a response to desiccation.

3.26 Pentastomida

Incidence of Parthenogenesis. Pentastomids are vermiform endoparasites of terrestrial vertebrates. They are gonochoric amphimicts; I have found no record of parthenogenesis.

3.27 Onychophora

Incidence of Parthenogenesis. The Onychophora are a phylum of terrestrial worms, in some respects intermediate between annelids and arthropods, living under cover in the humid tropics and the southern temperate zone. They are gonochores and appear not to reproduce parthenogenetically.

3.28 Arthropoda: Chelicerata

Reviews. Oliver (1971, 1977; Acari).

Incidence of Parthenogenesis.

MEROSTOMATA. The horseshoe-crabs, a class of large marine arthropods. They are gonochoric and are not known to reproduce parthenogenetically.

PYCNOGONIDA. The sea-spiders, a second class of marine chelicerates, again gonochoric and exclusively amphimictic.

ARACHNIDA. The scorpions, pseudoscorpions, harvestmen, mites, spiders and their allies; all are basically terrestrial, gonochoric and amphimictic. The Brazilian scorpion *Tityus serrulatus* is thelytokous: males are absent from natural populations, and isolated females can reproduce in the laboratory (Matthieson, 1962, 1971). Bisexual species of *Tityus* and another buthid, *Isometrus*, have an achiasmatic spermatogenesis (Piza, 1943, 1947). Parthenogenesis has been sketchily described in a very few opiliones (Gueutal, 1943; Briggs, 1971), whipscorpions (Lawrence, 1958) and spiders (Machado, 1964). With these exceptions, the only arachnids known to be parthenogenetic belong to the order Acari (Acarida), the mites and ticks. These comprise an enormous group of free-living and ectoparasitic arthropods; most are gonochoric amphimicts, but a considerable number of parthenogenetic forms is now known, and their distribution through the order is summarized in Table 3.7. Arrhenotoky is the most common mode of parthenogenesis, there being sound cytological evidence for male haploidy in many genera. Even in quite large taxa the majority of species may be arrhenotokous, as in some mesostigmatan families and in the prostigmatan Tetranychidae (spider mites). In the mesostigmatan Phytoseiidae, and in a few acarines from other families, unmated females will not lay eggs but, once mated, will produce both haploid and diploid embryos; this may indicate an arrhenotokous gynogenesis, but I do not think it is conclusively established that sperm penetration, rather than the act of mating alone, is required to elicit oviposition. Thelytoky is much less common, though it is often found occasionally amongst arrhenotokous species (see references in Helle and Overmeer, 1973). Obligate thelytoky occurs sporadically throughout the order, and is the only parthenogenetic mode of reproduction known from the orabatid mites. Little is known about the cytology of acarine thelytoky; the few cases which have been worked out suggest a good deal of diversity. The tenuipalpid *Brevipalpus obovatus* appears to be automictic, perhaps with a premeiotic doubling (see Pijnacker and Ferwerda, 1975). In the orabatids *Platynothrus* and *Trhypochthonious* automixis involves a suppression of the second meiotic division, diploidy being restored by the fusion of second anaphase plates (Taberly, 1958, 1960). Another type of automixis occurs in the prostigmatan *Cheyletus eruditus*, where diploidy is restored postmeiotically by the fusion of cleavage nuclei (Peacock and Weidmann, 1961, cited by Oliver, 1977). On the other hand, the thelytokous race of *Histiostoma feroniarum* is apomictic (Heinemann and Hughes, 1969). Thelytokous and amphimictic races of the same species may have different geographical distributions: *Macrocheles penicilliger* is thelytokous in Italy but amphimictic in the USSR (Filipponi, 1964; Bregetova and Koroleva, 1960, cited by Oliver 1971); *Typhlodromus guatemalensis* is thelytokous in California but amphimictic (or arrhenotokous) in Canada (Kennett, 1958; Putman, 1962); *Haemaphysalis longicornis* is thelytokous in the South Pacific, in Hokkaido and northern Honshu and in Primorye (northeastern USSR), but is amphimictic in southern Honshu and Kyushu, in Korea and in the extreme south of Primorye (Hoogstral *et al.*, 1968). In Japan the parthenogenetic races of *H. longicornis* seem to be triploid and the amphimictic races diploid (Oliver and Tanaka, 1969), but more generally the relationship between ploidy and reproductive mode in acarines is unclear, and most thelytokous forms may be automictic diploids. Deuterotoky, the parthenogenetic production of diploid individuals of both sexes, occurs in *Histiophorus* and perhaps elsewhere in the Listrophoridae (Dubinina, 1964, cited by Oliver, 1971), and may occur occasionally in some tetranychids (Jesiotr and Suski, 1970, cited by Helle and Overmeer, 1973). Finally, in some water mites (Hydrachnellae) the males have an achiasmatic meiosis (Keyl, 1957). Indeed, a low potential for the creation of recombinational diversity is characteristic of many acarines: even amphimictic species with a chiasmatic meiosis commonly have less than four or five chromosomes in the haploid set, although several chiasmata per bivalent may be formed.

3.29   Arthropoda: Crustacea Branchiopoda

Reviews. Longhurst (1954; Notostraca); Mathias (1937; phyllopods); Berg (1934), Hebert (1978; Cladocera).

Incidence of Parthenogenesis. The branchiopods are an almost exclusively freshwater group of crustaceans which can be conveniently divided into two series. The first includes the Anostraca, Notostraca and Conchostraca ('phyllopods' – but this term has also been used to denote the Branchiopoda other than Anostraca), nearly all of which are relatively large, benthic branchiopods inhabiting freshwater habitats which are unusually stressful in that they dry up, freeze solid or are highly saline; their distribution is often very local in space and sporadic in time, and it is unusual for any single locality to support more than one member of this series (Dexter, 1953). Their eggs, whether sexual or not, are normally resistant to freezing and drying (see short review of diapause in conchostracan eggs in Mattox and Velardo, 1950). The second series includes the Cladocera, which are typically small planktonic and epiphytic branchiopods inhabiting both temporary and permanent bodies of water.

ANOSTRACA. The fairy shrimps, most of which are gonochoric amphimicts, although there are unconfirmed reports of parthenogenesis in species of *Branchipus* and *Streptocephalus* (see Mathias, 1937). The best-known anostracan is the brine shrimp *Artemia salina*, which occurs naturally in saline lakes and has a broad but necessarily highly discontinuous distribution around the shores of the Mediterranean and in western North America. Most populations are amphimictic, and Bowen (1962) has ruled out the possibility of thelytoky or gynogenesis in samples from the Great Salt Lake of Utah. Some Mediterranean populations are parthenogenetic, but the nature of the parthenogenesis is still not entirely clear. In some localities there are polyploid cytotypes (triploids, tetraploids and pentaploids or hyperpentaploids) which are always parthenogenetic and

Table 3.7: Parthenogenesis in Acari. Families and genera arranged alphabetically within suborder.

Family	Ecology	Genus	Reproduction
MESOSTIGMATA			
Dermanyssidae	Ectoparasites of vertebrates, feeding on blood	*Dermanyssus*	Arrhenotokous gynogenesis
		Laelaps	Probably arrhenotokous
Haemogamasidae		*Eulaelaps*	Thelytoky inferred
Macrochelidae		*Areolaspis*	Arrhenotoky
		Geholaspis	Thelytoky
		Glyptholaspis	Arrhenotoky
		Holostaspella	Arrhenotoky
		Macrocheles	Arrhenotoky in most spp., thelytoky in some
Macronyssidae	Ectoparasites of birds and insects	*Ornithonyssus*	Arrhenotoky
		Ophionyssus	Arrhenotoky
Phytoseiidae	Predatory: feed on other mites, insects and pollen	*Amblyseius*	Arrhenotokous gynogenesis
		Iphiseius	Arrhenotokous gynogenesis
		Phytoseiulus	Arrhenotokous gynogenesis
		Typhlodromus	Arrhenotokous gynogenesis; one race of *T. guatemalensis* is thelytokous
Podocinidae		*Podocinum*	Arrhenotoky in one sp.; thelytoky in one sp.
Veigaiaidae		*Veigaia*	Thelytoky inferred
METASTIGMATA			
Argasidae	Ectoparasites of vertebrates	*Ornithodoros*	Tychoparthenogenesis
Ixodidae	Ectoparasites of terrestrial vertebrates, feeding on blood	*Amblyomma*	Sometimes produce many parthenogenetic offspring; *A. agamum* may be obligatorily thelytokous

	Boophilus		Tychoparthenogenesis
	Dermacentor		Some races or strains produce large proportion of parthenogenetic offspring
	Haemaphysalis		*H. longicornis* includes a mixture of amphimictic and thelytokous races
	Hyalomma		Tychoparthenogenesis
PROSTIGMATA			
Cheyletidae	*Cheyletus*	Predators of other mites	Automixis with fusion of cleavage nuclei
Cloacaridae		Endoparasitic in reptiles	Arrhenotoky inferred
Demodicidae	*Demodex*	Ectoparasites of vertebrates	Arrhenotoky
Eriophyoididae	*Aculus*	Plant parasites	Arrhenotoky
	Vasates		Arrhenotoky
	Phyllocoptruta		Arrhenotoky
Eylaidae	*Eylais*	Free-living predatory aquatic mites	Amphimictic with an achiasmatic meiosis in males
Harpyrhynchidae	*Harpyrhynchus*	Parasitic	Arrhenotoky
Hydrodromidae	*Hydrodroma*	Free-living predatory aquatic mites	Amphimictic with an achiasmatic meiosis in males
Pterygosomatidae	*Geckobiella*	Ectoparasitic	Arrhenotoky
	Pimeliaphilus		Probably arrhenotokous
Pyemotidae	*Pyemotes*	Parasitic on insect larvae	Arrhenotoky
	Siteroptes		Arrhenotoky
Syringophilidae	*Syringophiloidus*	Commensals of birds, feeding on other mites	Arrhenotoky
Tarsonemidae	*Hemitarsonemus*	Plant parasites	Arrhenotoky
	Steneotarsonemus		At least sometimes thelytokous, but males occur
	Tarsonemus		Arrhenotoky
Tenuipalpidae	*Brevipalpus*	Plant parasites	Thelytoky; perhaps automictic with a premeiotic doubling

Table 3.7 (*cont.*)

Family	Ecology	Genus	Reproduction
Tetranychidae	Plant parasites (spider mites)	*Raoiella*	Arrhenotoky in one sp.; thelytoky in three spp.
		Bryobia	Arrhenotoky in one sp.; thelytoky in two spp.
		Oligonychus	Arrhenotoky, with one species apparently thelytokous
		Petrobia	Arrhenotoky in one sp.; thelytoky in one sp.
		Tetranychus	Arrhenotoky; occasional thelytokous clones. One report of deuterotoky
		Tetranysopsis	Thelytoky
		Ten other genera	Arrhenotoky
ASTIGMATA			
Anoetidae		*Anoetus*	Arrhenotoky
		Histiostoma	Arrhenotoky, thelytoky and arrhenotokous gynogenesis. Thelytokous race of *H. feroniarum* is apomictic
Listrophoridae		*Histiophorus*	Deuterotoky
CRYPTOSTIGMATA			
(many families)	The orabatids: free-living terrestrial mites, usually detritivores, fungivores or carnivores	*Carnisia*	Thelytoky
		Dameobelba	Thelytoky
		Nanhermannia	Thelytoky
		Nothrus	Thelytoky
		Oppia	Thelytoky
		Platynothrus	Automixis with suppression of meiosis II
		Trhypochthonious	Automixis with suppression of meiosis II

Source: Oliver (1971, 1977).

which are probably apomictic, although a temporary synapsis of uncertain genetic significance has been observed in Dead Sea material (Barigozzi, 1944; Goldschmidt, 1952). More commonly the parthenogenetic cytotypes are diploid and automictic; the automixis is probably equivalent to a suppression of the second meiotic division, but the details are still in dispute (see Artom, 1931; Stefani, 1967; and reviews by Narbel-Hofstetter, 1964). Amphimictic strains are always diploid. Several different cytotypes — amphimictic and parthenogenetic, diploid and polyploid — may occur at the same locality.

NOTOSTRACA. The tadpole shrimps inhabit transient ponds, although *Lepidurus arcticus* is often found in large alpine lakes. Most species are gonochoric amphimicts, but some populations of *Triops cancriformis*, *T. longicaudatus*, *Lepidurus apus* and *L. arcticus* consist largely of internally self-fertilizing hermaphrodites, producing both oocytes and immotile spermatozoa in an ovotestis (Bernard, 1889; Longhurst, 1954, 1955). The reproductive mode of *T. cancriformis* varies regularly with latitude: North African populations are amphimictic, but the proportion of males declines to the north, and the northernmost populations, in Britain and Sweden, seem to be exclusively parthenogenetic (Mathias, 1937; Longhurst, 1955).

CONCHOSTRACA. The clam shrimps live in transient ponds and in more permanent bodies of water, including the littoral zone of lakes. Most are gonochoric amphimicts, but males are unknown in some species of *Limnadia* and *Cyclestheria*, and since large samples have been collected these species are almost certainly parthenogenetic (see Mattox and Velardo, 1950; Nair, 1968). *Limnadia lenticularis* is automictic, the pronucleus fusing with the first polar body and thus, in effect, suppressing the first meiotic division; the second division takes place normally, with a diploid polar body being extruded (Zaffagnini, 1969).

CLADOCERA. The water fleas often dominate the macroscopic zooplankton of lakes and ponds, besides including many epiphytic and semibenthic forms. A typical cladoceran life history comprises a variable number of parthenogenetic generations culminating in a sexual generation. Parthenogenetic eggs are produced by mitosis (Mortimer, 1936). Some authors (Ojima, 1958, cited by Hebert, 1978; Bacci *et al.*, 1961) have reported meiotic phenomena such as synapsis and disjunction in some strains of *Daphnia pulex*, but Hebert and Ward (1972) failed to detect any segregation amongst the progeny of individuals heterozygous for isozyme loci in *D. magna*, and it seems likely that parthenogenesis in cladocerans is thoroughly apomictic. The sexual generations are invariably amphimictic and gonochoric: diploid males and females are produced parthenogenetically, and go on to develop sperm and ova by a normal meiosis (Mortimer, 1936; Ojima, 1958, cited by Hebert, 1978). Parthenogenetic females usually produce single-sex broods (Cuvier, 1833). Males are smaller than females, but are not reduced to nearly the same extent as male rotifers. The fertilized eggs develop into embryos within a thick-walled structure called the ephippium, which is resistant to freezing, drying and digestive enzymes, and which normally undergoes a diapause of variable duration. The individuals hatching from ephippial eggs are always parthenogenetic females. The timing and regulation of sex in cladocerans has been a favourite research topic since Weismann's essay of 1879, and is reviewed at greater length below.

Discussion. I shall not spend much time in reviewing studies of sex in laboratory culture, since for once there is substantial agreement on the subject. In *Daphnia* sexual eggs can be produced at any time during adult life, but in some cladocerans their production is more restricted — in *Moina*, for example, only parthenogenetic eggs are produced unless the first are sexual (Banta and Brown, 1939). The stimuli for the production of males and of ephippial females are rather different, though they may well be connected in nature. The production of males can be affected by many aspects of the physical environment but, within generous limits set by nutrition and temperature, male production is elicited by crowding (Grosvenor and Smith, 1913; Banta and Brown, 1929a, b, 1939; Berg, 1931, 1934, and references therein). Banta and Brown found that ephippial females appear after males, when the food supply is low, but not so low as to prevent reproduction altogether. These results were confirmed by a detailed study of the dynamics of laboratory populations of *Daphnia* undertaken by Slobodkin (1954). He found that males were produced during or shortly after population maxima, just after the rate of parthenogenetic-egg production had peaked, whilst ephippial females appeared somewhat later when the rate of egg production had further declined. However, ephippial females disappeared from populations existing under persistently low food levels, with the remaining females producing small numbers of parthenogenetic eggs. In short, the sexual periods of laboratory populations are elicited by the nutritive stresses imposed by high population density.

For some years after the discovery that cladocerans alternate between sexual and asexual reproduction there was much debate about whether the reproductive cycle was regulated internally, with any given population producing sexual individuals after a certain fixed number of generations regardless of external conditions. This earliest literature has been reviewed by Wesenberg-Lund (1908). The notion of an inherent cycle was abandoned about the turn of the century — curiously, the controversy was one in which Herbert Spencer was right and August Weismann wrong — being finally discredited by the discovery that sex could readily be elicited in the laboratory by manipulating culture conditions. It should be remarked in passing, though, that there does seem to be some genetic variation in the ability of clones to produce male eggs (Berg, 1931; Banta and Brown, 1939; Banta and Wood, 1939). In more recent times the seasonal occurrence of sexuality in natural populations of cladocerans has been routinely documented by limnologists. Unfortunately, the innumerable reports in the literature have never been adequately collated; and rather than attempting the task myself I shall instead describe in some detail the results of studies on the well-known genus *Daphnia*, before dealing much more briefly with other genera. The discussion below is based on the classic papers by Wesenberg-Lund (1926) and Berg (1931), whose observations are summarized without further citation.

Species of *Daphnia* (including *Ctenodaphnia* s.l.) can be roughly arranged into an ecological series, with the inhabitants of small transient ponds at one end and the open-water plankters of large lakes at the other. *D. magna* lies at the former extreme, being most often found in small eutrophic ponds and ditches which are liable to dry up in summer and freeze solid in winter. During winter the population usually comprises last year's ephippia, but a few parthenogenetic females may survive mild winters. In spring, parthenogenetic reproduction quickens, and the population increases in density to a maximum. In small transient ponds ephippial females are almost always present, being more common if the water level is dropping and less common if the pond has been diluted recently by rainfall, so that pronounced sexual periods cannot be recognized. In more permanent ponds the spring maximum may be maintained through the summer, but the species usually decreases to a minimum in July before again increasing to a maximum in the fall. The maximal populations may be very dense; either the spring or the fall maximum may be the greater. First males and then ephippial females are found during or shortly after the maxima; if there is only one maximum there is only one sexual period. In dicyclic populations (those with two sexual periods) the spring sexual period is usually the more intense. Green (1955) found large numbers of males and ephippial females during population maxima in May and October in a very dense population inhabiting a sewage settling tank. At times of minimal population density, either in summer or in winter, only parthenogenetic females are present. If the spring maximum is prolonged into summer, males and ephippial females can be found throughout summer, but they become less numerous as the season advances. The fall sexual period may also be prolonged, with a few males and ephippial females being found as late as February.

D. atkinsoni is a southern species which Berg discovered in a few Danish localities. It has a large spring maximum during April–June, and often disappears entirely during the summer; sometimes, however, it may be very common in fall, and may even overwinter as parthenogenetic females beneath the ice. The sexual period occurs in spring at the time of the population maximum; very few males or ephippial females were seen in fall even when the species was abundant then.

D. pulex is a common and widely distributed species which shares village ponds with *D. magna* but is also found in a variety of unpolluted habitats from peaty forest pools to the littoral zone of lakes. After overwintering as ephippia or parthenogenetic females, it rises to a maximum in April–June, and then may or may not rise to a second maximum in September–November, following a summer minimum. The maxima may be very intense, colouring the water of shallow pools. When only a spring maximum occurs, there is only a single sexual period. Rammner (1932, cited by Berg, 1934) observed an intense sexual period of *D. pulex* in a pond near Leipzig when the species was extremely abundant in June, and I have found ephippial females in a temporary forest pool near Montreal in May–June. Proszynska (1962) found ephippial females in a permanent pond in May. Armitage and Smith (1968) found that ephippial females were produced at peak population densities during summer in two artificially fertilized fishponds, but in a third pond made the unusual observation of ephippial females appearing in June, before the population maximum in July–August. In ponds which dry up in summer but fill again in fall there is a second sexual period associated with a fall population maximum. The same locality may be monocyclic in some years and dicyclic in others; for instance, the populations of small shallow ponds which normally exist only between December and June are monocyclic in most years, with a sexual period in June, but may be dicyclic when there is enough rainfall to keep the pond filled. Wesenberg-Lund reviews a number of studies which describe monocyclic or dicyclic life histories; she also claims that polycyclic populations have been described, but Berg points out that there is no conclusive evidence that more than two distinct sexual periods ever occur. In small eutrophic ponds, however, ephippial females may be found in nearly every month of the year. In permanent ponds which freeze to the bottom Wesenberg-Lund found only parthenogenetic females between May and September, with a sexual period occurring during the fall and early winter maximum. However, Uéno (1934) found that the population of one large permanent pond in Japan had a maximum and a sexual period in March–May, whilst the population of another was dicyclic, with sexual periods in spring and late fall. In both cases the males and ephippial females were rather scarce. On the whole, a switch from predominantly vernal to predominantly autumnal sexuality appears to be characteristic of populations at high latitudes or high altitudes; for example, sexuality occurs only during fall in the mountains of northern Sweden and in Greenland. On the other hand, populations in the Lünersee at 2000 m elevation and in a small pond 1000 m above sea-level in Japan were dicyclic (Uéno, 1934). In the far north, males of *D. pulex* become extremely rare; Reed (1964) describes a few from the Adelaide Peninsula, but his records are clearly exceptional. P. D. N. Hebert (personal communication) has recently made the surprising discovery that populations of *D. pulex* throughout southern Ontario are entirely acyclic, with males absent and the ephippia being formed amictically. I have yet to find male *D. pulex* in the adjacent area of southwestern Québec and, in culture, females from these localities form ephippia in the absence of males. Hebert's discovery suggests that amictic ephippal eggs may be much more widespread than has hitherto been suspected.

D. schodleri resembles *D. pulex* and is also usually found in small ponds. In a shallow eutrophic lake in Alberta its ephippia hatched in May, and the population reached a maximum in June, with a sexual period in June–July. There was a second sexual period in September, perhaps associated with a weaker fall maximum, and the population disappeared in November (Lei and Clifford, 1974).

Curiously, when *D. pulex* and *D. schodleri* occur in the open water of lakes they behave as cold stenotherms (Birge, 1898; Hall, 1962). Birge found the

D. schodleri of Lake Mendota in deep water near the thermocline. The species disappeared from the plankton between August 1894 and July 1895, but it then began to increase and was quite abundant through the winter of 1895. After a short decline in March it reached a maximum in May 1896, before decreasing during the summer and eventually becoming scarce by October. Males and ephippial females were found abundantly only during June 1896.

D. ambigua is another cold-water form which usually has a spring maximum and reproduces parthenogenetically in deep water during the summer (see Tappa, 1965). In a small eutrophic lake in Michigan it had a distinct maximum and a strong sexual period in May (Allan, 1977).

D. longispina is notoriously variable both in its morphology and in its reproductive biology. In small ponds there are either one or two sexual periods, associated with population maxima; if there is only one sexual period it almost always occurs in the fall, and the fall period is usually the more intense in dicyclic populations. However, Elgmork (1964) found a rather different pattern in a set of shallow but permanent ponds formed by flooding from a river. *D. longispina* was present irregularly between April and December; there was always a maximum in spring, but the species was also sometimes abundant in October, after a summer minimum. Ephippial females were seen in June but disappeared later in the year, and there was no fall sexual period. In a permanent pond in Japan, Uéno (1934) observed a sexual period corresponding to the spring maximum in March–May, but none corresponding to the weaker fall maximum. In some localities in Saxony, three or four distinct sexual periods have been reported (Wagler, 1912; Berg is sceptical of the claim). Herr (1917, cited by Uéno 1934) found sexual individuals in every month between April and November. Populations may be monocyclic in one year and dicyclic in another; in one locality, Wesenberg-Lund observed one year that the number of individuals increased slowly through the season, culminating in a maximal population and a sexual period in the fall, whilst in the following year there were two large maxima, in June and in October, both associated with sexual periods and separated by minimal populations of parthenogenetic females. In larger lakes, planktonic populations of *D. longispina* may be monocyclic or dicyclic. In one Danish lake, Wesenberg-Lund (1908) found distinct maxima in both spring and fall, and caught ephippial females in July, when they were most frequent, and in October. In two other localities a few males were found between May and July but none in fall, and in a fourth locality a slight maximum in summer was followed by a sexual period in October–November. In all cases the sexual periods are inconspicuous, with only a small proportion of males and ephippial females; indeed, in some planktonic populations sexuality was never observed during several years of sampling. The same behaviour characterizes limnetic populations of *D. galatea*; in Lake Mendota, Birge (1898) found males both after the spring and, somewhat more abundantly, after the fall maxima, but the proportion of males in the population did not exceed about four per cent. In extreme climates in Iceland and Lapland *D. longispina* may have only two generations each year, the exephippial females being parthenogenetic and the second generation sexual.

D. cucullata is a fully planktonic daphnia, living in the open water of permanent ponds and lakes. In many localities no sexual period was observed; in others there was a sexual period in the fall, but this was usually inconspicuous, and only very rarely have males or ephippial females been found in the spring. In oligotrophic lakes in central Europe, *D. cucullata* is perennial and ephippial females are absent. Sebastyén (1948) did not find any males or ephippial females in Lake Balaton, though a few males appeared during October in a nearby marsh. *D. retrocurvata*, a plankter of large lakes, is a North American species whose ecology is similar to the European *D. cucullata*. It appears in June and reaches a maximum in August, after which the population declines before increasing to a second maximum in October, when the ephippial females appear. Wells (1960) found no ephippia in Lake Michigan, where the species may overwinter exclusively as parthenogenetic females. Birge (1898) found the species to be rare in Lake Mendota between November and June, with a maximum in October and perhaps sometimes also in August. Males were found very abundantly at the peak of the October maximum. An even more extreme form is *D. longiremis*, which inhabits ponds at high latitudes but is planktonic in the depths of large lakes to the south. It is perennial, and sexuality is almost unknown. The related northern species *D. cristata* has population maxima in June and December, often with two sexual periods of low intensity (Freidenfelt, 1913).

In high Arctic lakes, *D. middendorffiana* produces amictic ephippia, males being entirely lacking (Olofsson, 1913; Edmondson, 1955). Males are, however, common in the very dense populations of nearby tundra pools, which freeze solid in winter (Edmondson, 1955). More southerly populations also produce males, although Banta (1926; see also Schrader 1926) reported the amictic formation of ephippia in a Long Island population. I have already described the amictic production of ephippial eggs in the related *D. pulex*. The Australian *D. cephalata* is also obligately parthenogenetic, producing ephippia asexually (Hebert, 1977). Both *D. middendorffiana* and *D. cephalata* are able to produce males in the laboratory (Hebert, 1977, 1978). Furthermore, the fertilized eggs of *D. obtusa* may develop immediately instead of undergoing a diapause, just as parthenogenetic eggs normally do (Slobodkin, cited by Hutchinson 1967; see also Wood, 1932, on *Moina rectirostris*). We must conclude that whilst sexuality and the production of dormant eggs are normally associated, they are not necessarily associated; the two processes can be decoupled, and we must therefore seek an evolutionary explanation both for their normal association and for those exceptional cases in which this association breaks down.

I have outlined the reproductive periodicity of some north-temperate genera other than *Daphnia* in Table 3.8. This is not by any means an exhaustive survey, but I think that the data which it summarizes are representative. Taken together with the work on *Daphnia* it leads to some important generalizations about the timing and frequency of sex in natural populations of cladocerans.

The leading fact concerning the timing of sex is that the sexual period almost always occurs during or shortly after the period of maximal population density.

This agrees very well with the observation that sex can usually be elicited in laboratory culture by crowding. Moreover, it seems likely that high density in natural populations represents a real stress, since the sexual period often also coincides with a minimal rate of egg production (e.g., Berg, 1934; Uéno, 1934; Lei and Clifford, 1974; cf. Slobodkin, 1954). In populations which have two maxima in the same season, a sexual period may be associated with either or with both; if with both, the earlier sexual period is often the more intense. More generally, the frequency of sexual individuals seems to vary with the magnitude of the population maximum, although the relationship has never been adequately quantified (see, for example, Uéno, 1934, on *Daphnia longispina* and Elgmork, 1964, on *Bosmina obtusirostris*). When high population densities are maintained for some time after the maximum, males and ephippial females can be found for much longer periods of time than usual, but they become gradually fewer as the season grows later. Minimal populations consist exclusively of parthenogenetic females, no matter at what time of year they occur. Finally, sexuality seems to occur less frequently and with lower intensity in rarer than in more common species; this observation was made by Berg (1931), and seems to me so striking that I have reproduced his results as Figure 3.1. In short, it seems certain that sexual differentiation in natural populations is connected with fluctuations in population density, and not with any regular seasonal change in the physical environment.

The males of many cladocerans have never been described, and in some cases this may be the result of a genuine absence of sexuality, rather than being merely a reflection of inadequate sampling. In general, sex becomes less frequent and less intense towards the north and in larger bodies of fresh water; thus, males are rare among populations from the Arctic or from the open water of large lakes. On the other hand, there are obligately parthenogenetic populations of *Daphnia cephalata* along the southeastern coast of Australia and, despite an irritating paucity of data, males seem to be rare in large tropical lakes (see Jenkin, 1934).

3.30 Arthropoda: Crustacea non-Branchiopoda

Incidence of Parthenogenesis. The remaining crustacean groups are exceedingly diverse, but from our point of view can be dealt with quite briefly. There is no indication of any abbreviation of amphimixis in the Cephalocarida, Mystacocarida or Branchiura, nor in any of the malacostracan orders other than the isopods. I shall therefore confine this section to the four taxa in which some mode of parthenogenesis is known or has been suspected to occur.

OSTRACODA. The ostracods are small bivalve crustaceans, typically benthic organisms found in still and running freshwater and in shallow marine habitats. Marine ostracods are perhaps always amphimictic, but the fauna of brackish and

Table 3.8: Reproductive Periodicity in some Genera of North-temperate Freshwater Cladocerans. The final column of the table lists authorities adopting the abbreviations given in the list of sources, below. Other abbreviations: PF, parthenogenetic females; EpF, ephippial females; M, males; max, population maximum; min, population minimum; and names of months.

Genus	Ecology	Reproduction	Author
SIDIDAE			
1. *Diaphanosoma*	Pelagic in larger lakes; also found in ponds and littoral. Hatches from ephippia in spring, does not become abundant until July–Aug.; max in late Aug. to early Sept., disappears in Oct.	M, EpF in Sept., sometimes into October	Be, Bi
2. *Latona*	Rare; found near bottom along shores of small lakes and ponds. Does not overwinter. Max in fall	Sexual period late in fall, max during Sept.–Nov.	Be
3. *Sida*	Littoral of lakes and permanent ponds. Hatches from ephippia in April–May, common through summer with max in Aug.–Sept.	M, EpF in Sept.–Oct. or Nov.	Be, DD, Seb
DAPHNIIDAE			
4. *Ceriodaphnia*	Typically littoral or in small ponds; also in peatbogs and transient ponds; occasionally pelagic in large ponds and lakes. Does not overwinter; common through summer with max usually in fall but sometimes as early as June	Sexual period usually in fall at time of max. EpF of *C. reticulata* found during July max in fishponds, and in June–Oct. in small ponds with June–Aug. max. *C. pulchella* may have spring sexual period. *C. laticauda* may be sexual early in year in transient habitats. Males unknown in *C. acanthina* (found in Manitoba in weedy ponds) and *C. lacustris* (throughout USA, pelagic in lakes)	Be, Br, AS, Pr
5. *Moina*	Shallow muddy or eutrophic ponds. Does not overwinter; appears in July and disappears in October	Berg found M and EpF throughout season, but most abundantly in Sept.–Oct. AS found EpF at time of max in early Aug.	Be, Br, AS

6. *Scapholeberis*	Near surface along shores of lakes and ponds. Does not overwinter; appears in April–May and may be abundant by July, disappearing in Oct. or Nov.	Sexual period in Sept.–Oct. or Nov.; sometimes dicyclic with a second sexual period during June–July in permanent ponds and lakes	Be, Seb
7. *Simocephalus*	Typically along shores of lakes and ponds, also in marshes; rarely found in open water of lakes or in transient ponds. May overwinter as PF. Max may occur in June–July or in Sept.–Nov. or both. Often very abundant	Sexual period near time of max, in spring or fall; Berg described dicyclic populations of *S. vetulus*, *S. exspinosus* and *S. serrulatus* in which M and EpF were found in both spring and fall. Seb did not observe a sexual period in a littoral population of L. Balaton	Be, Pr, Seb
BOSMINIDAE			
8. *Bosmina*	Planktonic, in open water of permanent ponds and large lakes. May overwinter as PF. Often diacmic with spring max in May–June and fall max in Sept.–Nov., but much variation between years and localities; one max may be much weaker than the other, or even entirely absent, and it is sometimes abundant throughout summer	Diacmic populations are dicyclic; e.g., Berg found sexual periods of *B. longirostris* during June–July and Oct.–Nov. in smaller ponds; in same species WL found a few M and EpF during a strong spring max but none during a much weaker fall max in the same year. In *B. obtusirostris* Elgmork found a large population only once, in Nov., when it included a high frequency of males; a nearby pond had smaller but still appreciable numbers of the same species at this time, with a lower frequency of males. Sexuality occurs during pronounced fall max in *B. coregoni*, but not during weaker spring max. Pr says that maxima and sexual periods may occur at various seasons in small ponds. Planktonic populations of large lakes are often acyclic, with no M or EpF seen during long periods of intense sampling	Be, As, Pr, Elg, WL
MACROTHRICIDAE			
9. *Acantholeberis*	*Sphagnum* bogs. Max usually in fall	Sexual period shortly after fall max, in Sept.–Oct. (Be) or late Nov.–Dec. (DD). Berg reports that it is dicyclic in some localities with M and EpF in June and fall	Be, DD

Table 3.8 (*cont.*)

Genus	Ecology	Reproduction	Author
10. *Drepanothrix*	Locally abundant in forest pools. May over-winter but usually rare in spring, with max in Aug.–Oct.	M and EpF common in Oct.–Nov.	Be
11. *Ilyocryptus*	Near bottom of permanent ponds and lakes. Appears in April, with max in Sept.	Sexuality is rare. Berg found no M or EpF in *I. sordidus*, and only a few M during July–Aug. in *I. agilis*. Br says males are unknown in *I. acutifrons* and *I. spinifer*, but DD saw a few males of *I. spinifer* in Nov, without EpF. Seb did not observe sexuality in *Ilyocryptus* spp.	Be, Br, DD, Seb
12. *Lathonura*	In weeds at edge of permanent ponds and lakes. Max in Sept.	M and EpF in Oct.–Dec.	Be
13. *Macrothrix*	Benthic in small ponds	Males unknown in *M. montana* (Rocky Mountains) and *M. hirsuticornis* (generally distributed). Seb found no sexuality in Balaton population of *M. laticornis*	Br, Seb
14. *Ophryoxus*	Max during Aug. in acid pond population of *O. gracilis*	M during Sept.–Oct., EpF during Oct.–Nov.	DD
15. *Streblocerus*	Fall max in *Sphagnum* pool population of *S. serricaudatus*	M and EpF in Sept.–Nov. Males unknown in North American *S. pygmaeus*	Be, Br
CHYDORIDAE			
16. *Acroperus*	In permanent ponds and littoral of lakes; also in transient ponds and peatbogs. May overwinter as PF and is often abundant in spring and early summer, with a summer minimum and a second max in fall	M and EpF commonly found towards end of fall max in Oct.–Dec.; a weak sexual period has been observed in June	Be, DD, Seb

17. *Alona*	On submerged vegetation and near bottom in open water of permanent ponds and lakes; often abundant. Probably overwinters as PF; increases in spring and may remain abundant through summer, but there is often a second max in fall following a summer min	M and EpF most commonly reported in fall, during Oct.–Dec.; Berg mentions a weak sexual period in some species in June near end of spring max. In monacmic populations there is always a max and sexual period in fall	Be, DD, Pr, Seb
18. *Alonella*	In ponds, bogs and lake littoral, sometimes very abundant. Does not overwinter; appears in May and is usually scarce through summer with a max in fall; *A. nana* sometimes has a second max in spring	Sexual period in fall, usually Oct.–Nov.	Be, DD, Seb
19. *Alonopsis*	Weedy pools and lake littoral. May be abundant in late summer	M and EpF found in Sept.–Nov., probably near end of a fall max. Males unknown in *A. aureola* (rare, weedy pools in Maine)	Be, Br
20. *Camptocercus*	Permanent ponds and littoral of lakes. Does not overwinter; appears in April–May, becomes abundant during summer, reaches max in early fall	M and EpF found in Sept.–Nov.; sometimes dicyclic with a weaker sexual period in summer	Be, Pr
21. *Chydorus*	*C. sphaericus* is extremely abundant and widespread, and is often planktonic in the open water of lakes; it and other species may also inhabit ponds and the littoral. Often overwinters as PF; max may occur in June–July or in Sept.–Oct. or both	Berg found M and EpF in Oct.–Dec.; in one locality which dried out in July–Aug. there was a single sexual period in May–June, with neither M nor EpF seen in sparse fall population. DD found summer max in June–July with sexual period well afterwards in Oct.–Dec. Weismann said that the species can be polcyclic. Planktonic populations sampled by Birge and others are often acyclic, with no sexual individuals seen over long periods of time. Sebastyén found no sexual period in a marsh population. Males unknown in some species	Be, Bi, Br, DD, Seb
22. *Eurycercus*	In submerged vegetation in permanent ponds and the littoral of lakes. May overwinter; increases to max in summer	M and EpF in Nov., some time after max; but Sebastyén reports males in Aug.	Be, Seb

Table 3.8 (*cont.*)

Genus	Ecology	Reproduction	Author
23. *Graptoleberis*	Bottom-living in ponds and littoral of lakes. Usually scarce until late summer or fall; spring max observed in one locality	M and EpF during Sept.–Nov. or Dec. No spring sexual period recorded	Be
24. *Monospilus*	On muddy bottoms. Max in fall	M during Sept.–Oct. Sometimes sexual period lacking	Be, Seb
25. *Peretacantha*	In lakes and ponds; very common. Max in fall	M and EpF in Oct.–Nov., rarely in Sept. and Dec.	Be
26. *Pleuroxus*	Bottom-living in lakes, ponds and marshes. May be abundant from June till fall, but usually a more or less distinct fall max	M and EpF in Oct.–Nov.; EpF of *P. aduncus* also found in Aug. Males rare or absent in two North American spp.	Be, Seb, Br
27. *Rhynchotalona*	Near bottom, especially on sandy bottoms. Abundant in late summer and fall	M and EpF in Oct.–Nov. or Dec. Earlier sexual period observed in alpine lakes	Be
POLYPHEMIDAE			
28. *Polyphemus*	In marshes ponds and lake margins. May be abundant from May till Sept.	M seen in June and Sept.–Oct.	Be
LEPTODORIDAE			
29. *Leptodora*	Large predatory cladoceran, planktonic in lakes. Dynamics irregular, but max usually during Aug.–Oct., followed by decline and disappearance between Dec. and June	M in Oct., some time after peak of max; but Berg reports M during Aug.–Oct.	Be, Bi

Source: Principally Berg (1929; Be), but also Armitage and Smith (1968; AS), Birge (1898; Bi), Brooks (1978; Br), Daggett and Davis (1974; DD), Elgmork (1964; Elg), Proszynska (1962; Pr), Sebastyén (1948; Seb) and Wesenberg-Lund (1908; WL).

Figure 3.1: Abundance and Intensity of Sexual Reproduction in Cladocerans. The two curves are frequency distributions of abundance (as assessed by Berg, with *rrr* being very rare and *ccc* very common) for cyclic (solid line) and acyclic (broken line) cladocerans.

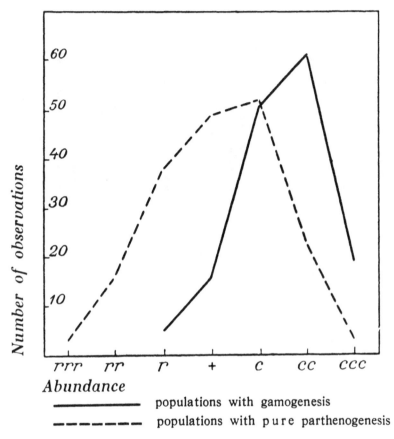

Source: Berg (1931), p. 185, Figure 14.

fresh water includes a large number of parthenogenetic species; Table 3.9 and Figure 3.2 show the patchy way in which these are distributed through the group. Bauer (1940) states that *Heterocypris incongruens* is apomictic, but the cytology of ostracods is little known, and apomixis may not be the general rule. So far as is known, all ostracods are diploid. In some species both amphimictic and parthenogenetic populations have been described, and it seems likely, though it has not been proven, that some populations include both sexual and parthenogenetic females. Weismann (1880) suggested that ostracods have a cyclical parthenogenesis similar to that of cladocerans, and Wohlgemuth (1914) claimed that sexuality could be elicited in laboratory populations of parthenogenetic

Table 3.9: Mode of Reproduction in some Genera of North-temperate Freshwater Ostracods.

Genus	Habitat	Reproduction
CYPRIDAE		
1. *Candocypria*	Small streams	Male described by Tressler
2. *Candonocypris*	Permanent and temporary still and running water	Males unknown in 2/6 N. American spp.
3. *Chlamydotheca*	Typically in small temporary ponds and ditches	Males unknown in 6/7 N. American spp.
4. *Cypretta*	Pools	Males unknown in 3/5 N. American spp.
5. *Cypricercus*	Small weedy ponds and swamps, lake littoral; *C. reticulatus* inhabits temporary ponds	A typically amphimictic genus. Males unknown in 4/15 N. American spp., and in all four Norwegian spp. described by Sars. Males rare in *C. reticulatus* but seem to become more common to north; Alm records them from Siberia and Greenland
6. *Cypridopsis*	Found in a variety of habitats and often very abundant. *C. vidua* is perhaps the commonest and most widespread Holarctic ostrocod	Many spp. are parthenogenetic, including *C. vidua*. Only 5/31 spp. listed by Ferguson (1959) which do not have type localities in N. America have males; all are from Africa. 5/16 spp. listed by Ferguson (1964) for N. America have males; all are from Yucutan or the Gulf States
7. *Cyprinotus*	Often in very small transient ponds and ditches; also in running water. *C. incongruens* is especially abundant and widespread. *C. salinus* is found in brackish water	Originally defined to include amphimictic spp. only. Males unknown in 6/15 N. American spp. (*inconstans, scytoda, pellucidus, salinus, fretensis, symmetricus*). Males of *C. incongruens* rare in many localities, and not found in Sweden by Alm. Sars found no males of *C. salinus* and *C. fretensis* in Norway
8. *Cypris*	Temporary ponds and swamps	Males unknown
9. *Dolerocypris*	Grassy pools	Males unknown in Norwegian *D. fasciatus*
10. *Eucypris*	Temporary ponds and swamps	Males virtually unknown
11. *Herpetocypris*	Small ponds, tarns and ditches	Males unknown in all four N. American spp. and in Norwegian *H. reptans*

	Genus	Habitat	Males
12.	*Heterocypris*	May be very abundant in small ponds and ditches	Males unknown in cosmopolitan *H. incongruens*; males recorded from two S. African spp.
13.	*Ilyodromus*	Weedy ponds and sluggish streams	Males unknown in sole N. American sp.
14.	*Prionocypris*	Small ponds, swamps and lake littoral	Males unknown in four Norwegian spp.
15.	*Prionocypris*	Ditches and small streams. *P. glacialis* is an arctic sp. living in water-holes in Spitzbergen	Males unknown in the three Norwegian spp. described by Sars, including *P. glacialis*. Males described for *P. longiforma* (USA) but not for *P. canadensis* (Canada)
16.	*Potamocypris*	Widespread, usually in permanent ponds and streams	Males unknown in 4/7 N. American spp. Hoff found no males in 44/50 samples of *P. smaragdina*; in remaining six samples males and females were equally abundant
17.	*Stenocypria*	Muddy bottom of lakes	Males unknown in sole species, *S. longicomosa*
18.	*Stenocypris*	Pools	Males unknown in both N. American spp.
19.	*Strandesia*	Pools; southern analogue of *Cypricercus*	Males very scarce, recorded for *S. intrepida* only

CYCLOCYPRIDAE

	Genus	Habitat	Males
20.	*Cyclocypria*	Permanent ponds and lakes	Males unknown
21.	*Cyclocypris*	Temporary and permanent still waters, from small vernal pools to several metres depth in lakes	Predominantly amphimictic; males usually described though sometimes uncommon. Sars does not remark any deficiency of males in four Norwegian spp.
22.	*Cypria*	Widespread in still and running water, often abundant. *C. turneri* may be abundant in temporary ponds, but *C. lacustris* is found in deep water in large lakes	Males usually abundant, but often uncommon in *C. opthalmica* and very uncommon in *C. turneri*
23.	*Physocypria*	Ponds and the littoral of lakes; some spp. in streams	Males unknown in 5/10 N. American spp. but, except for *P. dentifera*, these are little-known southern forms; reproduction probably amphimictic in most common northern spp.

Table 3.9 (*cont.*)

Genus	Habitat	Reproduction
CANDONIDAE		
24. *Candona*	A very large, diverse and widespread genus found in all sorts of freshwater habitats but especially common in temporary pools and streams	Many species are purely amphimictic, but males are unknown in 10/35 N. American spp., and are in addition often scarce in *C. albicans* and *C. candida*. Dobbin found males in only 2/5 southern US spp. Alm failed to find males in Swedish populations of *C. candida* and *C. parallela*. Sars found males in 8/9 Norwegian spp., but their frequency varied greatly: unknown in *C. lapponica*, very scarce in *C. candida*, rather scarce in *C. albicans*, as abundant as females in *C. compressa*, and actually more abundant than females in *C. rostrata*
25. *Candonopsis*	Muddy creeks	Males as as abundant as females in sole sp.
26. *Cryptocandona*	Sars records *Cryptocandona* from a shallow ditch, but also to 40 m in a lake	Males unknown in both spp. described by Sars
27. *Paracandona*	Lakes	Males abundant
ILYOCYPRIDAE		
28. *Ilyocypris*	Permanent and intermittent streams; also in temporary ponds and ditches	Males unknown in *I. bradyi*. Males of *I. gibba* reported from N. Africa, Spain, Hungary and south-eastern Russia, but apparently absent from Germany, Norway and N. America. Males of *I. biplicata* recorded from N. America and N. Africa but not from Norway
NOTODROMIDAE		
29. *Cyprois*	Typically in temporary pools; also found in streams	Males abundant
30. *Notodromas*	Widespread but scarce. Permanent weedy pools and littoral; weedy streams	Males abundant

DARWINULIDAE

| 31. | *Darwinula* | Bottom-dwelling at considerable depths in large lakes | Males scarce and often not found; reproduction thought to be predominantly parthenogenetic |

LIMNOCYTHERIDAE

32.	*Cyprideis*	Sars records *C. littoralis* from a brackish lake	Males abundant in *C. littoralis*
33.	*Cytherissa*	At considerable depths in large lakes	Males unknown in *C. lacustris*
34.	*Limnocythere*	Typically in permanent streams and lakes; occasionally in temporary ponds	Males abundant in most spp. Males of *L. inopinata* found in USA but not in Canada or Norway

ENTOCYTHERIDAE

| 35. | *Entocythere* | Ectocommensal on freshwater crayfish | Males abundant. Male of Canadian spp. *E. insignipes* not yet described |

Source: Alm (1916); Klie (1926); Furtos (1933, 1935, 1936); Dobbin (1941); Hoff (1942); Ferguson (1959, 1964); Delorme (1970a–d, 1971); Hart and Hart (1974).

Figure 3.2: Distribution of Amphimictic and Parthenogenetic Species Between Genera of North American Ostracods. Each point represents a genus: numbers refer to genera listed in Table 3.9. Note extreme positions of *Entocythere* (35), *Cyclocypris* (21), *Candona* (24) and *Eucypris* (10). Diagonal line indicates equal numbers of amphimictic and parthenogenetic species in genus.

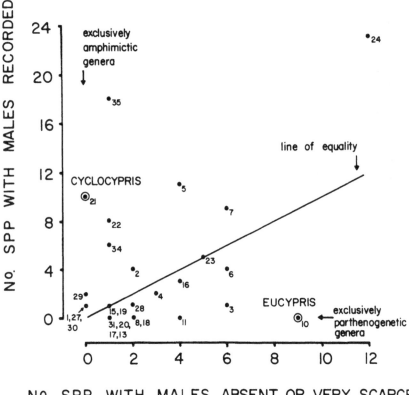

NO. SPP WITH MALES ABSENT OR VERY SCARCE

Source: Tressler (1959).

females. This opinion was accepted by Dobbin (1941). However, this view has been rejected by almost all other more recent authors, on the grounds that, in species which possess males, the males can be found throughout the year. The reproductive apparatus of amphimictic ostracods is extraordinary: the sperm may be ten times and the spermathecal duct a hundred times as long as the body of the adult. Lowndes (1935) claims that the sperm is functionless, and consequently that all ostracods are parthenogenetic, but his arguments are unsound. The resistant eggs produced by many ostracods may be either fertilized or unfertilized.

COPEPODA. The copepods are extremely abundant in the plankton of fresh-waters and the sea; the harpacticoids are benthic forms. Almost all copepods are gonochoric amphimicts; there are persistent references in the secondary literature (e.g., Charniaux-Cotton, 1960; Kaestner, 1970) to parthenogenesis in the harpacticoids *Elaphoidella* and *Epactophanes*, but I have been unable to locate a primary source.

CIRRIPEDIA. The barnacles include outcrossed hermaphrodites and gono-chores with dwarf males. The reproduction of isolated hermaphroditic individuals is normally limited by the maximum extension of the penis; individuals of *Balanus balanoides*, *B. crenatus* or *Elimnius modestus* which lie further than 5 cm from their nearest neighbour cannot copulate and are therefore almost always sterile. However, Barnes and Crisp (1956) found that isolated individuals of other species (*Chthalamus stellatus*, *Verruca stroemia* and *Balanus perforatus*) often did bear viable eggs, and concluded that they were capable of self-fertiliz-ation. The eggs of isolated individuals survived less well than those produced by contiguous individuals.

MALACOSTRACA: AMPHIPODA. There is a single unconfirmed report of parthenogenesis in some populations of the amphipod *Rhabdosoma* (Fage, 1954).

MALACOSTRACA: ISOPODA. The terrestrial, freshwater and marine wood-lice and sowbugs, almost all of which are amphimictic gonochores. Parthenogen-esis is known only from terrestrial isopods. The common and widespread species *Trichoniscus pusillus* has a parthenogenetic race, which has been studied by Vandel (1928, 1931) and is discussed further below. Parthenogenetic females are apomictic triploids, whereas other races of *T. pusillus*, and other species of *Trichoniscus*, are all amphimictic diploids. Occasional triploid males occur among the offspring of parthenogenetic females; they produce functionless triploid sperm. *T. pusillus* is also well known for its habit of producing offspring of one sex only (monogeny), and population sex ratios are often highly skewed, with an excess of females. *Nagara modesta* has a thelytokous, probably apomictic, race on Christmas Island, and an amphimictic race on the islands of Java, Sumatra and Salaja (Hill, 1948). Arcangeli (1925) has described an extraordinary (but so far unconfirmed) life history in Sardinian populations of *Philoscia elongata*. Some individuals are protandrous hermaphrodites, breeding as males in their first year of adult life and thereafter as females, whilst others are female through-out life; according to Arcangeli, members of the latter class of 'true females' reproduce amphimictically during their first and thelytokously during their second season.

Discussion. The North American ostracod fauna includes 79 species in 25 genera for which males are unknown or very scarce (see Tressler, 1959). The entire fauna comprises only 191 species in 30 genera, so very few genera consist wholly of amphimictic or wholly of parthenogenetic species; as Figure 3.2 shows, most include a mixture of the two modes of reproduction. Within this large natural group of organisms, therefore, almost every sexual species has a close asexual relative, and vice versa. Moreover, several species have both sexual and asexual races; and in some species sexual and asexual populations occur in the same geographical area. This is a most unusual situation, and one which offers an unrivalled opportunity to identify the ecological correlates of sexuality.

In the first place, parthenogenesis is much more common in fresh water than in the sea. Table 3.10 makes this point by summarizing data originally published by Sars (1928). Table 3.10 greatly understates the case, however, since many marine species are listed as having no males on the basis of very small samples: of the 21 species in this category, eight were represented by between one and three individuals, eight by 'a few' or 'only a few' and three by 'some', 'some few' or 'several', while for the other two species the sample size was not specified. Elofson (1938) also found that there were often no males in small samples of rare marine ostracods, and Poulsen (1962, 1965) has definitely established that females are usually more frequent than males in marine species. However, the preponderance of adult females seems to be adequately explained by their lesser rate of mortality (Thiesen, 1966) and so far as I know there is no compelling evidence for parthenogenesis in any marine ostracod.

Table 3.10: Distribution of Males Between Samples of Marine and Freshwater Ostracoda Podocopa.

	Marine	Freshwater
Males known	72 spp. in 26 genera	21 spp. in nine genera.
Males unknown	21 spp. in ten genera	33 spp. in 17 genera.

Source: Sars (1928).

In fresh water, parthenogenetic species tend to predominate in temporary, and sexual species in permanent habitats. Table 3.11 summarizes the habitat descriptions given by Tressler (1959) for those North American genera which contain five or more species. It is obvious that, as we pass from genera like *Cyclocypris*, *Cypria* and *Limnocythere*, which are made up largely of sexual species, to genera such as *Chlamydotheca* and *Eucypris*, where parthenogenetic species are in a majority, we pass from the communities of lakes, rivers and large ponds to those of transient pools and ditches. *Cyclocypris* and *Eucypris* represent extremes among free-living ostracods: *Cyclocypris* has ten species in North America, and all are amphimictic; *Eucypris* has nine, all of which are parthenogenetic. In Table 3.12 I have abstracted some representative descriptions of the habitats of these two genera from various authors. They scarcely overlap: *Cyclocypris* is largely restricted to permanent and *Eucypris* to temporary habitats. Table 3.13 lists the number of collections from different types of habitat in

Table 3.11: Habitats of Free-living North American Freshwater Ostracods. Only genera with five or more species included. Figures in cells are the numbers of times that a particular habitat type is mentioned in Tressler's descriptions (Tressler, 1959). The classification of habitats attempts to express degree of permanence, as follows: (1) lakes and rivers; (2) ponds, streams, canals, small shallow lakes and littoral of lakes; (3) pools, rock pools, marshes and small shallow ponds; (4) ditches and temporary ponds, marshes and streams. Some habitat types (fish ponds, brackish water, springs and caves) do not fit readily into the classification and have been excluded: they comprise only seven per cent of the habitats named by Tressler.

Genus	Parthenogenetic spp. (%)	Habitat type			
		1	2	3	4
Cyclocypris	0	5	9	0	0
Cypria	11	4	5	2	0
Limnocythere	14	5	2	0	0
Cypricercus	27	1	7	8	0
Candona	34	8	22	12	13
Cyprinotus	42	6	10	6	6
Physocypria	50	4	4	3	0
Potamocypris	56	0	1	4	1
Cypretta	60	0	0	3	0
Cypridopsis	60	4	2	10	2
Chlamydotheca	84	0	3	3	1
Eucypris	100	0	2	1	5

which Hoff (1942) found sexual and parthenogenetic ostracods in Illinois. Species in which males are unknown were collected twice as often from temporary as from permanent habitats, whilst species in which males are common were collected twice as often from permanent as from temporary habitats. Species in which males, although known, are very much less frequent than females can probably reproduce by amphimixis and by parthenogenesis, and are found equally often in permanent and in temporary habitats.

We can sharpen the focus of this sort of comparative analysis by concentrating on a single genus. The best for our purpose is *Candona*, which has 35 North American species, of which twelve are parthenogenetic. Broadly speaking, species in which males are rare or unknown inhabit temporary ponds, marshes and ditches – *uligninosa, parvula, biangulata, albicans, candida* and *stagnalis* all fit this generalization. Conversely, species in which males are common usually inhabit permanent bodies of water. However, there are exceptions: *caudata* is a parthenogenetic species which may be found, in Canada at least, to considerable depths in lakes, while *suburbana, indigena, inopinata* and *fossulensis*, although typically found in transient habitats, are amphimictic. The best comparison would lie between different populations of the same species, but at this point

Table 3.12: Habitats of an Amphimictic Genus, *Cyclocypris*, and a Partheno-genetic Genus, *Eucypris*.

Cyclocypris	Eucypris
1. *ampla*. Ponds, marshes, small lakes, cold streams (Tressler)	1. *affinis*. Temporary ponds and marshes (Tressler)
2. *cruciata*. Ponds, lakes (Tressler)	2. *crassa*. Pools in grassy swamps (Tressler). Shallow grassy swamp (Sars)
3. *forbesi*. Lakes, ponds (Tressler): Woodland pond (Sharpe). Mats of vegetation from lake (Hoff)	3. *elliptica*. Small pond (Sars)
4. *globosa*. Lakes (Tressler). Small temporary pools and ditches (Sars)	4. *fuscatus*. Temporary ponds (Tressler). Shallow grassy ponds and swamps (Sharpe)
5. *laevis*. Shallow ditches and pools (Sars). Weedy streams, ponds and swamps (Sharpe)	5. *hystrix*. Small weedy temporary ponds, in woods (Tressler)
6. *nahcotta*. Lakes, pools (Tressler)	6. *rava*. Small streams (Tressler)
7. *ovum*. Ponds, edges of shallow lakes, marshes (Tressler)	7. *reticulata*. Small temporary grassy pools (Tressler)
8. *serena*. Several metres depth in large lake (Sars)	8. *virens*. Weedy ponds (Tressler). Grassy ponds and swamps (Sars). Weedy ponds (Sharpe)
9. *sharpei*. Ponds, lakes, marshes (Tressler). Ponds and swamps (Sharpe). Ponds, marshes and lakes (Furtos)	
10. *Washingtonensis*. Ponds (Tressler)	

Source: Sharpe (1918); Sars (1928); Furtos (1933); Hoff (1942); Tressler (1959).

Table 3.13: Habitats and Genetic Systems of Illinois Ostracods. Entry in each cell gives the number of collections corresponding to the description: thus, asexual species were recorded 35 times in samples from temporary ponds. The data abstracted here excludes *Cypridopsis vidua*, which made up nearly 40 per cent of all collections, was found in all types of habitat, and whose inclusion would seriously bias the table.

Males	Still water: Temp.	Perm.	Running water: Temp.	Perm.	Total: Temp.	Perm.
Unknown	35	22	25	11	60	33
Rare or local	40	54	82	109	122	163
Always present	62	186	60	59	122	245

Source: Hoff (1942), Table 1 and list on pp. 36-7.

we run out of material. In 50 samples of *Potamocypris* taken from Illinois by Hoff (1942), 44 contained no males, while in the remaining six samples males and females were about equally abundant; but Hoff noticed no difference between the habitats occupied by sexual and asexual populations.

In short, the incidence of parthenogenesis in ostracods increases as we pass from the sea to freshwater, and as we pass from permanent to temporary bodies of freshwater. However, the habitat descriptions on which my analysis has been based are crude, and the opportunity offered by the ostracods for studying the relationship between habitat and genetic system deserves to be much more fully exploited.

I should mention in passing some unusual habitats occupied by ostracods. Six species are listed by Tressler (1959) from cenotes and caves in North America. Four are from Yucutan and all have males; the other two are species of *Candona* from Illinois in which males are unknown. Ganning (1971) found that all three species of ostracods inhabiting small brackish rock pools by the shores of the Baltic were parthenogenetic. Members of the large genus *Entocythere*, which are ectocommensal, or ectoparasitic, on crayfish, are probably all amphimictic. Of the five species which Tressler (1959) reports as having succeeded in invading the novel habitat provided by canals, four are parthenogenetic (*Cypricercus horridus, Candona caudata, Chlamydotheca speciosa* and *C. arcuata*) and only one amphimictic (*Candona distincta*). The only truly terrestrial ostracods are species of *Mesocypris* found in Africa (Harding, 1953) and New Zealand (Chapman, 1960, 1961). The New Zealand species, *M. audax*, is found in the forest-floor leaf litter, and occupies quite harsh environments — subantarctic *Notofagus* forest and even *Sphagnum* moss. Although it is quite plentiful, no males have ever been found.

Several authors state that the males of ostracods become less frequent to the north. In some cases this claim seems to be true. The males of *Ilyocypris biplicata* are absent from Scandinavia (Alm, 1914; Sars, 1928) and from most of North America, but they have been found in Texas, and Sars isolated some from a sample of dried mud from North Africa. The males of *Cypricercus reticulatus* are generally rare in North America, occurring only very occasionally in Illinois (Sharpe, 1908; Hoff, 1942), but they have been found in Newfoundland, Greenland and Siberia (Alm, 1914). In *Cyprinotus incongruens* the males are rare in Germany, are more common in central Europe, and are as abundant as females in North Africa (Klie, 1926). In *Heterocypris*, males are very rare in Europe and North America but have been described from two South African species (Sars, 1928). At the level of whole faunas, however, this generalization breaks down. In Table 3.14 I have detailed the latitudinal trend in the numbers of parthenogenetic and sexual species, both in the entire ostracod fauna and in the genus *Candona* alone. In neither case is there any convincing increase in the proportion of parthenogenetic species to the north; if anything, indeed, there seems to be the converse tendency.

The isopod *Trichoniscus pusillus* is a classic instance of 'geographical parthenogenesis'. Throughout much of its range, amphimictic and parthenogenetic females occur together in the same habitat, although they do not interbreed, but the frequency of the triploid apomicts increases to the north, and at high latitudes they completely replace the sexual form (Figure 3.3). Moreover, this latitudinal replacement is associated with an ecological separation. Along the northern shore of the Mediterranean the parthenogenetic cytotype occurs in

Table 3.14: Latitudinal Variation in the Occurrence of Males in Free-living North American Ostracods. Names of States of USA abbreviated.

Approx. limits N. latitude	Political divisions	All ostracods: Males	No males	Candona: Males	No males
15–25°	Yucutan, Jamaica, Trinidad, Peurto Rico	12	16	0	0
25–30°	Mexico, Fla	9	19	1	1
30–50°	SC, Ga, Ala, La, Ari, Tex, Miss	13	17	3	2
35–45°	NC, Va, Tenn, Mo, Ka, DC, Okla, Co, Utah, Ca, Dela, Md	9	14	3	0
40–5°	Mass, NY, Penn, Ohio, Ind, Mich, Ill, Wis, Neb, Wyo, Idaho, Ore, NJ	58	38	20	8
45–50°	Minn, Mont, Wash, Nova Scotia, Ontario, Newfoundland	17	17	3	2
50–65°	Alaska, Pribilofs, Greenland, Alberta, Yukon, NW Territories	8	5	2	2

Source: Tressler (1959).

arid garrigues, where conditions are unfavourable and the species is rare, whilst the amphimictic race is found in more humid habitats. For example, the only colony of *T. pusillus* which Vandel found on the dry limestone massif of La Montagne d'Alaric consisted exclusively of parthenogenetic females; and Vandel (1928) gives several similar instances. In Britain the asexual race replaces the sexual to the north and west, predominating first in open habitats, and further to the north in woodland too (Sutton, 1968, 1972). In transitional areas apomictic females are found in grassland and sexual females and males in the neighbouring woodland. Thus, the parthenogenetic form becomes more common as we pass from south to north, and from more sheltered to more open habitats.

The triploid apomictic females of *T. pusillus* are much larger than the diploid amphimictic females.

3.31 Arthropoda: Myriapoda

Review. Enghoff (1976a).

Incidence of Parthenogenesis. The myriapods are long-bodied, many-legged terrestrial mandibulates. Most are gonochoric amphimicts, but parthenogenesis has been reported recently from several species and may be quite common; the group is too little-studied to be certain. The cytology of parthenogenesis is not known for any myriapod.

Figure 3.3: Latitudinal Trend in the Frequency of Males in European Populations of the Isopod *Trichoniscus pusillus*. In Zone I, no males are found; in Zone II, no more than five per cent males; in Zone III, between five and 15 per cent males; in Zone IV more than 15 per cent males.

Source: Vandel (1928), p. 192, Figure F.

PAUROPODA. Large collections of *Allopauropus proximus* from Ceylon consisted exclusively of females, and the species is probably thelytokous (Scheller, 1970). Males are very scarce or absent in other species of *Allopauropus*, *Rabaudauropus* and *Scleropauropus*, but sample sizes are usually much too small for the paucity of males to be accepted as convincing evidence of thelytoky.

SYMPHYLA. No symphylan is known to be parthenogenetic.

DIPLOPODA. The millipedes. Thelytoky has been inferred from the absence or extreme scarcity of males in five millipedes: *Proteroiulus fuscus* (Brookes, 1974; Rantala, 1974), *Nemasoma (Isobates) varicorne* (Brookes, 1974; Enghoff, 1976b), *Polyxenus lagurus* (Enghoff, 1976c), *Cylindrodesmus laniger* (Enghoff, 1978) and *Archiboreoulus pallidus* (Palmen, 1961). The successful reproduction of isolated females in *Proteroiulus* and *Nemasoma* establishes the existence of thelytoky beyond doubt, but the genetic mechanisms involved remain unexplored.

CHILOPODA. The centipedes. Males are absent from large parts of the ranges of *Lamyctes fulvicornis, L. coeculus* and *Lithobius aulacopus*, which are inferred to be thelytokous in these areas (Eason, 1964; Enghoff, 1976a). Enghoff also ascribes thelytoky with somewhat less confidence to *Poratia digitata* and *Detodesmus attemsi*, two greenhouse centipedes which seem to lack males. Males are very often in a small minority among adult pauropods, millipedes and centipedes, and there is evidence that some millipedes are monogenous (Rantala, 1974).

Discussion. Enghoff (1976b) has published an excellent account of parthenogenesis in *Nemasoma varicorne*, a widely distributed European millipede which lives under the bark of dead trees. Populations may be exclusively amphimictic or exclusively parthenogenetic, but many are a mixture of the two types. Females from populations in which males are common have a well-developed seminal receptacle, whereas in those from populations in which males are rare the seminal receptacle is greatly reduced and very variable in structure, which indicates the derivation of the thelytokous from the amphimictic type. A few males occur even in samples from localities where all the females lack a functional seminal receptacle, and Enghoff speculates that these males may be the offspring of thelytokous females. These presumptively parthenogenetic males appear to be capable of fertilizing the eggs of amphimictic females in the laboratory, although they are less successful than indubitably amphimictic males, and Enghoff argues from this that the genetic isolation of parthenogenetic from amphimictic strains is incomplete. However, he observed no males among the offspring of parthenogenetic females in the laboratory, so the status of the occasional males in predominantly thelytokous populations is not finally settled. On the other hand, isolated females with fully formed seminal receptacles occasionally produce a very few offspring, presumably by parthenogenesis. Thelytokous females generally produce fewer eggs than amphimictic females in laboratory culture, perhaps only half as many, but this statistic is uncertain because of geographical variation in fecundity.

The geographical distribution of the thelytokous and amphimictic types is strikingly different, the thelytokous populations occupying the periphery of the species' range (Figure 3.4). Within Denmark the amphimictic form is largely restricted to old beech forests growing on clay soil, whilst the thelytokous form extends into less favourable, more open habitats on sandy soil. However, laboratory experiments have shown that, whilst males are less resistant to desiccation

Figure 3.4: Distribution of Thelytokous and Amphimictic Populations of the Millipede *Nemasoma varicorne*. Upper figure, in Europe; lower figure, in Denmark. Solid circles, amphimixis; open circles, thelytokous; half-filled circles, mixed population.

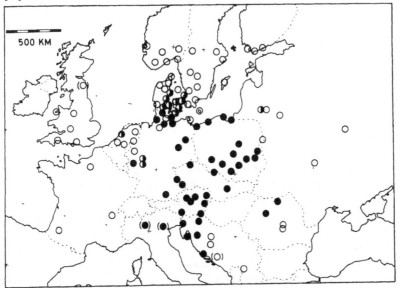

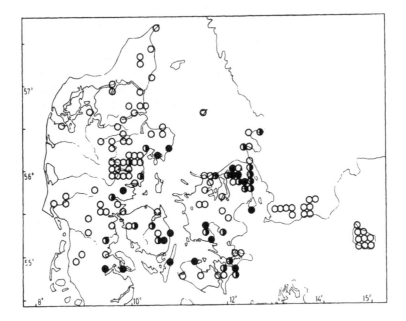

Source: Enghoff (1976b), p. 51, Figure 12 and p. 52, Figure 13.

and high temperature than females, there is no consistent difference in hardihood between amphimictic and thelytokous females. In *Polyxenus lagurus*, Enghoff (1976c) found no difference between the habitats occupied by thelytokous and amphimictic populations.

The millipede *Archiboreoulus pallidus* is parthenogenetic in Finland but amphimictic in central Europe.

In centipedes, too, the mode of reproduction may vary geographically. The males of *Lamyctes fulvicornis* are found only in the Azores and the Canary Islands, and those of *L. coeculus* only in Cuba. In *Lithobius aulacopus* males are absent from the northern part of the range, in Britain and Scandinavia.

3.32 Arthropoda: Insecta

Only a very small proportion of insects are parthenogenetic: and yet insects are so numerous that in absolute terms the number of parthenogenetic species is quite large. The huge size of the class, the sporadic occurrence of parthenogenesis and the diversity of cytological mechanisms make insects a special case, which will best be served by a departure from my usual format. I have listed all the instances of parthenogenesis that I have found in the literature in Tables 3.15, 3.16 and 3.17. Table 3.15 lists instances of thelytoky, and cites reviews down to the rank of family. Table 3.16 reorders this data for insects which have been investigated cytologically, to clarify the taxonomic distribution of different modes of thelytoky. Table 3.17 summarizes the occurrence of modes of parthenogenesis other than thelytoky. These tables form the basis for the discussion below, which is arranged under subheadings referring to those topics which seem to me the most germane to our eventual goal of understanding the functional significance of genetic systems. Fortunately, insect parthenogenesis has often been reviewed, which will make my review the briefer; it should be supplemented by reading the reviews listed in Table 3.15, especially those by Narbel-Hofstetter (1964), White (1973) and Suomaleinen *et al.* (1976).

3.32.1 Thelytoky

Taxonomic Distribution. Table 3.15 documents the existence of facultative or obligate thelytoky in over 200 genera from more than 80 families of insects, excluding the aphids and cynipids. A really thorough search of the entomological literature would doubtless add more instances; and many more must remain to be discovered, especially in the relatively little-studied apterygotes, thrips and psocopterans. Very roughly, about two per cent of the million or so known species of insects can reproduce by thelytoky, and about half as many are obliged to do so.

The first thing that strikes one about the distribution of thelytoky through the class is how sporadic it is; the second, how it is concentrated in certain taxa, where, though still exceptional, it is relatively frequent, whilst in other taxa it is entirely lacking. I have found no certain report of thelytoky from ten of the 30 orders of insects (Protura, Entotrophi, Odonata, Plecoptera, Gryllobattoidea, Zoraptera, Mallophaga, Anoplura, Mecoptera, Siphonaptera), and only vague or dubious reports from three others (Dermaptera, Neuroptera and Strepsiptera). There are a few authentic instances of thelytoky in the Collembola, Ephemeroptera, Embioptera, Isoptera, Hemiptera and Trichoptera, almost all of which refer to isolated species in families which are otherwise wholly amphimictic. In the four largest orders (Lepidoptera, Diptera, Coleoptera and Hymenoptera) many more cases of thelytoky are known, but this is no more than would be expected from the vast size of these groups. More interestingly, certain families within these orders include a disproportionate number of thelytokous species. Among lepidopterans obligate thelytoky seems to be almost exclusively confined to certain genera of bagworm moths in the family Psychidae. Among the dipterans, thelytoky is quite common in the Chironomidae, and especially in the subfamily Orthocladiinae. Weevils of the family Curculionidae include 42 thelytokous cytotypes in six subfamilies, as against 230 amphimictic species in 27 subfamilies (Takenouchi, 1970a). Nevertheless, the remaining moths, flies and beetles which are known to be thelytokous are for the most part taxonomically isolated; and thelytoky is also rather evenly scattered through the Hymenoptera, where the only large concentration of thelytokous forms occurs in the Cynipidae, most of which have a cyclical parthenogenesis. Thelytoky is most common in the orthopteroid orders, especially the Phasmatodea, in the Homoptera and in the Psocoptera.

Tychoparthenogenesis. A small proportion of eggs develop parthenogenetically in normally amphimictic species of orthopterans, cockroaches, phasmids, mantids, psocopterans, lepidopterans and dipterans (see Table 3.15). White (1973) has drawn attention to the remarkable fact that all such instances of tychoparthenogenesis (strictly, tychothelytoky) are automictic. He draws the conclusion that automixis may evolve gradually, through the accumulation of mutations which affect the fusion of meiotic products. Apomixis, on the other hand, seems to arise suddenly, without evolving from a prior automixis: amongst all the thelytokous weevils, for example, not a single automictic form is known, and it must be concluded that none ever existed. On the other hand, Porter (1971) argues that the existence of rudimentary first meiotic division in certain apomictic chironomids, embiopterans and weevils might indicate the existence of an automictic stage in the evolution of apomixis. Apparently meiotic phenomena have also been observed in maturation divisions of the apomictic eggs of aphids (see below).

Tychoparthenogenesis provides evidence for the existence of individual variation in the capacity for parthenogenesis, which, insofar as it is heritable, will respond to selection. Carson (1967) was able to increase the frequency of automictic parthenogenesis in *Drosophila mercatorum* from 0.1 per cent to over six per cent by artificial selection.

Table 3.15: Thelytoky in Insects. All orders are listed, including those which are not known to include thelytokous forms. Families are listed alphabetically within orders, and genera within families, except that the Homoptera is divided into suborders and superfamilies to separate aphids and coccids. The abbreviations used in the fourth column are as follows: Occ., occasional; Fac., facultative; Obl., obligate; Geog., geographical, i.e., allopatric sexual and thelytokous races; Sym., sympatric, i.e., mixed populations of sexual and thelytokous individuals; Cyc., cyclical parthenogenesis. Review articles are listed opposite the class, order, suborder or family to which they refer; only reviews recent enough to remain useful are included.

Family	Genus	Mechanism	State	Ploidy	Authorities
INSECTA					Vandel (1931)
					Grassé (1966)
					Suomaleinen (1950, 1962)
					Narbel-Hofstetter (1964)
					White (1973)
					Enghoff (1976a)
					Suomaleinen *et al.* (1976)
PROTURA					
THYSANURA					
Nicoletiidae	*Nicoletia*				Picchi (1972)
ENTOTROPHI					
COLLEMBOLA					Husson and Palévody (1967)
					Petersen (1971)
Isotomidae	*Folsomia*		Fac.		Goto (1960)
					Marshall and Kevan (1962)
	Hypogastrura				Green (1964)
					Grassé (1922)
Neanuridae	*Isotoma*				Christiansen (1964)
	Neanura				Petersen (1971)
					Cassagnau (1972)

				Reference
Onychiuridae	*Onychiurus*			Hale (1964)
	Tullbergia			Petersen (1971)
EPHEMEROPTERA		Degrange found facultative parthenogenesis in 26/51 European mayflies; only *Cloeon* was reared through two successive generations		Degrange (1960)
Baetidae	*Baetis*			Downes (1964)
Empherellidae	*Emphererella*			Dodds (1923)
				Needham (1924, cited by Vandel, 1931)
Heptageniidae	*Ephoron*		Fac.	Britt (1962)
	Ameletus		Fac. Obl.	Clemens (1922)
				Needham (1924, cited by Vandel, 1931)
				Needham *et al.* (1935)
	Cloeon	Four genera. Low hatch of unfertilized eggs in *Epeorus*, *Ecdyonurus*, *Rhitrogena* and *Heptagenia*	Geog.	Degrange (1960)
	Stenonema			McCafferty and Huff (1977)
ODONATA				
PLECOPTERA				
GRYLLOBLATTOIDEA				
ORTHOPTERA				Chopard (1948)
				White (1951a)

Table 3.15 (*cont.*)

Family	Genus	Mechanism	State	Ploidy	Authorities
Acrididae	Nine genera: *Arphia* *Chorthippus* *Dissoteira* *Locusta* *Melanoplus* *Pardolophora* *Romalea* *Schistocerca* *Trimerotropis*	Tychoparthenogenesis appears to be widespread. In cytologically investgated cases it is automictic, with fusion of cleavage nuclei. When fusion fails the embryos may be haploids or haploid/diploid mosaics. The embryos are usually inviable	Occ.	$2n$	Uvarov (1928) King and Slifer (1934) Creighton (1938) Creighton and Robertson (1941) Husain and Mathur (1945) Chopard (1948) Bergerard and Seugé (1959) Hamilton (1953, 1955)
Eumastacidae	*Warramaba*	Automixis: premeiotic doubling	Obl.	$2n$	White *et al.* (1963) White (1966) White and Webb (1968) Hewitt (1975) White *et al.* (1977) White and Contreras (1979)
Gryllidae	*Loxoblemmus* *Myrmecophila*		Fac.		Ohmachi (1929) Schimmer (1909) Chopard (1919)
Rhaphidophoridae	*Euhadenoecus* *Hadenoecus*		Geog. Geog.		Hubbell and Norton (1978) Hubbell and Norton (1978)
Tetrigidae	Three genera: *Apotettix* *Paratettix* *Tettigidea*	Widespread tychoparthenogenesis. Automixis, either with suppression of second meiotic division (Robertson, 1930) or with fusion of cleavage nuclei (White, 1973)	Occ.	$2n$	Nabours (1919) Nabours and Forster (1929) Robertson (1930, 1931) King and Slifer (1933)
Tettigoniidae	*Leptophyes* *Poecilimon*		Occ.		Cappe de Baillon (1939) Bei-Bienko (1954)

				References
Saga	Obl.	4n	Apomixis: one maturation division	Matthey (1941, 1946) Goldschmidt (1946) White (1978) Baccetti (1958)
Xiphidiopsis *Dolichoptera*				
PHASMATODEA				
Phasmidae				
Acanthoxyla				Cappe de Baillon *et al.* (1937b) Bergerard (1962) Bedford (1978)
Bacillus	Geog.	2n	Automixis: fusion of cleavage nuclei	Salmon (1955) Cappe de Baillon *et al.* (1937a) Bullini (1965, 1966) Pijnacker (1969) Scali (1970, 1972)
Bacteria	Geog.	2n	Most unfertilized eggs fail to hatch	Cappe de Baillon *et al.* (1934)
Baculum				Cappe de Baillon *et al.* (1934)
Carausius	Obl. Fac.	2n	Apomixis. Two maturation divisions. Pehani, however, stated that the two divisions were meiotic and synaptic. Cappe de Baillon claimed that *C. morosus* is triploid	Pehani (1925) Cappe de Baillon *et al.* (1934, 1935, 1937a) Bergerard (1962) Pijnacker (1966)
Clitarchus	Occ.		Clones deteriorate in laboratory; perhaps an automictic mechanism leading to extensive homozygosity	
Clitumnus	Fac.	2n	Automixis: cleavage nuclei fuse. Early embryo is haploid, later mosaic	Bergerard (1958)
Clonopsis	Obl.	2n		Cappe de Baillon *et al.* (1937b)

Table 3.15 (*cont.*)

Family	Genus	Mechanism	State	Ploidy	Authorities
	Didymuria				
	Epibacillus		Obl.		Cappe de Baillon and de Vichet (1940)
	Eurycnema				
	Extatosoma				
	Leptynia		Obl.	2n	Cappe de Baillon et al. (1937b)
					Cappe de Baillon and de Vichet (1940)
	Menexenus	Very few unfertilized eggs survive	Fac.		Cappe de Baillon et al. (1935)
					Bergerard (1954)
	Parasosibia	Most unfertilized eggs survive	Fac.		Cappe de Baillon et al. (1937a)
	Phalces	Almost all unfertilized eggs die	Fac.		Cappe de Baillon et al. (1937b)
	Phobaeticus	Almost all unfertilized eggs die	Fac.		Cappe de Baillon et al. (1937b)
	Podacanthus				
	Sipyloidea	Apomixis. Two maturation divisions	Fac. Geog.		Possompès (1956)
					Pijnacker (1968)
BLATTARIA					
Blaberidae	**Pycnoscelus**	Apomixis. Two maturation divisions	Obl. Occ.	2n Occ. poly-ploidy	Roth and Willis (1956)
					Matthey (1942, 1943, 1944, 1945)
					Roth and Willis (1961)
					Roth and Cohen (1968)
					Roth (1967, 1970, 1974)
					Stephan and Cheldelin (1975)
					Parker et al. (1978)

Family / Order	Genus	Type	Cytology	Notes	References
Blattidae	*Blatta*	Occ.			Roth and Willis (1956)
	Periplaneta	Occ.			Griffiths and Tauber (1942)
					Haydak (1953)
					Roth and Willis (1956)
	Other genera			Occasional hatching of eggs from unmated females in *Blatella, Nauphoeta, Supella*	Roth and Willis (1956)
MANTODEA Mantidae	*Brunneria*	Obl.	$2n$		White (1948)
	Miomantis	Occ.			Adair (1924, cited by White, 1951b)
DERMAPTERA Forficulidae	*Forficula*			Dubious; only female offspring produced in culture	Gunther and Herter (1974)
EMBIOPTERA Oligotomidae	*Haploembia*	Obl.	$2n$	Probably apomictic	Stefani (1956)
	Gynembia	Obl.	Occ. $3n$		Ross (1944)
					Ross (1961)
ISOPTERA Hodotermitidae	*Zootermopsis*				Light (1944)
PSOCOPTERA Atopidae	*Cerobasis*	Geog.			Mockford (1971)
					Pearman (1928)
					Badonnel (1943, 1951)
	Lepinotus	Obl.			Schneider (1955)
					Broadhead (1954)

Table 3.15 (*cont.*)

Family	Genus	Mechanism	State	Ploidy	Authorities
Caeciliidae	*Caecilius*		Obl. Geog.		Sommermann (1943) Schneider (1955) Mockford (1971)
Elipsocidae	*Elipsocus*		Geog.		Pearman (1928) Thornton and Broadhead (1954) Schneider (1955)
	Palmicola		Obl.		Mockford (1955)
	Pseudopsocus		Obl.		Schneider (1955)
	Reuterella		Geog.		Badonnel (1943, 1951) Schneider (1955)
Epipsocidae	*Epipsocus*				Pearman (1928)
Lachesillidae	*Lachesilla*				Mockford (1971)
Lepidopsocidae	*Echmepteryx*				Mockford (1971)
	Pteroxanium		Obl.		Pearman (1928)
Liposcelidae	*Liposcelis*	Probably apomictic; single maturation division with pronucleus remaining diploid	Obl.	$2n$	Broadhead (1947, 1954) Monterosso (1952) Goss (1954)
Peripsocidae	*Ectopsocus*		Fac. Obl.		Ribaga (1905) Schneider (1955) Thornton and Wong (1968)
	Peripsocus				Schneider (1955) Eertmoed (1966)
Philotarsidae	*Aaroniella*		Obl.		Mockford (1971)
	Philotarsus				Schneider (1955)
Psocidae	*Clematostigma*		Obl.		Schneider (1955)
	Psocidus		Geog.		Mockford (1971)

Family / Order	Genus	Notes	Type	References
	Psocus		Obl. Fac.	Schneider (1955)
	Trichadenotecnum		Obl. Fac.	Mockford (1971)
Psoculidae	*Psoculus*		Obl.	Schneider (1955)
Psyllipsocidae	*Psyllipsocus*			Badonnel (1951) Schneider (1955)
	Several general	Tychoparthenogenesis		Schneider (1955) Mockford (1971)
ZORAPTERA				
MALLOPHAGA				
ANOPLURA				
THYSANOPTERA				
Thripidae	*Anaphothrips*		Geog. Sym.	Pomeyrol (1928) Bournier (1956) Ananthakrishnan (1979)
	Heliothrips	Automixis: pronucleus fuses with second polar body	Obl.	Hinds (1903) Shull (1914)
	Physopus		?Fac. ?Cycl.	Buffa (1911) Pomeyrol (1928) Bournier (1956) See Pomeyrol (1928)
	Many genera	Males are very scarce or unknown in many genera of Aelothripidae, Thripidae and Phloeothripidae; some of these may be thelytokous		Uzel (1895) Hinds (1903) Priesner (1926, cited by Pomeyrol, 1928) Summary: Pomeyrol (1928)

Table 3.15 (*cont.*)

Family	Genus	Mechanism	State	Ploidy	Authorities
HOMOPTERA					
Sub. Auchenorrhyncha					
Cicadellidae	*Agallia*				Black and Oman (1947)
Delphacidae	*Muellerianella*	Gynogenesis	Sym.	$3n$	Drosopoulos (1976)
Sub. Sternorrhyncha					
Superf. Aleurodoidea					
Aleurodidae	*Aleurotulus*		Obl.		Thomsen (1927)
	Trialeurodes	Automixis: fusion of cleavage nuclei	Obl. Geog.		Schrader (1920, 1926) Thomsen (1927)
Superf. Aphidoidea					
Four families	All genera	Apomixis; maturation divisions are endomeiotic, probably with crossing-over. Alternates in most species with amphimixis. See text	Cyc.	Usually $2n$	Major papers include: Bonnet (1745) Stevens (1905) Tannreuther (1907) von Baehr (1909, 1920) Morgan (1909, 1915) Marcovitch (1923, 1924) Davidson (1929) Bonnemaison (1951) Kennedy and Stroyan (1959) Cognetti (1960, 1961a) Lees (1961, 1966) Pagliai (1961) Hille Ris Lambers (1966) Dixon (1977)

Superf.
Coccoidea

Diaspididae	Many genera	Apomixis in *Aspidiotus, Hemiberlesia, Howardia, Ichnaspis, Leucaspis, Phenacaspis* and *Poliaspoides*. Automixis in *Parlatoria, Pinnaspis* and *Targonia*. Undetermined thelytoky in *Chrysomphalus, Fiorinia, Kuwanaspis, Neoselenaspidus, Odonaspis, Selenaspidus* and *Xanthophthalma*	Obl. and Geog.	2*n*	Hughes-Schrader (1948) Brown (1965) Nur (1971) See Brown (1965)
Lecaniidae	*Eucalymantus*	Automixis: fusion of cleavage nuclei	Obl.		Nur (1971)
	Lecanium	Automixis and apomixis. In automictic forms, pronucleus fuses with second polar body	Obl. Fac.		Thomsen (1927, 1929) Suomaleinen (1940c)
	Pulvinaria	Automixis: fusion of cleavage nuclei	Obl.		Nur (1963)
	Gueriniella	Automixis: fusion of cleavage nuclei	Obl.		Hughes-Schrader and Tremblay (1966)
Margarodidae	*Icerya*	Autogamy. Unfertilized eggs yield occasional haploid males	Obl.	2*n*	Hughes-Schrader (1925) Schrader and Hughes-Schrader (1926) Hughes-Schrader and Monahan (1966)

Table 3.15 (*cont.*)

Family	Genus	Mechanism	State	Ploidy	Authorities
Phoenicoccidae	*Marchalina*	Apomixis	Obl.		Hovasse (1930)
	Thysanococcus		Obl.		Brown (1965)
Pseudococcidae	*Antonina*	Automixis and apomixis	Obl.		Nur (1971)
	Dysmicoccus	Apomixis	Geog.		Nur (1971)
	Phenacoccus	Automixis: pronucleus fuses with second polar body	Obl.		Nur (1971)
	Trionymus	Automixis	Obl.		Nur (1971)
HEMIPTERA					
Mesoveliidae	*Mesovelia*				Gagné and Howarth (1975)
Miridae	*Campyloneura*				Wagner (1968)
	Chlamydatus				Bocher (1971)
NEUROPTERA					
Chrysopidae					New (1967)
Hemerobiidae					New (1967)
MECOPTERA					
TRICHOPTERA					
Hydropsychidae	*Amphipsyche*				Corbet (1966)
Limnephilidae	*Apatania*				Corbet (1966)
					Lack (1933, 1934)
					Schmid (1954)
Psychomyiidae	*Psychomyia*				Marshall (1939)
LEPIDOPTERA					
Geometridae	*Alsophila*	Gynogenesis			Mitter *et al.* (1979)

Lasiocampidae					
Lymantriidae	*Mesocelis* *Orygia*	Var.	?2n		Taylor (1966) Eaton (1866) Rangnow (1912) Pictet (1924)
Psychidae	*Acanthopsyche* *Apterona*	Geog. Obl.	2n	Automixis: second division nuclei fuse in pairs to give a mosaic embryo	Trautmann (1909) Narbel (1946)
	Luffia		4n	Automixis: pronucleus fuses with first polar body, though details vary. *L. lapidella* is gynogenetic	Narbel-Hofstetter (1954, 1955, 1956, 1960, 1961, 1963)
	Solenobia			Automixis. *S. triquetrella* has a functionally apomictic mechanism involving the fusion of two central nuclei. In *S. lichenella* the pronucleus fuses with the first polar body	Narbel-Hofstetter (1950) Seiler (1923, 1927, 1936, 1959, 1961, 1963, 1964) Seiler and Schäffer (1960) Lokki *et al.* (1975)
DIPTERA					
Agromyziidae	*Phytomyza*	Geog.	2n, 3n		Hering (1926) Frick (1951) Block (1969)
Cecidomyiidae	Three genera: *Heteropeza*, *Miastor*, *Mycophila*; also *Tecomyia* and *Henria*	Obl.	See text	Highly aberrant cycles, described in text. Thelytokous reproduction is apomictic, with a single maturation division	Kahle (1908) Hegner (1912, 1914) Gabritschevsky (1928, 1930) Ulrich (1936) Reitberger (1939) White (1946a, 1950) Hauschteck (1959, 1962) Nikolei (1961) Wyatt (1961) Camenzind (1962, 1966) Panelius (1968, 1971)

Table 3.15 (*cont.*)

Family	Genus	Mechanism	State	Ploidy	Authorities
Ceratopogonidae	*Culicoides*				Williams (1961) Lee (1968)
Chamaemyiidae	*Ochthiphila*	Stalker suggests automixis, but this is not proven		$3n$	Sturtevant (1923) Stalker (1956a)
Chironomidae	*Chironomus*				Edwards (1919)
	Coryneura				Goetghebuer (1913) Edwards (1919)
	Limnophyes	Apomixis		$2n, 3n$	Scholl (1956, 1960)
	Lundstroemia	Apomixis, though with a rudimentary first division of meiotic appearance		$3n$	Lindeberg (1958) Edward (1963) Porter (1971)
	Monotanytarsus		Geog.		Lindeberg (1958)
	Pseudosmittia	Apomixis		$2n, 3n$	Scholl (1956, 1960)
	Smittia	Apomixis			Scholl (1956, 1960)
	Tanytarsus				Vandel (1931) Palmén (1961)
Culicidae	*Aedes*		Occ.		Enghoff (1976a)
	Culex		Occ.		Kitzmiller (1959)
Drosophilidae	*Drosophila*	Automixis, with a variety of fusion products	Occ. Obl.	$2n$	Stalker (1954) Carson *et al.* (1957) Murdy and Carson (1959) Carson (1961, 1962) Henslee (1966) Ikeda and Carson (1973) Templeton and Rothman (1973)
Empididae	*Platypalpus*				Chvala (1975)
	Tachydromia		Geog.		Tuomikoski (1935)

Family	Genus	Notes	Sym./Geog.	Ploidy	References
Lonchopteridae	*Lonchoptera*	Automixis: pronucleus fuses with second polar body			deMeijere (1906), Aldrich (1918), Stalker (1956b), Baud (1973)
Psychodidae	*Psychoda*				Séguy (1950)
Simuliidae	*Crephia*	Basrur and Rothfels suggest automixis, without cytological proof. There is no premeiotic doubling, but a fusion of central polar nuclei is conceivable	Sym.	$3n$	Davies and Petersen (1956), Basrur and Rothfels (1959)
	Gymnopais			$3n$	Basrur and Rothfels (1959)
	Prosimulium		Geog.	$3n$	Basrur and Rothfels (1959), Downes (1965)
SIPHONAPTERA					Suomaleinen (1940a), Smith (1971)
COLEOPTERA					
Chrysomelidae	*Adoxus*	Apomixis		$3n$	Jolicoeur and Topsent (1892), Suomaleinen (1965a), Lokki *et al.* (1976a)
	Altica			$3n$	Smith (1971)
	Calligrapha			$4n$	Robertson (1966)
Ciidae	*Cis*		Geog. Sym.	$2n$	Lawrence (1967)
Curculionidae	*Barynotus*	Apomixis		$3-5n$	Suomaleinen (1969), Suomaleinen (1940a, b), Takenouchi (1957a)
	Blosyrus			$6n$	Takenouchi (1970b)
	Brachyrhinus			$3n$	Takenouchi (1965)

Table 3.15 (cont.)

Family	Genus	Mechanism	State	Ploidy	Authorities
	Catapionus	Apomixis		4–5n	Takenouchi (1957a, b, 1966, 1970a)
	Cyrtepistomus			3n	Takenouchi (1970b)
	Eusomus	Apomixis		3n	Mikulska (1953) Takenouchi (1957a)
	Liophloeus	Apomixis		3n	Suomaleinen (1955)
	Listroderes	Apomixis		?3n	Sanderson (1953, 1956, 1973)
					Takenouchi (1957a, 1969)
	Otiorrhynchus	Apomixis	Obl. Geog.	3–6n	Andrewartha (1933) Mikulska (1940, 1960) Suomaleinen (1940a, b, 1947, 1948, 1954, 1955, 1961, 1966a, 1969) Seiler (1947) Suomaleinen and Saura (1973) Saura *et al.* (1976a, b)
	Peritelus	Apomixis		3–4n	Suomaleinen (1954) Takenouchi (1957a)
	Polydrosus	Apomixis		2n, 3n	Suomaleinen (1940a, b, 1954) Mikulska (1949) Takenouchi (1957a) Lokki *et al.* (1976b)
	Pseudoceneorhinus	Apomixis		4n	Takenouchi (1957a)
	Scepticus	Apomixis	Geog. Sym.	2n, 5n	Takenouchi (1961, 1968)

Family	Genus	Mode	Type	Ploidy	References
	Sciaphilus	Apomixis		$3n$	Suomaleinen (1940a, b, 1947) Takenouchi (1957a, 1964, 1965)
	Sciopithes	Apomixis		$4n$	Suomaleinen (1966a)
	Strophosomus	Apomixis		$3n$	Suomaleinen (1940a, b, 1947, 1954, 1966a) Takenouchi (1957a) Suomaleinen and Saura (1973)
	Trachyphloeus	Apomixis		$3n$	Suomaleinen (1940a, b) Takenouchi (1957a, 1964, 1965)
	Trophiphorus	Apomixis		$3-4n$	Suomaleinen (1954) Takenouchi (1957a, 1964, 1965)
Dermestidae	*Perimegatoma*		Obl.		Milliron (1939)
Hydrophilidae	*Helophorus*				Angus (1970)
Micromalthidae	*Micromalthus*	Complex cycle: see text	Cyc.		Barber (1913) Scott (1936, 1938) Pringle (1938)
Pselaphidae	*Bythinopsis*		Geog.		Reichle, cited by Dybas (1966)
Ptiliidae	*Eurygyne*		Obl.		Dybas (1966)
Ptinidae	*Ptinus*	Geometric automixis: premeiotic doubling, with single maturation division	Sym.	$3n$	Sanderson (1956, 1960) Moore *et al.* (1956)
Scolytidae	*Ips*	Gynogenesis	Obl.		Hopping (1962) Smith (1962) Lanier and Oliver (1966)
Staphylinidae	*Amischa*		Geog.		Strand (1951) Williams (1969)

Table 3.15 (*cont.*)

Family	Genus	Mechanism	State	Ploidy	Authorities
STREPSIPTERA					
Stylopidae					Kinzelbach (1971)
HYMENOPTERA					
Apidae	*Apis*	Automixis	Occ.	2*n*	Slobodchikoff and Daly (1971) Jack (1916) Mackensen (1943) Tucker (1958) Anderson (1963)
	Ceratina				Daly (1966)
Bethylidae	*Scleroderma*				Keeler (1929a, b)
Braconidae	*Bracon*				Speicher (1936)
	Perilitus				Balduf (1926)
	Pygostelus				Loan (1961)
Cephidae	*Cephus*				Farstad (1938) Smith (1938)
Cynipidae	All genera	Thelytoky alternates with arrhenotoky, except in species which have lost the sexual phase altogether. See text. In *Neuroterus* the thelytoky is automictic, probably with an endomitotic doubling	Cyc.		Schleip (1909) Doncaster (1910, 1911, 1916) Hogben (1920) Dodds (1937, 1939) Evans (1967)
Diprionidae	*Diprion*	Automixis: pronucleus fuses with second polar body	Geog.	4*n*	Smith (1941)
Encyrtidae	*Encyrtus*				Flanders (1943)
	Habrolepis				Flanders (1946)

Family	Genus		Notes	Reference
	Ooencyrtus			Wilson and Woolcock (1960)
	Trechnites			Slobodchikoff and Daly (1971)
Eulophidae	*Tropidophryne*			Doutt and Smith (1950)
	Aphytis			DeBach (1969)
	Prospaltella			Flanders (1953)
Formicidae	*Aphaenogaster*			Haskins and Enzmann (1945)
	Atta			Tanner (1892)
	Crematogaster		Automixis	Soulié (1960)
	Formica			Otto (1960)
	Lasius			Bier (1952)
	Oecophylla		Workers are apomictic, queens arrhenotokous	Ledoux (1949)
Ichneumonidae	*Nemeritis*	Cyc.		Speicher (1937)
Leucospidae	*Leucospis*	Geog.		Berland (1934)
Pelecinidae	*Pelecinus*	Geog.		Brues (1928)
Signiphoridae	*Signiphora*			DeBach (1969)
	Thysanus			DeBach (1969)
Tenthredinidae	*Empria*		Automixis	Doncaster (1906)
	Fenusa			Pieronek (1973)
	Nematus		Automixis	Doncaster (1907)
	Pristophora		Automixis: pronucleus fuses with second polar body	Comrie (1938) Heron (1955) Smith (1955)
	Thrinax		Apomixis	Peacock and Sanderson (1939)
Trichogrammatidae	*Trichogramma*			Bowen and Stern (1966) Stern and Bowen (1968)

Table 3.16: Parthenogenesis in Insects, other than or Supplemental to Thelytoky.

Mechanism	Taxa	Authorities
Arrhenotoky	Thysanoptera. At least five genera; perhaps all or most of order	Bournier (1956) Risler and Kempter (1962) Ananthakrishnan (1979)
	Homoptera: Margarodidae. All iceryine coccids	Hughes-Schrader (1948)
	Homoptera: Aleurodidae. *Trialeurodes*	Schrader (1920)
	Coleoptera: Scolytidae. *Xyleborus*	Entwhistle (1964) Takenouchi and Takagi (1967)
	Coleoptera. *Coccotrypes*. (Requires confirmation)	Herfs (1950)
	Hymenoptera. Entire order other than secondarily thelytokous forms	Many authors; see Schrader and Hughes-Schrader (1931); Whiting (1949); Slobodchikoff and Daly (1971); and White (1973)
Diploid arrhenotoky	Homoptera: Coccoidea: Lecaniidae. *Lecanium*. Male genome undergoes a postmeiotic doubling, after which one set becomes heteropycnotic	Nur (1971)
Deuterotoky	Phasmatodea: Phasmidae. *Ctenomorphodes*	Degrange (1960)
	Ephemeroptera. *Centroptilum*	Nur (1963)
	Homoptera: Coccoidea: Lecaniidae. *Lecanium* and *Pulvinaria*. Males nonfunctional in *Pulvinaria*	Phillips (1965)
	Lepidoptera. Tychoparthenogenesis in *Bombyx, Galleria, Lasiocampa, Lymantria, Phthorimaea* and *Porthesia* is deuterotokous. Females are heterogametic and arise from fusions between second division non-sister nuclei; males arise from fusions between cleavage nuclei	Goldschmidt (1917) Smith (1938) Astaurov (1940, 1962)
Arrhenotoky + apomictic thelytoky	Hymenoptera: Formicidae. *Oecophylla*. Workers are apomictic, queens arrhenotokous	Ledoux (1949)

Arrhenotoky + automictic thelytoky	Hymenoptera: Cynipidae. Some cynipids have lost the sexual phase	See Table 3.15
Amphimixis + apomictic thelytoky	Homoptera: Aphidoidea. Some recombination may occur during the maturation divisions of thelytokous eggs	See Table 3.15
Complex cycles (see text)	Diptera: Cecidomyidae. *Heteropeza*, *Miastor* and *Mycophila*	See Table 3.15
	Coleoptera: Micromalthidae. *Micromalthus*	See Table 3.15
Polyembryony	Hymenoptera. Reported from Braconidae, Dryinidae, Encyrtidae and Platygasteridae	Borror and DeLong (1971)

Table 3.17: Mechanisms of Thelytoky among Insects. For authorities, consult Table 3.15.

Order	Apomixis	Premeiotic	Automixis		Gynogenesis	Autogamy
			Intrameiotic	Postmeiotic		
Collembola					Onychiurus	
Orthoptera	Saga	Warramaba		Acrididae Tetrigidae		
Phasmatodea	Carausius Sipyloidea			Bacillus Clitumnus		
Blattaria	Pycnoscelus					
Embioptera	?Haploembia					
Psocoptera	Liposcelis					
Thysanoptera			Heliothrips			
Homoptera	Coccoidea (9 genera) Aphidoidea (all genera)		Lecanium Phenacoccus	Lecanium Eucalymantus Gueriniella Pulvinaria Trialeurodes		Icerya
Diptera	Cecidomyidae (three genera)	?Chironomidae (four genera)	Lonchoptera Drosophila			
Lepidoptera			Apterona Luffia Solenobia		Luffia Alsophila	
Coleoptera	Adoxus Cuculionidae (15 genera; 54 cytotypes)	Ptinus			Ptinus Ips	
Hymenoptera	Thrinax		Apis Diption Empria Formica Nematus Pristophora	Nemeritis ?Cynipidae		

Apomixis. Just as thelytoky itself is not distributed haphazardly amongst insect taxa, different mechanisms of thelytoky tend to be associated with different groups (see Table 3.17). Apomixis is the most common mode of thelytoky, and predominates among thelytokous aphids, Diptera (Cecidomyidae and – but see below – Chironomidae) and beetles (Curculionidae). The only thelytokous cockroaches, embiopterans and psocopterans which have been studied cyto-logically are apomicts. Both apomixis and automixis occur in orthopterans, phasmids and coccids, but automixis predominates among thelytokous Hymeno-ptera and Lepidoptera. There may be one (e.g., *Saga*, *Liposcelis*, Cecidomyidae, Curculionidae) or two (e.g., *Carausius*, *Sipyloidea*, *Pycnoscelus*) maturation divisions in apomictic eggs, but in either case, of course, the maternal genome is conserved.

The nature of the maturation divisions of the parthenogenetic eggs of aphids was questioned by Cognetti (1961a, b) and Pagliai (1961, 1962). There is only a single maturation division, but the chromosomes appear to undergo a peculiar sort of synapsis before falling apart into univalents. Cognetti believes that some crossing-over occurs during this process, which he terms 'endomeiosis', and claims to have demonstrated some response to selection for winglessness in thelytokous lines of *Myzodes persicae*. It will be recalled that he has made similar claims with respect to the maturation divisions of cladoceran eggs. His figures do not show unimpeachable chiasmata, and most authorities continue to regard aphids as thoroughly apomictic.

Automixis. The simplest automictic mechanism is a premeiotic doubling of chromosome number followed by a meiotic reduction; provided that only sister chromatids pair, crossing-over has no genetic significance, heterozygosity is conserved and the process is functionally apomictic. This has been found in only two insects: the eumastacid grasshopper *Warramaba* (formerly *Morabo*) and the beetle *Ptinus*. Equally simple, but with diametrically different conse-quences, is the fusion of haploid cleavage nuclei, which enforces homozygosity at all loci. This has been found in surprisingly many cases: the tychopartheno-genetic Acridiidae and Tetrigidae, the phasmids *Bacillus* and *Clitumnus*, four genera of coccids, the whitefly *Trialeurodes vaporariorum* and the ichneumonid *Nemeritis*. The mechanism of restitution in cynipid wasps is not entirely clear, but also seems to involve an endomitotic doubling. Nur (1971) observes that most of these cases involve arrhenotokous taxa, which might be preadapted to complete homozygosity by virtue of their hemizygous males. This generalization can be stretched to include all those intrameiotic mechanisms which have no special provision for the conservation of heterozygosity.

Intrameiotic mechanisms of restitution are more complicated. The more straightforward involve the fusion of the pronucleus with a polar body. The fusion of pronucleus and first polar body (suppression of first meiotic division) seems not to have been recorded from insects, though some psychid moths undergo a similar process (see below). The fusion of pronucleus and second polar body, on the other hand, has been reported from a number of very different insects: the thysanopteran *Heliothrips*, the coccids *Lecanium* and *Phenacoccus*, the dipteran *Lonchoptera* and the hymenopterans *Diprion* and *Pristophora*.

More complicated cases involve the fusion of haploid polar nuclei. The thelytokous Psychidae (Lepidoptera) are all automictic, and their extraordinary cytologies have aroused a good deal of interest (see references in Table 3.15). The problem that an automictic moth must overcome is that females are heterogametic, so that thelytokous lines can be propagated only if the automixis conserves heterozygosity. In *Solenobia triquetrella* this is achieved by centric fusion, the fusion of nonsister nuclei resulting from the second meiotic division. It is interesting that in the amphimictic race of *S. triquetrella* the two central polar nuclei also fuse, but this fusion is annulled by fertilization, the sperm nucleus fusing with the innermost of the two central nuclei to form a zygote nucleus. In *Apterona helix* both pairs of nonsister nuclei fuse, and both function as cleavage nuclei, so that the growing embryo is a mosaic for two genetically different diploid complements. The maturation divisions of *Luffia ferchaultella* vary from female to female, but all have in common the effective suppression of the first meiotic division, so that the second gives rise to a diploid zygote nucleus and a single diploid polar body. The cytology of these three genera is reviewed in much greater detail by Narbel-Hofstetter (1964).

The fusion of nonsister second division nuclei is also found in *Drosophila mangabeiri*, the only member of this large genus known to be obligately thelytokous, where it is such as to conserve structural heterozygosity. *D. parthenogenetica* is a facultative automict in which both terminal and central fusions occur, besides triple fusions which give rise to triploid zygotes.

There seems to be some doubt as to whether thelytokous Diptera such as *Ochthiphila*, *Cnephia*, *Phytomyza* and *Lundstroemia* are automicts or apomicts. Apomixis is indicated by their ploidy (all are triploids), but the much great structural heterozygosity of these thelytokous forms, especially in *Cnephia*, suggests that some segregation, presumably automictic, must occur. Stalker (1956) and Basrur and Rothfels (1959) suggest that a very regular meiotic orientation followed by centric fusion would make automixis compatible with triploidy; it is also possible that the triploids are apomicts with a highly polyphyletic ancestry. However, a suggestion made by White (1973) seems to me the most probable: a premeiotic doubling and subsequent reduction by a normal two-division meiosis, permitting the occasional pairing of nonsister chromatids between which crossing-over would generate some karyotypic variation. The origin, as opposed to the maintenance, of heterozygosity is also difficult to explain, since the presumptive amphimictic ancestors are chromosomally monomorphic; perhaps it arose from hybridization between a diploid automict and a related diploid amphimict.

Autogamy. Three species of *Icerya* (Homoptera: Coccoidea) are self-fertilizing hermaphrodites. They have an ovotestis whose core is testicular and haploid, one chromosome set having been eliminated during early gonadal development, whilst its cortex is ovarian and diploid; all somatic cells are diploid. They are thus more properly arrhenotokous than thelytokous. Rare males have also been found, and will copulate with the hermaphrodites; the behaviour of the hermaphrodites themselves is entirely female. According to Hughes-Schrader (1948) the males may, in some years and in some localities, be much more common than usual. Hughes-Schrader has also reported pure females of *I. bimaculata*.

Gynogenesis. The only insects known to be gynogenetic are the collembolan *Onychiurus*, the psychid *Luffia* and the beetles *Ptinus* and *Ips*; both *Luffia* and *Ptinus* are known to be automictic, though their restitution mechanisms are very different.

3.32.2 Other Mechanisms of Parthenogenesis

Deuterotoky. Deuterotoky is almost of necessity an occasional or facultative mechanism, and is found as such in a number of tychoparthenogenetic moths; it has also been described from a phasmid and — inevitably — from the coccid *Lecanium*. Nur (1963) claims that deuterotoky is obligate in the coccid *Pulvinaria hydrangeae*, the presumptively male embryos dying; a sort of inefficient thelytoky.

Arrhenotoky. Arrhenotoky is found throughout the Hymenoptera, except in the very few forms where it has been replaced partially or completely by thelytoky. It is also found among Homoptera in the iceryine coccids and the white flies (Aleurodidae); and isolated instances have been reported from Coleoptera — in two species of the scolytid *Xyleborus*, tentatively from *Coccotrypes*, and as part of the complex life cycle of the primitive monospecific family Micromalthidae (see below). Male haploidy has been confirmed cytologically in five genera of Thysanoptera, but whether these are exceptional, or the majority of the order reproduces by arrhenotoky, is not clear. I should stress the fact, so well known that it is liable to be overlooked, that whenever one sex is haploid that sex is male.

Diploid Arrhenotoky. In the coccid *Lecanium putmani* fertilized eggs develop into diploid females and unfertilized eggs into males in which diploidy is restored by the fusion of second cleavage nuclei. In later cleavage, however, one of the two haploid sets of male embryos becomes heterochromatic, so the males are effectively haploid.

Polyembryony. Polyembryony occurs in a few hymenopterans: the dryinid *Aphelopus theliae*, corn borers (Braconidae) of the genus *Macrocentus*, and a few platygasterids and encyrtids. In some braconids each fertilized egg may yield up to a thousand embryos.

Achiasmatic Meiosis. It is convenient to note here the occurrence of an achiasmatic meiosis in a number of insects: in the mecopteran *Panorpia* (Ullerich, 1961); in two eumastacid grasshoppers (White, 1965) and in fourteen genera of mantids (White, 1938, 1951a; Hughes-Schrader, 1943, 1950, 1953; Gupta, 1966), as well as in the wholly aberrant chromosome cycles of Cecidomyidae (see below), Sciaridae and various lice and coccids (see White, 1973). Meiosis is achiasmatic in certain families of the Diptera Nematocera (Phryneidae, Bibionidae, Scatopsidae, Thaumalidae, Blepharoceridae and Mycetophilidae; see Wolf, 1941, 1946, 1950; Le Calvez, 1947), but not in others; in the Tipulidae, meiosis is chiasmatic in 15 members of the genus *Tipula* but achiasmatic in two (see Bayreuther, 1955). Finally, virtually all the Diptera Brachycera have an achiasmatic meiosis. In all these cases meiosis is achiasmatic in only one sex, and that sex is always the male: there is thus a striking parallel between such species and arrhenotokous insects. It is presumed that the absence of visible chiasmata implies the complete, or almost complete, absence of genetic recombination in spermatogenesis, but definitive proof of this is available only for *Drosophila*. The cytology of achiasmatic meiosis is reviewed by White (1973), and I shall discuss the topic further in section 5.2.

3.32.3   Cyclical Parthenogenesis

Thysanoptera. Several early authors (e.g., Jordan, 1888; Shull, 1914) believed that some thrips alternate between sexual and parthenogenetic generations. This has never been confirmed; but neither, in my view, has it been decisively refuted, and the matter deserves reinvestigation.

Aphidoidea (Homoptera). The life histories of aphids have often been reviewed (for recent treatments, see Kennedy and Stroyan, 1959; Engelmann, 1970; Dixon, 1977), and this account will be correspondingly brief. The fertilized egg always undergoes a diapause, and usually represents the dormant overwintering stage of the life history. In some species, adults or nymphs hibernate or aestivate. The individual which hatches from the fertilized egg at the start of the growing season is a female which reproduces by apomictic thelytoky, bearing live young. After one or more parthenogenetic generations, diploid sexual males and females appear, copulate and produce mictic eggs to complete the cycle. The number of parthenogenetic generations in each cycle varies greatly between species: there are typically several, but in some species the founding female lays eggs which develop immediately into sexual males and females. By contrast, in some Pemphiginae the sexual generation has been lost altogether, and reproduction is indefinitely thelytokous (anholocyclic) (see Steffan, 1961). I know of no case in which the parthenogenetic generation has been lost and the sexual generation retained. Sex usually occurs late in the growing season; virtually the whole population may suddenly switch from parthenogenetic to sexual reproduction, but the change-over is often more gradual, with the proportion of sexual individuals increasing from mid-summer on.

The parallel between the life cycles of aphids and cladocerans is striking, and has often been remarked on. In aphids the cycle is sometimes complicated by heteroecy, with winged individuals migrating from the primary to the secondary plant hosts and back again. The migratory individuals seem always to belong to parthenogenetic lineages; the sexual individuals lay eggs which, after diapausing, hatch and reproduce on the same host as that of their parents.

The extensive variation both within and between species in the number of parthenogenetic generations intervening between episodes of sexuality has aroused a great deal of interest in the proximate factors governing the appearance of sexuality. The topic has been reviewed recently by Bonnemaison (1951), Hille Ris Lambers (1966) and Lees (1966). In some aphids the cycle is short and fixed, with sexual individuals always appearing after a certain number of parthenogenetic generations. In most species it is labile to some extent, but even so there seems to be a strong inhibition of sexuality for the first few partheno-genetic generations (Marcovitch, 1924; Wilson, 1938; Bonnemaison, 1951; Lees, 1960a), so that sex can be elicited by an appropriate manipulation of the environment only after this refractory period has passed. In *Aphis fabae*, Marcovitch (1923, 1924) discovered that sex could be elicited by short photoperiods, imitating the fall conditions under which sexual individuals appear in natural populations. This result was subsequently confirmed for several other aphids (Shull, 1928, 1929; Davidson, 1929; Wilson, 1938; Bonnemaison, 1951, 1958; Kenten, 1955; Lees, 1959, 1960a, b). Conversely, by using long photoperiods, aphids can be made to reproduce entirely by thelytoky for scores or even hundreds of generations. The effect of photoperiod is modulated by that of temperature, so that the effect of short photoperiod can be annulled by high temperature (Bonnemaison, 1958; Lees, 1963); indeed, Lees (1959) managed to elicit male production over a wide range of photoperiods by manipulating temperature alone. The effect of photoperiod and temperature on sexual development may be direct, or it may act indirectly through the nutritive status of the host plant. In *Brevicoryne* (Shull, 1928, 1929; MacGillivray and Anderson, 1964) and *Megoura* (Lees, 1961, 1964, 1966) there is good experimental evidence that the effect is direct, but this can hardly apply to subterranean aphids which feed on roots, and in *Eriosoma* the nutritive status of the host may control the timing of sex (Sethi and Swenson, 1967).

Winged individuals, whether thelytokous females or sexual males and females, develop in crowded populations (see Hille Ris Lambers, 1966; Lees, 1966). In *Brevicoryne* and *Dysaphis*, crowding will cause the appearance of sexual aphids, even under long photoperiods. *Pemphigus bursarius* is normally heteroecious and holocyclic, but under uncrowded conditions becomes monoecious and anholo-cyclic; the parthenogenetic females overwinter on the primary host, forgoing the production of fertilized eggs (Judge, 1968). It is possible to imitate the tactile stimulation caused by individuals crawling over one another in crowded populations by gentle brushing, and this treatment elicits the production of winged offspring even in isolated individuals of *Aphis craccivora* (Johnson, 1965, cited by Dixon, 1977). High densities of conspecifics are most effective in eliciting the

appearance of winged and sexual individuals, of related species less so, and high
densities of unrelated species are entirely ineffective (Bonnemaison, 1958).

Cecidomyidae (Diptera). Some of the more primitive members of the midge family
Cecidomyidae have very complicated modes of reproduction (for references see
Table 3.15). The larvae are paedogenetic and reproduce by apomictic thelytoky:
their eggs undergo a single mitotic maturation division and commence cleavage,
to produce embryos which parasitize and eventually kill their mother. During
cleavage there is a peculiar process of chromosome elimination. The zygote
nucleus contains two different sorts of chromosomes, which have been called
E- and S-chromosomes. The S-chromosomes are eliminated from the somatic
line in late cleavage, resulting in an embryo which has diploid somatic nuclei
and 'polyploid' nuclei in the germ line (the homology of E- and S-chromosomes
is questionable). This regular elimination of chromosomes during development
is also found in sciarids and in some chironomids. Larval thelytoky may continue almost indefinitely in the laboratory, but adverse conditions elicit sexuality.
This happens in the laboratory when cultures become stale or crowded; in natural
populations it occurs in summer, when the substrate dries out and the fungal
hyphae on which the larvae feed become scarce. Under such conditions the
paedogenetic larvae of *Miastor*, *Heteropeza* (= *Oligarces*) and *Mycophila* give
rise, by thelytoky, to sexual larvae, which pupate and develop into winged
imagos. The somatic nuclei of the male larvae are haploid, relative to the somatic
nuclei of the female larvae; nevertheless, oogonia and spermatogonia have the
same number of chromosomes. The male larvae of *Miastor* have an asynaptic
spermatogenesis in which the 48 chromosomes of the primary spermatocyte are
shared unequally between two secondary spermatocytes, the smaller receiving
a haploid set of six S-chromosomes and the larger a mixture of 36 E- and six
S-chromosomes. The larger cell degenerates; the smaller undergoes a second
maturation division to give rise to two haploid sperm. The female larvae produce
imagos bearing a small number of large haploid eggs, which will not develop
without fertilization; they give rise to a new generation of paedogenetic larvae.
In *Heteropeza* the situation is further complicated by the existence of three
types of paedogenetic larvae: a thelytokous paedogenetic larva which gives rise
to individuals of the same type, and which can also produce female imago-
larvae and thence female imagos; an 'androgenic' larva which gives rise to male
imago-larvae and thence to male imagos; and an 'amphogenic' larva which may
give rise to thelytokous larvae or to male imago-larvae. The latter two types
have been called 'arrhenotokous' and 'deuterotokous', respectively, because of
the haploidy of male somatic cells, but since the germ line of the males contains
a full chromosome complement this usage is illegitimate. The eggs of androgenic
larvae undergo a meiosis which halves the chromosome number and produces
three polar bodies, which persist as supplementary cleavage nuclei. The chromosome complement of the pronucleus is restored by fusion with one of the two
small somatic nuclei which persist in the egg cytoplasm, an aberrant process of

automixis. The eggs of the female imago, following meiosis, fuse with several sperm nuclei and with several small somatic nuclei, an even more aberrant halfway house between amphimixis and automixis. Moreover, some eggs develop parthenogenetically after fusion with somatic nuclei alone, although such eggs are usually inviable. In *Tecomyia populi* and *Henria psalliotae* the pupae rather than the larvae reproduce parthenogenetically (see Wyatt, 1961).

Micromalthidae (Coleoptera). The cecidomyid life cycle described above could, perhaps, be dismissed as a chapter of accidents, were it not that a very similar cycle has been found in a wholly unrelated group, the single species, *Micromalthus debilis*, of the taxonomically isolated family Micromalthidae. The main reproductive phase of *Micromalthus* is a viviparous paedogenetic larva which produces other larvae by thelytoky (Barber, 1913). The mechanism of thelytoky is not known. Some larvae, however, produce a single egg which after hatching devours its mother and develops into a haploid adult male beetle (Scott, 1936, 1938); the diploid adult females are produced by the thelytokous larvae. There are also deuterotokous larvae which give rise both to males and to females. In nature the adult beetles appear in August; in the laboratory, the production of sexual adults can be elicited by warming or drying the rotten wood in which the beetles live. However, their role in reproduction remains questionable: Scott even argued that the adults are sterile, in which case reproduction is effectively thelytokous. Males are very rare in North America, and have never been found in South Africa (Pringle, 1938).

Formicidae (Hymenoptera). A sort of cyclical parthenogenesis appears to occur in some ants, with the queens reproducing by arrhenotoky and the workers by deuterotoky or by apomictic thelytoky. The details are obscure and debatable, and I refer the reader to Wilson (1971) for an authoritative discussion.

Cynipidae (Hymenoptera). The cynipid wasps have a simple alternation between thelytokous and arrhenotokous generations. The thelytokous females emerge from plant galls in autumn and lay eggs which hatch the following spring into sexual males and females. This strict succession seems to be genetically fixed: no exceptions are known from nature, and the cycle has never been manipulated in the laboratory. A number of cynipids, however, have lost the sexual generation and become indefinitely thelytokous.

3.32.4 Ploidy and Reproductive Isolation

Ploidy. The usual association of parthenogenesis with polyploidy holds good for insects. The classic instance is the Curculionidae, where of the 44 apomictic cytotypes known to 1970 only two are diploid, 26 being triploid, eleven tetraploid, four pentaploid and one hexaploid (Takenouchi, 1970a). Other apomictic polyploids are the tetraploid tettigoniid *Saga pedo* and the triploid chrysomelid *Adoxus obscurus*. All obligately thelytokous species of *Calligrapha* are tetraploid, the amphimicts being diploid (Robertson, 1966; see Table 3.21). On the other hand, the phasmid *Carausius* and the psocopteran *Liposcelis* seem to be diploid apomicts, and polyploidy is exceptional in apomictic cockroaches (*Pycnoscelus surinamensis*) and embiopterans (*Haploembia* cf. *solieri*). Automicts are usually diploid, the only exceptions being some psychids and chironomids, the beetle *Ptinus* and the hymenopteran *Diprion*; most of these exceptions have either a premeiotic doubling or some means of intrameiotic restitution which conserves heterozygosity.

Reproductive Isolation. The polyploidy of most thelytokous biotypes enforces their isolation from closely related bisexual species. But even unsuccessful attempts at fertilization will waste eggs, and the problem is the more serious since thelytokous females usually retain an apparently functional seminal receptacle (in the gryllid *Myrmecophila*, Schimmer, 1909; in psocopterans, Jordan, 1888, and Buffa, 1911; in the mayfly *Apatania*, Corbet, 1966; in the chrysomelid *Calligrapha*, Robertson, 1966; in the ptiliid *Eurygyne*, Dybas, 1966). Robertson, indeed, has seen copulation between males and thelytokous females in *Calligrapha*. However, it seems that thelytokous females can acquire means of avoiding the unwelcome attentions of males. In *Cnephia mutata*, diploid amphimictic and triploid parthenogenetic strains live side by side in the same stream without interbreeding (Basrur and Rothfels, 1959). Lawrence (1967) was unable to obtain crosses between thelytokous female *Cis fuscipes* and males of the nominate sexual species *Cis impressus*, even though the two taxa cannot be distinguished through larval or female characters and are probably synonymous. In psocopterans, there are behavioural barriers to copulation. Schneider (1955) found that males of *Psocus bipunctatus* will not attempt to copulate with thelytokous females, unless they are simultaneously exposed to material which has been in contact with amphimictic females – in which case the males do attempt to copulate but are avoided by the females. Thelytokous females of *Bertkauia lucifuga* flee from males, and males will not court thelytokous females of *Peripsocus quadrifasciatus* (Eertmoed, 1966). In *Drosophila* it has been possible to select for reproductive isolation in thelytokous females (Ikeda and Carson, 1973).

Ploidy and Size. Ploidy also has an effect on body size, as it does in other animals. In *Otiorrhynchus rugifrons* and *Liophloeus tessulatus* the triploid parthenogenetic race is larger than the diploid amphimictic race (Suomaleinen, 1954, 1955), and the hexaploid thelytokous *Blosyrus japonicus* is larger than its diploid amphimictic relative, *B. falcatus* (Takenouchi, 1970b). Moreover, tetraploid individuals are larger than triploids in thelytokous races of *O. scaber*, *O. subdentatus* and *Peritelus hirticornis* (Suomaleinen, 1954, 1961; Mikulska, 1960). Tetraploid parthenogenetic species of *Calligrapha* are usually larger than diploid amphimictic species, although the largest species of all, *C. pnirsa*, is amphimictic (Robertson, 1966). On the other hand, pentaploid *Catapionus gracilicornis* are not larger than tetraploid individuals, and the triploid *Lonchop-tera furcata*, the only parthenogenetic species of its genus in the European fauna, is also the smallest. Nor are the eggs of polyploid parthenogentic weevils larger than those of related diploid amphimicts in *Otiorrhynchus* and *Solenobia* (Seiler, 1936; Suomaleinen, 1947). The somatic cells of the tetraploid *Saga pedo* are said to be no larger than those of its diploid amphimictic relatives.

3.32.5 Geographical Parthenogenesis

The fact that thelytokous and amphimictic animals referred to the same species often have quite different geographical distributions was first made widely known by Vandel (1931). He used the term 'geographical parthenogenesis' as a category in a generally unsatisfactory classification of modes of parthenogenesis; I shall retain it as a descriptive term because of its familiarity. The major work on geographical distribution since Vandel has been done by Seiler (1943, 1946) on the psychid *Solenobia*, and by Suomaleinen and his group on European weevils (Curculionidae) (see Suomaleinen, 1950, 1969; Suomaleinen *et al.*, 1976). A very useful compilation of further examples from the entomological literature has been published by Glesener and Tilman (1978). In Table 3.18, I have collated all the instances of geographical parthenogenesis in insects known to me. Most of these instances refer to the different distribution of thelytokous and amphi-mictic 'races' of the same Linnean species; some to thelytokous and amphimictic species of the same genus; a few to general trends in higher taxa. In the remainder of this section I shall use the data summarized in Table 3.18 to document patterns in the geographical distribution of thelytokous insects.

Allopatry. In the classical instances on geographical parthenogenesis, for example the European weevils, the ranges of thelytokous and amphimictic races of the same species are completely disjunct (Figure 3.5). This is not invariably or even usually the case; perhaps the most common pattern is that of partially separate but overlapping distributions. In some cases, thelytokous and amphimictic forms appear to be completely sympatric: the thelytokous and amphimictic biotypes of *Cnephia mutata* may inhabit the same stream (Basrur and Rothfels, 1959); those of *Calligrapha alni* may occupy adjacent bushes, whilst the amphimictic *C. philadelphica* and the closely related but thelytokous *C. vicina* may live side by side on the same host plant (Robertson, 1966). However, although extreme cases are easy to recognize it is difficult to decide how generally applicable the concept of geographical parthenogenesis is, not only because of a paucity of adequate information, but also because of ambiguities in terms like 'distribution' and 'range' (see MacArthur, 1972) and because of the possibility of small-scale spatial separation even between forms which occupy the same restricted geo-graphical area.

Latitude. Amongst forms whose geographical distributions are obviously separate, the best-supported and most widely accepted generalization is that

Table 3.18: Geographical Parthenogenesis in Insects. Initial letter abbreviations used for 'central' and points of the compass. Cf. Vandel (1931); Glesener and Tilman (1978).

Taxon	Species	Distribution of: sexual type	Distribution of: parthenogenetic type	Authorities
Ephem: Baetidae	*Baetis feminum*	S. Canada	Arctic Canada	Downes (1964)
Ephem: Heptageniidae	*Ameletus* spp.	Other spp: Pacific Coast of N. America	*ludens*: throughout N. America	Needham (1924, cited by Vandel, 1931)
Orth: Gryllidae	*Myrmecophila* spp.	*subdula*: Italy *surcoufi*: Africa	*acervorum*: C. Europe	Schimmer (1909) Chopard (1919)
Orth: Rhaphidophoridae	*Euhadenoecus insolitus*	N. America	N. limit of distribution	Hubbell and Norton (1978)
Orth: Rhaphidophoridae	*Hadenoecus cumberlandicus*	N. America	N. limit of distribution	Hubbell and Norton (1978)
Orth: Tettigoniidae	*Poecilimon intermedius*	Mediterranean region; S. European USSR	E. European USSR, N. to Siberia	Bei-Bienko (1954) White (1973)
Orth: Tettigoniidae	*Saga* spp.	About 15 spp. in Balkans, Turkey, Palestine, Iran	*pedo*. Most widespread sp.; Spain E. to S. USSR	White (1951a)
Phasmatodea	*Acanthoxyla* spp.	Eight sexual spp., throughout New Zealand	One parthenogenetic sp., in extreme northernmost part of New Zealand	Salmon (1955)
Phasmatodea	*Bacillus rossius*	S. Italy, Mediterranean coast, N. Africa	France, NW Italy. Sympatric with sexual race in C. Italy	Bergerard (1962)
Phasmatodea	*Didymuria violescens*	S. New South Wales	N. New South Wales	
Phasmatodea	*Leptynia* spp.	*attenuata*: Spain and Portugal	*hispanica*: Spain, Portugal and French Mediterranean coast	Cappe de Baillon and de Vichet (1940)
Phasmatodea	*Podacanthus wilkinsoni*	S. New South Wales	N. New South Wales	

Mantodea	*Brunneria* spp.	Tropical and subtropical S. America	*borealis*: S. United States	White (1948)
Embi: Oligotomidae Embioptera	*Haploembia* cf. *solieri* New World spp.	Mediterranean coast Tropical and subtropical	Mediterranean islands *Gynembia tarsalis*: S. United States	Stefani (1956) Ross (1944)
Psoc: Atropidae	*Cerobasis guestfalicus*	Germany, England	France	Badonnel (1943, 1951)
Psoc: Caeciliidae	*Caecilius auranticus*	Peripheral: N. Mexico; SE United States; N. Pacific coast of United States; N. Great Lakes and St Lawrence R.	Eastern N. America; scattered records in NW	Badonnel (1943) Thornton and Broadhead (1954) Smithers (1969)
Psoc: Elipsocidae	*Elipsocus hyalinus*	European mainland	British Isles	Badonnel (1943, 1951)
Psoc: Lepidopsocidae	*Echmepteryx hageni*	S. Florida; also at two localities in S. Illinois	Eastern N. America	Mockford (1971)
Psoc: Peripsocidae	*Peripsocus quadrifasciatus*	Peripheral: chiefly SE United States and NW Pacific coast	Eastern N. America, scattered records in W	Mockford (1971)
Psoc: Psocidae	*Psocidus pollutus*	S. Carolina and Gulf states	Eastern N. America	Mockford (1971)
Psoc: Psocidae	*Trichadenotecnum alexanderae*	New Jersey	Eastern N. America	Mockford (1971)
Thys: Aelothripidae	*Aelothrips fasciatus*	Europe	United States	Williams (1917)
Thys: Thripidae	*Anaphothrips striatus*	Michigan	Massachusetts	Shull (1914)
Thys: Thripidae	*Taeniothrips pyri*	Europe	California	Foster and Jones (1915) Williams (1917)
Homo: Aleurodidae	*Trialeurodes vaporariorum*	N. America, Britain, Denmark	Britain, Denmark	Schrader (1920, 1926) Thomsen (1927)
Homo: Aphidoidea			Anholocycly seems to be most common in arctic and tropical aphids	See text under 'Heteroecy'

Table 3.18 (*cont.*)

Taxon	Species	Distribution of: sexual type	Distribution of: parthenogenetic type	Authorities
Homo: Coccoidea	See text and Table 3.20			
Homo: Coccoidea			*Pulvinaria ellesmerensis* one of the two most northerly coccids, is thelytokous	Richards (1965)
Homo: Diaspididae	*Phenacaspis pinifoliae*	C. and S. California	N. and E. California, Utah, Arizona, Idaho	Brown (1965)
Homo: Lecaniidae	*Lecanium*, two spp.	Germany, C. and S. Europe	Facultatively parthenogenetic females found only in Denmark	Thomsen (1927)
Hemi: Miridae	*Chlamydatus pullus*	S. of range	Arctic Greenland	Bocher (1971)
Trich: Limnephilidae	*Apatania* spp.	*fimbriata*: S. Europe	*muliebris* and *arctica*: N. Europe	McLachlan (1874–80); Despax (1928)
Trich: Psychomyiidae	*Psychomyia flavida*	E. Asia	Michigan	Marshall (1939); Schmid (1965)
Dipt: Agromyzidae	*Phytomyza crassiseta*	Romania	Germany, N. Europe	Hering (1926)
		Predominates in S. of S. Sweden	Frequency increases to N. in S. Sweden; all-female samples from Finland and Norway	Block (1969)
Dipt: Agromyzidae	*P. plantaginis*	Eastern N. America, Europe	C. California	Frick (1951)
Dipt: Ceratopogonidae	*Culicoides* spp.	Most spp. are amphimictic	Tundra spp. are parthenogenetic	Downes (1964)
Dipt: Chironomidae			Predominance of parthenogenetic spp. at high latitudes in Arctic and Antarctic	Downes (1964)
Dipt: Chironomidae	*Monotanytarsus boreoalpinus*	S. of range	Finland	Lindeberg (1958)

Order: Family	Species	Distribution	Notes	Reference
Dipt: Chironomidae	*Tanytarsus norvegicus*	Mainland and in S. of range	Islands off coast of Finland	Palmén (1961)
Dipt: Culicidae	*Aedes* spp.	Most spp. are obligately amphimictic	Arctic spp. are often facultatively thelytokous	Corbet, cited by Downes (1964)
Dipt: Emphididae	*Tachydromia*, four spp.	S. Scandanavia, Denmark and further S. in Europe	Finland	Tuomikoski (1935)
Dipt: Simuliidae		S. spp. almost all obligate amphimicts	Thelytokous spp. (*Prosimulium ursinum*, *Gymnopais* sp.) occur in in Arctic	Basrur and Rothfels (1959); Downes (1964)
Lepid: Psychidae	*Acanthopsyche villosella*	Italy	N. Germany	Trautmann (1909)
Lepid: Psychidae	*Apterona helix*	Mediterranean	C. Europe	White (1973)
Lepid: Psychidae	*Solenobia triquetrella*	Unglaciated areas of C. European mountains	Diploid: C. European mountains, especially in glaciated areas. Tetraploid: widespread in Europe, N. to C. Finland	Seiler (1943, 1946, 1961)
Col: Chrysomelidae	*Adoxus obscurus*	Europe		Smith (1971)
Col: Chrysomelidae	*Calligrapha* spp.	Canada	Species diversity highest in central N. America, where both parthenogenetic and amphimictic spp. occur. In E. and W. coastal provinces of Canada fewer spp. are found and all appear to be amphimictic. Most N. record is for Manitoba at Lat. 54° N, for an amphimictic sp., *C. alni*; most S. record is for *C. scalaris* in Kansas, where both parthenogenetic and amphimictic types occur	Robertson (1966)
Col: Ciidae	*Cis fuscipes*	N. America, broadly overlapping range of parthenogenetic type	N. America; only parthenogenetic type found in central states. *C. fuscipes* is only sp. of Ciidae found in northern USA	Lawrence (1967)

Table 3.18 (*cont.*)

Taxon	Species	Distribution of: sexual type	Distribution of: parthenogenetic type	Authorities
Col: Curculionidae	European spp.	Generally speaking, the range of the parthenogenetic types is larger and situated to the north of that of the amphimictic types, which live in the mountains of C. Europe. Where they are sympatric the parthenogenetic type usually occupies glaciated and the amphimictic type unglaciated areas. See text and Table 3.19		Reviewed recently by Suomaleinen (1969), Suomaleinen *et al.* (1976)
Col: Curculionidae	*Scepticus griseus*	S. Japanese islands	Hokkaido	Takenouchii (1961)
Col: Pselaphidae	*Bythinopsis tychoides*	New York, New Jersey	West of Allegheny Mountains	Reichle, cited by Dybas (1966)
Col: Ptiliidae	*Eurygyne* spp.	Tropical and subtropical	Five spp. at N. limit of range of genus in S. United States	Dybas (1966)
Col: Staphylinidae	*Amischa* spp.	European mainland	British Isles	Williams (1969)
Hymen: Leucospididae	*Leucospis gigas*	Mediterranean	C. France	Berland (1934)
Hymen: Pdecinidae	*Pelecinus polyturator*	Tropical S. America	N. America	Brues (1928)

Figure 3.5: Distribution of Apomictic and Amphimictic Races of the European Weevil *Otiorrhynchus dubius*. Diploid amphimictic race mapped in black, tetraploid apomictic race cross-hatched.

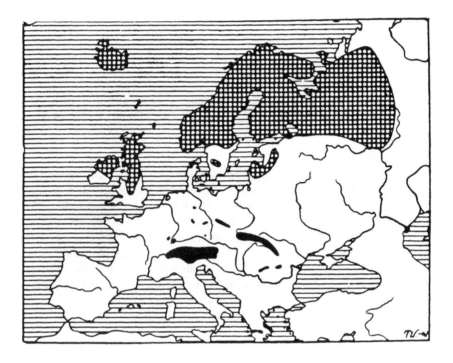

Source: Suomaleinen (1969), p. 277, Figure 23.

parthenogenetic biotypes occupy areas at higher latitudes. Of the 56 instances in Table 3.18, 13 are immaterial because they concern longitudinal rather than latitudinal separation, and in addition the North American Psocoptera are a special case that will be discussed further below; of the remaining instances, 30 conform to the pattern, without including the many weevils in which apomictic races are found in northern and amphimictic races in central Europe; these are noted separately in Table 3.19. The aphids are ambiguous; parthenogenesis seems to increase both towards high and towards low latitudes (see below, under 'Heteroecy'). There are only four clear counterexamples. the chrysomelid *Calligrapha*, in which the most northerly species for which sex-ratio data are available is amphimictic, whilst the most southerly is a mixture of amphimictic and parthenogenetic races; two European psocopterans which are said to be amphimictic in Germany and Britain but thelytokous in France; the phasmid *Acanthoxyla*, where a parthenogenetic species inhabits high latitudes in New Zealand whilst all its amphimictic relatives live further south; and the

Table 3.19: Geographical Parthenogenesis in European Curculionidae. The trend towards parthenogenesis at higher latitudes is real but perhaps overstated by the table: samples from the Austrian Alps were taken only from unglaciated areas, where sexual forms predominate.

| | | Number of species | |
	Sexual	Parthenogenetic	Total
Fennoscandia	2	7	9
Poland	10	5	15
Switzerland	8	11	19
Austrian Alps	13	5	18
Total	33	28	61

Source: Suomaleinen (1969), p. 280, Table 3.

genus *Drosophila*, where the only obligately thelytokous species, *D. mangabeiri*, is neotropical. I have also examined the latitudinal distribution of sexual and thelytokous coccids, using the extensive data reviewed by Brown (1965). Surprisingly, there is no tendency for parthenogenetic species to become relatively more abundant at higher latitudes; if anything, indeed, a weak trend in the reverse direction might be inferred (Table 3.20a). It will be recalled that a similar result was obtained for North American ostracods, where the correlation between parthenogenesis and high latitude seems to hold good at low but not at high phyletic rank. Of two species of coccid from very high latitudes in the Canadian Arctic, one is sexual and one thelytokous (Richards, 1965).

Size of Range. It has often been said that parthenogenetic forms occupy larger ranges than their nearest sexual relatives. This is the case in the mayfly *Ameletus ludens*, in the orthopteroids *Poecilimon intermedius*, *Saga pedo* and *Leptynia hispanica*, in most psocopterans, in *Solenobia triquetrella* and in most of the European Curculionidae. However, it appears not to be the case in the whitefly *Trialeurodes vaporariorum* (where the contrast is between thelytoky and arrhenotoky), in the dipterans *Tanytarsus norvegicus* and *Tachydromia* spp. and in the beetles *Calligrapha* spp., *Cis fuscipes* and *Amischa* spp.; it is probably not the case in the coccid *Lecanium* and in the dipteran *Phytomyza plantaginis*. In addition, I suspect that in several other forms, for which adequate distributional data are not available, parthenogenesis occurs only in a relatively narrow northern fringe. The proposed generalization is therefore not securely founded.

Peripheral and Central. It has also been said that parthenogenetic forms tend to occupy the periphery and amphimictic forms the centre of their joint range; obviously, this is difficult to distinguish from the previous case. In fact, I can find no well-documented instances to substantiate the claim, although some might exist at certain phyletic levels amongst aphids and chironomids. It seems certain, however, that in several North American psocopterans and in the beetle

Table 3.20: Geographical Parthenogenesis in Coccids.

(a)

Latitude.

Degrees latitude North or South	Number of localities for:		Total
	sexual spp.	parthenogenetic spp.	
0–10	18	8	26
10–20	36	10	46
20–30	38	15	53
30–40	40	9	49
40–50	16	4	20
Total	148	46	194

Source: Brown (1965)

(b)

Mainland and Island Distribution. Islands are: Jamaica (17 records), Trinidad (ten), Oahu (Hawaii) (seven), Sicily (seven), Hokkaido (seven; including one locality given as 'Japan'); Mauritius (six), Bermuda (three); and Ceylon, Java, Fiji and Guadalupe Island, Baja California (one each).

	Number of localities for:		Total
	sexual spp.	parthenogenetic spp.	
Mainland localities	99	34	133
Island localities	49	12	61
Total	148	46	194

Source: Brown (1965)

(c)

Occurrence on Extremely Isolated Islands. Three of the records of parthenogenetic species on these islands refer to *Howardia biclavis*.

	Number of species recorded:		Total
	sexual	parthenogenetic	
Mauritius	4	2	6
Bermuda	2	1	3
Oahu	4	3	7
Fiji	1	0	1
Total	11	6	17

Source: Brown (1965).

Cis fuscipes the reverse is true: the parthenogenetic forms occupy the centre and the sexual forms the periphery of the range (see the figure in Lawrence, 1967, p. 5). This also holds for different species within the genus *Calligrapha*. I hesitate to claim this as a generalization on such slender evidence; but the original generalization is certainly untenable at present for insects.

Island and Mainland. Several recent authors (e.g., Cuellar, 1977; Glesener and Tilman, 1978) have claimed that parthenogenetic forms tend to occupy islands, whilst related amphimicts occur on the nearby mainland. This holds for *Haploembia* cf. *solieri* and for *Tanytarsus norvegicus*; and in *Elipsocus hyalinus* and *Amischa* spp., males seem to be common on the European mainland but very rare in the British Isles. Specimens of *Cis fuscipes* collected on Madeira, Maui (Hawaii) and Cuba were all females, whereas males are known from about 20 per cent of mainland localities. The thelytokous coccid *Howardia biclavis* might be cited as a further instance, since the only localities given by Brown (1965) are Bermuda, Mauritius, Hawaii and Jamaica. However, the coccids serve better to illustrate the danger of citing examples without any more rigorous comparative analysis. The locality records given by Brown (1965) have been collated in Table 3.20b, and fail to show any correlation at all between parthenogenesis and island life. Moreover, Table 3.20c, taken from the same data set, shows no notable excess of parthenogenetic coccids even on the very isolated islands of Mauritius, Bermuda, Oahu and Fiji. Dybas (1966) described the North American distribution of eight species of the ptiliid genus *Eurygyne*, of which five are probably thelytokous and two amphimictic (the remaining species is known from only a very few specimens). Their distribution is less likely to have been affected by human activity than that of coccids. Of the three species found on South Bimini Island, relatively close to the mainland, two were thelytokous and one amphimictic; the only species recorded from the much more isolated island of Bermuda was thelytokous. The domiciliary cockroach *Pycnoscelus surinamensis* is usually apomictic, but specimens from Hawaii were found to be amphimictic (Roth and Willis, 1961). In short, the correlation between parthenogenesis and island life, if not illusory, is at all events elusive.

Maritime and Continental. A related assertion, made by the same authors, is that parthenogenetic forms tend to occupy continental and amphimictic forms maritime regimes. Since islands have a maritime climate, this must be carefully separated from the previous case. Support for the generalization is given by *Cis fuscipes*, where central North America is occupied exclusively by thelytokous populations, and by *Calligrapha*, in which parthenogenetic species are common in central Canada but unknown from the coastal provinces. Amphimictic populations of Psocoptera also tend to occur chiefly on the southern and maritime fringes of their species' distribution in North America.

Longitude. In a number of cases, thelytokous and amphimictic populations are separated by longitude rather than by latitude or, at least, are separated by so many degrees of longitude that any difference in latitude seems accidental: for example, the caddis fly *Psychomyia flavida* and the chrysomelid *Adoxus obscurus*. In some cases these longitudinally disjunct distributions can be explained by human introduction (see below); in the rest, no explanation or generalization has been proposed.

Ploidy and Distribution. Suomaleinen has often maintained that a major proximate cause of geographical parthenogenesis is the higher ploidy of parthenogenetic forms; that is, that cytotypes with different ploidies have different distributions, their modes of reproduction being more or less incidental. In support of this claim he has cited an observation made by Linderoth (1954) that the lower limit of thermal activity in tetraploid parthenogenetic *Otiorrhynchus dubius* from Scandinavia is about $2°C$ lower than that of the diploid amphimictic race from the Alps. This is doubly unsatisfactory: first, because the experiment cannot distinguish between the effect of ploidy and the effect of parthenogenesis; and secondly, because any form living at higher latitudes might be expected to evolve a capacity to function at lower temperatures, irrespective of its ploidy or mode of reproduction. To investigate the assertion we must compare like with like: first by comparing diploid parthenogenetic taxa with their diploid amphimictic relatives, and then by comparing parthenogenetic races with different ploidies. Neither of these comparisons yields very encouraging results. The diploid parthenogenetic biotypes of *Bacillus*, *Brunneria* and *Apterona* show the usual tendency for parthenogenesis to predominate at higher latitudes, though *Drosophila mangabeiri* can stand as an exception. The summary table in Suomaleinen (1969) shows no clear difference between the distribution of parthenogenetic weevils with different ploidies, except in the case of *Polydrosus mollis*, where the diploid race is found in Finland and Poland, to the north of the triploid race in Germany and Switzerland. I conclude that the mode of reproduction rather than ploidy is instrumental in causing differences in geographical distribution.

Automixis and Apomixis. The classical instances of geographical parthenogenesis refer both to automicts (*Bacillus*, with postmeiotic restitution; psychids, with intrameiotic restitution) and to apomicts (curculionid weevils). Although one would welcome more evidence, there is at present no reason to believe that the tendency for thelytoky to occur at high latitudes depends on the cytological mechanism involved.

3.32.6 Ecological Parthenogenesis

'Ecological parthenogenesis', by analogy with 'geographical parthenogenesis', is intended as a convenient if inelegant term to refer to correlations between parthenogenetic modes of reproduction and various ways of life.

Elevation. There is a weakly supported tendency for parthenogenetic forms to occupy higher elevations. The aphid *Anuraphis subterranea* is holocyclic at low elevations in central Asia but anholocyclic at high elevations (Narzikulov, 1970). Thelytokous populations of *Cis fuscipes* are found only at high elevations (Lawrence, 1967). Among weevils, *Tropiphorus terricola* is exceptional in living at high altitudes, and is parthenogenetic; but *Liophloeus tessulatus* is a counter-example, since the triploid parthenogenetic type is found at lower elevations than the diploid amphimicts (Horion, cited by Suomaleinen, 1969). The statement by Glesener and Tilman (1978) that thelytokous populations of the beetle *Orthotomicus* live at high altitudes is mistaken: *Orthotomicus*, like *Pityophthorus* and *Dendroctonus*, is an amphimictic beetle which often produces all-female broods, because of a sex-linked lethal which aborts male embryos (see Lanier, 1966; Lanier and Oliver, 1966; Lanier and Wood, 1968).

Exposure. When it is possible to classify habitats into open and closed, or xeric and mesic, parthenogenetic insects tend to be found in open, xeric habitats. This is reflected in the correlations of parthenogenesis with geography and elevation; to obtain closely comparable material in which these variables can be discounted is very difficult. Perhaps Mockford's (1971) observation that the thelytokous race of the psocopteran *Echmepteryx hageni* lives in trees, whilst amphimictic males and females at the same localities live on rocks, belongs here. More convincingly, the habitat descriptions given by Baud (1973) make it clear that *Lonchoptera dubia* occupies more open habitats than most of its amphimictic congeners.

Disclimax. There is a tendency for parthenogenetic insects to occupy disclimax habitats. Takenouchi (1965) examined 34 species of Canadian weevils, of which four turned out to be thelytokous. These four were found on vacant grassy lots (three species) or on dandelion, a colonizing herb of early succession (one species). Only eight of the 30 amphimicts were collected on vacant lots, and only one on dandelion; three others were collected on a lawn, in a park and on strawberry, respectively, but a majority (18/30) of amphimictic species was found only on trees or bushes. The parthenogenetic water-beetle *Helophorus orientalis* is a colonist of temporary ponds (Angus, 1970). In two cases, parthenogenetic species are known to occupy recently burned areas. White *et al.* (1963) found *Warramaba virgo* only in an area which had recently suffered a bush fire, and the samples taken by White and Contreras (1979) were from land subject to burning, tree-felling and sheep-grazing. The principle food plant of the apomictic chrysomelid *Adoxus obscurus* in northern Europe is fireweed (*Chaemaenerion*), a plant of very early succession which appears in places where soil nitrogen has been elevated, especially by fire (Lokki *et al.*, 1976a).

Glaciation. A dozen or so species of *Otiorrhynchus* (Curculionidae) have both sexual and apomictic races in the eastern Alps, where Jahn (1941) made the remarkable observation that the amphimictic races occur only in areas which were ice-free during the last glaciation, whilst the apomicts have extended their distribution into glaciated areas. This has since been confirmed by Suomaleinen (1947). Nor is this an isolated case. The diploid amphimictic race of *Solenobia triquetrella* has been found only in an unglaciated region north of the Alps, and in areas of the Swiss Mittelland which were raised above the glaciers, whilst the diploid automictic race is also found in glaciated areas of the Alps and the Swiss Mittelland (Seiler, 1961). According to Cuellar (1977), the distribution of the tettigoniid *Saga pedo* suggests the early colonization of glaciated areas, with the species now becoming steadily rarer as these habitats approach a post-glaciation climax. We may also place here the curious distribution of certain psocopterans, worked out by Mockford (1971), where amphimictic populations are usually found on the periphery of the range of the species or, if centrally located, occupy unusual habitats. These amphimictic populations are most often found in Florida and in the Mexican Highlands, both of which remained unglaciated during the Pleistocene. Exceptional amphimictic populations of *Echmepteryx hageni* in southern Illinois occupy sandstone outcrops just below the limit of glaciation. Amphimictic populations of *Peripsocus quadrifasciatus* in Indiana and Tennessee occupy old native stands of hemlock on bluffs overlooking streams, probably also a refugial habitat. In short, it seems that parthenogenetic races tend to occupy recently glaciated areas whilst closely related amphimicts occupy nearby glacial refugia. I know of no clear counterexamples.

Caves. The orthopterans *Dolichoptera*, *Hadenoecus* and *Euhadenoecus* and the thysanuran *Nicoletia* belong to families which are principally cavernicolous in habitat. This may be too slight a basis for generalization, but seems too substantial for coincidence (see Section 4.4).

Domiciliary Species. Many of the insects which have succeeded in colonizing human habitations and storehouses seem to have been already thelytokous, for example, the thysanuran *Nicoletia meinerti*, the tropical thrips *Heliothrips haemorrhoidalis* and the coccid *Howardia biclavis*. What is more interesting is that in some cases parthenogenetic domiciliary insects have been derived from an amphimictic wild stock, showing that the mode of reproduction is somehow critical to success in the new way of life. Wild populations of the beetle *Ptinus clavipes* are diploid and amphimictic, but in warehouses a triploid gynogenetic form (*mobilis*) is also found (Sanderson, 1960). The apomictic domiciliary cockroach *Pycnoscelus surinamensis* seems to have arisen from a wild amphimictic stock, the tropical *P. indicus* (Roth, 1970; Roth and Willis, 1956). The coccid *Aspidiotus hederae* is thelytokous in greenhouses, but males occur in wild populations (Schrader, 1929). The only species of the beetle *Eurygyne* to occur north of Florida is the thelytokous *E. fusca*, which in northern localities lives exclusively in the sawdust piles of lumber mills (Dybas, 1966). The only counterexample I have found is a rather curious one: on the basis of sex ratios and breeding experiments, Schrader (1926) concluded that a native thelytokous English race of the greenhouse whitefly *Trialeurodes vaporariorum* was largely displaced between 1917 and 1920 by an arrhenotokous race introduced from America.

Introduction. Where comparative material is available, parthenogenetic insects seem to invade natural habitats as successfully as they infest human habitations. Takenouchi (1965, 1970a) found that five species of weevil have successfully invaded North America (four from Europe and one from Japan); all are apomictic polyploids. The phasmid *Sipyloidea sipylus* is amphimictic (though with some facultative parthenogenesis) in the Malay Archipelago, where it is native, but is thelytokous in Madagascar, where it has been introduced (White, 1973).

Abundance. Parthenogenetic insects may attain high local population densities, as the simuliids of the Canadian Arctic do, but there is some indication that they are usually less abundant than closely related amphimicts. *Saga pedo* is both by far the rarest and the only parthenogenetic member of its genus. When the arrhenotokous race of *Trialeurodes vaporariorum* replaced a thelytokous race in English greenhouses, population density increased to a point where the insect became for the first time a serious pest (Schrader, 1926). Table 3.21 presents some semiquantitative data for *Calligrapha* from Robertson (1966): parthenogenetic species tend to be locally somewhat less abundant than amphimicts, and Robertson remarks that their colonies, as well as being smaller, are also less numerous.

Population Structure. Arctic simuliids not only reach high local population densities: they may also form large continuous populations extending over a considerable area. In more temperate regions, however, a quite different sort of population structure seems to be characteristic of parthenogenetic insects: a pattern of small, isolated populations each occupying an island-like patch of habitat. This pattern is common to the orthopterans *Saga pedo* (White, 1951a) and *Warramaba virgo* (White, 1973), the mantid *Brunneria borealis* (White, 1973), the dipteran *Drosophila mangabeiri* (Carson, 1962), and the beetles *Cis fuscipes* (Lawrence, 1967) and *Eurygyne fusca* (see above) (Dybas, 1966), whereas the nearest amphimictic relatives of these forms have much more continuous distributions.

Dispersal. It has often been remarked that, while parthenogenesis has evolved in a few flightless or feebly flying psychids, mantids, phasmids, grasshoppers, scale insects and so forth, it is very scarce or entirely absent in such highly mobile forms as butterflies and dragonflies. Among the orthopteroids, *Warramaba virgo* and *Brunneria borealis* are both apterous or nearly so, though most amphimictic forms are quite powerful fliers. The same is true of thelytokous psychid moths. The majority of parthenogenetic weevils, in contrast to

Table 3.21: Reproduction in *Calligrapha* (Coleoptera: Chrysomelidae). Material from Canada and north-eastern United States. 'Abundance' is a subjective assessment expressed on a logarithmic scale to the base two; no estimate is available for *verrucosa* and *pruni*, which nevertheless had rather large sample sizes. The figures in the third column are expressed as an arithmetic mean over all samples, with range for different samples given in parentheses. The fourth column gives the minimal number of chiasmata at diplotene or, when this information is not available, at first metaphase ('at MI'). The fifth column gives the average number of eggs laid by females in the laboratory; figures in parentheses are judged to be unreliable because specimens may have laid some eggs before capture.

Species	Abundance	Percent males	Chiasmata	Fecundity	Ploidy
Amphimictic					
rowena	1	31 (0–78)	14–16		$2n$
philadelphica	8	40 (0–50)	15–18	308	$2n$
phirsa	2	50	14	(253)	$2n$
amator	5	42 (36–53)	13 (at MI)		$2n$
confluens	5	25 (0–50)	18	(93)	$2n$
californica	4	56 (50–64)	11		$2n$
bidenticola	1	48 (47–50)	14		$2n$
multipunctata	10	47 (low–54)	16	359	$2n$
verrucosa	(high)	34 (20–46)	14 (at MI)		$2n$
pruni	(high)	56 (50–83)	13 (at MI)		$2n$
Facultatively parthenogenetic					
alni	3	29 (50–83)	16	422	$2n$
scalaris	4	27 (7–48)	15 (at MI)	148	$2n$
Obligately parthenogenetic					
vicina	5	0		(79)	$4n$
virginea	2	0		301	$4n$
apicalis	4	0		168	$4n$
alnicola	1	0		(117)	$4n$
scalaris		0.3			$4n$
ostryae	2	0			$4n$

Source: Robertson (1966).

amphimictic species, are flightless or have only vestigial wings, though *Polydrosus mollis* and *Listroderes costirostris* are exceptions to the rule (Takenouchi, 1970). At high latitudes there is a very pronounced tendency towards the reduction of wings, with or without the evolution of parthenogenesis. The two northernmost coccids are *Pulvinaria ellesmerensis*, which is thelytokous, and *Pseudococcus altoarcticus*, which has wingless males (Richards, 1965). In high Arctic aphids there is not only a tendency to anholocycly, especially when the species extends to the north beyond the range of its woody host (Hille Ris Lambers, 1955), but also to the development of a peculiar brachypterous male (Hille Ris Lambers, 1960; Richards, 1963). The Arctic simuliids generally have morphologically reduced nonfeeding females, but in the parthenogenetic *Prosimulium ursinum* and *Gymnopais* spp. this reduction is carried to an extreme: eggs are deposited, without dispersal, in the first day or two of the short life of a structurally degenerate female (Downes, 1964). In northern Norway, most *P. ursinum* release their eggs by rupture of the pupal wall, and the adults never appear (Carlsson, 1962). The limit of this tendency is reached in the paedogenesis of cecidomyids and *Micromalthus*, where a nondispersing larva reproduces by thelytoky and a winged adult by amphimixis or arrhenotoky. In short, there seems to be a sound general rule that parthenogenesis is associated with a reduction of the ability to disperse in space.

Heteroecy. In aphids there seems to be a correlation between holocycly and the habit of alternating between two host species. The adelgid *Sacchiphantes viridis* is a particularly clear example. Like most Adelgidae it alternates between two species of conifer, producing galls on the primary host. The fertilized egg is laid on spruce, and gives rise to a female whose parthenogenetic offspring are winged and migrate to larch. Here the species goes through several more parthenogenetic generations, and in the following year the final parthenogenetic generation migrates back to spruce, where sexual individuals are produced, whose fertilized eggs complete the cycle. Two monoecious species are thought to have been derived from *S. viridis*: *S. abietis*, which lives on spruce, and *S. segregis*, which lives on larch. Both monoecious species are anholocyclic, the sexual generation being entirely suppressed (Steffan, 1961, 1962). Now, heteroecy is most common in temperate regions, and becomes less common towards the poles and towards the equator: for example, of 35 species of aphid in Iceland only six are heteroecious (Prior and Stroyan, 1960), whilst only monoecious species are known from Greenland (Hille Ris Lambers, 1960) and Arctic Canada (Richards, 1963). I suggest that the prevalence of holocycly in temperate aphids reflects the frequency of heteroecy, whilst the increase of anholocycly at higher and at lower latitudes is associated with an increase in the monoecious habit.

Among biting flies there may be a correlation between thelytoky and autogeny (absence of feeding in the adult). This is certainly true in the ceratopogonid genus *Culicoides*, where *C. bermudensis* and *C. bambusicola* are thelytokous and autogenous (Williams, 1961; Lee, 1968); it may also be true for the ceratopogonid

Dasyhelea (where evidence of thelytoky is not conclusive; see Downes, 1955, 1958) and the simuliid *Prosimulium ursinum* (where amphimictic races in North America are also autogenous; see references in Table 3.15 and Davies, 1954).

Fecundity. There seems to be a tendency, although only a weak and erratic one, for parthenogenetic insects to have relatively low fecundities. Typical biting simuliids mature about 200–500 eggs during their first reproductive attempt, and may go on to make further attempts; in nonbiting Arctic forms, the average number sinks to about 150, and only one reproductive attempt is made; and the lowest fecundity of all is that of parthenogenetic species, which produce only about 20 eggs (Downes, 1964). However, this is difficult to disentangle from the generally lower fecundity and larger eggs which are characteristic of many wholly sexual taxa of Arctic organisms. In *Calligrapha* the parthenogenetic species seem to be somewhat less fecund (see Table 3.21), but no stronger statement can be justified. White (1973) has remarked that orthopteroids which have an achiasmatic meiosis are all relatively small, and infers that they produce relatively few eggs. On the other hand, since most parthenogenetic weevils are polyploid, and polyploidy is associated with an increase in body size but not with any change in egg size (see above), they may illustrate the reverse trend.

Degenerate Males. Amphimictic coccids and Arctic dipterans have highly specialized, structurally degenerate males; Downes (1964), Nur (1971), Porter (1971) and White (1973) have suggested that in these cases thelytoky provides a means of eliminating a 'biologically unsatisfactory' male sex. Stefani (1956) has made the same suggestion with respect to *Haploembia* cf. *solieri*, where the males, but not the females, of the amphimictic race are heavily parasitized by a gregarine. As an hypothesis, it is radically unsound; as a generalization it holds well enough, though it must be borne in mind that structural degeneracy is a common adaptation of high-latitude organisms, connecting with the quasipaedogenetic modes of reproduction described above.

3.33 Chaetognatha

Reviews. Alvarino (1965); Ghirardelli (1968); Reeve and Cosper (1975).

Incidence of Parthenogenesis. The chaetognaths are the dominant predators of the marine zooplankton; one genus is benthic. All are simultaneous hermaphrodites. No means of nongametic reproduction has been reported, and the authorities disagree about the importance of self-fertilization. In the benthic chaetognath *Spadella cephaloptera* the spermatophore must be placed in a certain position on the dorsal surface of the animal to ensure fertilization (Ghirardelli, 1968); self-spermatophores are as effective as foreign spermatophores, but Ghirardelli concludes that the site of spermatophore placement acts as a morphological barrier to self-fertilization. In the planktonic *Sagitta hispida* the spermatophore is placed laterally, and self-fertilization is easily induced by experimentally rupturing the seminal vesicles (Reeve and Walter, 1972); nevertheless, only a very small fraction of isolated unmanipulated individuals produced viable eggs. In other species of *Sagitta*, however, self-fertilization has been known for a long time (Stevens, 1910). In *S. setosa*, Jägersten (1940) observed apparently spontaneous self-fertilization, and Dallot (1968) found that about 50 per cent of isolated individuals spawned viable eggs; neither author observed copulation or cross-fertilization. It seems, then, that most chaetognaths are capable of self-fertilization, and that in some species it may be a regular mode of reproduction.

3.34 Pogonophora

Review. Southward (1975).

Incidence of Parthenogenesis. Pogonophores are benthic tubiculous marine worms. *Sclerolinum brattstromi* is an elongate species which fragments readily; the fragments regenerate into complete small worms which, at least for some time, share the same tube. The same process of architomy may occur in other species of *Sclerolinum*, but otherwise asexuality and parthenogenesis are unknown; since pogonophores are gonochoric, there is no possibility of self-fertilization.

3.35 Echinodermata

Incidence of Parthenogenesis. Echinoderms are an exclusively marine group of benthic carnivores and suspension feeders, usually most abundant in the littoral. Almost all echinoderms are gonochoric amphimicts. Artificial parthenogenesis was demonstrated in sea-urchin eggs by Loeb (1899), and was subsequently investigated by many workers (see Tyler, 1941). In some starfish and sea-urchins unfertilized eggs may develop spontaneously (see MacBride, 1896; Viguier, 1900; Lyon, 1903; Newman, 1921), but the distinction between a spontaneous process and artificial parthenogenesis caused, for example, by temperature shock is not often clear, and only Newman's observations on *Asterina miniata* suggest the frequent occurrence of parthenogenesis in natural populations. Self-fertilization does not occur; the hermaphroditic holothurian *Labidoplax buski* is self-compatible but eggs and sperm are not shed simultaneously (Nyholm, 1921). Thus, the only regular mode of reproduction, other than amphimixis, is the transverse fission of certain holothuroids, asteroids and ophiuroids.

HOLOTHUROIDEA. Synaptid holothurians may undergo multiple fission, but apparently only the most anterior fragment can regenerate a complete new individual (Pearse, 1909; Domontay, 1931). However, *Cucumaria* and *Holothuria* have been seen to reproduce by a more or less equal binary fission in the labora-tory (Chadwick, 1891; Crozier, 1917), and *Holothuria surinamensis* does so in nature. Deichmann (1921) found that many individuals in samples of *Holothuria parvula* and *Actinopyga difficilis* were regenerating, some orally and some anally, and concluded that fission was a regular mode of reproduction. Both *Holothuria* and *Actinopyga* form dense aggregations in shallow subtropical seas, but their clonal structure has not been investigated.

ASTEROIDEA. Asexual reproduction by fission has been reported from the asteriids *Coscinasterias*, *Sclerasterias*, *Stephanasterias*, *Allostichaster* and *Linckia* (see Fisher, 1925; Yamazi, 1950; Hyman, 1955), and from the asterinids *Nepan-thia* and *Asterina* (Bennett, 1927; Edmondson, 1935). Fission is often an import-ant mode of reproduction and, in Greenland populations of *Stephanasterias albula*, sexual reproduction is a rare phenomenon (Thorson, 1936). Fission occurs during summer in *Coscinasterias tenuispinus*, being replaced by amphimixis in January and February (Crozier, 1920), but in *C. acutispina* fission continues throughout the year (Cognetti and Delavault, 1962). Fission is confined to younger individuals in *Stephanasterias* and *Coscinasterias*, sexual reproduction occurring later in life (Fisher, 1925; Cognetti and Delavault, 1962; Delavault, 1966). In most cases, regeneration is successful only if the fragment includes part of the disc, but a more extreme architomy is practised by *Linckia*, where an arm alone, or even part of an arm, can give rise to a complete new individual (Kellogg, 1904, cited by Hyman, 1955; Monks, 1904; Clark, 1913). Thus, where *Linckia* is common, cast-off arms with four small regenerating arms are often found; they are called 'comets'.

OPHIUROIDEA. Fission seems to be confined to small six-armed forms. It is especially common in *Ophiactis* – almost all specimens of *O. virens* collected by Simroth (1877) had recently undergone fission – and has also been reported from *Ophiothela*, *Ophiocoma*, *Astrogymnotes*, *Astrocharis* and *Astroceras* (see Mortensen, 1933; Yamazi, 1950). As in asteroids, fission seems to be characteristic of younger individuals. The hundreds of juvenile *Ophiactis virens* which can be found within a single large sponge (Simroth, 1877) may all have descended from the same founder by fission.

3.36 Hemichordata

Review. Hadfield (1975).

Incidence of Parthenogenesis.

PTEROBRANCHIA. Pterobranchs are sessile tentaculate marine animals which for convenience I shall unite with enteropneusts as Hemichordata. There are three genera. *Atubaria* is a solitary naked zooid, found only once, on a hydroid off the coast of Japan. Only juveniles and females were found; none were budding. *Cephalodiscus* occurs in rather large aggregations of unconnected zooids which inhabit a common casement, the coenecium. The zooids in each colony may be male, or female, or a mixture of both, or hermaphrodite; nothing is known about fertilization. Each colony is believed to arise by basal budding from a single sexually produced larva. As many as a dozen buds may be borne at the same time by a single zooid. According to Hyman (1959), the buds do not disperse; the dispersal stage is the ciliated, planula-like sexual larva. I have found no report of a dormant stage. *Rhabdopleura* consists of a branching prostrate stolon giving rise at frequent intervals to erect tubes, each of which houses a zooid. The zooids are connected by a mesenteric strand of tissue which passes through the interzooecial walls. As in *Cephalodiscus* the colony arises by asexual budding from a single ovoid, ciliated sexual larva, which represents the dispersal stage. Some buds contain food reserves and lie dormant rather than differentiating immediately into a feeding zooid; they appear to represent an asexually produced resistant stage. A good account of the reproduction of *R. compacta* has been published recently by Stebbing (1970).

ENTEROPNEUSTA. Enteropneusts are solitary burrowing marine worms, found in the upper littoral. All are gonochoric. Many enteropneusts are known to possess considerable powers of regeneration, but in many species posterior fragments regenerate anteriorly only very slowly, if at all. Regular asexual reproduction by a complicated process of architomy has been described in two genera of Ptychoderidae, *Balanoglossus* (Gilchrist, 1923; Packard, 1968) and *Glossobalanus* (Petersen and Ditardi, 1971); it may also occur in *Ptychodera* (Crozier, 1920). The adult worm breaks into two large fragments; the anterior half then divides into two pieces, the more posterior of which breaks into several small fragments ('regenerands'). The original worm has thus given rise by vegetative fission to two large and several small fragments. The two large fragments regenerate anteriorly or posteriorly; the small fragments differentiate into morphologically complete worms before growing to adult size. This process was first adequately described by Gilchrist (1923) in *Balanoglossus*, who discovered that an asexually reproducing summer form (*proliferans* s.l.) gave rise to a sexually reproducing winter form (*capensis*). Sexual reproduction gives rise to a small ciliated tornaria larva, which is probably responsible for dispersal (but see Packard, 1968); small vegetatively produced individuals are often found in the same burrow as the parental worms, or in nearby burrows, and at least in some localities the existence of large isolated clones of vegetatively produced individuals has been inferred (Packard, 1968). There is no dormant stage.

PLANCTOSPHAEROIDEA. A class erected for large planktonic tornaria-like organisms, perhaps the larvae of enteropneusts. Their reproduction is unknown.

3.37 Tunicata (Urochordata)

Reviews. Berrill (1935, 1951, 1975); Brien (1958).

Incidence of Parthenogenesis.

   ASCIDIACEA. Sedentary benthic tunicates. Asexual budding is universal amongst colonial ascidians, but does not occur in solitary forms. Ascidian colonies range from highly integrated complexes, in which the zooids are connected by a vascular system, to loose groups of physically isolated, autonomous zooids, and budding may thus represent anything between a nearly pure process of growth and a nearly pure process of reproduction. In no case, however, do the budded offspring disperse actively away from the site where they are produced. Budding in ascidians has been intensively studied because of the variety and complexity of the morphogenetic processes involved. I shall give only a very brief account of these, family by family; the reviews by Berrill and Brien should be consulted for details and for references to the earlier literature.

Diazonidae. A series of transverse fission zones appears in the abdomen, cutting off up to eight cell masses supplied with food reserves ('strobilation'). The parental zooid degenerates anteriorly, and the buds regenerate new, physically isolated zooids, increasing the size of the colony.

Polyclinidae. A large number of buds are formed by strobilation of the post-abdomen.

Didemnidae. A pair of external buds form on the oesophageal region, the anterior developing into the abdomen and the posterior into the thorax of a new zooid; physical continuity with the parental zooid is retained.

Clavelinidae (Polycitorinae). Abdominal strobilation.

Clavelinidae (Holozoinae). Buds are formed both by the larva and by the zooid, through the strobilization of an elongate stolon.

Clavelinidae (Clavelininae). Buds are formed by the accumulation of trophocytes in stolonic ampullae. These are dormant structures which germinate after the degeneration of the colony.

Perophoridae. Buds appear distally on the upper surface of the stolon, remaining in communication with the parental zooid through a strand of mesenchymatous tissue, the whole colony sharing a common vascular system. Budding continues throughout the active life of zooids, not causing their dissolution, as it does in diazonids and polyclinids.

Styelidae (Botryllinae). An outgrowth of the thoracic wall becomes either a bud, or a stolon on whose tip a bud rudiment forms. The bud may be paratomical, remaining attached to its parent until fully grown (*Botryllus*), or it may detach and become secondarily attached to the colony vascular system (*Metandrocarpa*) or, well supplied with reserve materials, it may become detached and grow independently (*Stolonica*). In brief, then, asexual reproduction in ascidians may involve budding to form a highly integrated colony in a manner reminiscent of hydrozoans or entoprocts (as in perophids) or, at the other extreme, transverse fission to form a loose aggregation of individuals united only by a common matrix, in a manner which approaches that of annelids (as in polyclinids). In all cases the dispersal phase is a sexually produced appendicularia (tadpole) larva. The dormant phase is a strobilated asexual bud, rich in food reserves and somewhat reminiscent of an ectoproct sessoblast.

THALIACEA. Pelagic tunicates.

Pyrosomida. The fertilized egg develops into an oozooid, which proliferates by strobilization of a stolon. Further generations of zooids are added to the colony by subsequent stolonic budding. The zooids are functionally cooperative – their atrial siphons provide the current which moves the colony through the water – but anatomically separate, being merely embedded in a common test.

Doliolida. The fertilized egg is liberated by a solitary sexual individual (the gonozooid), and develops into a tadpole larva. This loses its tail and by the strobilization of a ventral stolon becomes colonial. The colony comprises not only permanently attached feeding zooids but also a special class of phorozooids, which after a short period of attachment are released and become free-swimming. Before detachment, however, a wandering bud from the persistent oozooid has settled on the phorozooid, and this bud, during the free-swimming life of the phorozooid, elongates and contricts off a series of small buds, which develop into the gonozooids. These are eventually liberated and reproduce sexually to close the cycle.

Salpida. All salps exist as an oozooid, which has a budding stolon but no gonads, and as a blastozooid, which has gonads but no budding stolon; the succession of sexual and asexual phases was observed very early by Chamisso (1819). The stolon of the oozooid strobilates to form a chain of zooids, each of which is eventually released to become a solitary free-swimming blastozooid. This produces a single egg (in *Salpa*; more are produced by other genera), which develops within the blastozooid until liberated by rupture of the body wall. The salps thus exemplify a life cycle in which increase in number may be wholly asexual, with sexuality replacing a parental individual by a single, genetically different offspring.

LARVACEA. Small pelagic tunicates resembling the tadpole larva of ascidians. They are not known to reproduce asexually.

Discussion. The great majority of tunicates either have hermaphroditic zooids or live in monoecious colonies, and are therefore liable to be self-fertilized. Sabbadin (1971) has published an exceptionally careful study of self-fertilization in *Botryllus schlosseri*, a colonial ascidian in which the zooids are united by a common vascular system. The colonies are slightly protogynous, the eggs ripening a few days before the sperm. This appears to be effective in reducing selfing in natural populations to a negligible level, since colonies maintained in the laboratory will not fertilize one another when their gonadal cycles are in phase. However, by manipulating conditions so that sperm ripens in one detached fragment of a colony at the same time that eggs ripen in another, there is no difficulty in achieving self-fertilization in the laboratory, nor is there any preference for foreign over self sperm. The barrier to effective selfing acts after fertilization, in the form of an inbreeding depression which becomes steadily more severe in later growth stages; selfed eggs hatch as successfully as cross-fertilized eggs, and produce nearly as high a proportion of morphologically normal larvae, but selfed offspring are much less competent at metamorphosis and colony-formation. I have abstracted some of Sabbadin's quantitative data in Table 3.22; they show that each cross-fertilized egg gives rise on average to about ten times as many fifth-generation blastozooids as a self-fertilized egg, reflecting such a large inbreeding depression of fitness that habitual self-fertilization in natural populations seems very improbable.

Table 3.22: Performance of Self-fertilized and Cross-fertilized Progeny of *Botryllus schlosseri*. Provenance of progeny inferred from the segregation of alleles at two loci determining colouration. 'Field colonies' are freshly collected; 'Lab. colonies', maintained for several generations in the laboratory, suffer some depression. Figures in columns are: A, fraction of eggs developing; B, fraction of viable (morphologically normal) larvae; C, fraction of metamorphosing larvae which become functionally oozooids; D, fraction of young colonies reaching the fifth blastogenic generation; E, mean number of zooids per colony in the fifth blastogenic generation; F, a measure of fitness in the field colonies, the product of the five preceding columns. There is considerable variation for each measurement between clones and between crosses, and the estimate in column F has a large standard error.

		A	B	C	D	E	F
Field colonies	Selfed	0.956	0.769	0.469	0.502	2.74	0.47
	Crossed	0.965	0.984	0.873	0.887	7.54	5.55
Lab. colonies	Selfed	0.630	0.796	–	0.430	2.96	–
	Crossed	0.628	0.854	–	0.809	6.11	–

Source: Sabbadin (1971).

The solitary ascidian *Ciona intestinalis* is self-sterile, the barrier to selfing occurring before fertilization (Morgan, 1905, 1923, 1938, 1942, 1945, etc.).

On the other hand, Dhandapani (1975) has pointed out that in most doliolids the testes and ovary ripen simultaneously, and their openings often face one another at a very small distance. In *Doliopsoides horizoni* and *Doliopsis* sp. the vas deferens of fully mature animals thickens and grows upwards towards the ovary. Furthermore, isolated gonozooids of *Dolioletta gegenbauri* have reproduced successfully in the laboratory. Dhandapani also claims that pyrosomids are selfed (no authority given), and that the presence of a developing embryo in the newly released blastozooids of some salps suggests self-fertilization. I note, however, that in thaliaceans the male structures are usually bulky, with no sign of the extreme reduction to be expected if selfing is habitual or obligate.

Some ascidians, such as *Distaplia* and *Diplosoma*, have several cycles of sexual reproduction and budding in the course of a single season. More commonly there is only a single cycle, with sexual reproduction occurring in the summer, asexual budding in winter and further colony growth in spring; Berrill (1935) records this pattern for *Diazona, Morchellium, Sidynum, Euherdmania, Eudistoma, Clavelina* and *Pycnoclavella*, and for various perophorids and styelids. Detailed accounts of seasonal variation in the sexual reproduction and budding of the clavelinid *Metandrocarpa taylori* have been published by Abbott (1953) and Haven (1971). Both sexual reproduction and budding occurred throughout the year, but sexuality was most intense from June to September, whilst budding declined through the summer to a minimum in September before rising abruptly to a maximum in October.

Sexuality is often a terminal condition, the zooid or colony degenerating after spawning.

3.38 Cephalochordata (Acrania)

Incidence of Parthenogenesis. The lancelets live in shallow seas, from the tropics to the temperate zone. They are gonochoric, and are not known to reproduce except by amphimixis.

3.39 Vertebrata

Reviews. Uzzell (1970; cytogenetics); Beatty (1967; chiefly experimental); Schultz (1971; fish); Maslin (1971; reptiles). See also Cuellar (1977).

Incidence of Parthenogenesis.

PISCES. In 1932 Hubbs announced that *Poecilia formosa* (Poeciliidae) exists as all-female populations in nature and, after mating with males of *P. mexicana* or *P. latipinna*, exhibited strictly matroclinous inheritance in the laboratory. It was inferred that *P. formosa* was a stable diploid gynogenetic hybrid between *P. mexicana* and *P. latipinna*, its two sexual hosts. Despite some early scepticism (Howell, 1933), this inference has been supported by all subsequent investigations (Meyer, 1938; Hubbs and Hubbs, 1946; Kallman, 1962). Triploid individuals also occur, sometimes in considerable numbers (Rasch *et al.*, 1965; Prehn and Rasch, 1969; Rasch *et al.*, 1970), but their reproductive status is not clear: triploids produced by hybridization in the laboratory are sterile (Rasch *et al.*, 1965; Schultz and Kallman, 1968), but triploid individuals from natural populations produce both diploid and triploid offspring (Prehn and Rasch, 1969). There are at least three gynogenetic cytotypes of *Poeciliopsis* (Poeciliidae), all of which are triploids derived by hybridization between the diploid amphimicts which now act as sperm donors – *P. monacha, P. lucida* and *P. viriosa* (Miller and Schultz, 1959; Schultz, 1961, 1966, 1967, 1969, 1971). These triploids are automictic, with premeiotic restitution (Cimino, 1971, criticized, rather opaquely, by Cuellar, 1974); premeiotic doubling has also been suggested, but not confirmed, for *Poecilia formosa* (Schultz, 1967). Cimino and Schultz (1970) have described an exceptional diploid male, whose triploid mother bore two genomes from *monacha* and one from *lucida*, and which itself bore one of each kind, suggesting that chromosomes from different sexual ancestors do not segregate independently. Five other cytotypes of *Poeciliopsis* are hybridogenetic diploids, transmitting a maternal genome from *monacha* and utilizing a paternal genome from *lucida, occidentalis* or *latidens* (Schultz, 1966, 1969, 1971, 1973). A gynogenetic strain (*gibelio*) of the goldfish *Carassius auratus* (Cyprinidae) has been found in stock ponds in Russia (Berg, 1961). It is triploid, with an auto-mictic restitution involving the suppression of the first meiotic division through the collapse of a peculiar tripolar spindle on which all the chromosomes are present as univalents (Cherfas, 1966); however, in diploid gynogenetic goldfish from European ponds, Lieder (1955, 1959) has concluded that restitution is postmeiotic. There are also reports in the Russian literature (Taliev, 1950; Novikov, 1962) of gynogenesis in *Atheresthes* (Pleuronectidae) and *Comephorus* (Comephoridae); I have not read them, but Schultz (1971) doubts their validity. *Rivulus marmoratus* (Cyprinodontidae) is an internally self-fertilized simultaneous hermaphrodite (Harrington, 1961; Kallman and Harrington, 1964; Harrington and Kallman, 1968). Cross-fertilization between hermaphrodites in captivity was never observed (Harrington, 1961), but some crossing may be achieved by oc-casional male individuals. Primary males can be produced at will in the laboratory by exposure of the eggs to low temperature during extraparental brooding (Harrington and Kallman, 1968; Harrington, 1971); only one primary male has ever been found in natural populations in Florida, but they are rather more common in the Caribbean region (Kristensen, 1970; Harrington, 1971). Second-ary males, developing from hermaphrodites by conversion of the ovotestis into a testis, have been produced in the laboratory but not yet found in the field (Harrington and Kallman, 1968; Harrington, 1971). Both types of male are capable of fertilizing the unfertilized eggs occasionally emitted by females. Pure females do not occur. A laboratory strain of *Poecilia reticulata* (Poeciliidae), typically a gonochoric species, developed into simultaneous hermaphrodites which produced viable self-fertilized eggs (Spurway, 1957; Comfort, 1961; these impaternate offspring were wrongly attributed to thelytoky by Spurway, 1953). Apart from *Rivulus*, the only fish known to be true simultaneous hermaphrodites are the Serranidae, where self-fertilization has been obtained experimentally in *Serranus subligarius* (Clark, 1959) and *S. scriba* (Reinboth, 1962; Salekhova, 1963, cited by Kallman and Harrington, 1964). In *S. subligarius* isolated gravid fish will release eggs and sperm which give rise to viable selfed larvae, but whether self-fertilization is a frequent occurrence in natural populations of any serranid remains highly questionable. Finally, Kilby and Kallman (unpublished, cited by Kallman and Harrington 1964) 'have observed several cases of parthenogenesis in the mosquito fish, *Gambusia affinis*' (Poeciliidae), leading to the production of viable but sterile offspring.

AMPHIBIA. The occurrence of parthenogenesis in the *Ambystoma jeffersonianum* species group (Ambystomidae) has long been suspected, both because of highly skewed sex ratios and because of the demonstration of matroclinous inheritance by Clanton (1934), and was confirmed by Uzzell (1963, 1964; McGregor and Uzzell, 1964). The group includes two diploid amphimicts, *A. jeffersonianum* and *A. laterale*, and two triploid gynogenetic hybrids, *A. platineum* and *A. tremblayi*. Restitution is by premeiotic doubling, with a very high frequency of chiasmata (which presumably have no genetic effect) between sister chromatids (McGregor and Uzzell, 1964). *Rana esculenta* (Ranidae) is hybridogenetic, its hosts being *R. ridibunda* and *R. lessonae* (Uzzell and Berger, 1975).

REPTILIA.

Agamidae. No males of *Leiolepis belliana* have been described, and its karyotype suggests triploidy (Hall, 1970).

Chamaeleonidae. The subspecies *affinis* of *Brookesia spectrum* is known only from females (Hall, 1970).

Gekkonidae. The subspecies *ogasawarisimae* of *Gehyra variegata* is probably triploid and all-female (Makino and Momma, 1949; Hall, 1970). *Hemidactylus garnotii* is known only from females, and is probably triploid (Smith, 1935; Kluge and Eckardt, 1969). Males are very rare in *Lepidodactylus lugubris*, a diploid, and sperm is never found in the reproductive tract of females during the breeding season; furthermore, tissue grafting has confirmed the genetic uniformity of sibships (Cuellar and Kluge, 1972).

Lacertidae. Parthenogenesis has been proven in four biotypes (*armeniaca, dahli, defilippi* and *rostombekovi*) of the *Lacerta saxicola* group by the existence of all-female populations (Lantz and Cyren, 1936, and subsequent authors), the absence of sperm in the reproductive tract of ripe females (Darevsky, 1958), the production of all-female progeny and the completion of an entire reproductive cycle in the absence of males (Darevsky, 1966), and the histocompatibility of sibships (Eckhardt and Whimster, 1971). The mechanism of restitution is uncertain; Darevsky and Kulikova (1964) suggest that it proceeds either via the suppression of the second meiotic division or via the fusion of cleavage nuclei (see Cuellar, 1971, who states that some confusion has been created by a mistranslation of the original Russian paper).

Teiidae. The best-known parthenogenetic lizards are the dozen or so thelytokous biotypes of the genus *Cnemidophorus*. The evidence for thelytoky includes the observation of all-female populations (Minton, 1958, Tinkle, 1959; Maslin, 1962; and many subsequent authors), the occurrence of triploidy (Pennock, 1965), the absence of sperm from the reproductive tracts of ripe females (Cuellar, 1968), the production of all-female progeny in the laboratory (Maslin, 1966), the completion of an entire reproductive cycle in the absence of males (Maslin, 1971) and the histocompatibility of sibships (Maslin, 1967). Restitution is by premeiotic doubling (Cuellar, 1971). Males are very rare in another teiid, *Gymnophthalmus underwoodi* (Thomas, 1965).

Xantusiidae. Males are very rare in *Lepidophyma flavimaculata* (Bezy, 1969; Telford and Campbell, 1970), which produces all-female broods in the laboratory (Telford and Campbell, 1970).

Typhlopidae. There is no certain instance of parthenogenesis among snakes. Many species are able to store sperm for long periods of time, but the occasional production of young by long-isolated females may indicate tychoparthenogenesis (see Fox, 1977; Magnusson, 1979). The best candidate for an obligately parthenogenetic snake is *Typhlina bramina*, where all of 114 specimens examined by McDowell (1974) were female.

AVES. Natural parthenogenesis is unknown in birds. Olsen (1960, 1965; Olsen and Marsden 1954) found low levels of parthenogenetic development in domestic turkeys, which he succeeded in raising substantially by selection. Those which reached maturity were fertile diploid males, which suggests a restitution mechanism acting to suppress the second meiotic division; since the female is heterogametic, any parthenogenetic offspring would then be male or inviable. The fusion of cleavage nuclei is excluded because of evidence that some heterozygosity is retained by parthenogenetic male offspring (Poole *et al.*, 1963).

MAMMALIA. No mammal is known for certain to have produced full-term parthenogenetic offspring. The spontaneous development of unfertilized eggs to early cleavage stages is occasionally reported, and impaternate blastocysts have been produced by mice and rabbits in the laboratory (see review by Beatty, 1967). However, armadillos of the genus *Dasypus* (Dasypodidae) regularly reproduce by polyembryony. The process was described in detail by Patterson (1913) for *Dasypus novemcinctus*, which usually produces identical quadruplets; other species produce as many as twelve embryos from a single fertilized egg. The subject has been reviewed recently by Galbraeth (1980). Polyembryony occurs occasionally in many other mammals, including man.

Discussion. The origin of parthenogenetic vertebrates has been the subject of some controversy. The prevailing view has been that they arise as a consequence of interspecific hybridization: hybridization yields allodiploid hybrids, which reproduce parthenogenetically, and which by back-crossing to the ancestral amphimicts may in turn give rise to parthenogenetic allotriploids. This view has been urged for fish by Hubbs and Hubbs (1932), Meyer (1938), Schultz (1961,

1966, 1969, 1971, 1973), Abramoff *et al.* (1968), Vrijenhoek (1972) and Vrijenhoek and Schultz (1974); for amphibians by Uzzell (1964), Uzzell and Goldblatt (1967) and Uzzell and Berger (1975); for reptiles by Lowe and Wright (1966), Wright and Lowe (1967), Neaves and Gerald (1968, 1969), Neaves (1969), McKinney *et al.* (1973), Uzzell and Darevsky (1975), Cole (1975), Parker and Selander (1976) and Brown and Wright (1979). The lonely opposition to the hybridization theory has been conducted by Cuellar (1974, 1977), whose articles have recently raised some dust in the correspondence columns of *Science* (Cole, 1978; Cuellar, 1978; Wright, 1978). Cuellar maintains that spontaneous genetic change within an amphimictic diploid may produce a parthenogenetic diploid, which by crossing with the same or another amphimict in turn gives rise to a parthenogenetic autotriploid or allotriploid. To decide between these conflicting opinions we must first recognize their common ground. In the first place, the controversy is restricted to vertebrates; it is not alleged that hybridization is usually or commonly involved in the origin of parthenogenetic invertebrates. Secondly, it is restricted to diploids: both parties agree that triploidy is likely to arise only through hybridization. The controversy therefore concerns the origin of: *Poecilia formosa*; the diploid gynogenetic race of *Carassius auratus*; *Rivulus marmoratus*; the five hybridogenetic taxa of *Poeciliopsis*; four taxa of the *Lacerta saxicola* complex; four taxa of *Cnemidophorus*; and the gecko *Lepidodactylus lugubris*. Of these, *Rivulus marmoratus* is a self-fertilized hermaphrodite, irrelevant to the present dispute; and too little is known of *Carassius auratus gibelio* or *Lepidodactylus lugubris* to permit any conclusion to be drawn; these three are therefore excluded, and the issue must be decided on the merits of the remaining fourteen diploid taxa.

Poecilia formosa is morphologically intermediate between the amphimictic species *P. latipinna* and *P. mexicana*; crosses between these two species give hybrids with the appearance of *P. formosa* (Hubbs and Hubbs, 1932). *P. formosa* is heterozygous for two serum albumen bands for which its presumptive sexual ancestors are both homozygous, and possesses an albumen phenotype identical with that of hybrids produced in the laboratory (Abramoff *et al.*, 1968). The five hybridogenetic biotypes of *Poeciliopsis* are, by universal consent, hybrids from which the paternal genome is eliminated in every generation. Schultz (1973) crossed *P. monacha* with *P. lucida* in the laboratory to produce hybrids which not only reproduced by hybridogenesis but were also indistinguishable from wild *P. monacha-lucida*. Darevsky (1962) has claimed a hybrid origin for three of the thelytokous biotypes of *Lacerta* (*L. defilippi* is described as having both parthenogenetic and sexual races), but the situation is not entirely clear because of the karyotypic uniformity of the complex. There is both karyotypic and biochemical evidence that *Cnemidophorus tesselatus* is a hybrid between *C. tigris* and some species in the *sexlineatus* group (Wright and Lowe, 1967; Neaves and Gerald, 1968), and karyotypic evidence that *C. cozumelae* is a hybrid between *C. deppii* and *C. angusticeps* (Fritts, 1969). The very high levels of heterozygosity discovered in some of these diploids (see below) adds a further dimension to the evidence for their hybrid nature.

In almost every relevant case, therefore, there is evidence, ranging from the suggestive to the convincing, that interspecific hybridization is involved in the genesis of parthenogenetic diploid vertebrates. The only possible exceptions seem to be *Lacerta defilippi*, and perhaps also *Lepidodactylus lugubris*, which is the only member of its genus throughout most of its geographical range. To this strong empirical support for the hybridization theory, Cuellar opposes the argument that hybridization does not necessarily or usually or even very often lead spontaneously to parthenogenesis. Diploid '*Poecilia formosa*' produced in the laboratory by crossing *P. latipinna* with *P. mexicana* reproduce amphimictically, not parthenogenetically (Hubbs, 1955), and laboratory hybrids between amphimictic species in the *Lacerta saxicola* group are sterile (Darevsky, 1966). In fact, with the exception of the highly aberrant *P. monacha-lucida*, there is no example among vertebrates of an interspecific cross yielding parthenogenetically reproducing hybrids. The force of this argument must be admitted; the relationship between hybridization and parthenogenesis is entirely obscure. Nevertheless, the existence of a relationship seems undeniable; and Cuellar's objection loses much of its force when we recall that nobody has ever claimed that interspecific crosses always or even very often give rise to parthenogenetic hybrids. It also seems significant to me that the restitution mechanism in almost all cytologically known cases (*Lacerta armeniaca* may be an exception) is such as will permit the regular segregation of two non-homologous sets of chromosomes.

Little is known for certain about the antiquity of parthenogenetic vertebrates. There is some slender circumstantial evidence which suggests a late Pleistocene date for the origin of some parthenogenetic taxa of *Ambystoma*, *Lacerta* and *Cnemidophorus* (Uzzell, 1964; Axtell, 1966; Uzzell and Darevsky, 1975), but Parker and Selander (1976), on equally slight grounds, think that *Cnemidophorus tesselatus* may be no more than two hundred years old.

Apart from the diploids I have listed above, all parthenogenetic vertebrates whose karyotypes are known are triploid: some *Poecilia formosa*, three gynogenetic biotypes of *Poeciliopsis*, the two gynogenetic *Ambystoma*, at least six taxa in *Cnemidophorus*, the gecko *Hemidactylus garnotii*, and two lizards which are presumed to be parthenogenetic, *Leiolepis belliana* and *Gehyra variegata*. Diploidy and triploidy are thus about equally common; amongst amphimictic vertebrates, of course, polyploidy is extremely rare.

Partly because of triploidy, parthenogenetic vertebrates are usually reproductively isolated from their nearest amphimictic relatives. Matings between diploid thelytokous and amphimictic members of the *Lacerta saxicola* group occur in nature, but the offspring are sterile triploids (Darevsky, 1966). Male specimens of thelytokous lizards are occasionally found (e.g., Maslin, 1962; Taylor and Medica, 1966; Taylor *et al.*, 1967) and Wright and Lowe (1967) claim that at least some of these rare males are allotriploids; Lowe *et al.* (1970) have found what they think to be allotetraploid hybrids between triploid parthenogenetic *Cnemidophorus sonorae* and diploid amphimictic *C. tigris*. These males usually possess morphologically normal reproductive structures, but often have less sperm than male amphimicts, or none at all, and what they do have may include many abnormal gametes (Christiansen and Ladman, 1968). They are therefore unlikely to constitute a frequent or effective channel of genetic communication between parthenogenetic and amphimictic species.

In gynogenetic fishes the situation is somewhat different for, whilst thelytokous females should normally repel the advances of sexual males, gynogenetic females are obliged to solicit them in order to reproduce at all; at the same time the males themselves make no genetic contribution to the progeny in either case, and should therfore prefer to mate with normal sexual females. The preference of males for sexual over parthenogenetic females has been demonstrated in the laboratory by Hubbs (1964) for *Poecilia* and by McKay (1971) for *Poeciliopsis*; and both authors add, as confirming the existence of a similar preference in natural populations, that wild amphimictic females are almost always pregnant, whilst gynogenetic females are often not. This phenomenon is of critical importance in the dynamics of gynogenesis, and I defer a more detailed discussion to a later chapter.

There is some slight evidence that parthenogenesis is associated with lowered fecundity in vertebrates. *Rivulus marmoratus* produces far fewer eggs per lifetime than *Oryzias latipes*, the only other cyprinodont for which adequate data are available (Harrington, 1971). The gynogenetic salamanders of the *Ambystoma jeffersonianum* group are only about two-thirds as fecund as their sexual hosts (Uzzell, 1964). I have already referred to the lower fertility of gynogenetic fishes. On the other hand, there is no consistent difference either in egg number or in egg size between thelytokous and amphimictic species of *Cnemidophorus* (Schall, 1978).

Maslin (1968) observed that parthenogenetic vertebrates have weedy tendencies, and succeed best in novel or disturbed habitats. This generalization has been denied by White (1970), and there are instances in which it does not hold: for example, Telford and Campbell (1970) found thelytokous populations of *Lepidophyma flavimaculata* in undisturbed tropical forest, though it may be worth remarking that these populations occupy the southernmost limit of the range of the genus. This is an exception, however, to what seems to be a solidly established general rule. *Rivulus marmoratus* is a brackish-water species which inhabits 'zones of alternate desiccation and tidal-pluvial flooding' on the coast of the southern United States (Harrington, 1961). The gynogenetic *Poeciliopsis 2 monacha-lucida* inhabits small bedrock pools in the headwaters of mountain streams in northwestern Mexico (Moore, 1975), a habitat which Vrijenhoek *et al.* (1978) describe as 'unpredictable'. The populations are small and isolated, and local extinctions often occur during the annual dry season (Vrijenhoek *et al.*, 1978). Cuellar (1977) observes that parthenogenetic geckos (species of *Lepidodactylus, Hemidactylus* and *Gehyra*) are primarily insular, in contrast to most of their close amphimictic relatives. According to McDowell (1974), *Typhlina bramina* is a successful colonist of isolated islands. *Hemidactylus garnotii* has successfully colonized Miami, Florida and Hawaii, but appears eventually to be replaced, through subsequent introductions, by amphimictic species (Kluge and Eckardt, 1969). Cuellar (1977) has reviewed the evidence that thelytokous biotypes of *Cnemidophorus* are distributed primarily along rivercourses and floodplains which are liable to sudden catastrophic inundation. He makes the interesting suggestion that the control of flooding by dams in the Rio Grande Basin may have been responsible for the disappearance of some parthenogenetic populations. Cuellar (1980) has recently reviewed the occurrence of sympatry between thelytokous and amphimictic species of *Cnemidophorus*. The neotropical *C. lemniscatus* also has a riparian distribution along the Amazonas, but according to Vanzolini (1978) it is never found in undisturbed forest, but instead is following the spread of human habitation along the river and settling in 'places of maximum disturbance'. Two other thelytokous lizards, *C. cozumelae* and *C. rodecki*, inhabit open beaches on the Yucutan peninsula (Fritts, 1969), which Cuellar (1977) suggests, I know not with how much justice, are subject to frequent inundation by tropical storms; other thelytokous species of *Cnemidophorus*, *Lacerta* and *Lepidophyma* also inhabit open lacustrine beaches (Cuellar, 1977). Thelytokous species of *Lacerta* live along the headwaters of streams in the Minor Caucasus, primarily along recent glaciated ridges (data of Darevsky, 1966, interpreted by Cuellar, 1977). It may be asserted with confidence, therefore, that parthenogenetic vertebrates most often succeed in establishing themselves in disclimax habitats.

4 A COMPARATIVE AND EXPERIMENTAL CRITIQUE OF THE THEORIES

4.1 A Scientific Method. (2) Comparisons

Whether we should use the experimental or the comparative method depends on the question we ask. Our problem concerns the function of sex; we ask how the distribution of sexuality among multicellular animals is to be explained. This offers little purchase for the experimental method: we can perform experiments to elicit sexuality in the laboratory, but these will identify only proximate causes, while we are interested in function, or ultimate causation. More generally, we can sometimes perform experiments which test the validity of the axioms on which rival hypotheses rest, but it is difficult in this way either to exclude hypotheses or to investigate their generality. The more powerful technique for these purposes is the comparative method, which allows us to examine the relationship between theoretical predictions and the patterns in nature whose explication is our ultimate goal.

The comparative and experimental methods are similar, of course, in that we use both to scrutinize the consequences of different treatments. However, the development of the comparative method has fallen behind that of the experimental method, because the rigorous statistical technique that can be applied to the design and analysis of experiments on sets of individual organisms is simply not available for comparisons between taxa. The fundamental problem here concerns the units of comparative analysis, and is best explained through an example. Suppose that we have reason to believe that amphimixis will evolve in acidic and parthenogenesis in alkaline environments. We must first choose a group of organisms on which to perform the analysis: we should like this group to be small enough to minimize as far as possible the effects of taxonomic (historical) effects, but large enough to include a range of variation sufficient to decide the issue in question. We choose, say, an order of aquatic invertebrates, knowing in advance that some members of the group inhabit acidic and some alkaline environments, while some are amphimictic and others parthenogenetic. By casting the data in the form of a contingency table we discover that no amphimictic species inhabits alkaline environments and no parthenogenetic species inhabits acidic environments. This clearly supports our hypothesis: but how powerful is the support? In acidic environments we find a single family containing two genera and five species of amphimictic animals; in alkaline environments there is again only a single family, but one which contains ten genera and a hundred species. What numbers should appear in the appropriate cells – one and one; two and ten; or five and a hundred? The confidence that we place in the result will depend critically on our choice of units, but there is no widely accepted criterion that we can use to guide this choice.

Ghiselin (1974b) has argued that species are not sets but individuals, in the logical sense of having no defining properties. I am not concerned, or competent, to debate the validity of this notion here, but if it is accepted it immediately suggests a protocol for comparative analysis: comparisons should be made only between sets of taxa each of which is individual at the same level. In practice, I shall be forced to interpret this as meaning that comparisons should be made only between taxa of equal phyletic rank. Naturally, any statistical analysis of such comparisons should not rest on assumptions, such as normality, which are unlikely to be met by the data. However, I shall often ignore even these rudimentary rules, foregoing statistical analysis altogether or even making comparisons between taxa of different rank, when more suitable comparative material is not available.

The purpose of these comparisons is to identify the correspondence between the theories of sex described in Chapter 2 and the facts set out in Chapter 3. The most efficient method of procedure is a pairwise comparison of rival hypotheses, with one alternative being eliminated by each comparison. We can start by observing that the entire set of theories can be divided into two groups by either of two criteria. In the first place, we can separate a set of functional from a set of nonfunctional hypotheses: the functional set includes all those hypotheses which attribute patterns of sexuality to some identifiable consequence of sexual reproduction, whilst the nonfunctional set includes all those hypotheses which assert that the distribution of sex is essentially independent of its function. The nonfunctional set comprises the historical hypothesis, and the functional set all other hypotheses. The alternatives can easily be distinguished since, while nonfunctional hypotheses predict that the correlates of sex will be taxonomic rather than ecological, the functional hypotheses make the converse prediction that the correlates of sex will be ecological rather than taxonomic. The occurence of broad ecological generalizations which transcend phyletic boundaries is therefore sufficient to falsify the nonfunctional alternative. The existence of such generalizations will be demonstrated below.

The second dichotomy separates a set of balanced from a set of nonbalanced hypotheses. The balanced set comprises those hypotheses (best man, tangled bank, Red Queen, hitch-hiker) which operate through individual selection and thus demand a substantial amount of contemporary genetic variance for the mode of reproduction. The unbalanced set comprises the nonfunctional hypotheses, together with hypotheses of group selection (Vicar of Bray, ratchet), both categories requiring that contemporary genetic variation for the mode of reproduction be virtually absent. If the nonbalanced hypotheses are rejected, we can proceed to a pairwise comparison of the balanced hypotheses. Here the most important dichotomy lies between the best man, with its emphasis on temporal change in the physical environments, and the tangled bank and Red Queen, operating in spatially complex environments whose principle features are biotic rather than physical. If the best man be rejected, then the tangled bank and Red Queen can be separated by searching for correlations between sex and either intraspecific or interspecific stress. On the other hand, if the balanced hypotheses are rejected, then we would compare the nonfunctional (historical) with the group selection hypotheses and, if the historical hypothesis were rejected, would ask whether sex were associated with constant or with frequently disturbed environments in order to reject either the Vicar of Bray or the ratchet.

At this point I should re-emphasize the nature of the question being asked and the techniques employed to find an answer. I am concerned to find a broad general principle through which to interpret patterns in nature. To apply any such principle to any particular situation will demand a detailed knowledge of the natural history of the situation. It is through such applications that the usefulness of general principles is established, but this falls beyond the scope of the present work; and I shall attempt only the bluntest explanation of the broadest patterns.

The definition of the problem and the choice of a method of analysis are of crucial importance, because they immediately rule out certain approaches to an explanation of sexuality. They rule out that category of hypotheses which ascribe a function to sex but are incompetent to predict its distribution in nature. This is because any consequence of sex can be judged to be a function of sex, so that no hypothesis is excluded which correctly identifies a consequence of sex; whereas the more interesting problem is to discover a consequence of sex which leads to a sufficient explanation of its occurrence. In particular, the hitch-hiker hypothesis is excluded. Its fatal flaw is stated succinctly by Maynard Smith (1978, p. 123): 'This model . . . has the universality we seek. It must operate in all real populations at all times.' Quite so: it must operate in all real populations at all times, and to about the same extent. It cannot tell us why bdelloid rotifers should lack recombination, or why monogonont rotifers should usually reproduce sexually towards the end of the growing season, or whether resistant and dispersive propagules should be produced sexually or asexually; so that for all its ingenuity it gives us no purchase on the problem at hand. This is not to say that it must be devoid of all empirical support, for it might still be possible either to demonstrate the truth of its axioms or even to perform comparative tests on a small scale.

In the first place, there is an appreciable increase in the frequency of alleles for higher recombination only if the locus controlling recombination is rather tightly linked to the pair of loci undergoing selection. This requirement provides a stringent axiomatic test of the hypothesis, and is readily shown to be false. In *Neurospora crassa*, 'the targets are usually not close to the *rec* locus (the locus which regulates recombination at a given target) and are commonly on different chromosomes' (Catcheside, 1977, p. 73). Simchen and Stamberg (1969), drawing on their work with *Schizophyllum commune* but referring to microorganisms generally, wrote that: 'It is clear . . . that recombination is regulated in short segments by controlling genes unlinked or loosely linked to the regions they control'.

Secondly, the hypothesis explains why an allele causing some recombination should be favoured over an alternative causing none, but not why an allele causing a higher should be favoured over an alternative allele causing a lower rate of recombination. This might be interpreted to mean that the hypothesis requires or predicts a complete absence of genetic variance for intermediate rates of recombination, which is demonstrably false (see Section 5.2). Alternatively, the hypothesis must be viewed as a theory of obligate sexuality rather than as a theory of recombination.

Thirdly, selection for higher rates of recombination in the hitch-hiker model is more effective if alleles directing higher rates of recombination are recessive. Treating this as a prediction, we find that high-recombination alleles are recessive in *Schizophyllum commune* (Simchen, 1967), *Neurospora crassa* (Smith, 1975; Catcheside, 1977), *Saccharomyces cerevisiae* (Roth and Fogel, 1971; Fogel and Roth, 1974) and *Drosophila melanogaster* (Kidwell, 1972a, b), but show a tendency to dominance in *Schizocerca gregaria* (Shaw, 1974) and, more strongly, in *Hordeum* spp. (Gale and Rees, 1970). If these facts be interpreted as favouring the hitch-hiker, two difficulties remain. The first is that since the hypothesis contrasts zero recombination with some degree of recombination, it does not properly predict that high rates of recombination should be recessive to low, but rather that any non-zero rate of recombination should be recessive to complete suppression of recombination. This is false, since the major genes controlling asynapsis and desynapsis in higher organisms are usually recessive (review by Riley and Law, 1965; see also Sjödin, 1970; Parker, 1975; Tease and Jones, 1976). Secondly, the test is a weak one since, even if the prediction were verified, it would neither provide sufficient grounds for believing the hypothesis to be true, nor would it exclude any well-defined alternative hypothesis.

Finally, Charlesworth *et al.* (1977) and Maynard Smith (1978) have argued that selection for higher recombination in hitch-hiker models is stronger if the organisms concerned are self-fertilized, so that the usual observation that rates of recombination are greater in selfing than in outcrossing plants lends empirical, and indeed comparative, support to the theory. However, imagine an outcrossing population in which some equilibrium level of recombination between any given pair of loci has been attained through selection, by whatever means. This equilibrium level is defined in the usual way: rare genotypes which increase or decrease the rate of recombination do not spread. We next imagine that the population switches to self-fertilization. The effect of selfing is to decrease the efficacy of recombination, since loci are now more likely to be homozygous. Since there is no reason to suppose *a priori* that the equilibrium rate of recombination under selection has been changed by the switch to selfing, it follows that selection will favour alleles which increase recombination, until the inhibitory effect of selfing is balanced by a greater rate of crossing-over. This conclusion is even more obvious if, instead of a switch to selfing, some event occurs which arbitrarily increases linkage, the effect of tighter linkage being analogous to that of selfing. This merely represents a disturbance of the original equilibrium, and from the definition of an equilibrium it follows that selection will favour alleles causing higher rates of recombination until equilibrium is restored. In this way, all hypotheses which for any reason attribute the function of sex and recombination to the increased diversity of progeny predict the observed result; and this category includes all hypotheses of interest. Maynard Smith (1978, p. 120) objects to this rather straightforward conclusion because an innovation like selfing may itself alter the position of the equilibrium. If not actually a *petitio principii*, this objection at least seems arbitrary, and the onus of proof rests on those who assert such to be the case. The simulations of Charlesworth *et al.* (1977) show that selfing favours alleles for high recombination both in the hitch-hiker and in a best-man model of a capricious environment, and do not include any such effect on the equilibrium. I conclude that the proposed comparative test fails to exclude any of a large number of rival hypotheses, and is therefore null.

In short, the hitch-hiker is only by courtesy a scientific hypothesis. It may be true: the disentangling of stochastically created linkage disequilibria might indeed be an important factor in the evolution of sex and recombination. But the axiomatic analysis above shows that whenever the hypothesis makes exclusive predictions they are false, and whenever it makes true predictions they are not exclusive. Nor can I think of any satisfactory way in which the competence of the hypothesis to explain the distribution of sex amongst organisms can be exposed to empirical test. When such a way is devised, I shall retract my opinion; until then the hitch-hiker must be dismissed from contention.

4.2   Group Selection, Historicity and the Balance Hypothesis

Is sex functional in the short term? Two influential ideas answer that it is not. The first is the conventional group-selection interpretation of sex, which holds that sexual populations evolve more rapidly, or deteriorate less rapidly, than asexual populations, and thereby become extinct less often. The second is the historical hypothesis, which interprets sex in the majority of animals as a mal-adaptive character which persists only because asexual individuals either arise only very rarely or are unable to reproduce effectively. We are led to attribute the scarcity of asexual animals to their low rate of origination and their high rate of extinction: group-selection arguments stress the high rate of extinction, historical arguments the low rate of origination. Opposed to both arguments is the balance hypothesis, which states that sexual and asexual reproduction are usually in short-term balance, there being an adequate amount of genetic variation to ensure that asexuality will spread whenever it creates an individual selective advantage. Both the group-selection and the historical hypotheses posit a very small amount of genetic variation in the mode of reproduction; the group-selection hypotheses also require that the great majority of presently existing parthenogenetic animals should be of very recent origin, and that they should evolve only very slowly. If these statements can be shown to be true, we shall be compelled to accept the existence of a group effect in the evolution of sex.

The discussion in the remainder of this chapter will be based on the data already collated and documented in Chapter 3, and authorities will be cited only when new material is introduced. Tables and figures will be cross-referenced, but otherwise the authority for any particular statement can be ascertained by referring to the appropriate section of Chapter 3.

4.2.1 Origins and Antiquity of Parthenogenetic Animals

Origins. Parthenogenesis might be acquired either gradually or abruptly. In the first case, we are led to imagine that a female in an amphimictic population acquires the heritable capacity to produce a certain proportion of unreduced eggs; that this character spreads through the population; that selection then favours an increase in the proportion; so that by a process of gradual change the population eventually reaches an equilibrium at which the proportion of parthenogenetic eggs produced by each female represents an optimum. Alternatively, we may choose to imagine that a rare female type arises, all of whose eggs are produced parthenogenetically, and that this type proceeds wholly or partly to replace the amphimictic type. At this point, we must anticipate a generalization that I shall prove in Section 4.3: parthenogenesis is usually associated with polyploidy. The problem of the origin of parthenogenesis in animals is thus intimately connected with the problem of the origin of polyploidy.

It seems likely that automicts with intrameiotic or postmeiotic restitution usually evolve gradually, according to the first model (see Section 3.32). Such animals are usually (but not invariably) diploid, so that polyploidy is not a serious complication. In phasmids, in psychid moths and in *Drosophila* there are species which regularly produce a small proportion of unreduced automictic eggs, and in which this proportion can be increased by artificial selection. The hypothesis of gradual change offers an attractive model for the evolution of parthenogenetic species in these taxa, and by extension for similar automicts in other taxa. In the psychid *Solenobia*, Seiler has argued from strong circumstantial evidence that automictic diploids arose first from an amphimictic diploid stock, later giving rise to the automictic tetraploid race.

On the other hand, apomicts, and automicts with premeiotic restitution, are usually polyploid. To anticipate again the discussion in Section 4.3, it seems unlikely that polyploidy arose first, since the first polyploid individuals would be sexually sterile and, having arisen from an amphimictic lineage, would be unlikely to bear alleles directing parthenogenetic reproduction; but if parthenogenesis arose first, we would expect the majority of apomicts to be diploid, which is not the case. It follows that polyploidy and apomixis arise together; or alternatively that parthenogenesis evolves first in diploids, but that only those taxa which succeed in acquiring polyploidy soon thereafter survive for any appreciable length of time. In either case, the joint probability that apomixis and polyploidy should both arise in the same lineage within a very short span of time must be very small if they are independent events. Apomixis will in that case arise only very infrequently.

A parthenogenetic line might acquire polyploidy in any of three ways: by an internal chromosomal event, such as endomitosis; by the fertilization of an unreduced egg by a conspecific male; or by interspecific hybridization. The first mechanism is presumably implicated in the origin of even-ploid automicts such as *Solenobia*, but can scarcely explain the prevalence of triploidy amongst apomictic animals. In vertebrates stable parthenogenetic triploids have arisen through interspecific hybridization, but there is little evidence for a rôle of hybridization in other taxa, and I suspect that the second explanation is the more general. A female which produces unreduced eggs but lives in a predominantly amphimictic population runs a very high risk of having most of these eggs fertilized, giving rise to embryos which are both triploid and which carry alleles predisposing them to parthenogenetic reproduction. These will usually be sterile if the unreduced eggs are produced by automixis, but not if they are apomictic. Since polyploidy is then both compatible with, and a natural consequence of, apomixis, their joint occurrence is much less improbable than appeared at first glance.

Apomixis will therefore arise abruptly, with the creation of stable triploids, and at a rate which depends primarily on the amount of genetic variation for reproductive mode. This abruptness has been held to be proven by the lack of intermediate forms (diploid apomicts) in taxa such as curculionid weevils. Parthenogenetic ostracods, which are said to be diploid apomicts, present an obvious difficulty; their cytology needs further investigation.

Antiquity. It is not possible to date with certainty the origin of any parthenogenetic taxon. Taxa in which the sexual types are glacial relicts (e.g. curculionid weevils, psychid moths, some vertebrates and psocopterans) are supposed to have acquired parthenogenesis at the end of the last glacial period; by the same reasoning, parthenogenetic insects which are domiciliary or introduced must be of very recent origin. At best, these arguments establish only a minimum period for the duration of parthenogenetic taxa, which may have arisen at any time before the events which are presumed to have increased their abundance and widened their distribution.

More generally, it is widely accepted that thelytokous taxa are the short-lived offshoots of amphimictic (or arrhenotokous) lineages. This idea is usually supported by the observation that the taxonomic distribution of obligate thelytoky is much more sporadic than that of the genetic systems with which it can be legitimately compared. Amphimixis, arrhenotoky and heterogony often characterize large phyletic groups, whereas obligate thelytoky typically occurs only in a few distantly related species of any given major taxon, the closest relatives of most thelytokous species being amphimictic, arrhenotokous or heterogonic. The truth of this assertion is beyond doubt − almost all the major taxa discussed in Chapter 3 show this pattern − and it forces the conclusion that either a group or historical effect is implicated in the evolution of sex, or that the circumstances in which parthenogenesis is advantageous are both very rare and largely uncorrelated with taxonomic similarity. Nevertheless, one major reservation must be expressed: the taxonomic distribution of obligate thelytoky may be sporadic, but it is certainly not random. At the highest phyletic level there are two whole classes of multicellular animals which are exclusively thelytokous, the apomictic bdelloid rotifers and the automictic chaetonotoid gastrotrichs. Among the insects (Table 3.15), obligate thelytoky is very rare or unknown in some orders (e.g., Odonata, Hemiptera), but relatively common in others (e.g., Orthoptera, Homoptera). Thelytoky is rare among Coleoptera, but relatively common in a single family, the Curculionidae and, within this family, much more common in some subfamilies than in others. The distribution of obligately thelytokous species between taxa of ostracod, cladoceran and coccid faunas is analyzed quantitatively in Tables 4.1, 4.2 and 4.3. The distribution of acyclic species between both genera and families of North American cladocerans seems to be entirely random, as does the distribution of thelytokous species between genera of coccids (with the exception of *Hemiberlesia*, with four thelytokous and only two amphimictic species). However, the distribution of thelytokous species between tribes of coccids, and between genera of North American ostracods, is certainly not haphazard; rather, some taxa have a much higher and others a much lower proportion of parthenogenetic species than would be expected if the origin of each were an independent event. The nature of this heterogeneity is discussed below. Such aggregation is not necessarily fatal to an historical or a group-selectionist interpretation of sex. It might be argued that thelytokous forms, for unknown reasons, arise more frequently in some taxa than in others or, more plausibly, persist longer if they belong to taxa which inhabit unusually stable environments (the Vicar of Bray) or unusually unstable environments (the ratchet). These latter predictions march with those made by hypotheses of individual selection, and will be examined at length in Section 4.3.

It remains a general rule that, in most large taxa, obligately thelytokous species, whether reproducing by apomixis, nongametic automixis, autogamy or vegetative proliferation, are taxonomically isolated, presumably because they frequently become extinct, at least in the genetic sense of failing to give rise to new lines of descent before the disappearance of the parental lineage. Extinction has never been observed directly. The asexual anemone *Haliplanella luciae* behaves in the manner envisaged by the theory, with shifts in physical factors such as salinity or temperature beyond a certain sharply defined lethal threshold precipitating mass extinction; related sexual anemones have a much shallower response, with a few individuals surviving even under very unfavourable conditions.

4.2.2 Heritable Variation in the Mode of Reproduction

Heterogony. The balance hypothesis holds that amphimixis and thelytoky occur at equilibrium in the same populations as the result of short-term individual selection. In the extreme cases of obligate amphimixis or obligate thelytoky, the alternative genetic system is eliminated by selection, but maintained at very

Table 4.1: Relationship Between Size of Genus and Proportion of Partheno-genetic Species in North American Freshwater Ostracods. Parthenogenetic species are those for which Tressler states that the male is unknown.

(a) Tabulation of all Genera According to the Number of Species Per Genus. Kolmogorov-Smirnov one-sample test statistic $D = 0.1719$, with associated $P < 0.01$.

Species per genus	Number of genera	Number of species	Parthenogenetic species
1	7	7	4
2	5	10	7
3	4	12	7
4	1	4	4
5	1	5	3
7	3	21	11
9	2	18	10
10	3	30	11
15	2	30	10
19	1	19	1
35	1	35	12

Source: Tressler (1959).

(b) Two-by-two Contingency Table for Large and Small Genera. This has $\chi^2 = 9.94$, $df = 1$, $0.01 > P > 0.001$. The Spearman rank correlation coefficient between the number of species per genus and the proportion of parthenogenetic species, corrected for ties, is $r_s = -0.261$, which yields $t = 1.431$ with $df = 28$, and thus $0.20 > P > 0.10$ for a two-tailed test; it thus approaches significance at the five per cent level if a one-tailed test is allowed.

	Number of species:		
	amphimictic	parthenogenetic	Total
Twelve large genera (> seven species)	98	55	153
18 small genera (< five species)	13	25	38
Total	111	80	191

Source: Tressler (1959).

low frequency by recurrent mutations which affect the fate of meiotic products. In all other cases two alternative genetic systems continue to coexist as the result of a balance between short-term advantage and disadvantage. It is this latter class of phenomena which is critical to the hypothesis. If it is a large class, it is likely to be generally true that there is abundant genetic variation in the mode of reproduction available to short-term individual selection, and the hypotheses of history and of long-term group selection become extremely onerous. There are five relevant categories of evidence.

Table 4.2: Relationship Between Size of Taxon and Proportion of Parthenogenetic Species in Diaspidid Coccids. The number of species per genus is the number studied by Brown. Four species with amphimictic and parthenogenetic races are scored as being parthenogenetic.
(a) Tabulation of Species by Genus. There is no sign of heterogeneity in the data, other than that contributed by the four parthenogenetic species of *Hemiberlesia*.

		Number of species in genus:						
		1	2	3	4	5	6	T
	0	25	16	4	3	1	0	49
	1	4	5	1	2	1	0	13
	2		2	0	1	0	0	3
Number of	3			0	0	0	0	0
parthenogenetic	4				0	0	1	1
species:	5					0	0	0
	6						0	0
	T	29	23	5	6	2	1	66

Source: Brown (1965).

(b) Tabulation of Species by Tribe. Kolmogorov-Smirnov $D = 0.1328$, with $0.05 > P > 0.01$. The related family Phoenicococcidae, also studied by Brown, has nine cytologically known species of which one is parthenogenetic. Its inclusion in (b) slightly weakens the result.

Tribe	Number of amphimictic species	Number of parthenogenetic species	Total
Xanthophthalmini	0	1	1
Odonaspidini	1	2	3
Parlatorini	5	2	7
Diaspidini	48	8	56
Aspidiotini	53	10	63
Total	107	23	130

Source: Brown (1965).

Firstly, in some heterogonic taxa a succession of generations of females committed to asexual reproduction is followed by a generation committed to sexual reproduction. This life history occurs more or less commonly in Mesozoa, Hydrozoa, Scyphozoa, Rhabdocoela, Tricladida, Rhynchocoela, Monogononta, Polychaeta, Cladocera, Enteropneusta, Thaliacea, and some insect groups (Aphidoidea, Cecidomyidae, Micromalthidae, Cynipidae). Since the sexual generation is produced asexually from the final asexual generation it follows that the same genotype is capable of directing either sexual or asexual reproduction, according to circumstances. There would therefore be directional selection in favour of more frequent or less frequent sexuality, if the observed situation did not represent a balanced equilibrium.

Table 4.3: Relationship Between Size of Taxon and Proportion of Acyclic Species in North American Cladocerans. Only those species in which Brooks explicitly states the male to be unknown are scored as acyclic.
(a) Tabulation of Species Within Families. Kolmogorov-Smirnov D = 0.084, with $P > 0.20$.

	Number of species:		
Family	cyclic	acyclic	Total
Leptodoridae	1	0	1
Polyphemidae	1	0	1
Holopedidae	1	1	2
Bosminidae	3	0	3
Sididae	8	0	8
Macrothricidae	12	6	18
Daphnidae	30	4	34
Chydoridae	48	9	57
Total	104	20	124

Source: Brooks (1978).

(b) Pooled Data for Large and Small Families. χ^2 = 0.72, df = 1, $0.50 > P > 0.30$. Different ways of pooling do not alter this result much.

	Cyclic	Acyclic	Total
Small families	26	7	33
Large families (Daphnidae and Chydoridae)	78	13	91
Total	104	20	124

Source: Brooks (1978).

(c) Tabulation of Species Within Genera. Kolmogorov-Smirnov D = 0.108, with $P \simeq 0.10$.

Species per genus	Number of genera	Number of species	Parthenogenetic species
1	22	22	3
2	11	22	4
3–5	5	17	5
6–8	3	21	2
9–13	4	42	6
Total	45	124	20

Source: Brooks (1978).

Secondly, in most other heterogonic taxa each individual female may reproduce either sexually or asexually according to circumstances, and may produce both sexual and asexual broods during her lifetime. Life histories of this sort occur in Demospongiae, Hydrozoa, Anthozoa, Ctenophora, Rhabdocoela, Digenea, Endoprocta, Bryozoa, Sipunculoidea, Oligochaeta, Cladocera, Echinodermata, Pterobranchia and Ascidiacea. The inference drawn is the same as in the previous case: either sexual or asexual broods would tend to become more frequent if short-term selection were not maintaining an equilibrium.

Thirdly, females are in some cases able to produce both sexual and asexual progeny in the same brood. This 'facultative thelytoky' (or 'amphotery') is conventionally distinguished from heterogony, or cyclical parthenogenesis, but the genetic difference is trivial. It shades into heterogony in the case of taxa which reproduce more or less continuously and release both sexual and asexual propagules throughout the year, as with some marine Demospongiae. Facultative thelytoky is known from: *Asplanchna, Sinantherina* and *Conochiloides* (Monogononta); *Potamopyrgus* (Prosobranchia); *Dendrobaena* (Oligochaeta); several arrhenotokous mites; some mayflies; *Loxoblemmus* (Orthoptera); seven genera of phasmids; *Psocus* and *Trichodenotecnum* (Psocoptera); *Physopus* (Thysanoptera); and *Lecanium* (Homoptera). Since facultatively thelytokous females can produce offspring whether mated or not, they are unlikely to be recognized without painstaking research, and the phenomenon is doubtless more common than these scanty data suggest. Wherever facultative thelytoky occurs there is a presumption that the production of reduced and unreduced eggs is in balance, since selection would otherwise favour a greater or lesser proportion of each kind.

Fourthly, obligately thelytokous and obligately amphimictic females may coexist in the same population or at least in the same local habitat. Apart from the enforced sympatry of gynogenetic and hybridogenetic forms, the coexistence of sexual and asexual strains has been reported from: *Dendrobaena* (Oligochaeta); several tardigrades, e.g., *Macrobiotus hufelandi*; *Artemia* (Anostraca); *Trichoniscus* (Isopoda); *Nemasoma* (Myriapoda); *Anaphothrips* (Thysanoptera); *Cnephia* (Diptera); *Cis, Scepticus* and *Calligrapha* (Coleoptera). It probably also occurs among freshwater ostracods. Like facultative thelytoky, it is difficult to detect, and may be much more widespread than the short list of known instances suggests. It is presumed that, if either the thelytokous or the sexual strain had a net selective advantage, the one would eventually eliminate the other, so that their continued coexistence is evidence of an equilibrium maintained by short-term selection.

Finally, different populations of the same morphospecies may comprise exclusively thelytokous or exclusively parthenogenetic individuals. Geographical parthenogenesis is known from a number of triclads, oligochaetes, prosobranchs, tardigrades, mites and myriapods, as well as from insects (Table 3.18); it also occurs in such heterogonic taxa as Monogononta, Phylactolaemata and Cladocera. It is the weakest category of evidence, since a pattern of this sort might be taken to indicate the operation of group selection. However, thelytokous and sexual populations are often so close to one another, relative to the active or passive dispersal of which the animals are capable, that there must be a substantial amount of migration from thelytokous to sexual populations and vice versa. If short-term selection did not favour different genetic systems in different places, migration would eventually create uniformity across the whole geographical range of the species.

I have excluded from these lists cases in which the alternative to amphimixis is autogamy. A heterogonic alternation between amphimixis and autogamy is known in typhloplanid rhabdocoels and the mollusc *Omalogyra*; facultative autogamy occurs in nematodes, and probably in pond snails; amphimictic and autogamous strains coexist in the mollusc *Bulinus* and in the notostracans *Triops* and *Lepidurus*; and a geographical separation of amphimictic from autogamous populations is well known in *Triops*.

These examples take in a large fraction of the animal kingdom, and provide compelling evidence of the validity and generality of the balance hypothesis. Maynard Smith (1978, p.62) argues that the evidence is inconclusive, since sexual and parthenogenetic females or propagules are often not only genetically but also ecologically distinct; thus, sex may be retained in cladoceran populations not because sexuality itself is functional, but because it is the ephippia which are produced sexually, and without ephippia the population or lineage would be unable to survive adverse conditions. As I have said before, this argument seems to me quite unacceptable. If there is a necessary (nonfunctional) correlation between sex and the production of a resting stage, or between sex and anything else, then any comparative analysis will be meaningless; since, if the correlates of sex are unrelated to its function, then its function cannot be discerned. Fortunately, we know that the correlation between sexuality and ephippial eggs is not a necessary one: first, because the ephippial eggs of cladocerans can be formed asexually; secondly, because the sexual eggs of cladocerans are sometimes subitaneous; finally, in other organisms with both sexual and asexual eggs either may be dormant and either subitaneous. The occurrence of heterogonic or facultative life cycles is therefore a valid argument in support of the balance hypothesis.

Obligate Sexuality. However, it is not a conclusive one. It could be maintained that, whilst sex and thelytoky may be in short-term balance in the animals listed above, they are not in obligately sexual animals, which comprise the bulk of such large taxa as insects, echinoderms, molluscs and vertebrates. The balance hypothesis interprets the genetic system of these taxa as being maintained by selection despite a thin but constant stream of thelytokous mutants. It rests on two lines of evidence.

First, although there is almost always a postmeiotic block to the further development of unfertilized eggs, this block can often be removed by simple chemical or physical treatments. Artificial parthenogenesis was a favourite research topic around the turn of the century, and has often been reviewed (e.g., Delage and Goldsmith, 1913; Tyler, 1941). The force of this argument derives from the general rule that there is likely to be substantial genetic variance for any character which may readily be altered by manipulating the environment.

Secondly, many normally amphimictic animals produce a very small proportion of unreduced eggs, which develop more or less normally without fertilization. Tychoparthenogenesis, which obviously cannot be sharply distinguished from facultative parthenogenesis, has been recorded from various Metastigmata, Orthoptera, Blattoidea, Phasmatodea, Mantodea, Psocoptera, Lepidoptera, Diptera, Asterioidea and Echinoidea. Since it can be detected only by rather careful and prolonged work in the laboratory, this list will no doubt become longer with time. It seems unlikely, however, that it will be extended indefinitely. A good deal of effort has been put into obtaining parthenogenetic development in some mammals, but unreduced eggs never seem to develop beyond blastocysts (see Beatty, 1967). In addition, it has been objected by Maynard Smith (1978, p.63) that where a few parthenogenetic eggs are produced they are usually automictic and are likely to have very low viability. While not always true, it is certainly a general rule that apomixis tends to arise abruptly, and presumably through the rapid selection of a rare and extreme mutation, rather than through the gradual selection of an increased proportion of parthenogenetic eggs per female. Nevertheless, the viability of automictic eggs has been increased substantially by only a few generations of artificial selection in *Drosophila*, and even in phasmids, which have a postmeiotic restitution and thus completely homozygous parthenogenetic offspring. If this can be achieved so easily in the laboratory, perhaps it may also occur in natural populations.

Arguments such as these become conclusive only when the lists of examples become exhaustive. It seems very probable that sexual and parthenogenetic reproduction are in short-term balance amongst a large fraction of multicellular animal taxa, and amongst obligately sexual taxa there is some direct evidence for tychoparthenogenesis. But in such well-studied and thoroughly amphimictic animals as mammals and birds there is still no single well-authenticated case of a viable, fecund female developing from an unfertilized egg, and we must conclude that in such taxa the capacity for a radical shift in the genetic system is extremely limited, and that any parthenogenetic offspring which are produced are likely to be so sickly as to negate the theoretical cost of sex. This does not prove that sexuality is maladaptive in these taxa: it is a fallacy to suppose that invariable characters are necessarily functionless, and entirely plausible to argue that their lack of variation indicates the importance of their function (see Cain, 1965). But it does imply that, within such taxa, comparative analysis is impossible, and therefore that the axiom of perfection cannot be sustained nor the historical hypothesis rejected.

4.2.3 The Rate of Evolution in Parthenogenetic Animals

Variation and Selection in Parthenogenetic Animals. The amount of morphological and genetic variation in parthenogenetic animals was discussed in Chapter 1.

Two facts stand out clearly. First, almost all parthenogenetic animals form polyclonal populations, the only consistent exceptions being the automictic lizard *Cnemidophorus*, the apomictic beetle *Adoxus* and the autogamous snail *Rumina* (see Section 1.3.2). Secondly, variation is almost always much less than in related amphimicts: the exceptions are shell colour in the facultatively parthenogenetic prosobranch *Potamopyrgus anitopodarium* and the chromosomal polymorphisms of certain Diptera, especially *Cnephia mutata*. It should follow that, whilst selection will be effective in producing genetic change in parthenogenetic animals, this change will occur very slowly.

The work of Hebert and of Young has shown that there may be appreciable selection during the parthenogenetic generations of heterogonic cladocerans. However, although genetic change may be directional over short periods of time – for example, in the shifts of genotype frequency observed in permanent ponds by Hebert or the elimination of homozygous clones early in the growing season observed by Young – the main result of selection is the maintenance of a considerable amount of genetic variation, presumably by some frequency-dependent process. It is not the intense directional selection envisaged by the Vicar of Bray.

Suomaleinen and his colleagues have shown considerable geographical variation in morphological and isozymic characters in apomictic weevils, and interpret this as representing adaptation to local conditions, and thus as showing continued evolution even in obligate apomicts. M. J. D. White, on the other hand, is inclined to think that it may merely reflect multiple ancestry, as is almost certainly the case in *Solenobia*. There is no conclusive evidence on the point, since the distinct tendency for populations in nearby localities to resemble one another in both morphometric and isozymic characters is predicted by both hypotheses. However, the fact that apomictic populations of the weevil *Otiorrhynchus dubius* from Scandinavia can remain active at lower temperatures than sexual Alpine populations proves that evolution has occurred, if it is granted that the apomicts arose in central Europe from the resident amphimictic stock and spread northward as the ice retreated. There is also strong circumstantial evidence that new alleles have arisen in the beetle *Polydrosus mollis*, and new karyotypes in the orthopteran *Warramaba virgo* and the dipteran *Cnephia mutata*, since the divergence of the parthenogenetic from the parental amphimictic stock.

Parthenogenetic insects are often flightless or nearly so, and this suggests that selection has been able to act effectively on morphological characters even in strictly parthenogenetic lineages. This does not necessarily follow: parthenogenesis and aptery tend to vary together, with wings being reduced even in sexual forms at high latitudes, for example, so that the immediate sexual ancestors of wingless parthenogenetic insects may have been wingless themselves. However, this does not convincingly account for the rather strong tendency for apomictic weevils to be wingless or nearly so, when their nearest amphimictic relatives are usually strong fliers.

The entirely thelytokous classes Bdelloidea and Chaetonotoidea have undergone considerable adaptive radiations in fresh water. In bdelloids this radiation has occurred in the absence of any known mechanism for recombination whatsoever. This fact is sufficient to prove the occurrence of evolution in obligately apomictic lineages, since an hypothesis of multiple ancestry would be incredible, but by no means excludes the possibility that such evolution may have taken place extremely slowly over a very long period of time. Moreover, the bdelloids are much less diverse than the heterogonic monogonont rotifers: in particular, the whole class is characterized by a single arrangement of the trophic structures, whereas in monogononts at least eight basic types exist, and this lesser morphological diversity is associated with a much more restricted ecological range.

Conservation of Accessory Sexual Structures. Accessory sexual structures such as seminal receptacles or a bursa copulatrix are useless or even dangerous to parthenogenetic females, and if selection is effective it will inevitably tend to procure their reduction or elimination. The status of such structures therefore supplies a very sensitive and reliable measure of the efficacy of selection in parthenogenetic animals. In the enchytraeids *Achaeta affinis* and *Fredericia maculata* the spermathecae are absent, and in the diplopod *Nemasoma varicorne* females from populations in which males are rare have greatly reduced and very variable seminal receptacles. On the other hand, the enchytraeid *Mesenchytraeus pelicensis* possesses morphologically normal spermathecae, and many automictic enchytraeids even still produce sperm. The apomictic prosobranch *Potamopyrgus* has both a bursa copulatrix and a seminal receptacle. Among insects, a functional seminal receptacle is found in several parthenogenetic psocopterans, in *Myrmecophila* (Orthoptera), in *Apatania* (Ephemeroptera) and in *Eurygyne* and *Calligrapha* (Coleoptera). These seem to be clear instances in which a functionally optimal character state has not been attained under selection because of a lack of relevant genetic variation. But those cases in which accessory structures are reduced or absent seem equally to prove that selection can occur, if only slowly, and this conclusion is supported by the evolution of mating reluctance in some parthenogenetic insects (see Section 3.32.4).

I should also point out what may be obvious, that the persistence of accessory reproductive structures in many thelytokous animals is good evidence, independent of their taxonomic distribution, that they evolved secondarily from amphimictic ancestors.

Rate of Speciation. Let us grant that thelytokous species evolve more slowly, and in consequence have a higher rate of extinction or a lower rate of speciation than related amphimicts. Let us further assume that the characters used by taxonomists bear no systematic relation to the mode of reproduction, and that the rate of formation of new species is much greater than the rate of formation of new genera. It then follows that thelytokous species will be more frequent in small than in large genera. In North American ostracods it seems quite certain that the expected correlation occurs: there is a pronounced tendency for genera which contain few species to include a larger than expected proportion of thelytokous species, and vice versa (Table 4.1). In coccids the situation is less straightforward (Table 4.2). There is no tendency for thelytokous species to be disproportionately frequent in small genera, but there are more thelytokous species than expected in small tribes. The expected effect thus occurs at a higher but not at a lower taxonomic level. Among cladocerans (Table 4.3) the effect does not occur at any level, and species which are thought to be acyclic occur indiscriminately among small and large genera and families. It is interesting that whilst ostracods are thought to be either wholly thelytokous or wholly amphimictic, at least four species of coccids have both thelytokous and amphi-mictic races, and cladocerans, of course, usually exhibit both thelytokous and amphimictic reproduction. The evidence for the proposed effect is strongest where the comparison is purest.

The two assumptions involved in this analysis are probably sound. First, the major characters used by ostracod taxonomists are the shape and ornamentation of the shell, and the form and processes of the appendages and the caudal furca (Tressler, 1959); none bear any obvious functional relationship to the mode of reproduction. Secondly, if new genera arose at a rate comparable to that of speciation, we would expect to find a large number of small genera and only a very few large genera. This is not the case: 40 per cent of the genera of ostracods considered have seven or more species.

Heterogonic taxa which alternate between amphimictic (or arrhenotokous) and apomictic reproduction may evolve either towards obligate amphimixis or towards obligate apomixis. If obligately apomictic taxa suffer a higher rate of extinction, then they should be much less frequent than obligately amphimictic species in heterogonic taxa. The reverse is usually the case: apparently obligately apomictic forms are quite common among monogonont rotifers, cladocerans and aphids, but obligately sexual species are unknown. In freshwater sponges, on the other hand, a few species are obligately amphimictic but none obligately apomictic. The test is a weak one, however, since it could be argued that those taxa in which males or mictic females have never been observed may nevertheless experience very infrequent sexual episodes which exert a disproportionate effect on the probability of extinction.

4.2.4 Conclusions

An advocate of the balance hypothesis would argue that heterogony is sufficient to establish its validity for a large fraction of the Earth's biota; that despite technical difficulties it has been established in a large number of obligate amphi-micts by the facts of artificial parthenogenesis and tychoparthenogenesis; and that, by extension, it is generally true. An opponent of the hypothesis would point out that for certain groups, such as mammals and birds, the cumulation of negative evidence has become conclusive; that a number of lines of evidence, especially the restricted genetic variation of parthenogenetic populations, the limited adaptive radiation of the bdelloids and the conservation of accessory sexual structures, support the belief that, although evolution occurs in parthenogenetic lineages, it does so very slowly; and that the high rate of extinction suggested by these facts, with the low rate of speciation for which there is some inferential evidence, is precisely what is anticipated by group-selection theories of sex. We may take it that both proponents are right. The evidence that the mode of reproduction will have an effect on the fate of taxa is compelling, and it would be difficult to interpret the sporadic taxonomic distribution of parthenogenesis other than as an effect at the level of the group. At the same time, it is clear that there will often by ample scope for selection at the level of the individual. It follows that either of two opinions may be true. The first is that group selection has caused the present distribution of sexuality, and that the Vicar of Bray or the ratchet or the historical hypothesis, or some combination of these, will provide an adequate explanation of this distribution. The second is that the group effect is merely a consequence of individual natural selection, which is alone involved in the causation of any particular instance of parthenogenesis. To distinguish between these two opinions, I shall first accept Williams' view that adaptation should always be interpreted at the lowest level compatible with the facts since, when both are acting, individual selection will almost always be more powerful than group selection. We can then search for ecological correlates of sex, and ask whether these are predicted by any theory of individual selection. If the correlates do not exist, we shall accept the historical hypothesis. If they do exist, but are not predicted by any theory of individual selection, we shall accept one or other of the group-selection hypotheses. If they exist, and are accurately predicted by an hypothesis of individual selection, in such a manner as to exclude rival hypotheses, we shall accept that hypothesis, and then enquire how the function thus ascribed to sex could lead to consequences at the level of the group.

4.3 Sex in Space and Time

In recent years the best-man hypothesis has been the interpretation of sex received and accepted by most evolutionary biologists. But, for all its popularity, and for all the effort devoted to patching and propping its logical structure, there has been no sustained attempt to evaluate its merit as a description of the world. The purpose of this section is to supply this deficiency, by exposing the theory to as searching a comparative scrutiny as I can devise. In order to do so, I shall first identify the axioms on which the theory rests, and comment on their validity. I shall then use the theory to generate predictions concerning the occurrence of sex in multicellular animals, and attempt to state these predictions so that they might be falsified by an appropriate and accessible comparison. There is, of course, no difficulty in constructing a theory couched in terms so abstract, or involving a logical machinery so complicated, that its value as a general account of nature can never be established. For this reason I shall be chiefly concerned with the simplest version of the best-man hypothesis, which states that sex is adaptive in uniform environments whose physical parameters often change in time. My justification for doing this is set out below; it does not, of course, exclude the testing of any more sophisticated derivative theory when an opportunity occurs.

The best-man hypothesis involves temporal heterogeneity as an explanation for sex. It is appropriate, therefore, to rest this general proposition by contrasting it with a rival hypothesis, the tangled bank, which operates through spatial heterogeneity — through the diversity of ecological niches existing at any given point in time. By applying the data assembled in Chapter 3 to this fundamental dichotomy, I shall attempt to reach a decision on the basic functional significance of sex.

4.3.1   The Axiomatic Bases of the Rival Hypotheses

The Nature of Selection and Variation. The best-man hypothesis is an offshoot of the classical theory of population structure, according to which most loci are homozygous for wild-type alleles, with evolution occurring by the rapid fixation of beneficial alleles or the rapid loss of deleterious alleles whenever the environment changes or a novel mutation appears. The tangled bank relates to the balance theory of population structure, according to which high levels of genetic variation are maintained by balancing selection of some sort, and evolution consists in the main of a 'shifting balance' between the frequencies of common alleles at different loci. For present purposes, we can ignore that proportion of genetic variation accounted for by drift, and concentrate on those genetic polymorphisms which are protected by selection. The best-man hypothesis leads us to interpret these polymorphisms as being protected by temporal variation in selection coefficients; the tangled bank, as being protected by spatial differences in selection.

To fix ideas, imagine an environment made up of a number of ecological niches. A range of genotypes is defined, each of which is assigned a fitness in each of the niches. Each individual hatches and develops to maturity within a niche, and then leaves the niche to mate at random, each of its offspring being assigned at random to a new niche to complete the cycle. The adults surviving to maturity within any given niche comprise some constant fraction of the total population. Given these assumptions, we can imagine two extreme cases. On the one hand, the fitness of any given genotype might be the same in all niches, whilst being liable to change from generation to generation; this would be a model of a uniform environment which changes in time. On the other hand, genotypic fitnesses might differ between niches but remain the same in any given niche from generation to generation; this would be a model of a complex environment which does not change in time. We can then ask whether spatial or temporal heterogeneity is the more effective in maintaining genetic variation. I have no space to review the subject thoroughly, but there is a convenient summary of the more important theoretical results in Hedrick *et al.* (1976). Purely temporal heterogeneity is rather ineffective in conserving genetic variation. If two alleles are segregating at a locus, and selection coefficients are chosen randomly in each generation from a normal distribution with mean zero and small variance, the polymorphism persists in an infinite population only if there is marginal overdominance for geometric mean fitness. Fine-tuning the model usually makes this condition more restrictive, except that negative serial auto-correlations — a capricious tendency for high and low fitnesses to alternate in time — relaxes it to some extent. Purely temporal models thus have an element of structural instability which does not recommend them as a general explanation for allelic diversity. Purely spatial models are more robust. The condition for stable polymorphism is somewhat less restrictive, and becomes even more favourable when such biologically reasonable assumptions of habitat choice or limited movement between niches are added.

The greater robustness of the spatial model derives from the frequency dependence of genotypic fitnesses, which in turn is a consequence of the assumption that the population within each niche is regulated independently by some density-dependent factor. If this assumption is discarded, the conditions for polymorphism become as restrictive as they are in the temporal model, where genotypic fitnesses are not functions of genotype frequencies. The clash between the best-man and the tangled-bank hypotheses thus reflects the difference between two opinions of environment, one of which is related to classical ideas of genetic population structure and holds that environments are uniform in space but vary in time, whereas the alternative holds that environments are complex but change only slowly. The reflection is imperfect, since the more sophisticated versions of either theory admit both temporal and spatial variation, but the fundamental dichotomy must be admitted.

One cannot at present make an unambiguous choice between the two hypotheses, and presumably neither is wholly true. If I incline towards the spatial model, it is because many of the most fundamental ecological processes necessarily generate frequency-dependent patterns of fitness: population regulation, interspecific competition, predation and parasitism are ubiquitous population phenomena which seem very likely, both on theoretical and on empirical grounds, often to generate frequency- or density-dependent fitnesses. Lewontin (1974), impressed by the allelic diversity of laboratory populations of *Drosophila*, has expressed what is probably a widely felt doubt that the very simple environment provided by culture vessels can possibly be sufficiently complex to maintain a large number of genetic polymorphisms. The notion is not absurd, however: under most experimental regimes differences between jars will contribute to selection, and the spatial segretation of different species within a single jar seems in any case to prove the existence of some level of complexity. Moreover, genetic variation may be maintained even in bacterial populations grown under conditions which are as uniform as human ingenuity can devise (Levin, 1972).

Comparative evidence can be cited on either side of the argument. For example, Gooch and Schopf (1973) found rather high levels of genetic variation in elements of the abyssal fauna, which is thought to live under very constant physical conditions; on the other hand, Bryant (1974) obtained an astonishingly high correlation between the coefficient of variation of certain environmental variables and the level of genetic variation among a number of unrelated taxa. A more promising line of attack is through the study of variation in organisms which are either asexual or haploid, since in neither case will a purely temporal model yield any point of stable genetic equilibrium. The polyclonal nature of most parthenogenetic animals is therefore a severe blow to the temporal hypothesis; it does not represent unequivocal support for the spatial hypothesis, however, since the variation might equally be maintained by heterosis, at least in apomicts. A decisive test is likely to come from haploid organisms, and particularly from haploid asexual eukaryotes in which even a transient or dormant diploid zygote does not occur. According to the temporal model, such organisms will be genetically uniform, since the genotype with the highest geometric mean fitness will eventually be fixed; according to the spatial model, as many genotypes can be maintained as there are niches (Strobeck, 1974). In this case the comparison is not complicated by heterosis, which will be rejected if the temporal model is rejected. Astonishingly, this critical experiment has never been done but, since Milkman (1973) found a great deal of genetic variation in the prokaryote *Escherischia coli*, I would expect it to reveal substantial genetic variation among haploid eukaryotes, thus discrediting both temporal heterogeneity and heterosis as major determinants of heterozygosity in diploids.

The Intensity of Selection. A feature of almost all best-man models, from the simplest to the most sophisticated, is the intense hard selection supposed to be acting in each generation to pick out those rare genotypes which alone have an appreciable chance of reproducing. In lottery models, for instance, the differences in fitness within a habitat during a growing season must be great enough to permit the total elimination of all its competitors by the one best-adapted clone. This is possible only if the clones have an enormous capacity for increase between sexual generations — Williams' so-called zygote-to-zygote increase, or ZZI. Williams buttresses this assumption by remarking that a large ZZI is possible even with quite moderate fecundity: a single parthenogenetic female cladoceran producing 40 eggs could be replaced by 40^4, or about 2.5×10^6, descendents after only four asexual generations, even if each individual dies after reproduction. However, this calculation makes the rather improbable assumption that all individuals survive to reproduce. If each female produces b eggs of which a fraction s survive to reproduce, then after n generations there will be not b^n but $(sb)^n$ descendents; Williams' calculation thus overestimates the number of descendents by a factor s^{-n}, which may be large if s is small or n large — for example, if $s = 0.1$ then the ZZI of the previous example is not several millions but only 341.

Very high rates of increase are most effectively procured by very short generation times; very short generations are characteristic of very small organisms (see Bonner, 1965). Thus, if sex is likely to evolve only when ZZI is very great, we expect it to predominate among very small organisms but to be rare among large animals. The reverse is the case. The smallest metazoans — rotifers, gastrotrichs, kinorhynchs and tardigrades, with the smallest members of such taxa as hydrozoans, nematodes and oligochaetes — exhibit a great variety of modes of reproduction, but are especially remarkable for a high frequency of parthenogenesis. The only large taxa in which amphimixis is entirely unknown — for example, euglenoid flagellates, bdelloid rotifers and chaetonotoid gastrotrichs — invariably comprise very small organisms. By contrast, larger organisms such as molluscs, echinoderms, vertebrates and large arthropods are monotonously amphimictic. This difficulty is admitted by Williams, who, pursuing his advocacy of best-man models to its logical conclusions, has despaired of obtaining a functional interpretation of sex in large organisms.

Since lottery models require a drastic reduction of genotypic diversity during each growing season, they seem inconsistent with the polyclonal nature of most parthenogenetic populations. They are directly refuted by the work of Hebert and Young, who have both found that natural populations of cladocerans comprise several or many clones at the end of the period of asexual reproduction.

Reproductive Isolation. By far the most restrictive assumption made by tangled-bank models is that sexual and parthenogenetic populations should experience total reproductive isolation. If this requirement is not met, the models become much less plausible. However, this apparently improbable assumption turns out to have a sound basis in fact. Polyploid animals are normally unable to reproduce sexually, since their gametes, on fusing with the normal haploid gametes of diploid individuals, give rise to inviable or sexually sterile zygotes. In automicts, however, egg maturation will occur normally with any even ploidy; and apomicts (or automicts with premeiotic doubling) are able to produce balanced eggs of any degree of ploidy. In this way, polyploidy, whilst not hindering reproduction, effectively closes all channels of genetic communication between a parthenogenetic cytotype and its nearest amphimictic relatives. This appears to be the case, not merely in a few instances, but in the majority of obligately parthenogenetic animals. The relevant facts are set out in Table 4.4. Only two caveats need be added. The first is that, whilst apomicts are almost invariably polyploid, automicts are often diploid. In tardigrades and in *Artemia*, for instance, apomictic cytotypes are polyploid but automictic cytotypes tend to be diploid. Secondly, there are isolated exceptions to the general rule amongst apomicts, such as the prosobranch *Campeloma* and the weevil *Polydrosus*; parthenogenetic ostracods may comprise a large group of diploid apomicts, but their cytology is very poorly known.

These exceptions do not impeach a remarkably consistent general rule, that parthenogenetic animals are usually polyploid. It is possible to interpret this

Table 4.4: Digest of Parthenogenesis and Ploidy in Higher Animals. The ploidy given refers only to those members of a taxon which are obligately thelytokous. A quantitative treatment for oligochaetes is given in Table 3.4, and can be inferred for nematodes (Table 3.2), tardigrades (Table 3.6), mites (Table 3.7) and insects (Table 3.15). In all cases it can be assumed that related amphimicts are diploid. This table is intended only to indicate relevant sections in the text of Chapter 3.

Taxon	Reproduction and Ploidy
Turbellaria: Planariidae	Vegetative strains of *Dugesia* often polyploid or aneuploid
Nematoda: Heteroderidae	Mostly high-polyploid apomicts (see Table 3.2)
Prosobranchia	*Potamopyrgus* and *Melanoides* are polyploid apomicts, *Campeloma* a diploid apomict
Oligochaeta: Enchytraeidae	Predominance of polyploid automicts (Table 3.4)
Oligochaeta: Lumbricidae	Predominance of polyploid automicts (Table 3.4)
Tardigrada	Apomicts polyploid; automicts diploid or polyploid (Table 3.6)
Acari	Most may be automictic diploids; some polyploidy (Table 3.7)
Anostraca: *Artemia*	Apomictic cytotypes polyploid, automicts diploid
Conchostraca: *Limnadia*	Automictic diploid
Ostracoda	Perhaps apomictic diploids
Isopoda: *Trichoniscus*	Apomictic cytotype triploid
Insecta	For insects, see text and Table 3.15
Orthoptera	*Warramaba* is an automictic diploid, *Saga* an apomictic triploid
Phasmatodea	Several genera of automictic diploids
Blattaria: *Pycnoscelus*	Apomictic diploid
Mantodea: *Brunneria*	Thelytokous diploid
Embioptera: *Haploembia*	Probably apomictic: diploid, occasionally triploid
Psocoptera: *Liposcelis*	Probably diploid apomict
Homoptera: Coccoidea	See text; broadly, thelytokous diploids
Lepidoptera: Psychidae	Diploid and polyploid automicts
Diptera: Agromyziidae	Diploid and triploid cytotypes
Diptera: Chaemomyidae	Automictic triploids
Diptera: Chironomidae	Diploid and triploid apomicts
Diptera: Drosophilidae	Automictic diploids
Diptera: Simuliidae	Thelytokous, perhaps automictic, triploids
Coleoptera: Chrysomelidae	Apomictic polyploids
Coleoptera: Ciidae	Thelytokous diploids
Coleoptera: Curculionidae	Mostly polyploid apomicts
Coleoptera: Ptiniidae	Automictic gynogenetic triploids
Hymenoptera: Diprionidae	Automictic tetraploids
Vertebrata: Poeciliidae	Gynogenetic triploids and hybridogenetic diploids

Taxon	Reproduction and Ploidy
Vertebrata: Cyprinidae	Gynogenetic triploids
Vertebrata: Ambystomidae	Gynogenetic triploids
Vertebrata: Agamidae	Thelytokous, perhaps triploids
Vertebrata: Chamaeleonidae	Thelytokous, perhaps triploids
Vertebrata: Gekkonidae	*Hemidactylus* is probably triploid; *Lepidodactylus* is diploid
Vertebrata: Lacertidae	Automictic diploids
Vertebrata: Teiidae	Automictic triploids

pattern in either of two ways. Darlington, a botanist, emphasized the fact that polyploidy often makes mixis more difficult, so that apomixis becomes the most efficient mode of reproduction for organisms which have, for whatever reason, increased their ploidy (Darlington, 1939). For example, an individual whose germ-line cells are tetraploid but which nevertheless undergo a normal meiosis will produce diploid gametes; on fusing with the haploid gametes produced by normal diploid individuals these will give rise to triploid zygotes, which will be sexually sterile even if they are not inviable. Apomixis, automixis and selfing are other ways in which this difficulty can be circumvented, making possible the production of tetraploid progeny. However, for this process to work it is necessary that genes coding for parthenogenesis should often occur in newly arisen polyploid genomes, and in a population with an unbroken history of amphimictic reproduction this seems implausible. This snag is not perhaps as apparent, or as serious, in higher plants, where polyploidy is very common, as it is in higher animals, where polyploidy in the germ line is very exceptional. Zoologists (e.g., White, 1946b; Suomaleinen *et al.*, 1976) have therefore tended to distrust Darlington's explanation, and have instead preferred the diametrically opposed view that the evolution of polyploidy is cotingent on the prior evolution of parthenogenesis. The rationale behind this hypothesis is that polyploidy is usually disadvantageous when outcrossing is the rule, because it gives rise to sterile odd-ploid progeny, but this disadvantage disappears when syngamy is autogamous or else is dispensed with altogether. Although this view seems to me the more reasonable, it presupposes some general advantage to polyploidy, whose evolution is held in check only by the formation of odd-ploid zygotes in outcrossed species. Such an advantage has been sought in the greater heterozygosity of polyploid apomicts (Lokki, 1976b). I would also point out that polyploidy is associated with increased body size in oligochaetes (Table 3.5), tardigrades and insects, and perhaps in other groups: the same phenomenon is well known in plants.

Whatever the mechanism involved, there is certainly a strong correlation between parthenogenesis and polyploidy, and this leads to the crucial consequence that thelytokous and amphimictic populations of the same species will commonly experience complete reproductive isolation, even if they are sympatric. Imagine a gene which codes for parthenogenesis, but which has incomplete penetrance. Normally, we would expect individuals bearing this gene to produce a certain proportion of reduced gametes and thus to cross with amphimictic individuals, so that the parthenogenetic and amphimictic types comprise a single interbreeding population. But if the parthenogenetic individuals are also polyploid, they will be unable to produce balanced gametes, or else the zygotes into which their gametes enter will be odd-ploid, and therefore inviable or sterile. The result will be at least a drastic diminution and very likely a complete cessation of gene flow between the parthenogenetic and amphimictic 'races', so that there is no longer a single Mendelian population segregating for genes which determine the production of a certain fraction of unreduced eggs, but rather a Mendelian population and a set of parthenogenetic clones which have similar ecologies but which are no longer able to re-establish genetic contact.

The Real Cost of Sex. Polyploidy is a very efficient primary isolating device, but it has the drawback, from the point of view of parthenogenetic females, that it acts after fertilization. It therefore becomes imperative for them either to prevent sperm penentration or to annull its genetic consequences. Neither amphimictic males nor parthenogenetic females have anything to gain from mating together, and both will be selected to avoid one another; but the females, having the more to lose, will undergo much the more stringent selection, and it is therefore the females which should be the more likely to evolve a reluctance to mate, unless they lack sufficient genetic variation. Among psocopterans, female mating reluctance has evolved in *Bertkauia lucifuga*, male mating reluctance in *Peripsocus quadrifasciatus*, and a combination of both in *Psocus bipunctatus*. It would be interesting to know how widespread these phenomena are. It is possible to select for female mating reluctance in thelytokous *Drosophila*.

Postmating devices for the destruction of sperm have not been described, but might be sought in the labrinthine reproductive tracts of parthenogenetic gastropods and ostracods. I shall discuss later the possibility that pseudogamy evolves as a device for the genetic destruction of sperm.

The initial evolution of parthenogenesis is thus associated with a cost, in terms of lethally fertilized eggs, which may be severe, and whose existence is proven by the evolution of mating reluctance among psocopterans. In effect, this lowers the cost of sex, and makes the evolution of parthenogenesis more difficult.

The production of a substantial proportion of inviable eggs by automicts is an additional burden which has been documented in orthopterans, phasmids and dipterans. It will no doubt lessen with time, since the viability of automictic eggs has been increased by artificial selection in phasmids, but observations on dipterans suggest that there may often be an obdurate upper limit to the fertility of automictic females.

I have previously documented a weak trend among insects for parthenogenetic species to be somewhat less fecund; on the other hand, any such trend might be opposed by the greater size of parthenogenetic polyploids.

The evolution of parthenogenesis is thus conditioned historically – by the fact that in most cases (those other than acyclic representative of heterogonic taxa) it must arise secondarily from a strictly amphimictic lineage. Tychothely-tokous animals are almost always automictic, and at least in some taxa (e.g., phasmids) usually have a postmeiotic restitution; one would expect partheno-genetic eggs produced in this manner to have low viability, and this has often been observed to be the case. Obligate apomicts arising abruptly from an amphi-mictic lineage should not face the same problem (though the low viability of the eggs of *Pycnoscelus surinamensis* shows that they sometimes do), but they run the risk of lethal fertilization. All this suggests that the twofold cost of sex is a theoretical limit which is rarely attained or even approached in natural populations; indeed, what quantitative evidence there is seems to show that novel or occasional parthenogenesis in most cases scarcely increases the effective rate of reproduction.

Low Density. Amphimicts have one other potential handicap: they must be able to find a mate, and this may be an expensive, risky and time-consuming process. Several authors (the most recent and thorough treatment is that due to Gerritson, 1980) have advanced the hypothesis that low population density is for this reason a critical factor in the evolution of parthenogenesis. This is an admirably simple idea, invoking a selection pressure that must always operate to some extent and may sometimes be powerful. Nevertheless, I have not been able to accept that it contributes substantially to our understanding of parthenogenesis.

In the first place, it is emphatically a theory of parthenogenesis, and not of amphimixis or sexuality. It leads straightforwardly to the predictions that parthenogenesis should predominate in novel or disturbed habitats, that repro-duction following long-distance dispersal should be parthenogenetic and that the sexual periods of heterogonic animals should occur at maximum population density; and in subsequent sections I shall establish the truth of these predictions. What it does not do is to explain why sex should occur at all. It follows that, if any rival hypothesis should generate the same predictions, whilst at the same time providing an explanation for the occurrence of amphimixis, that hypothesis will be preferred.

Further, there are several lines of evidence which lead one to doubt that low population density usually or even often facilitates the evolution of partheno-genesis by raising the cost of amphimictic sex. First, many animals live at high density but nevertheless reproduce by obligate parthenogenesis — simuliids are a dramatic example. Furthermore, whilst parthenogenetic insects may generally be somewhat less numerous than related amphimicts, they very often live in small patches of high local population density. Secondly, the hypothesis predicts a correlation between the timing of sex and the absolute value of population density in any given species, and it seems legitimate to infer that this level should be very similar for organisms of similar form; say, for free-swimming cladocerans of comparable size. But sexual periods occur at maximum density even though the absolute numerical value of the maximum varies greatly between comparable species and, in the same species, between different habitats (see

Section 4.3.4). Thirdly, the hypothesis predicts that parthenogenesis will predominate among animals which live at chronic low density, for instance in the deep sea or in habitats which are difficult of access, like caves. The anticipated correlation does not exist (see Section 4.4.2). Finally, if nongametic parthenogenesis is a device to ensure reproduction at chronic low density, it should be part of an adaptive complex which also includes hermaphroditism, self-fertilization, sperm storage, parasitic males and so forth. In deep-sea faunas and among parasites (Section 5.4.1), these devices occur virtually to the exclusion of thelytoky; conversely, in free-living heterogonic animals which reproduce by eggs, the adults are almost invariably gonochoric.

4.3.2 The Method of Prediction

At the core of the best-man hypothesis lies the notion that sex is a device for ensuring preadaptation to changeable, disturbed or novel environments. This may or may not be true; it may or may not be theoretically plausible; it may or may not rest on soundly demonstrated axioms. It is at any rate a clearly stated and empirically vulnerable hypothesis. More careful theoretical work has led to the conclusion that sex will be selected, not in environments which are merely mutable, or even unpredictable, but which are capricious. But how is capriciousness to be measured? Clearly, it cannot be measured without prior knowledge of which environmental factors influence fitness, and of how the phenotypes which determine competence in different environments are inherited. An operational definition of capriciousness, therefore, concerns the genotype rather than the environment: an organism inhabits a capricious environment if the population experiences frequent changes in the sign of linkage disequilibrium between given elements of its genome. Now, this is technically quite a difficult measurement to make, especially in natural populations, and it is inconceivable that it should ever be made for more than a minute fraction of all the organisms of interest. This introduces a dilemma. On the one hand, we might stake the whole future of the best-man hypothesis on the criterion of capriciousness. It is then for all practical purposes invulnerable: there is little relevant information (what there is, is reviewed below), nor any reasonable prospect of ever performing a comprehensive and decisive test. The theory has become little more than an interesting exercise in applied mathematics, and must be abandoned as a general interpretation of patterns of sexuality in nature. On the other hand, we might fall back on more subjective criteria to assess the frequency and severity of environmental change. I shall choose this latter alternative, not only because it enables the theory to be tested, but also because it provides the test which the best modern authorities have regarded as decisive. Thus, G. C. Williams (1975, p.3) writes that: 'sex is a parental adaptation to the likelihood of the offspring having to face changed or uncertain conditions . . . sexual reproduction will occur where ecological differences will be greatest between two successive generations.' In a similar vein, M. J. D. White (1970) remarks that: 'Because of their ability to respond to ecological change by alterations in the genetic composition of the population due to recombination, bisexual forms would seem far more likely to be able to colonize new, man-made or man-disturbed habitats.' These are not self-evident truths, however: they are specific and empirically vulnerable predictions from general theory. Before the theory can be accepted, the prediction must survive as rigorous a comparative scrutiny as we can devise.

The conceptual bases of the best-man hypothesis that I shall examine are due to Williams: they state that sex should occur when the heritability of fitness is least. Thus, obligate sexuality should predominate in environments whose physical characteristics are subject to frequent severe fluctuation, whilst sex in organisms which can also reproduce parthenogenetically should coincide with or anticipate these fluctuations. The conceptual basis of predictions from the rival tangled-bank hypothesis is due to Ghiselin: sex should occur where or when intraspecific competition is most intense. This leads to predictions which are diametrically opposed to those of the best-man hypothesis: in environments which are kept permanently undeveloped by frequent and intense disturbance, we expect to find parthenogenesis, not amphimixis, whilst among heterogonic animals sex should occur when competitive stress is most intense, or in general near to population maxima. In the following three sections I shall try to give these general predictions particular form, and decide which of the two rival hypotheses reflects nature the more faithfully. The first section will compare sexually reproducing with obligately parthenogenetic animals; the next two will deal with heterogonic life histories.

4.3.3 *The Ecology of Obligate Sexuality*

Sex in Fresh Water and the Ocean. With the exception of the largest lakes, the physical features of fresh waters vary more frequently and more severely than those of the ocean. The best-man therefore predicts that sex will predominate in fresh water and parthenogenesis in the sea. By the same token, fresh water will often offer vacant habitats or niches for colonization, whereas the marine environment will be more fully and continuously exploited. The tangled bank, therefore, predicts that parthenogenesis will predominate in fresh water and sex in the ocean.

The relevant data are presented in Table 4.5, where, for convenience, I have included heterogonic as well as obligately sexual or parthenogenetic taxa. The result is one of the soundest generalizations in evolutionary biology: wherever a comparison is possible, with the solitary and equivocal exception of Rhynchocoela, parthenogenesis predominates in fresh water and sexuality in the sea. The best-man hypothesis is rejected.

Ostracods offer an excellent opportunity to compare the modes of reproduction of the same taxon in different freshwater habitats. The results (Tables 3.11-3.13) contradict the prediction of the best-man hypothesis: parthenogenetic species predominate in small temporary ponds and streams, and sexual species in larger, more permanent bodies of water. Parallel cases occur in other taxa, but they are too little-studied to be trustworthy. For example, the triclad

Table 4.5: Modes of Reproduction in Freshwater and Marine Fauna. For further details, see appropriate section of Chapter 3.

Taxon	Reproduction in freshwater	Reproduction in ocean
1. Demospongiae	Generally heterogonic; asexual reproduction by gemmules	Generally amphimictic, without asexual reproduction
2. Hydrozoa	Hydras. Paratomical budding to form isolated polyps. Also often sexual	Budding usually leads to colony formation; unequivocal reproduction is amphimictic
3. Catenulida	Asexual, by paratomy. Sex very rare	Exclusively sexual
4. Tricladida	Paludicola. Sexual or by architomical fission. One genus obligately selfed. Terricola. Often by fission	Maricola. Exclusively sexual; regular selfing unknown
5. Rhynchocoela	*Prostoma* is sexual but may self	Sexual. Some species of *Lineus* reproduce by fragmentation
6. Rotifera	Bdelloidea. Exclusively apomictic. Monogononta. Heterogonic or apomictic	Seisonidea. Exclusively amphimictic
7. Monogononta	Heterogonic or apomictic	Heterogonic, sexuality more pronounced
8. Gastrotricha	Chaetonotoidea. Exclusively thelytokous	Macrodasyoidea. Exclusively amphimictic
9. Chaetonotoidea	Thelytokous	Two marine genera are amphimictic hermaphrodites
10. Nematoda	Males unknown amongst free-living mematodes in 53% of freshwater and 63% of soil species. In remaining species males often rare	Males unknown in 20% of free-living marine nematodes. In remaining species males usually as frequent as females
11. Endoprocta	Urnatellidae. Sexual, other than colony budding; dormant phase a reduced stalk	Loxosomatidae. Solitary zooids reproduce by budding. Pedicellinida. Sexual, other than colony budding
12. Bryozoa	Phylactolaemates and freshwater gymnolaemates are heterogonic; asexual reproduction by statoblasts	Stenolaemates and marine gymnolaemates exclusively sexual. Colony formation by budding in freshwater and marine bryozoans
13. Gastropoda	Apomixis confined to three prosobranch genera, two freshwater and one terrestrial. Autogamy may be common among freshwater opisthobranchs	Marine gastropods are all sexual and probably always cross-fertilized
14. Oligochaeta	Paratomy usual in naiads and aelosomatids, sex rare or very rare. Tubificids usually amphimictic, sometimes selfed. Apomixis and automixis widespread among terrestrial enchytraeids and lumbricids	Apparently always amphimictic

Taxon	Reproduction in freshwater	Reproduction in ocean
15. Polychaeta	*Nereis limnicola* is selfed	Amphimixis usual; vegetative fission exceptional but widespread
16. Tardigrada	Parthenogenesis common among freshwater and moss-dwelling tardigrades	Parthenogenesis unknown
17. Cladocera	Heterogonic; many species solely apomictic	Heterogonic; males usually common and sex more frequent
18. Ostracoda	Cypridae. High proportion of parthenogenetic species	Cytheridae. Perhaps always amphimictic

Phagocata velata is said to reproduce by asexual fission in small, temporary ponds, but to be amphimictic in permanent bodies of water; the sponge *Heteromeyenia baileyi* gemmulates in seasonal but not in permanent habitats.

The most extreme freshwater habitats are small rock-pools and the water film on mosses and lichens. These are ephemeral habitats, created suddenly by rainfall and persisting for only a few hours or days. They are also very fragile habitats: for instance, the character of a rock-pool holding half a pint or so of water may be changed radically by the chance arrival of a leaf or a pine-cone. According to the best-man, such habitats as these provide ideal conditions for the evolution of sex. In fact, the dominant multicellular animals are usually the wholly apomictic bdelloid rotifers, whilst the other important elements of the fauna – chaetonotoid gastrotrichs, tardigrades and nematodes, with a few turbellarians and aelosomatid oligochaetes – are also wholly or largely parthenogenetic. The shift from amphimixis to parthenogenesis as the predominant mode of reproduction on passing from lakes to small ponds and swamps and finally to rock-pools and water-films is most remarkable; it can be verified by anyone with access to a microscope, but to illustrate the point I have reproduced in Table 4.6 some data collected by students on a field course. It parallels the shift from sex to parthenogenesis between marine and freshwater habitats, and leads to the same conclusion, that the best-man hypothesis is to be rejected.

Sex in Long and Short Growing Seasons. As one passes from the tropics to the poles the physical features of the environment fluctuate with increasing magnitude. For instance, McNaughten's study of the rush *Typha* showed how reproductive success becomes more variable and intraspecific competition less intense at higher latitudes (McNaughton, 1975). The best-man hypothesis therefore predicts that sex should predominate at higher latitudes and parthenogenesis at lower latitudes, whilst the tangled bank leads us to expect the converse pattern.

The best comparative data are supplied by insects, where it is well known that parthenogenesis generally predominates at high latitudes (Table 3.18). The same pattern is shown by some mites, anostracans, ostracods, isopods, millipedes, centipedes and asteroids, and among heterogonic taxa by hydras, phylactolaemate bryozoans, monogononont rotifers and cladocerans. There are some exceptions to the rule: in particular, I have found no evidence that the proportion

Table 4.6: Major Taxa of Freshwater Invertebrates Collected by an Undergraduate Class at Mont St Hilaire, Québec. The taxa are arranged into four groups according to the frequency of amphimictic reproduction: absent (bdelloids and gastrotrichs); very occasional in all species (turbellarian and oligochaete species collected) or absent in many species (tardigrades, nematodes, ostracods); intermittent, with a regular cyclical parthenogenesis (monogononts, cladocerans, sponges); and obligate (copepods, water mites, amphipods, leeches and snails). An attempt has been made to rank habitats from most transient and mutable (small rock pools and the water-film on mosses), through intermediate states (habitats offered by small woodland ponds) to most permanent and stable (habitats in a small lake). Some features of the table — for example, the relatively small number of monogonont rotifers counted from a plankton haul taken in the open water of the lake — are undoubtedly due to the inexperience of the observers, but the general tendency for an increase in the frequency of amphimixis from low-ranked to high-ranked habitats is beyond dispute.

HABITAT DESCRIPTION	OBLIGATE ASEXUALITY		REDUCED OR VARIABLE DEGREE OF SEXUALITY					INTERMITTENT SEXUALITY			OBLIGATE SEXUALITY					Total individuals	Total major taxa
	Bdelloidea	Chaetonotoidea	Tardigrada	Turbellaria	Nematoda	Oligochaeta	Ostracoda	Monogononta	Cladocera	Demospongiae	Cyclopoidea	Hydracarina	Amphipoda	Hirudinea	Mollusca		
1. Exposed dry moss, lichen and litter	91		17		25											133	3
2. Rain pools on rocks	1974		7	5	77											2063	4
3. Damp moss on rocks	263		23		169			2								457	4
4. Damp moss on logs	134		7	14	159	4		26			11					355	7
5. Damp soil and litter	156		3	30	215	3	3	100			71					581	8
6. Puddles in bed of dried-up pond	30	11	40		43	1	3	35			21					185	8

7.	Standing water at edge of small pond	11	4		11	8	3	3	35	2		22	1	1	1	102	12
8.	Submerged detritus in pond	25	4	2	10	58	3	5	36			24				167	9
9.	Submerged detritus at margin of lake	15	4		8	25	7		35	3	1	15	3	1		117	11
10.	Macrophyte bed in shallow bay of lake		2		3	15	1	3	53	21		13	2	6	2	3	124	12
11.	Open water of lake			1				1	121	19		304				3	445	4

of parthenogenetic species varies with latitude in coccid and ostracod faunas (Table 3.20a and 3.14). At lower taxonomic levels the general rule is clear, however, and provides a convincing refutation of the prediction made by the best-man hypothesis.

Sex and Size. Very small multicellular animals have short generation times and high rates of increase, and are thereby able to persist in uncertain and transient environments through their ability quickly to recolonize any newly available habitat or niche. Theories such as the best man which view sex as a preadaptation to an uncertain future therefore predict that, as a general rule, sex should predominate among very small and parthenogenesis among large animals. As I have already pointed out, the reverse is the case.

Sex in Constant and Disturbed Habitats. The best man interprets sex as an adaptation to environments which are novel or which frequently undergo severe disturbance; the tangled bank leads us to expect parthenogenesis in such habitats.

Among the Demospongiae, *Plakina, Tethya, Aaptos* and *Halichondria* are mostly asexual in the intertidal, with *Halichondria* occupying the highest position on the shore and being almost exclusively amictic, whilst they are sexual in the subtidal. *Heteromeyenia* forms asexual gemmules in seasonal but not in permanent habitats. *Laxosuberites* gemmulates in shallow water but is exclusively sexual in deeper water. Most freshwater sponges gemmulate, but *Ochridaspongia* is a deep-water species which seems to be exclusively sexual. Similar patterns are found among Zoantharia. *Polythoa* lives higher in the intertidal, colonizes sandflats where mass extinction is common and reproduces asexually; *Zoanthus* lives lower in the intertidal, does not colonize and reproduces sexually. *Haliplanella luciae* is an ephemeral species with frequent mass extinctions, occupies exposed habitats in the intertidal and reproduces asexually by fission or pedal laceration; the sympatric *Diadumene leucolena* is found in similar but more sheltered habitats, does not undergo periodic mass extinctions and is exclusively sexual. *Anthopleura elegantissima* reproduces by longitudinal fission in the intertidal, giving rise to clones of small individuals, but larger solitary specimens in the subtidal reproduce sexually. *Metridium senile* undergoes active pedal laceration when part of the fouling community, but reproduces sexually in natural subtidal habitats. *Cereus pedunculatus* is parthenogenetic in heavily polluted water of variable salinity, but elsewhere sexual. Thus, these two groups of sessile marine invertebrates suggest a pattern which is the reverse of that predicted by the best man.

The same pattern can be traced in other taxa. Among triclads, asexuality predominates in unproductive lotic habitats and sexuality in productive lentic habitats. Parthenogenetic oligochaetes occur in the upper, more disturbed soil layers, while sexual species characterize the lower, less disturbed horizons. One of the only three known asexual gastropods, the prosobranch *Potamopyrgus jenkinsi*, is a recent invader of European fresh waters. A parthenogenetic race of the isopod *Trichoniscus pusillus* occurs in arid, open habitats and a sexual race in more favourable, humid, protected habitats. A parthenogenetic race of the myriapod *Nemasoma varicorne* is found in less favourable open habitats on sandy soils, with a sexual race occupying more favourable habitats in old beechwoods.

I have collated the ecological correlates of parthenogenesis for insects in Chapter 3. Parthenogenetic insects tend to live at higher elevation; tend to occur in more exposed, xeric habitats; tend to occupy the disclimax; tend to be found in recently glaciated areas; and tend to be domiciliary or recently introduced forms which are successful in colonizing novel habitats. Comparable sexual forms are generally found in lower, less exposed, unglaciated climax habitats, and are less successful in colonizing novel habitats.

Similar trends have been identified among vertebrates. The autogamous *Rivulus marmoratus* inhabits an unstable coastal habitat; gynogenetic *Poeciliopsis* live in unpredictable mountain streams; *Hemidactylus garnotii* is a successful colonist; thelytokous *Cnemidophorus* are concentrated along floodplains and other habitats subject to frequent inundation, with *C. lemniscatus* occupying man-disturbed localities; thelytokous *Lacerta* are usually found on recently glaciated ridges.

The prevalence of parthenogenesis in freshwater, at high latitudes and among the smallest metazoans can now be seen as aspects of a more general phenomenon: parthenogenesis tends to predominate in all sorts of novel or disturbed habitats. Other geographical patterns can be interpreted by the same rule. For instance, we would expect physical conditions to be perceived as more equable near the centre of a species' range, and as less equable towards its periphery, so that parthenogenesis should be more frequent in peripheral populations. This is the case in the millipede *Nemasoma varicorne* (Figure 3.4), where the peripheral habitats happen also to be the drier and more exposed. It is not true for North American psocopterans, where it is the sexual populations which occur peripherally; but in this case the peripheral populations occupy maritime or unglaciated climax habitats, and so do not contradict the generalization.

In short, the overwhelming concensus of the comparative evidence is that obligate sexuality is not associated with novel or disturbed habitats; quite to the contrary, for these are precisely the habitats in which parthenogenesis prevails. We are led to reject the best-man hypothesis.

4.3.4 The Ecology of Intermittent Sexuality

The Frequency and Timing of Sex in Natural Populations. The conventional interpretation of heterogonic life histories given by the best-man hypothesis is that several or many generations of asexual reproduction occur during the growing season, with a single sexual generation at the close of the growing season producing dormant mictic eggs. The heritability of fitness is thought to be least at the end of the growing season, because the environment will change more between the last date on which reproduction is possible in one growing season and the first date on which it is possible in the next than it will between any two successive generations within the same season. In north-temperate environments we expect to find asexual reproduction during spring and summer and sexuality in the fall, just before reproduction shuts down for the winter. This is, indeed, a reasonable description of the life histories of many aphids, cladocerans and monogonont rotifers, the groups on which most theoretical arguments have drawn for their examples. The competence of the theory to explain the finer details of such life histories is examined below. First, however, it must be pointed out that in other heterogonic taxa the supposed general rule is not obeyed.

The freshwater sponges *Spongilla* and *Tubella* reproduce sexually in May-June and produce asexual gemmules in October-November. The marine *Haliclona* also produces sexual larvae in summer and gemmules in late summer or fall; or there may be two periods of sexual reproduction, in early summer and early fall, punctuated by periods of gemmulation. In other marine Demospongiae, gemmulation and sexual reproduction occur together throughout the year. In the freshwater hydrozoan *Hydra oligactis* sexuality occurs in late summer and fall, just after the period of maximum budding rate, but in *H. viridissima* the sexual period occurs early in the growing season. In the marine hydrozoan *Coryne* the medusae are budded in midwinter, with sexual reproduction in the following spring; in *Rathkea* the medusae are produced in fall, reproduce asexually by manubrial budding during the winter and switch to sexual reproduction in spring. The scyphistoma larvae of the scyphozoan *Aurelia* reproduce asexually by budding or strobilation from fall to spring, the medusae reproducing sexually in midsummer; other scyphozoans reproduce sexually in midsummer; and still other scyphozoans reproduce sexually at various times during the year. The ctenophore *Coeloplana* reproduces asexually by pedal laceration during summer, then disappears from the littoral; when it reappears in late summer it reproduces sexually. Freshwater rhabdocoels usually reproduce by fission, but the rare episodes of sexuality seem to occur in fall. Typhloplanid rhabdocoels produce subitaneous eggs by selfing early in the growing season, and dormant eggs by cross-fertilization at the end of the season. Freshwater triclads are variable; sex may occur early in the growing season and fission later, or vice versa. In general the cocoons of sexual eggs are laid during winter in unproductive habitats, but in spring or early summer in productive habitats. The intertidal rhynchocoel *Lineus* reproduces asexually by fragmentation in summer but sexually in winter. Phylactolaemate bryozoans produce sexual larvae in late spring or early summer, but asexual statoblasts in fall; in the tropics statoblasts are produced most abundantly just before the hot season. The asexual resting stages of gymnolaemate bryozoans are also produced in fall, though they may exceptionally be produced in summer as a response to adverse conditions. The gastropod *Omalogyra* is cross-fertilized in spring but selfs in summer, during the most favourable part of the season. The polychaete *Dodecaceria* reproduces by fission during

September–November and gives rise to sexual epitokes in spring and summer. The sexual adults of Cecidomyidae appear in summer, when the substrate dries out and food becomes scarce. The asteroid *Coscinasterias tenuispinus* reproduces by fission in summer but sexually in January–February; in other asteroids, fission may occur throughout the year. The enteropneust *Balanoglossus* reproduces by fission in summer and sexually in winter. Amongst ascidians, the usual pattern is asexual budding in winter and spring, followed by a summer sexual period.

Doubtless further examples could be adduced, but they will not upset the conclusion that, among heterogonic animals, sex may occur at any time during the year, and is not related in any straightforward way to a period of maximum environmental change or unpredictability. To emphasize this point: if we confine ourselves to the major heterogonic taxa of freshwater, we find that cladocerans and monogonont rotifers usually obey the supposed general rule, with a sexual period late in the growing season; but that sponges and bryozoans have the converse pattern, with a sexual period during the growing season and asexual reproduction at its close; whilst turbellarians and hydroids are intermediate between these two extremes. The failure of the generalization is not fatal to the best man, since sexual propagules may disperse in space as well as in time; this is a topic I shall take up below. It is sufficiently disconcerting, though, to tempt us to turn to the rival predictions of the tangled bank. This makes no general prediction about the time of year at which the sexual period should occur, but instead predicts that sexuality should be associated with high population density. I shall now turn to a detailed analysis of the life cycles of cladocerans and rotifers. To substantiate the points to be made, I shall nominate examples in parentheses, rotifers being listed before a semicolon and cladocerans after, in order to facilitate reference to Tables 3.1 and 3.8 and to the relevant text.

In the first place, the prediction that sex should occur at times of high population density is well-supported, since the broadest generalization that can be made about the sexual periods of rotifers and cladocerans is that they occur at or shortly after the population maximum. There seem to be very few exceptions to this rule (*Cephalodella, Euchlanis dilatata*; *Chydorus sphaericus, Leptodora*). Since this maximum often occurs near the end of the growing season, however, the predictions of the best-man and tangled-bank hypotheses overlap to a considerable extent.

Secondly, sexual reproduction occurs during the spring maximum rather than during the fall minimum in aestival species inhabiting permanent bodies of water (*Rhinoglena, Notholca, Synchaeta, Gastropus, Conochilus*; *Daphnia, Chydorus*). If the best-man hypothesis be true, it is not easy to see why sex should not be deferred until the end of the growing season.

Next, species or populations which have two population maxima in the same growing season often have a sexual period associated with each of these maxima (*Asplanchna, Synchaeta*; *Daphnia, Scapholeberis, Simocephalus, Bosmina, Acroperus, Camptocercus*. The rotifer *Polyarthra platyptera* may have three sexual periods, and some *Daphnia* populations in permanent habitats are said to be polycyclic). The best-man hypothesis is forced to interpret the two maxima as indicating the existence of two growing seasons within the same calendar year, with a low heritability of fitness during the intervening period. Besides undercutting the basis on which the best man is usually understood to interpret rotifer and cladoceran life cycles, this seems improbable, because the generation period and the rate of increase of cladocerans are known to be determined primarily by temperature (e.g., Hall, 1962).

Next, if a diacmic population has only one sexual period, this may be associated either with the earlier or with the later population maximum. (The earlier in *Brachionus, Synchaeta*; *Daphnia magna, D. atkinsoni*. The later in *Polyarthra vulgaris*; *Daphnia longispina*. The earlier in the phylactolaemate *Plumatella repens*.) The best-man hypothesis predicts that the single sexual period should always be associated with the later maximum.

Next, if a diacmic population has only one sexual period, this is usually associated with the larger of the two population maxima (*Polyarthra vulgaris, Synchaeta tremula, Asplanchna*; *Daphnia longispina, Bosmina longirostris, B. coregoni, Alona*). This receives a ready explanation from the tangled bank, but the best man offers none.

Next, minimal populations are exclusively amictic, whenever they occur during the growing season. I know of no exception to this rule, which offers strong support for the tangled bank; for, if the best-man hypothesis were true, why would not females always reproduce sexually at the end of the growing season, without reference to population density?

Next, the intensity of the sexual period may vary from year to year according to the height of the population maximum, with larger maxima inducing the appearance of a larger proportion of males and mictic females (*Keratella, Polyarthra major, Filinia longispina*; *Daphnia schødleri, D. longispina*). This receives no explanation from the best-man hypothesis.

Next, the sexual period tends to be less intense in localities where a species is less abundant (*Gastropus minor*; *Bosmina obtusirostris*). This seems allied to the previous point, and again resists interpretation by the best man.

Next, sexual periods seem to be less frequent among species which are less abundant. This is not well-established among rotifers (*Platyas, Trichotria* and *Macrochaetus* might be taken to be examples, but *Colurella* is a very common genus comprising many species, few of which have sexual periods at all frequently, whilst males and resting eggs have been described for all species of the rather scarce *Asplanchnopus*), but is generally true among cladocerans (Figure 3.1). The best-man hypothesis is again in difficulties.

Finally, there are several aspects of the population genetics of cladocerans which seem incompatible with the lottery versions of the best man that were devised to explain heterogenic life histories. Firstly, Hebert has concluded from the small-scale spatial heterogeneity of allele frequencies that most populations are founded by only one or a very few ephippia. This offers little scope for the lottery principle. Secondly, there is the apparently decisive evidence of clonal diversity at the end of the growing season, to which I have already referred. Thirdly, the linkage disequilibrium and large fluctuations of genotype frequency, which are prerequisites for any attempt at a best-man explanation in terms of capriciousness, are characteristic of populations whose sexual periods are weak or infrequent, whilst populations whose sexual periods are frequent and intense are close to linkage equilibrium and show less change in genotypic composition.

Considered separately, none of these generalizations are conclusive, and in some cases the comparative material is sufficiently scanty to nourish hopes that they will be overturned by further work. Taken together, they supply a convincing refutation of the predictions made by the best-man hypothesis.

The life cycles of heterogonic organisms in different habitats offer a further opportunity for comparison. It has often been stated that sexual periods are more frequent among the inhabitants of small ponds than among the pelagic species of large lakes, and that this demonstrates the function of sex in environments subject to abrupt change.

There appears to be little support for this proposition at the generic level among monogonont rotifers. Sexual periods do not occur, or are exceedingly infrequent, in *Platyas*, *Mytilina*, *Macrochaetus*, *Colurella*, *Lecane* and *Horaella*. Most of these are benthic or littoral rotifers, often found in small ponds; the exception is *Horaella*, a plankter of small ponds. Sexual periods are rare in most species of *Anuraeopsis*, *Trichotria*, *Lepadella*, *Squatinella*, *Cephalodella*, *Scaridium* and *Testudinella*. All of these are littoral or small-pond species, often benthic. The typical plankton rotifers of the open waters of large lakes are *Rhinoglena*, *Keratella*, *Kellicottia*, *Gastropus*, *Asplanchna*, *Synchaeta*, *Ploesoma*, *Polyarthra*, *Pompholyx*, *Filinia*, *Hexarthra* and *Conochilus*. Males and resting eggs have been described for almost all of the species of these genera (a prominent exception is *Polyarthra minor* — which lives in small temporary ponds and wet moss). Whilst the lake plankters may have received more attention, if we are to keep within the data we must conclude that sexuality is less pronounced among littoral and small-pond rotifers than among the pelagic species. This is not surprising, since it merely extends the generalization that parthenogenesis is more frequent among the taxa of disturbed habitats, and is consistent with the rules that sex is more frequent in marine than in freshwater cladocerans and rotifers, and more frequent at low than at high latitudes.

Nevertheless, it is well-established that when we compare different populations of the same species those living in small ponds have more intense and more frequent sexual periods than those living in large lakes (species of *Keratella*, *Kellicottia*, *Trichocerca*, *Synchaeta*; *Daphnia*, *Bosmina*, *Chydorus*). This is a puzzling exception to an otherwise well-established general rule. Other than accepting it at face value, I can only point to the fact that lake populations are usually several orders of magnitude less dense than pond populations of the same species. In large lakes the growing season may simply not afford enough time in which to saturate the available niches, while the enormous populations commonly observed in small ponds may experience intense competition. If this is so, one would expect fecundity (the rate of production of amictic eggs) to fall much more sharply towards the end of the growing season in pond populations than in pelagic populations of the same species. I have not found the data with which to test this prediction directly; it seems compatible with the general rule that the sexual period, regardless of its intensity, tends to switch from earlier to later in the growing season as one passes from lower to higher latitude or from lower to higher elevation. It also seems compatible with the observation, made both by Hebert and by Young, that in cladoceran populations ephippial females tend to have a genotype which is declining in frequency in the population as a whole.

4.3.5 The Elicitation of Sex in the Laboratory

Most heterogonic animals — cynipid wasps are an exception — can be induced to switch from parthenogenetic to sexual reproduction by an appropriate manipulation of culture conditions in the laboratory. The proximate factors responsible for this switch may tell us something about the function of sexuality: if sex occurs in anticipation of future change, it should be elicited by an abrupt change in culture conditions, or by culture conditions which, if encountered in nature, would signal the imminence of change; if sex acts to reduce competition between progeny it should be elicited by crowding (when food supply is held constant) or by starvation (when population density is held constant). However, it is unlikely that such experiments will yield a conclusive result, since it can often be argued — for example, in a species with a population maximum and a sexual period in the fall — that low temperature and short photoperiod act as cues which signal the approach of high population density or, alternatively, that high population density signals the approach of a period during which the heritability of fitness will be low. Nevertheless, the data are instructive.

Changes in temperature, photoperiod or food quality seem to elicit sexuality in different genera of plankton rotifers. Change of temperature, especially exposure to low temperature, induces gametogenesis in *Hydra* and *Protohydra*, medusoid budding in the hydrozoan *Coryne*, gametogenesis in another hydrozoan, *Rathkea*, and gonad development in asexual strains of the turbellarian *Dugesia*. Sexual individuals appear in aphid cultures when these are maintained under a temperature and photoperiod régime simulating that of normal fall conditions. Conversely, in the freshwater sponge *Spongilla*, which reproduces asexually at the end of the growing season, gemmulation is triggered by cues associated with the onset of autumn. Other than a few dubious observations on *Hydra* there is no evidence that simply bringing about an abrupt change in culture conditions will induce sexuality.

More generally, however, sexuality can be elicited by crowding or starvation. These are by far the most potent factors known to elicit sexuality in the freshwater coelenterates *Hydra* and *Craspedacusta*, and in cladocerans; but they also contribute largely to the switch to sexuality in plankton rotifers, aphids and

Cecidomyidae. The frequency of males in several plant-parasitic nematodes increases with crowding and starvation, perhaps implying a lower frequency of parthenogenesis under these conditions. Conversely, asexual reproduction can be triggered by a temporary superabundance of food, as it is in the scypho-zoan *Aurelia*. The generality of this phenomenon is impressive: apart from the complicated situation reported from the rotifer *Notommata*, I know of no well-studied case in which crowding or starvation have been eliminated as factors effective in eliciting sexuality.

It is true that the dynamics of natural populations of these animals implies that high population density will often signal the onset of autumn, and so indicate the approach of a period of low fitness heritability in which sex might be adaptive. But this best-man interpretation of the facts seems to assert that high population density or low food availability are more reliable cues than physical variables such as temperature and photoperiod. It seems more reason-able, and less onerous, to suppose that cause and effect are more directly related, and that the nearly universal efficacy of crowding as a trigger for sexuality reflects the adaptiveness of sexuality at high population density in nature. The more straightforward interpretation of events in laboratory culture, therefore, is that provided by the tangled-bank hypothesis.

4.3.6 Dispersal and Dormancy

The life cycles of heterogonic organisms often include two stages which are of great importance to our argument: a dormant stage, resistant to hostile environ-mental conditions such as freezing or drying, which serves to disperse progeny in time; and an active motile stage which disperses progeny in space. Many authors (e.g., Williams, 1975, pp. 4-7) have pressed the point that dispersal, whether in time or in space, will be associated with a low heritability of fitness, and therefore that the dispersive propagules will always be sexually produced if the best-man hypothesis is correct. I agree with this reasoning; I would also point out that the tangled-bank, with its emphasis on the importance of spatial as opposed to temporal heterogeneity, leads us to expect a difference between propagules designed to disperse in time and those designed to disperse in space. A correct prediction of the mode of production of dispersive propagules is, I think, crucial to the plausibility of the best-man hypothesis, and it has been asserted by previous authors (e.g., Bonner, 1965; Williams, 1975) that the best-man predictions are indeed well-supported. I shall now examine the evidence for this assertion.

It is convenient to begin with a brief review of vegetative reproduction. Radial or asymmetrical animals (Porifera, Hydrozoa, Scyphozoa, Anthozoa, Endoprocta, Bryozoa, Pterobranchia, Tunicata) commonly reproduce by budding, the process by which an individual, usually a tentaculate polyp or zooid, is formed as an outgrowth from the body wall of its parent. Asexual reproduction by fission is characteristic of elongate bilateral animals which lack a rigid skeleton. Paratomical fission, which is morphologically so similar to the budding of radial animals that it often goes by the same name, is found in Rhabdocoela, Polychaeta and Oligochaeta. Architomical fission is found in Tricladida, Rhynchocoela, Phoronidea, Sipunculoidea, Polychaeta, Oligochaeta and Enteropneusta. Some radial and asymmetrical animals (Mesozoa; some Hydrozoa, Scyphozoa, Anthozoa, Ctenophora, Holothuroidea, Asteroidea, Ophiuroidea) reproduce by fission, usually architomical in nature, or (Porifera, Bryozoa) by colony fragmentation. At its limit, architomical fission in some polychaetes, which are able to regenerate from single cast segments, approaches reproduction by eggs. Despite the diversity of vegetative reproduction, and the fact that it characterizes so many taxa of aquatic animals, it has yet to receive a theoretical treatment comparable to the theory of life histories in egg-laying animals. For present purposes, however, the pertinent hypothesis is that in most cases the function of vegetative reproduction is to occupy as fully as possible a restricted local habitat. This is obvious enough in the case of animals which bud to form a highly integrated colony, and in which budding partakes more of growth than of reproduction, but it is also true of those animals, such as phylactolaemate bryozoans and some ascidians, in which the colony comprises a group of almost autonomous zooids. More generally, even animals which reproduce by fission to create a group of unequivocally independent individuals are almost always sedentary benthic forms which live in large, semi-isolated, probably clonal aggregates (Anthozoa, Rhynchocoela, Phoronidea, Polychaeta, Pogonophora, Holothuroidea, Asteroidea, Ophiuroidea; perhaps Sipunculoidea). Some motile benthic animals – the architomical triclads and the paratomical rhabdocoels, aelosomatids and naiads – may be exceptions to this rule; but in freshwater hydras, which are capable of considerable dispersal, I have found that the polyps in natural populations show a highly clumped distribution, and the same might be true of benthic worms. In general, it can scarcely be doubted that the primary function of vegetative reproduction is to ensure as complete an occupancy as possible of a favourable patch in a spatially heterogeneous environment. This conclusion receives strong support from the correlation between age and the mode of reproduction. In Trachylina, Siphonophora, Hydroida, hydrozoan medusae, Scyphozoa, Tricladida, Endoprocta, Bryozoa, Polychaeta, Oligochaeta, Asteroidea, Ophiuroidea, Pterobranchia and Ascidiacea, vegetative asexual reproduction occurs in younger and sexual reproduction in older individuals; indeed, sexual reproduction is often a terminal phase which is followed by the death of the individual. It is, therefore, a very sound general rule that colonial, sessile and sedentary animals reproduce vegetatively at first, until the local habitat is saturated. At this point they must produce dispersive propagules, which should be designed to ensure continued reproductive success by colonizing new habitats in space and time. How are these dispersive propagules produced?

If we consider first the mode of formation of propagules which disperse in space, we at once encounter the difficulty that if the function of a propagule were to be to achieve dispersal in time through prolonged dormancy, it would be likely also to be dispersed passively in space. Indeed, propagules which experienced long periods of dormancy may become dispersed far more widely than highly motile but short-lived propagules. In the first place, then, let us consider only those propagules which possess locomotory structures, such as cilia, and whose activity seems to be directed towards at least a local dispersal in space. The critical distinction we wish to make is between the mode of reproduction within local habitats and the mode of reproduction responsible for the production of propagules which disperse between local habitats. The correlation between age and the mode of reproduction, established in the previous paragraph, seems to guarantee a strong correlation between sexuality and active dispersal in space, and the comparative data, summarized in Table 4.7, bear this out very fully. It seems to be an almost invariable rule that, in sedentary and colonial animals, active dispersal away from the local habitat is associated with a switch from vegetative to sexual reproduction. The only clear exceptions known to me are the supposedly amictic larvae of certain ceractinomorph sponges, tentacle constriction in the scyphozoan *Chrysaora* and the peculiar motile buds of the endoproct *Loxosomella*. If we restrict the comparison to animals which reproduce only by eggs the data are less extensive but point in the same direction: where one stage in the life cycle reproduces sexually and another asexually, the stage which reproduces sexually is capable of a greater degree of active dispersal (Cecidomyidae, Micromalthidae). The reduction of flight ability in adult parthenogenetic insects also conforms with the generalization. All this is perfectly consistent with the conventional statement of the best-man hypothesis. It is also consistent with the tangled bank, or with some lottery versions of the best man, but only if active dispersal really is local, so that nearby habitats are likely to be stocked with several individuals from the same sibship. When the dispersing individual is an adult, itself produced asexually, which reproduces sexually after dispersal (for instance, hydrozoan medusae or the winged adults of cecidomyids or aphids), this does not seem implausible. When the dispersing individual is a motile larva, much depends on the function of the organs of locomotion; if they are indeed primarily responsible for dispersal, then limited movement and consequently competition within sibships may be the rule; but if dispersal is largely passive, with the locomotory structures being used only for detailed manoeuvring near the site of settlement, then this need not hold. In short, whilst the correlation between sex and active dispersal is not sufficient to falsify either tangled-bank or lottery models, it is more obviously favourable to a conventional version of the best man.

Sooner or later, conditions may deteriorate to the point where dispersal in time becomes crucial. Instead of examining the mode of reproduction within and between local habitats, we must now examine the mode of reproduction within and between growing seasons. The best-man hypothesis again makes the unequivocal prediction that dispersal in time should be achieved by sexually diversified propagules. The comparative data, which are summarized in Table 4.8, do not support this prediction. With some exceptions, dormant and resistant propagules are produced sexually among Hydrozoa, Tricladida, Monogononta, Cladocera and Aphidoidea; but are produced asexually among Demospongiae,

Table 4.7: Comparison of Modes of Reproduction Within and Between Local Habitats in Heterogonic Animals with an Actively Dispersing Stage.

Taxon	Reproduction between local habitats	Reproduction within local habitats
1. Porifera	Sexual (parenchymella larva). Asexual parenchymellae may be produced by some marine ceractinomorph sponges	Asexual, by budding, gemmulation or fragmentation
2. Hydrozoa	Medusa disperses and gives rise sexually to polyps. Sexual reproduction by polyps produces planula larva. Asexual frustule may disperse	Asexual, by polyp budding. Frustules are produced asexually by polyp. Medusae often bud
3. Trachylina and Siphonophora	Sexual	Asexual; non-colonial budding confined to young, often brooded, larvae
4. Scyphozoa	Sexual	Usually sexual, but asexual budding or strobilation proliferates a polypoid phase
5. Anthozoa	Sexual (planula larva). Some dispersal probably achieved by pedal laceration	Asexual budding, fission or pedal laceration to form colony or clone
6. Ctenophora	Sexual (cydippid larva)	Presumably less dispersal achieved by pedal laceration
7. Endoprocta	Sexual larva. In *Loxosomella* the liberated buds can disperse actively	Asexual budding to form colonies or clones
8. Phylactolaemata	Sexual larva	Asexual budding to form colony; asexual statoblasts
9. Gymnolaemata	Sexual larva	Asexual budding to form colony; asexual hibernacula and other structures
10. Polychaeta	Sexual epitokes and larvae	Asexual, by vegetative fission
11. Insecta	Parthenogenetic species and life stages tend to have reduced dispersal ability; see text	
12. Echinodermata	Sexual larva. Parthenogenetic larvae may be produced regularly by some asteroids	Vegetative fission in some asteroids, ophiuroids and holothuroids
13. Pterobranchia	Sexual, planula larva	Asexual budding to form colony
14. Enteropneusta	Sexual, tornaria larva	Asexual, by architomical fission
15. Tunicata	Sexual, appendicularia larva. In salps and doliolids the phorozooids and gonozooids may be dispersive, but all stages in life cycle are pelagic	Asexual budding to form colony or local group of zooids

Table 4.8: Comparison of Modes of Reproduction Within and Between Growing Seasons in Heterogonic Animals which have a Dormant Stage.

Taxon	Reproduction between growing seasons	Reproduction within growing seasons
1. Demospongiae	Asexual, by gemmules. Some species form reduction bodies	Asexual, by budding; subsequently sexual, by larvae
2. Hydrozoa	Encysted sexual egg. The polyp may form a reduction body	Asexual, by budding; subsequently sexual
3. Scyphozoa	Asexual, by podocysts. Scyphistomae, ephyrae and medusae may form reduction bodies. In *Cyanea* sexual planulae encyst	Usually sexual, by medusa; polypoid phases may bud asexually
4. Tricladida	Sexual, by cocoons of mictic eggs. In *Phagocata* the dormant phase is an encysted embryo produced by sexual fragmentation	Sexual or by asexual fission
5. Typhloplanidae	Cross-fertilization	Self-fertilization
6. Monogononta	Sexual, by amphimictic resting eggs. Asexual resting eggs produced by some species	Asexual, by subitaneous apomictic eggs
7. Endoprocta	Asexual, by reduced stalks	Asexual, by budding; subsequently sexual
8. Phylactolaemata	Asexual, by statoblasts	Asexual, by budding; subsequently sexual
9. Gymnolaemata	Asexual, by ancestrulae, hibernacula or dormant stolons	Asexual, by budding; subsequently sexual
10. Nematoda	Only dormant phase in free-living nematodes is anabiotic adult	Parthenogenetic or sexual eggs
11. Tardigrada	Anabiotic adult or a resting egg; cytology of resting egg is unknown	Amphimictic, automictic or apomictic eggs
12. Cladocera	Sexual resting eggs. Asexual resting eggs produced by some species	Subitaneous apomictic eggs. Rarely by subitaneous sexual eggs
13. Aphidoidea	Dormant sexual eggs	Subitaneous apomictic eggs
14. Pterobranchia	Dormant asexual bud	Asexual colony budding; subsequently sexual
15. Tunicata	Dormant asexual bud	Asexual budding; subsequently sexual

Scyphozoa, Monogenea, Phylactolaemata, Gymnolaemata, Endoprocta, Pterobranchia and Tunicata. In nematodes and tardigrades the dormant phase is often an anabiotic adult, and it can be argued that this too represents an asexual means of dispersal in time. In short, there are at least as many exceptions to the proposed rule as there are cases which obey the rule; and the prediction made by the best man stands refuted.

The general pattern has, then, been imperfectly perceived by previous authors. Active dispersal in space is indeed almost the exclusive province of sexual propagules in heterogonic animals, but dispersal in time by dormant or resistant propagules may be achieved either sexually or asexually, with sharp differences between major taxa and some variation within these taxa. The primary function of sexual propagules is thus to achieve short-range dispersal in space and not to achieve dispersal in time, and hypotheses which assert that sex is a preadaptation to an uncertain future are in this way falsified.

Short-range dispersal in space through a specialized propagule is essential to sessile benthic animals; motile animals will need a specialized propagule only for long-distance dispersal. Dormant asexual propagules, which disperse in time but may also be very widely dispersed in space by currents or animal vectors, seem to be produced mainly by sessile animals, whilst dormant sexual propagules are produced mainly by motile animals. This rule is illustrated nicely by the dichotomy among heterogonic freshwater animals to which I have already referred, with the sessile sponges and bryozoans producing dormant propagules asexually and the motile cladocerans and monogonont rotifers producing them sexually. This pattern can be interpreted tentatively by the tangled-bank principle. Sessile animals germinate in some locally favourable patch of a two-dimensional environment which they are able to saturate rather early in the season by continued colonial growth or by the proliferation of autonomous but nondispersing zooids. They must then colonize unoccupied local habitats nearby, and do so with sexually diversified propagules which germinate immediately, without diapause, in order to exploit as fully as possible a spatially heterogeneous environment. Motile planktonic animals live in a three-dimensional habitat which they do not saturate by clonal reproduction until relatively late in the growing season. They then make resistant propagules which will recolonize the environment in the following season, and these propagules are sexually diversified in order to reduce competition between sibs when they germinate.

That dormant propagules often serve to achieve passive dispersal is proven by the fact that they often have structures — such as the floats and circumferential hooks of some bryozoan statoblasts, or the dorsal spines of some cladoceran ephippia — which can be easily interpreted only as aids to dispersal in space. The very long diapause of which some such propagules are capable (see text for sponge gemmules, bryozoan statoblasts and reduced adults of bdelloids and nematodes) implies that they must often be passively transported great distances from their parents. It is, then, likely that each propagule produced by a given parental individual or clone will germinate in a different local habitat, and the tangled bank then predicts that such propagules will often be produced asexually, which is the case. More specifically, one might expect that dormant propagules which are produced asexually would remain viable for longer periods, would contain more stored food reserves and would be more difficult to germinate in the laboratory than those which are produced sexually. It is my impression from the literature that this is true — gemmules and statoblasts, for instance, contain bulky reserves of stored food and remain viable when air-dried for many years — but I have found no quantitative data adequate for a comparison.

In short, the phenomena of dispersal and dormancy are largely consistent with the tangled bank, but we know too little of their ecology to decide whether or not they give strong support to it. Two objections should be particularly stressed. The first is that, whilst the amictic production of dormant propagules falsifies most best-man models, it is consistent with some lottery versions of the best man; for example, the aphid-rotifer model yields more or less the same predictions as the tangled bank, since it too predicts a possible advantage for sexuality when two or more individuals from the same sibship grow up together in the same local habitat. The strawberry-coral model, which is designed to deal with the situation in which all the survivors of a given sibship grow up in different local habitats, is obviously inconsistent with the amictic production of dormant propagules, and must therefore be rejected as a general interpretation of the life cycles of sessile animals. The second objection is that, if the tangled-bank interpretation is correct, it is not clear to me why rotifers and cladocerans should not often produce subitaneous sexual eggs at high population density, before they produce resting eggs. By doing so, a female would produce new clones which, experiencing relatively little competition, might give her greater eventual reproductive success than females which either continued to reproduce asexually or which produced resting eggs. Perhaps genotypic diversity is sufficiently high to make such a strategy unprofitable; or perhaps subitaneous sexual eggs are more common than is usually thought. But stating these possibilities does not remove a worrying inconsistency.

4.3.7 Conclusions

Concluding that the unities of time and place are neither necessary nor desirable criteria of drama, Johnson, in his *Preface to Shakespeare*, added:

> Yet when I speak thus slightingly of dramatic rules, I cannot but
> recollect how much wit and learning may be produced against me;
> before such authorities I am afraid to stand, not because I think the
> present question one of those that are to be decided by mere authority,
> but because it is to be suspected, that these precepts have not been so
> easily received but for better reasons that I have yet been able to find.

I find myself in a similar situation ('in which' — to continue the quotation — 'it would be ludicrous to boast of impartiality') with respect to the best-man hypothesis. Accepted, at least as a conceptual foundation, by the best minds which have contemplated the function of sexuality, it seems utterly to fail the test of comparative analysis; not, that is, to fail a few of the many possible tests, or to fail them but partially, but to fail with respect to every major particular in which it is examined. Sex does not predominate in freshwater rather than in the sea, at high rather than at low latitudes, among small rather than among large organisms, and in general in disturbed rather than in equable environments, but contrariwise; it does not invariably occur at times of minimum fitness heritability, but rather at times of high population density; it is not elicited by abrupt environmental change, or specifically by cues indicating the imminence of change, but by crowding and starvation; and if it is associated with dispersal in space, it is not generally associated with dispersal through time. To these criticisms one may add the several inadequacies of the theory's axiomatic structure, besides its failure to anticipate the details of heterogonic life histories. In short, its failure is so general and so comprehensive that it must be judged to be void of explanatory or predictive power; and if it were a new theory, advanced without such weight of past authority, I would not have discussed it at such length, but would instead have dismissed it after slight consideration.

If my conclusion is accepted, as I think it must be, there are three roads open to us. In the first place, we might tinker with the basic theory, adjusting it so as to produce a different model for each particular situation. Human ingenuity is such that there is little doubt that this could be done successfully; yet I take it as a sign of radical weakness that the theory should require such piecemeal renovation to save it from rejection. Secondly, we might argue that the comparative analysis is itself unsoundly based; that we must look to capriciousness, and not to the mere magnitude and frequency of environmental disturbance, for the true correlates of sexuality; and — inescapably — that those environments are the most capricious which appear to change the least. This would seem to require more courage than wit. The third road, and the one which I propose to take, involves rejecting the best-man hypothesis as having been comprehensively discredited, and searching for an alternative. The most general conclusion that one can draw from the comparative evidence is that sexuality is normally associated with spatial (ecological) rather than temporal heterogeneity, and this immediately suggests the tangled bank as the preferred alternative. Indeed, where I have contrasted the best man and the tangled bank in the analysis above, the tangled bank has almost invariably given a more satisfactory interpretation of the facts. Whilst sufficient to falsify the best man, however, this evidence is not sufficient to establish the truth of the tangled bank, since it is not proven that some other hypothesis might not be equally or more satisfactory. This is the possibility that I shall examine in the following section.

4.4   The Red Queen

The Red Queen hypothesis proposes that the function of sex is related to the frequency and intensity of interactions between species, from which it seems legitimate to infer that the occurrence of sex will somehow be related to patterns of species diversity. But if this reasoning is correct we expect to find thelytoky where species diversity is low; and low species diversity is often associated with undeveloped habitats, so that, for instance, we find fewer species at high latitudes or in fresh water than we do at low latitudes or in the sea; so that the Red Queen leads us towards the same predictions that are made by the tangled bank. By no means all the patterns reported in the previous section are equally compatible with both theories, but the overlap is large enough to make it desirable to distinguish between them more sharply. I shall attempt to do so in two ways: first, by analyzing the life cycles of parasites and mutualists; and secondly by comparing habitats in which population density is low and species diversity high with those in which population density is high and species diversity low.

4.4.1 The Lives of Symbionts

According to the Red Queen, sex may be favoured by any negative interaction between species. Sex will be particularly important when the interaction is intense, so that the magnitude of the effect is large; and when the interaction is specific, so that the effect is not buffered by selective forces acting in different directions. The parasite-host interaction, which is usually more intense and more specific than predator-prey or competitive interactions, is therefore a particularly appropriate testing ground for the theory.

Parasitism is found in most of the major protostome taxa. A great many parasites are obligately amphimictic (Aspidobothria, Monogenea, Cestodaria, Acanthocephala, Gordiacea, Seisonidea, Myzostomida, Pentastomida, Siphonaptera, Mallophaga, Anoplura; parasitic members of Ostracoda, Copepoda, Cirripedia, Isopoda, Amphipoda, Decapoda, Tardigrada, Hemiptera, Diptera, Coleoptera, Hymenoptera, Strepsiptera). The Eucestoda and some parasitic nematodes are habitually self-fertilized, but otherwise obligate thelytoky is restricted to a few mesostigmatan mites, some plant-parasitic nematodes, a few blood flukes and a cestode. Indeed, it would scarcely be an exaggeration to say that there is no well-authenticated instance of a multicellular animal parasitic on other animals which is obligately thelytokous. On the other hand, a great many species of Mesozoa and Digenea, together with a few nematodes, have heterogonic life cycles in which thelytoky alternates with amphimixis.

These heterogonic cycles are all organized in the same way. First, thelytokous reproduction always occurs within the host: the agamogony of mesozoans, the vegetative proliferation of cercariae in digenetic trematodes and the production of parthenogenetic eggs by nematodes; one might add the regular polyembryony of some hymenopteran parasitoids. Secondly, and consequently, sexual reproduction occurs between hosts, with dispersive larvae and resistant eggs being always sexually produced. Finally, in forms with an indirect life cycle, sexual reproduction always occurs in the final host, the sexual products infesting the primary host.

The first generalization is easy to understand. The life cycles of most parasites, including those in heterogonic taxa, is governed by the difficulty of passing from one host to the next. One expression of this is the enormous fecundity of many parasites. Another is the array of devices to ensure fertilization: hermaphroditism, self-fertilization, dwarf males, sperm storage and permanent associations between males and females are all often encountered among parasites. Indeed, nothing is more impressive in the biology of parasites than the lengths to which they will go in order to retain amphimixis, or at least cling to some remnant of sexuality. These devices would not be needed if host individuals were commonly infested with a large initial number of parasites. Thus the body of the host will usually constitute an undeveloped habitat, and asexual reproduction will be favoured just as vegetative proliferation is favoured among sessile animals colonizing favourable environmental patches.

The usual interpretation of the other features of heterogonic life cycles among parasites is that offered by simpler versions of the best man. Sexual reproduction occurs between hosts because it is then that the heritability of fitness is least; it occurs in the final host because the final host is usually larger and more motile than the primary host, dispersing the propagules more widely and thus making less predictable the conditions in which they will germinate (Williams, 1975). The failure of such an hypothesis to explain the life cycles of free-living organisms, however, inclines one to turn to it as an explanation of parasite life cycles only as a last resort.

A rival interpretation can be developed from tangled-bank principles. If the hosts differ in their ability to resist infection by parasites with a given genotype, then the host population will constitute a complex environment, and sex might be selected in the same way as in an environment whose physical features are heterogeneous: by diversifying the infective propagules, a female will ensure that her progeny are not all competing to infest the same subset of the host population. Among parasites with an indirect cycle, the passage from primary (or intermediate) to final host is accomplished either by the production of dormant or dispersive propagules, or through the primary host being eaten by the final host. In either case the final host is likely to accumulate parasites from several primary hosts. Thus, while sex would be pointless in the primary host, whose parasite burden has usually been developed through the asexual proliferation of a single parasite individual, it will be functional in the genetically diverse parasite population accumulated by the final host (M. T. Ghiselin, personal communication).

However, the analogy between a population of hosts and a physically complex environment is incomplete in one major respect: the relationship between parasite and host is mutual, so that the host population can respond adaptively to any selective stress imposed on it by the parasite, and vice versa. This immediately suggests the Red Queen as a possible interpretation of parasite life cycles. The first two generalizations present little difficulty: asexual reproduction within hosts is explained as before, whilst sexual reproduction between hosts is explained as a device to keep up with host counteradaptation. The third generalization, that sex occurs during the passage from the final to the primary host, requires a little more thought. The frequency of sex in the parasite should be related to two features of its host. First, if the host is relatively small and short-lived, the parasite will need to undergo more recombination in unit time to maintain its adaptation. Secondly, if the parasite is highly host-specific the interaction between parasite and host will be the more intense, and will thereby generate more powerful selection for recombination in the parasite. We therefore expect to find sexual reproduction associated with the passage from large, long-lived, slowly reproducing hosts, to which the parasite is less specific, to small, short-lived rapidly reproducing hosts, to which it is more specific. It is generally true, both that the final host is larger and longer-lived than the primary host, and that in parasites with an indirect life cycle there is less specificity for the final than for the intermediate or primary hosts. The Red Queen hypothesis thus offers a possible explanation of the life cycles of heterogonic parasites.

With sufficient ingenuity, therefore, all three major hypotheses of individual selection are capable of providing a solution to the problem. The purpose of the present section is to discover whether the Red Queen has a broader explanatory range than the tangled bank, so I shall proceed by attempting to separate Red Queen from tangled-bank predictions.

First, the Red Queen predicts that obligate thelytoky should be extremely rare among parasites. I have already established that this is the case: despite a variety of reproductive adaptations to low density, parasites almost invariably retain a gametic sexuality of some sort. It may also be significant that most parasites which are obligately amphimictic have direct life cycles, and generally display greater host specificity than heterogonic parasites with indirect life cycles.

Secondly, in comparable taxa the Red Queen predicts that thelytoky should be more common in free-living than in parasitic forms. Among freshwater ostracods, the large genus of ectoparasites *Entocythere* is exclusively amphimictic (with the dubious exception of one little-known Canadian species), whilst nearly 50 per cent of free-living species are parthenogenetic. The ectoparasitic seisonids are the only exclusively amphimictic class of rotifers, but since they are also marine the Red Queen and tangled-bank predictions are identical. There seems to be much less parthenogenesis in Siphonaptera, Mallophaga and Anoplura than in most orders of small free-living insects. More satisfactory comparative material exists for Nematoda (Table 3.2) and Acari (Table 3.7), but the situation is complicated by the occurrence of autogamy in the former and arrhenotoky in the latter. Briefly, thelytoky seems to be most common among nematodes which are free-living in soil or in freshwater, or which are ectoparasites of plant roots; it is much less common among nematodes which inhabit marine sediments or which are parasites of animals. Among nematodes which have an indirect life cycle in which a vertebrate is the final host, thelytoky is known only from *Strongyloides*, where it alternates with amphimixis. With the possible exception of some Rhabditidae, all parasitic nematodes with a direct life cycle are amphimictic or, less often, autogamous. Most Acari are amphimictic or arrhenotokous; among bloodsucking ectoparasitic genera there are only a few unconfirmed reports of thelytoky. Cytologically confirmed cases of thelytoky are known only from the plant-parasitic Tenuipalpidae and Tetranychidae, and from the free-living prostigmatan Cheyletidae and Cryptostigmata. Thelytoky appears to be most common in the Cryptostigmata, the free-living orabatid mites.

Finally, there should be a very powerful comparison between parasitic and mutualistic taxa, the negative interaction of parasitism promoting selection for higher rates of recombination and the positive interaction of mutualism favouring the suppression of recombination. The most highly evolved mutualisms involve protists which live within the cells or tissues of body cavities of protostomes, and the Red Queen predicts that these protists will have evolved genetic systems which minimize recombination. Unfortunately, the genetic systems of the group most obviously useful, the endozooic algae, seem to be very poorly known. Mutualistic blue-green and green algae (zoocyanellae and zoochlorellae) seem to reproduce wholly asexually, by fission or through zoospores; I have come across no reference to sexual phenomena. On the other hand, the nearest free-living relatives of these forms are also invariably asexual. The zooxanthellae are related to free-living algae in the Pyrrhophycophyta and other taxa in which sexual reproduction is common. Endozooic forms normally reproduce by fission or asexual zoospore formation, but a meiotic process has been reported from *Gymnodinium microadriaticum*, the only form whose life cycle has been studied carefully.

The life cycles of the mutualistic flagellates (Polymastigina) of the cockroach *Cryptocerus* have been the subject of a classic series of papers by L. R. Cleveland (summarized and cited by Cleveland, 1956). In the vegetatively haploid genera *Trichonympha* and *Eucomonympha*, and in the vegetatively diploid *Macrospironympha*, there is a normal two-division meiosis and amphimictic syngamy. In other genera, however, the amount of recombination is reduced either by an automictic syngamy or by the adoption of a one-division meiosis: *Leptospironympha* is amphimictic but has a one-division zygotic meiosis; *Rhynchonympha* is a vegetatively diploid form in which the first meiotic division is followed by cytokinesis, after which the second division is effectively suppressed by an automictic fusion of the gametic pronuclei; in *Oxymonas* and *Saccinobaculus* there is a one-division zygotic meiosis followed by either automictic or amphimictic fusion. In *Barbulonympha* fusion may be either automictic or amphimictic, with a two-division meiosis, but the zygote may also be formed by endomitosis. In *Notila* each sexually competent cell produces two diploid gametes or gametic pronuclei. After cytoplasmic fusion each pronucleus undergoes a one-division meiosis to produce four haploid pronuclei. These fuse in pairs, with subsequent cytokinesis leading to the creation of two diploid vegetative cells. Some populations of *Notila* are vegetatively tetraploid. Finally, in *Urinympha* the diploid vegetative cell goes through a one-division meiosis to form two haploid pronuclei, which promptly fuse again to restore the original diploid condition. It seems clear, then, that these mutualistic flagellates have gone to the most extraordinary lengths to reduce the amount of recombination within the context of what in the case of *Urinympha* can only by courtesy be called a sexual process. It is very tempting to conclude that their mutualistic way of life has caused selection for the suppression of recombination, especially since the mutualism is both obligate and highly host-specific. However, the appropriate comparison seems to lie between the mutualistic Polymastigina on the one hand, and the parasitic Protomonadina and Opalinina on the other. The Protomonadina, which includes the trypanosomes, are well-known to be thoroughly asexual: they reproduce by binary or multiple fission, and so far as I know no meiotic process has ever been described. The Opalinina reproduce by binary fission, but also by amphimixis (El Mofty and Smyth, 1959; Wessenberg, 1961).

In short, the comparative evidence from animal symbioses is inconclusive. The tangled-bank interpretation is not excluded, nor is it very convincing: it is capable of explaining the sequence of events in heterogonic life histories, but the obvious direct test – thelytoky is to be expected among parasites whose hosts suffer frequent population crashes – is vitiated by the rarity of obligate thelytoky among parasites. This rarity is itself an argument for the Red Queen, which is strengthened by what appears to be a greater frequency of thelytoky in comparable free-living taxa. However, the equivocal result of what should be a very powerful comparison between parasitic and mutualistic protists is disappointing. My judgement is that much more research needs to be directed towards this last point, including the acquisition of new data, before any final decision can be reached on the operation of Red Queen mechanisms in symbionts.

4.4.2 Life in Extreme Environments

The tangled-bank and Red Queen hypotheses are difficult to resolve because environmental stability and complexity may favour sexuality either through competition within species in a fully developed economy (tangled bank) or through competition between species under conditions of high species diversity (Red Queen). However, the rival hypotheses will generate different predictions for environments which are stable but have low species diversity. An excess of parthenogenetic modes of reproduction in such environments would provide support for the Red Queen.

Very few organisms can withstand extremes of temperature, salinity and acidity and, in such habitats as hot springs and saline lakes, species diversity is usually very low. The fauna and flora of hot springs have been reviewed by Brues (1927), Kahan (1969) and Brock (1978). Above about 50°C only prokaryotes survive, the exception being the eukaryotic alga *Cyanidium*, which is restricted to very acid waters, where blue-green algae do not grow. It appears to be exclusively asexual, but its debatable taxonomic position inhibits a comparison with related forms growing elsewhere. At about 40°C *Cyanidium* is joined by an acidophilic *Chlorella*; this is also asexual, but this very common alga is everywhere asexual. A few protozoans have also been reported from hot springs. Among the most tolerant is *Cyclidium citrullus* (Ciliata, Hymenostomatida), which lives at 50-58°C in hot springs in Israel. Kahan (1972) did not observe conjugation, and concluded that it was apomictic; however, its nearest relatives in normal habitats are also apomictic (Jankowski, 1964). The multicellular animals listed from waters at 40-50°C by Kahan (1969) are a nematode (*Dorylaimus*), two bdelloid rotifers (*Philodina* and *Adineta*), two ostracods

(*Cypris* and *Potamocypris*), a mite (*Thermacarus*) and a gastropod ('*Paludestrina*'). I have found no explicit description of their modes of reproduction in hot springs, but the bdelloids will certainly be apomictic and the ostracods probably are too.

The fauna of highly saline waters has been reviewed recently by Cole (1968), Bayly and Williams (1973, Australia) and Beadle (1974, Africa). It is only at very high salinities that species diversity is depressed to the levels observed in hot springs. The most tolerant metazoan is the anostracan *Artemia*, which is often but not always thelytokous. The permanent inhabitants of highly saline (>100 parts per thousand) waters in Australia are the calanoid *Calamoecia*, the ostracods *Diacypris*, *Platycypris* and an unidentified large cyprid, the anostracan *Parartemia*, the isopod *Haloniscus* and the gastropod *Coxiella*; the monogonont rotifers *Brachionus* and *Hexarthra* live in waters only very slightly less saline. Again, the crucial reproductive data are lacking, though the anostracan and the ostracods are probably parthenogenetic.

To summarize an inconclusive set of observations, it appears that extreme environments are liable to be colonized by asexual taxa, but there is no evidence that normally sexual taxa become asexual under these conditions. Moreover, the comparison is greatly weakened by the fact that these extreme environments do not have the constancy required for a definitive test: hot springs are liable to considerable short-term changes in temperature and water chemistry, salt lakes vary in salinity through time, and both types of habitat seem to disappear altogether quite often (see Brock, 1978; Cole, 1968).

Caves and phreatic waters provide physically rather constant environments with low species diversity. Moreover, an alleged overrepresentation of parthenogenetic species in cavernicolous taxa has been used to support hypotheses allied to the Red Queen by Glesener and Tilman (1978) and by Hamilton *et al.* (1981). The basis for this claim is the occurrence of thelytoky in three genera of Rhaphidophoridae (Orthoptera) and one genus of Nicoletiidae (Thysanura) (see text and Table 3.15). We should note, however, that the parthenogenetic cave crickets *Hadenoecus* and *Dolichopoda* are not truly troglodytic species, since they leave their caves to forage at night (Nicholas, 1962; Chopard, 1932). The Nicoletiidae are all unpigmented and anophthalmic, but most species live in the nests of ants or termites. I would add that, according to Badonnel (1943), the parthenogenetic cavernicolous psocid *Psyllipsocus troglodytes* is only an environmentally induced variety of the common thelytokous species *P. ramburi*; and that Henry (1976) described a large excess of females in many populations of the phreatic isopod *Proasellus walterii*, without, however, raising the possibility of thelytoky. These fragmentary data are scarcely sufficient to support a generalization; moreover there are two counterexamples. Of six cavernicolous ostracods found in North America, the four species from the Yucutan are amphimictic and the two from Illinois are parthenogenetic. Among triclads, both Planariidae and Dendrocoelidae have cavernicolous representatives; but whilst asexual reproduction is known only from the former, there is a much larger number of cave species in the latter, which includes the Kenkiids, formerly separated as a distinct family of wholly cavernicolous triclads in which asexual reproduction was unknown. Finally, I have noticed that A. Vandel, the author of distinguished works both on parthenogenesis and on cave faunas, does not mention caves in his book on parthenogenesis (Vandel, 1931), and does not mention parthenogenesis in his book on caves (Vandel, 1965).

It has also been contended that support for the Red Queen is given by the disproportionate occurrence of parthenogenesis on islands. As a general rule, this seems dubious for insects as a whole (see text) and is not supported by quantitative data for coccids (Tables 3.20b and 3.20c). Parthenogenetic geckos do seem to have a predominantly insular distribution, but this serves only to highlight the absence of good examples elsewhere.

I conclude that there is no obvious tendency for stable environments with low species diversity to develop faunas with a disproportionate element of parthenogenetic species.

4.4.3 Conclusions

The area in which the tangled bank and Red Queen intersect comprises their prediction of the mode of reproduction in novel and disturbed habitats. When we turn to relatively stable environments with chronically low species diversity, the two hypotheses make different predictions, but there appear to be no well-established correlations which would extend the explanatory range of the Red Queen beyond that of the tangled bank. Quite on the contrary: the timing of sex in free-living heterogonic animals, the elicitation of sex in the laboratory and the mode of production of resistant or dispersive propagules are areas in which the tangled bank can usually make clear predictions, which are supported by the evidence, whereas the Red Queen principle, more often than not, leads to no clear prediction at all. In the central problem of the timing of sex in heterogonic life cycles, the Red Queen anticipates that sex should occur when the stresses imposed by predators, parasites and competitors are most intense, whilst the field and laboratory evidence all point unequivocally to the overwhelming influence of the population density of conspecifics; where the effect of related or unrelated organisms has been tested, it has been found (in hydras and monogonont rotifers, for instance) to be relatively slight. Only in the special case of parasitism does the Red Queen yield an interpretation which seems to be superior to that of the tangled bank, and even here the evidence is far from conclusive. This is a topic I shall take up again in the section on recombination in the following chapter. For the present, it must be concluded that the Red Queen fails to add to the generality of the tangled bank, and in consequence must be rejected as a general theory of sexuality.

4.5 The Group Effect

I have previously inferred that a group effect of some sort is involved in the evolution of sex, with asexual taxa evolving and speciating more slowly, and perhaps going extinct more frequently, than amphimictic lines of descent. This section will discuss possible explanations of this process.

The Vicar of Bray. The effect is usually explained by asserting that asexual taxa are unable to adapt quickly enough to keep pace with a changing environment. The correlates of sexuality predicted by the Vicar of Bray are the same as those predicted by the corresponding hypothesis of individual selection, the best man. Consequently, the best man having been rejected, the Vicar of Bray must also be rejected as an explanation of the group effect.

The Ratchet. Manning (1976) has advanced two lines of comparative evidence in favour of the ratchet. First he points out that the ratchet will operate in environments which are constant in time, and uses this fact quite properly to exclude the alternative Vicar of Bray model of group selection. He goes on to argue that animals with highly developed behavioural and physiological homeostasis (vertebrates, insects and cephalopods) will suffer more from a given load of deleterious recessives than will less well-buffered organisms (prokaryotes, protists and lower protostomes), and for this reason are obliged to be sexual. Secondly, he explains the timing of sex in heterogonic animals by arguing that the ratchet mechanism will create the greatest advantage for mixis at times of high mortality, during dispersal by larval or juvenile phases or during crashes following population maxima.

Both of these arguments can be challenged. The first seems too vague to pin down empirically; it can easily be inverted, to argue that taxa with well-developed homeostasis can for that reason afford a greater load of mutations and should therefore tend to be asexual; and it makes only the vaguest predictions about what sort of animals are expected to be sexual. The second is similar to the best-man prediction, except that it predicts the occurrence of sex when the intensity of selection increases rather than when its direction changes. It is therefore falsified by the observation, extensively documented above, that parthenogenesis prevails in environments which are subject to frequent and intense disturbance.

Perhaps an even more serious source of scepticism is that the ratchet operates in environments which are not only constant in time (in that the direction of selection does not change) but which are also uniform in space. This is an outcome of its descent from classical genetic theory, and makes it incompetent to deal with complex environments. The simple concept of deleteriousness invoked by the ratchet is simply irrelevant in a complex environment where alternative alleles are favourable or deleterious according to the niche they experience, and in which evolution proceeds by a shifting balance between allele frequencies at loci held near equilibrium by frequency-dependent selection. I have not attempted to prove the assertion, but it seems likely to me that an appropriate simulation would show that the ratchet becomes much less effective as one moves from uniform to ecologically complex environments. If this is granted, the analytical problem is still difficult, for no objective way of measuring environmental complexity is available. Still, few would care to argue that a lake is not a much more complex habitat than a rock-pool or a puddle in a wheel-rut; and yet it is in these latter, less complex environments that parthenogenesis is more frequently found. Species diversity might provide at least a rough measure of complexity, in that the presence of a greater number of species will indicate the existence of a greater number of niches; the ratchet then predicts that sex will be less frequent in habitats whose species diversity is greater, an assertion which is clearly falsified by the evidence.

The ratchet operates through the accumulation of deleterious recessive mutations in the germ line, and is intended to be a theory of sex amongst organisms which reproduce by eggs. It will operate with less force when the comparison is made between sexual egg-producers and organisms which produce asexual embryos, such as the gemmules of sponges, which are formed by the accumulation of somatic cells. It will operate with least force when sexual egg-producers are compared with organisms which reproduce by fission, and especially by paratomical fission; deleterious mutations will in this case occur only in individual somatic cells, or small groups of cells, and the loss of function will be very slight. Therefore, the ratchet predicts that the correlates of these modes of asexuality will differ from those of parthenogenesis, when either is compared with amphimixis. More specifically, it predicts that if we compare viviparous amphimicts to asexual organisms producing progeny of comparable size by fission or by a process akin to gemmulation, the correlations we have previously noted between environment and mode of reproduction will be greatly reduced; conversely, a comparison of brooding parthenogenetic animals with animals reproducing by fission would restore these correlations, with fission being found in the more stable environments characteristic of sexual organisms. The comparative evidence deployed in the middle part of this chapter gives no support to either prediction.

Finally, there are cytogenetic objections to the ratchet. For instance, when comparing apomicts with amphimicts we are usually comparing polyploids with diploids, and, for obvious reasons, polyploidy will blunt the teeth of the ratchet. More seriously, if populations prosper according to the efficiency with which they can prevent the accumulation of deleterious recessives, then we expect arrhenotokous and automictic animals to be extraordinarily successful, since deleterious recessives are continually exposed to selection in hemizygous condition in the one and in homozygous condition in the other. Quite apart from the observation that amphimixis is the rule, even in those taxa for which the ratchet claims to explain the occurrence of sex, this argument leads us to expect a profound dichotomy between automictic and apomictic parthenogenesis. I shall take up this topic in the next chapter, merely stating here that I can find no evidence that the ecological correlates of automixis and apomixis are radically different.

The Tangled Bank. In this way we are brought to an impasse; neither of the two theories of group selection is competent to explain the group effect. It is necessary to retrace our steps a little way.

It seems certain that parthenogenetic taxa evolve less rapidly; it is probable that they speciate less frequently. The arguments which lead to this conclusion form the conceptual foundation of the Vicar of Bray hypothesis, and the foundation seems solid. It is not the assumptions of the hypothesis which are contradicted by the facts, but rather the inference drawn from these assumptions, that extinction is the result of a failure to adapt quickly enough to a changing environment. We must therefore accept the assumptions, while rejecting the inference. This at once gives us half a solution: the sporadic taxonomic distribution of thelytoky, which I take to be the most important fact requiring explanation, will be generated by a low rate of speciation, which is in turn the expression of a low rate of evolution, provided that thelytokous taxa readily became extinct. The other half of the solution is, then, a correct theory of extinction.

According to the tangled-bank hypothesis, a sexual population will not be supplanted by a competing but reproductively isolated set of clones if the clones have the narrower ecological range. Now, suppose that extinction is caused, not by any failure to adapt to a changing niche, but by the disappearance of that niche. Because they exploit fewer niches, the clones will be more sensitive to such catastrophes than the sexual population, the more so since they are unable to recolonize the niche if its disappearance should have been only temporary. This corresponds to the theory of extinction advanced by Williams (1975), transcribed from a best-man to a tangled-bank context; Williams speaks of the imprecise adaptation of sexual populations as buffering them against extinction, whilst I see the concept of imprecise adaptation as masking the ability of sexual populations to exploit a wider variety of niches.

The tangled bank therefore provides a unified explanation of the ecological and taxonomic distribution of sexuality. In the short term, obligately parthenogenetic taxa can hold their own only in simple environments which are kept permanently undeveloped by frequent severe disturbances. In the long term they are unable to persist in the circumstances to which they are restricted by short-term selection, since the niches they occupy must sooner or later cease to exist; and having given rise to no new lines of descent, they disappear without trace. All the major ecological and taxonomic correlates of parthenogenesis can now be understood as consequences of a single general principle.

4.6 The Function of Sex

The comparative analysis has led me to the conclusion that the tangled bank is by far the most broadly supported of all the empirically vulnerable theories which attempt to identify the function of sex. All other theories have been decisively falsified, except the possible implication of the Red Queen in highly evolved symbioses. It would be superfluous to repeat all the relevant arguments

Figure 4.1: Logical Scheme for Testing Theories of Sex. This diagram is intended to summarize – very roughly and therefore with too great an appearance of certitude – the structure of my reasoning in distinguishing between, testing and rejecting rival theories of sex.

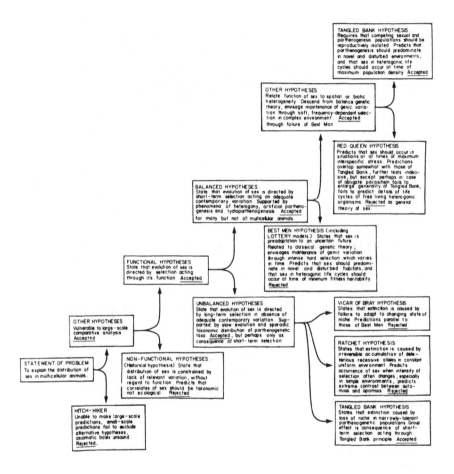

here; instead I have drawn up a chart (Figure 4.1) to summarize the structure of my reasoning.

This conclusion is not entirely palatable. Although based on a simple and compelling analogy, the tangled bank is by no means the most elegant theory at our disposal, nor does it seem to be more than a weak evolutionary force, even when hedged around with all sorts of apparently improbable restrictions. However, it is worth reiterating the most important of these restrictions. In the first place, the tangled bank, at least in its simplest form, demands a strict reproductive isolation between amphimictic and parthenogenetic types. Now, this is precisely the effect of the almost universal polyploidy of parthenogenetic animals; or to put this in another way, it seems to be very difficult for a stable parthenogenesis to evolve without polyploidy acting as a primary isolating mechanism. Far from being a weakness, the restriction becomes a source of strength; if we had decided in favour of a theory in which alleles causing successively greater investments in unreduced eggs were successfully substituted into an amphimictic population, we should be wholly at a loss to explain the leading cytogenetic correlate of parthenogenesis. Moreover, the requirement that selection should take place in the context of competition between reproductively isolated groups of sexual and asexual individuals immediately suggests an explanation for the indubitable group effects associated with parthenogenesis.

Secondly, the tangled bank seems to be such a weak force that it could not procure the initial evolution of sex in an asexual stock unless this earliest sex were not costly; and once sex has become established it seems unable to do more than to protect a sexual population against invasion, either by a constant and therefore moderately diverse stream of reproductively inefficient parthenogenetic mutants, or by extremely rare and therefore genetically uniform mutants whose parthenogenesis is efficient. But the world is such that these properties seem sufficient. We can only guess at the origin of sex, but it seems eminently plausible that the first sexual organisms were haploid isogametic protists whose sex was not costly. In the modern world, the description of what the tangled bank is able to do is also a good description of all that it is required to do in order to prevent an amphimictic population from being invaded by automicts or apomicts. Tychoparthenogenetic animals in which parthenogenetic individuals are constantly present at low frequency are almost always automictic, with high embryonic mortality and thus only a small cost of sex, whilst reproductively efficient apomicts arise very rarely and are in consequence extremely depauperate or even monoclonal. The restrictions which seem so crippling from a mathematical point of view are thus firmly based in the comparative biology of sex.

And that, I think, is what will be found unpalatable or even indigestible in my conclusion. We are taught to respect the simple and elegant theory, which requires no special or onerous assumptions, and whose consequences can be fully and unequivocally established by straightforward mathematical analysis. The reasons for this preference, as I have stressed, are aesthetic rather than scientific. Now, the tangled bank is not an elegant theory; on the contrary, it is (at least in my unskilful hands) cumbersome to the point of being baroque. But nature, it seems, is baroque also; and, where neater and more pleasing concepts have failed, the tangled-bank principle of imperfect competition leads us to an easy and natural interpretation of the major correlates of sex. In short, it seems certain – and here we must simply grit our teeth and swallow – that it is impossible to understand the function of sex without both a firm grasp of the rival hypotheses, their assumptions, logical development and consequences, and also a reasonably comprehensive knowledge of the nature, occurrence and correlates of sexual systems in nature. I do not think it would be possible to understand the role of sex in the natural economy merely by reasoning from general principles; on the contrary, any attempt to do this would be very misleading. Biology is not physics; our abstractions must not refer to supposed universals, but must instead be tailored quite precisely to the curious details of what, from an infinity of possibilities, has actually evolved.

All my efforts so far have been directed towards explaining the crucial but crude distinction between amphimixis and thelytoky. It does not necessarily follow that the success of the tangled bank in providing an explanation will be repeated when we turn to related phenomena. In particular, whilst the restrictions implicit in the tangled bank turn out not to handicap it as a theory of sex, it seems likely that they will disqualify it as a theory of recombination. All previous authors have taken for granted that, once a general principle governing the evolution of sex had been found, it could be applied indiscriminately to the evolution of recombination. Sex, so to speak, acts as a switch and crossing-over as a rheostat, but the underlying function of switch and rheostat has always been assumed to be the same. Now it seems that the rules governing whether or not the switch should be thrown incorporate a variety of special considerations which simply do not apply to fine tuning through the rheostat. The appalling possibility that sex and recombination may have to receive quite different explanations is taken up in Section 5.2.

A second qualification is that I have been much more concerned with the overall correlates of parthenogenesis than with differences between the correlates of alternative modes of parthenogenesis. The tangled bank gives no obvious interpretation of the differences between automixis and apomixis; and it is entirely possible that, whilst one theory explains why a certain set of animals are parthenogenetic, a quite different theory is required to explain why, within this set, some are automictic and others apomictic. This topic is addressed in Section 5.1.

Finally, I have not found it possible to construct decisive comparative tests other than by opposing the underlying principles of rival hypotheses in order to exclude one of them. No doubt this is unrealistic, and in nature sex might often serve other functions in addition to reducing competition between sibs. For instance, some flavour of the runt model has been admitted into the tangled bank in order to explain the group effect. By admitting some degree of temporal change in a model of a complex environment it might be possible to fashion a theory of recombination from the tangled bank. And if the resources for which sibs compete are biotic, this competition might generate a Red Queen effect, with the frequency dependence of fitness becoming lagged in time. I have chosen not to pursue these mixed hypotheses further, because I am afraid that a judicious blend might be able to explain everything, without being vulnerable to falsification.

The vagaries of nature have thus forced me to retreat from a starkly functionalist position to one where ancestry and circumstance cannot be ignored. However, none of these reservations affect the main point, that the problem originally set has been solved, by the discovery of a principle which correctly predicts the major correlates of sexuality among multicellular animals. Although this is as far as I intend to go, it is a beginning rather than an end. I am conscious that the formal theory in which I have tried to encapsulate the principle deserves a much more thorough investigation. If the principle is universal, it should lead to an explanation of patterns of sex among protists, algae, fungi and higher plants; if it is operationally useful, it should make possible the detailed understanding of particular heterogonic life histories. There is much still to be done.

5 EPIPHENOMENA OF SEXUALITY

5.1 Automixis

So far I have been more concerned to identify the common function of the various modes of cheap, conservative reproduction than to distinguish between them. But if each mode of reproduction is indeed selected to serve a different function then it should be easy to distinguish between automixis and apomixis because of their contrary effects on heterozygosity. Most previous authors seem to have burked the issue. The most popular line of argument, to be found in most textbooks of botany, is that automicts (in this context, obligate selfers such as cleistogams) lose the capacity for evolutionary change, even though automixis creates a short-term advantage, so that an optimal frequency of selfing will be set by the balance of short-term and long-term selection (see Mather, 1973). On the other hand, Haldane (1932) imagined that selection will proceed faster in selfed than in outcrossed organisms, because genes are selected more rapidly when homozygous. In the case of facultative or cyclical automixis or autogamy, it has often been argued (following Darwin, 1888) that the ability to self-fertilize is favoured because it ensures reproduction when mating partners are not available. These theories have a crucial failing in common, which is that if they are theories of automixis they are equally theories of apomixis. Indeed, they are not really theories of automixis at all, but are rather theories of any conservative, economical mode of reproduction. Because none of them distinguish adequately between automixis and apomixis, they can all be regarded as special cases of the historical hypothesis: granted that selection will favour parthenogenesis, none of them provides any grounds for predicting whether automixis or apomixis should evolve. Williams (1975) has explicitly defended this historical interpretation of thelytoky.

It is obvious enough that both automixis and apomixis will be autoselected relative to amphimixis. The outcome of competition between alternative and otherwise neutral alleles determining automixis and apomixis is less clear. On the one hand, some resources must be used by autogamous individuals to elaborate sperm and accessory male structures and, even in strictly automictic individuals, the extra time required to make gametes through meiosis might represent a cost. On the other hand, if the alleles concerned have only partial penetrance, automixis will be selected more efficiently than apomixis because the differential loci will quickly be made homozygous. I would therefore be reluctant to erect a theory which distinguished between automixis and apomixis on the basis of reproductive economy. Instead we must turn to the more obvious difference between the two systems, and ask under what conditions an allele embedded in a genome which it has made largely homozygous will outcompete an alternative allele whose context is highly heterozygous.

Ecology of Homozygosity. In discussing the relationship between sex and intraspecific competition I assumed as a simplifying device that the intensity of competition between individuals with different genotypes was a constant, and did not vary with the nature and magnitude of the difference between the genotypes. In an attempt to extend the same principle of imperfect competition to an interpretation of automixis, I shall analyze the more general case in which the intensity of competition between two individuals is proportional to the difference in gene dosage. For simplicity, consider a single diallelic locus. The two homozygotes differ genetically more than either homozygote differs from the heterozygote, since each shares one allele with the heterozygote but has none in common with the other homozygote. If this genetic difference is reflected in some aspect of the phenotype which influences resource utilization, then the two homozygotes may compete less intensely than either does with the heterozygote. This immediately suggests a possible advantage for automixis: the progeny as a whole may suffer less competition if they are segregated into the two most dissimilar phenotypic classes.

More formally, consider interactions between genotypes at an autosomal locus which is segregating for the alleles $A1$ and $A2$. The intensity of competition between individuals which have both alleles in common ($A1A1$ and $A1A1$; $A1A2$ and $A1A2$; $A2A2$ and $A2A2$) is expressed by a competition coefficient u; that between individuals which share one allele ($A1A1$ and $A1A2$; $A2A2$ and $A1A2$) by v; and that between individuals which have no alleles in common ($A1A1$ and $A2A2$) by w. The general scheme of fitnesses is then:

Genotype	Fitness
$A1A1$	$R - uf_{11} - vf_{12} - wf_{22}$
$A1A2$	$R - uf_{12} - vf_{11} - vf_{22}$
$A2A2$	$R - uf_{22} - vf_{12} - wf_{11}$

Here, R is a 'baseline' fitness, chosen so as to keep fitnesses positive, and f_{ij} is the frequency of the genotype A_iA_j. The frequency dependence of the fitness scheme is self-evident, and ensures the maintenance of genic variance whether the population is amphimicitic, automictic or apomictic. We postulate that an otherwise neutral allele causing automixis will be favoured if $u > v > w$.

The simplest and most realistic comparison lies between automixis and apomixis, the cost of sex being small or zero in either case. Since homozygotes replicate the maternal genome whether they reproduce by automixis or by apomixis, only heterozygotes need be considered; and the problem reduces to finding the conditions under which it is preferable to segregate progeny into the two homozygote classes rather than making them all heterozygous. From the fitness scheme, it follows immediately that the mean fitness of a mixture of equal numbers of each homozygote class will exceed that of a pure culture of heterozygotes if

$$w < [2v(1 - 2f_{12}) - u(1 - 3f_{12})]/(1 - f_{12}).$$

As we anticipated, automixis may be favoured if the interaction between homo-zygotes is sufficiently weak. Another way of expressing this result is to say that there is a value of $f_{12}, \hat{f}_{12}$, at which the progeny of automictic and apomictic heterozygous parents have the same mean fitness. This equilibrium value is

$$\hat{f}_{12} = (2v - u - w)/(4v - 3u - w).$$

This equation enables us to discover the dynamics of selection for all values of u, v and w, and these are illustrated in Figure 5.1. So long as $v < u$ automicts are not completely excluded from the population at equilibrium, irrespective of the value of w. If w is fairly large, such that $w > 2v - u$, the population will eventually come to consist of a mixture of automicts and apomicts. If w is rather small, such that $w < 2v - u$, selection will eliminate the apomicts entirely to produce a pure automictic population. Thus, when the intensity of competition is directly proportional to gene dosage — or, in the language of the model I have used, when $u > v > w$ — a mixed population of automicts and apomicts either approaches a stable equilibrium at which both types persist, or else the apomicts are completely eliminated.

Figure 5.1: Dynamics of Competition Between Automixis and Apomixis. The terms and the argument are explained in the text. Automixis is fixed for par-ameter combinations falling within the hatched zone.

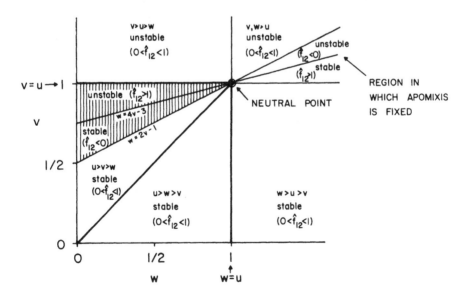

This is an encouraging result, since it shows how automixis may succeed given a simple and ecologically plausible set of conditions. At the same time it appears to introduce a difficulty, since apomixis can become fixed only if the intensity of competition is *inversely* proportional to gene dosage ($w > v > u$) – surely not a very attractive conjecture. However, we have so far described selection as being frequency-dependent, whilst ignoring the effect of population density. A more general fitness scheme would posit a different baseline fitness (R) for each genotype. Suppose that the locus in question were heterotic, in that the heterozygote can exploit a wider range of niches, but exploits each less efficiently than the best-adapted homozygote. At low population density, competition in a mixed population would be minimal, and if the heterotic effect were sufficiently strong the apomictic clone would increase in frequency. As the population grew, the intensity of competition would increase so that, as the carrying capacity was approached, the automicts might not be eliminated and might even become fixed. This is the formal realization of a concept I have already mentioned, that heterozygosity and homozygosity are different ways of packaging genetic variations. In an unsaturated economy it pays to produce generalist heterozygous progeny, each of which can cope, more or less, with a wide range of niches. As competition becomes more intense, the advantages of homozygous specialization increase, until eventually the generalists may be excluded altogether. Apomixis might even then be expected to be more characteristic of density-independent, novel or disturbed regimes and automixis of more strictly density-regulated conditions.

It is possible, but less illuminating, to compare automixis with amphimixis in the same way. Using the same fitness scheme, imagine for simplicity that sexuality is controlled at a haploid locus in some unpaired part of the genome, or else that the allele for automixis is rare. The fitness scheme can then be written as follows.

Sexuality	Genotype	Frequency	Fitness
	$A1A1$	$f_{11} = rp^2$	$R - u(f_{11} + g_{11}) - vf_{12}$ $- w(f_{22} + g_{22})$
Amphimictic $B1$	$A1A2$	$f_{12} = 2rpq$	$R - uf_{12} - v(f_{11} + g_{11})$ $- v(f_{22} + g_{22})$
	$A2A2$	$f_{22} = rq^2$	$R - u(f_{22} + g_{22}) - vf_{12}$ $- w(f_{11} + g_{11})$
Automictic $B2$	$A1A1$	$g_{11} = sm$	$R - u(f_{11} + g_{11}) - vf_{12}$ $- w(f_{22} + g_{22})$
	$A2A2$	$g_{22} = sn$	$R - u(f_{22} + g_{22}) - vf_{12}$ $- w(f_{11} + g_{11})$

Here, p is the frequency of the $A1$ allele in the amphimictic section of the population ($q = 1 - p$); m is its frequency in the automictic section ($n = 1 - m$); r is the frequency of the $B1$ allele, the allele which determines amphimixis ($s = 1 - r$). The algebra is more tedious than in the previous case; at equilibrium the frequency of amphimixis is

$$\hat{r} = -(Au + Cv + Ew)/(Bu + Dv + Fw),$$

using the definitions: $A \equiv (m^2 + n^2) - (mp^2 + nq^2); B \equiv 2(mp^2 + nq^2 - p^2q^2)$ $- (m^2 + n^2) - (p^2 + q^2); C \equiv - 2pq; D \equiv 8p^2q^2; E \equiv 2mn - (mq^2 + np^2);$ $F = 2(np^2 + mq^2 - p^2q^2 - mn)$. For the special case in which the $A1$ and $A2$ alleles are equally frequent among both automicts and apomicts (p and $m = 1/2$) this cumbersome expression simplifies, as it should, to:

$$\hat{r} = (2v - u - w)/(4v - 3u - w).$$

This is precisely the same expression as that describing the outcome of competition between an automict and an apomict, and shows that automixis may coexist with or exclude amphimixis even when autoselection is neglected. The analysis is not worth pursuing further, because it refers to a single locus. This is excusable when comparing automixis with apomixis, since all apomictic progeny will be identical, and genetic variation amongst automictic progeny will be sufficiently restricted to allow us to imagine without too much impropriety that the properties of many loci can be summed up at a single locus. This procedure is plainly unrealistic when applied to the extremely diverse progeny of an amphimictic female, where it is easily conceivable that greater genetic variety will more than compensate for any effect of gene dosage.

These arguments seem to me vague and unsatisfactory, because at bottom they refer to ecological processes which are very difficult to measure, and about which we have no great fund of knowledge. I offer them because I know of no better alternative. If they have any substance, they imply that the genetically intermediate strategy of automixis may also be ecologically intermediate. Starting with the assumption that the intensity of competition is proportional to gene dosage, I have shown that automixis is expected often to replace apomixis, and may do so most readily at high population density. A formally similar result is obtained for the case in which automicts compete with amphimicts, but the much greater genotypic diversity of amphimictic progeny may mask any effect due to gene dosage at one or a few loci. I infer that automixis will be favoured by relatively high levels of intraspecific competition when matched against apomixis, but by relatively low levels when competing with amphimixis. Automixis should be specially common amongst organisms living in environments which are neither so stressful as to elicit selection for apomixis, nor so crowded and complex as to ensure the success of amphimixis.

Correlates of Automixis. I have searched for ecological correlates of automixis, in the hope that they would shed light on the rather dim predictive scheme developed above. This search has been monotonously unsuccessful. In most cases this can be attributed to the lack of adequate information on the ecology of comparable automicts and apomicts in such otherwise suitable taxa as Tardigrada and Oligochaeta. Among nematodes the cytology of parthenogenesis is still poorly understood, but we can at least separate autogamous forms from those that practise a nongametic thelytoky which is in some cases known to be apomictic; we can also distinguish monogenetic and heterogenetic parasites of animals, plant parasites and free-living genera inhabiting soil and freshwater and marine sediments (see Table 3.2). No obvious pattern emerges: indeed, autogamous and agametic forms occur in the same genus (e.g., *Criconemoides*), and even in the same species (e.g., *Meloidogyne hapla*), and share closely similar ways of life. It might be possible to achieve more with higher plants, where the colonists of waste places are predominantly autogamous but the flora of high latitudes tends to be amictic, but direct comparison is difficult because of the scarcity of nongametic automixis in plants and because most amictic propagation is vegetative.

While the paucity of the evidence makes it impossible to reject any particular functional hypothesis, the absence of any obvious ecological correlates led me to test the general validity of the functional approach by attempting to falsify an historical interpretation. Before doing so, however, we must define the three possible hypotheses more carefully.

First, the pure historical hypothesis: history exclusively determines the mode of parthenogenesis, and any functional differences between modes cannot be detected by comparative analysis.

Secondly, the pure functional hypothesis: the optimal functional state is always attained, regardless of the history of the line of descent under consideration.

Thirdly, an hypothesis in which both function and history play a part: the fitness of any particular mode of parthenogenesis is modulated by the nature of the prior genetic system of the population. This hypothesis states, in effect, that history and function are not independent, and therefore that the optimal functional state cannot be defined without a knowledge of the previous state.

The data that I shall use to test these hypotheses is summarized in Table 5.1.

The basic prediction of the historical hypothesis is that the correlates of automixis should be largely taxonomic; i.e., that the parthenogenetic relatives of automicts should themselves be automictic and those of apomicts apomictic. It is obvious that this is broadly true, since in many large taxa (e.g., Demospongiae, Bdelloidea, Monogononta, Phylactolaemata, Cladocera, Curculionidae) only a single mode of parthenogenesis has ever been described. However, this test can be faulted on two counts. First, there are a number of exceptions to the rule (Tardigrada, Nematoda, Oligochaeta), and even a few species in which both automixis and apomixis have been described (e.g., the nematode *Meloidogyne hapla* and the tardigrades *Hypsibius oberhauseri* and *Macrobiotus hufelandi*; cf. also the coccids *Lecanium* and *Antonina*). Secondly, the test is open to the objection that closely related animals may have very similar ways of life, so that the test is bound to be indecisive.

A more promising line of approach is to compare the parthenogenetic relatives of gonochoric and hermaphroditic amphimicts. In the first place, if history is decisive then we should certainly expect to find that obligate autogamy is much more frequent in generally hermaphroditic than in generally gonochoric taxa, and in support of this argument we could point to the turbellarian *Cura foremanii*, the pulmonate *Rumina decollata* and aphallic individuals of *Bulinus*. However, while casual selfing is probably very common among hermaphroditic animals, and while there are dozens of instances in which selfing occurs in isolated individuals (e.g., in Cirripedia) or at certain times of the year (e.g., in Typhloplanidae), obligate selfing is very unusual except in the parasitic Eucestoda, and no single case has yet been proven from such large hermaphroditic taxa as Ctenophora, Oligochaeta, Hirudinea, Bryozoa and Chaetognatha. Moreover, a comparable number of cases is known from gonochoric taxa: the polychaete *Nereis limnicola*, the conchostracans *Triops* and *Lepidurus* and the fish *Rivulus marmoratus*. The many genera of autogamous nematodes also contradict the proposed rule, since obligately cross-fertilized nematodes are always gonochoric.

If the effect of history is not to compel the adoption of a particular genetic system but rather to modulate the fitnesses of alternative genetic systems in the manner suggested by the third hypothesis, then we can reason that the casual or occasional self-fertilization which is probably widespread among hermaphroditic animals will lead to a certain tolerance of homozygosity, and that a nongametic automixis should therefore be more frequent in hermaphroditic than in gonochoric taxa. A glance at Table 5.1 shows that this is indeed the case, but nevertheless we cannot accept the hypothesis without serious reservation. In the first place, the only hermaphroditic taxon from which there is sound evidence of the cytology of parthenogenesis is the Oligochaeta, so the taxonomic basis of the comparison is narrow. Secondly, whilst apomixis in the strictest sense is very rare among hermaphroditic taxa, the gemmulation of sponges and the statoblast formation of phylactolaemate bryozoans approach apomixis very nearly and are included in Table 5.1 on this basis; while if we cast the net wider and include extreme cases of architomy we find many instances in hermaphroditic taxa (e.g., the fragmentation of the turbellarians *Phagocata* and *Bipalium*, the rhynchocoel *Lineus* and the enchytraeids *Buchholzia*, *Cognettia* and *Enchytraeus*) but rather few instances from gonochoric taxa (the schizometamery of certain polychaetes and polydisc strobilation in scyphozoan scyphistomae).

But we may still argue that the prevalence of automixis in hermaphroditic taxa is a purely historical phenomenon which has no reference to function at all: autogamy will evolve readily, and from autogamy a nongametic automixis. To settle the question we have to focus the enquiry a little more sharply. If the optimal functional state in some defined set of circumstances is apomixis, but in hermaphroditic taxa a history which includes autogamy strongly predisposes parthenogenetic populations to be automictic, then what we expect to observe is an automixis with premeiotic restitution: a mode of parthenogenesis whose cytology is meiotic but whose consequences are as nearly mitotic as possible. This prediction is supported by the evidence: premeiotic restitution is much more common, relative to other modes of automixis, in hermaphroditic than in gonochoric taxa. However, the taxonomic basis for the comparison remains

Table 5.1: Taxonomic Distribution of Cytologically Confirmed Instances of Obligate and Cyclical Thelytoky, Arranged According to the Genetic System Prevailing among other Members of the Same Taxon. Columns are different modes of thelytoky; genetic system of nearest relatives of parthenogenetic genera shown in left-hand margin. By genus; for economy, where all or most thelytokous genera within a larger taxon have the same cytology the larger taxon is printed in capitals. Cases in which there is some doubt about the mode of thelytoky are indicated by a single asterisk*. Cases in which there is some doubt as to the genetic system of related forms are indicated by a double asterisk**. Taxa with cyclic parthenogenesis are enclosed in parentheses. The number preceding the name of the taxon refers to the appropriate section of Chapter 3. This table incorporates Table 3.17.

	APOMIXIS	PREMEIOTIC	INTRAMEIOTIC	POSTMEIOTIC	AUTOGAMY
GONOCHORIC	4 Margelopsis*	28 Brevipalpus**	14 Meloidogyne	32 Bacillus	24 Nereis
AMPHIMIXIS	14 Meloidogyne	32 Warramaba	25 Hypsibius	32 Lecanium**	29 Triops
	14 Meloidodera	32 Ptinus	28 Platynothrus	32 Eucalymantus**	29 Lepidurus
	20 PROSOBRANCHIA	39 Poeciliopsis	28 Trhypochthonius	32 Pulvinaria**	39 Rivulus
	25 Hypsibius	39 Poecilia*	29 Artemia*		
	25 Macrobiotus	39 Ambystoma	29 Limnadia		
	29 (CLADOCERA)	39 Cnemidophorus	32 Parlatoria*		
	30 Heterocypris		32 Pinnaspis*		
	30 Trichoniscus		32 Targonia*		
	30 Nagara*		32 Lecanium		
	32 Saga		32 Phenococcus		
	32 Carausius		32 Ochthiphila*		
	32 Sipyloidea		32 Drosophila		
	32 Pycnoscelus		32 Lonchoptera		
	32 Haploembia		32 Cnephia*		
	32 Liposcelis		32 PSYCHIDAE		
	32 (APHIDOIDEA)		39 Carassius		
	32 DIASPIDIDAE		39 Lacerta*		
	32 Marchalina				
	32 Antonina				
	32 Dysmicoccus				

	32 CHRONOMIDAE			
	32 (CECIDOMYIDAE)			
	32 Adoxus			
	32 CURCULIONIDAE			
HERMAPHRODITIC AMPHIMIXIS	1 (DEMOSPONGIAE)	24 TUBIFICIDAE*	24 Mesenchytraeus	6 Cura
	17 (PHYLACTOLAEMATA)	24 Eisenia	24 Cognettia	20 Rumina
	24 Dendrobaena	24 Eiseniella	24 Fredericia	20 Bulinus
		24 Allolobophora	24 Lumbricillus	
		24 Bimastus		
		24 Octolasium		
ARRHENOTOKY	10 (MONOGONONTA)	32 Heliothrips	28 Cheyletus	32 Icerya
	28 Histiostoma**	32 Diprion	32 Trialeurodes	
	32 Marchalina	32 Empria	32 Gueriniella	
	32 Oecophylla*	32 Formica	32 Lecanium	
	32 Thrinax	32 Nematus	32 Nemeritis	
		32 Pristophora	32 (Neuroterus)	
THELYTOKY	10 BDELLOIDEA	11 Lepidoderma		11 Lepidoder-mella

narrow, since for hermaphroditic taxa the comparison reduces to comparing lumbricid (and perhaps tubificid) with enchytraeid oligochaetes.

At the other extreme, automixis may involve a postmeiotic restitution. If the third hypothesis is correct, we expect a system of this sort to represent a functional optimum only in taxa which are predisposed to tolerate high levels of homozygosity. This hypothesis therefore predicts that postmeiotic restitution should be much more common, relative to other modes of automixis, in arrhenotokous than in amphimictic lines of descent. The existence of such a pattern has previously been noticed by Nur (1971). The few counterexamples of amphimicts with postmeiotic restitution are not very convincing: a dubious report from the fish *Carassius*, some but not all populations of the phasmid *Bacillus* and three genera of coccids, some of whose more distant relatives are arrhenotokous.

Conversely, the same reasoning leads one to expect that apomixis and premeiotic restitution will be rare among thelytokous members of arrhenotokous taxa. This is certainly true for arthropods, but unfortunately is just as certainly false for the other major centre of arrhenotoky, the monogonont rotifers.

An important corollary of the hypotheses under test is that only certain combinations of modes of reproduction are permitted in systems of cyclical parthenogenesis. The functional hypothesis permits the widest latitude, since it predicts the occurrence of an alternation between any two modes of parthenogenesis whose genetic consequences differ. Only the combination of apomixis with premeiotic automixis seems to be excluded. Any hypothesis which grants a large role to history, on the other hand, will exclude the alternation of apomixis with any mode of automixis. The latter view seems the more satisfactory, since I know of no case in which a female may produce both apomictic and automictic eggs.

It will be objected, and justifiably, that there seem to be uncomfortable numbers of exceptions to almost any proposed rule. In particular, there are many instances of automixis from taxa of gonochoric amphimicts. One of these instances, however, turns out to confirm rather strongly the rule which at first glance it appears to contradict. The peculiar mechanisms of automictic restitution found in three genera of psychid moths, which despite being automictic nevertheless conserve heterozygosity, are made possible by a sequence of cytological events (Section 3.32 and Narbel-Hofstetter's review should be consulted for the details) which also occur in the amphimictic races from which the automicts have descended.

The final line of argument that I shall present is that. if a purely historical hypothesis were correct, facultative parthenogenesis and tychoparthenogenesis should have the same cytology as obligate parthenogenesis. This is clearly untrue among insects, the only taxon for which we have any substantial knowledge of the cytology of facultative thelytoky. It is invariably automictic, and in Acrididae, Tetrigidae and Phasmatodea usually involves postmeiotic restitution. Since in these cases thelytoky is associated with high levels of embryonic lethality, it is difficult to argue that they represent early stages in the evolution of obligate thelytoky.

In the previous chapter I have shown that although the occurrence of thelytoky is influenced by history this influence is relatively small, and the major correlates of thelytoky are ecological rather than taxonomic. The facts that I have set out above show that the particular mode of thelytoky which evolves is much more tightly constrained by historical factors. Nevertheless, while these facts rule out the rigidly functional philosophy with which I opened this essay, they rule out with equal firmness a purely historical approach, and the only hypothesis still holding the field is one in which neither history nor function can be ignored. This hypothesis is composite, and states that the evolution of thelytoky occurs in two stages. The first is simply the input of mutations which determine the production of unreduced eggs. These characteristically result in facultative automixis in diploids or obligate apomixis in polyploids; I have discussed this topic in Section 4.2. This process is purely historical and, if such mutations were very rare, and if the mode of thelytoky adopted were always that which arose the most frequently, functional hypotheses would be superfluous. But neither of these statements is true. Instead it seems that selection provides a filter which screens out some genetic systems, whilst allowing others to evolve; a filter, moreover, which perceives the optimal genetic system differently according to the ancestry of the population undergoing selection. This hypothesis has the merit of being compatible with all the evidence that I have presented. It has two substantial drawbacks. The first is the usual failing of mixed hypotheses, that it does not specify precisely the magnitudes of the relative effects of history and function, and therefore cannot predict precisely what is to be expected in any particular case. The second is that, whilst the decisive falsification of a purely historical hypothesis through several lines of evidence forces us to grant a significant role to the functional difference between automixis and apomixis, we are no closer to understanding where this function might reside. A detailed study of the ecology and genetics of some taxon in which both apomixis and a nongametic automixis occur would be a formidable task, but it is clearly one which needs to be undertaken.

Arrhenotoky. I have shown (in Chapter 1) that arrhenotoky will be selected relative to amphimixis, but that apomixis will be autoselected relative to arrhenotoky. If the function of arrhenotoky resides principally in the reproductive economy implicit in the more efficient transmission of genes through the male line, it follows that exceptional genetic systems in otherwise arrhenotokous taxa should be thelytokous and not amphimictic. This prediction is supported by hymenopterans and monogonont rotifers: in the former, all species which are not obligately arrhenotokous have either obligate thelytoky or an alternation between thelytoky and arrhenotoky, while in the latter the only exceptions to the general rule of an alternation between apomixis and arrhenotoky seem to be obligately apomictic. The same argument can be made for margarodid coccids and Micromalthidae, but the interpretation of Acari is not as clear: amphimixis, arrhenotoky and thelytoky all occur, but it has been suggested (e.g., by Nur,

1971) that there have been several independent origins of arrhenotoky, which may still be spreading at the expense of amphimixis.

Arrhenotoky is not a conservative mode of parthenogenesis, and therefore should not produce the high rates of extinction and low rates of adaptation and speciation which seem to be characteristic of thelytokous systems. Consequently, we expect arrhenotoky to occur among the majority of members of those taxa in which it occurs at all, in contrast to the sporadic taxonomic distribution of thelytoky (Maynard Smith, 1978). This is broadly true, though the exceptional nature of the Acari must again be emphasized.

It has occurred to me to wonder why, in arrhenotokous animals, it is always the male and never the female which is haploid. The answer may be absurdly simple: a system with haploid females could not tolerate even the temporary absence of males. If the males were to disappear entirely for a generation or more, as they do in monogonont rotifers and cynipid wasps, the females could reproduce successfully through haploid apomixis. However, in the absence of sperm they would then be unable ever again to produce diploid male progeny, unless indeed we choose to imagine that some automictic doubling device arose, in which case the males would be completely homozygous. In obligately arrheno-tokous systems with haploid females this objection is not fatal, but the temporary local absence of males might create an analogous difficulty in viscous populations.

Much more extensive theoretical discussions of the origin and evolution of arrhenotoky, stressing the role of inbreeding, are given by Hartl and Brown (1970) and Borgia (1980).

Pseudogamy. In gynogenetic and hybridogenetic animals, sexual and partheno-genetic reproduction are necessarily in balance. The nature of this balance in the hybridogenetic guppy *Poeciliopsis monacha-lucida* has been the subject of an elegant study by Moore (1976). He reasoned that the overall fitness of either a parthenogenetic or an amphimictic female would be the product of three components: a 'primary fitness', a cost of sex and a probability of being pregnant. If the two types have the same primary fitness and sex always incurs a full twofold cost, then the dynamics of the system will depend only on rates of pregnancy, which are in turn determined by competition for males. Being unable to sample the males efficiently, Moore argued that, if the sex ratio among the amphimicts were constant, the frequency of males in the population as a whole would vary directly with the frequency of amphimictic females. He then measured the rate of pregnancy of parthenogenetic females as a function of the frequency of amphimictic females, and found it to be an S-shaped curve with a slope parameter $b = 27.6$ and an asymptote at unity. The amphimictic females them-selves were almost always pregnant, even when rare. Clearly, parthenogenetic females will tend to increase when they are rare, since they are not then limited by sperm availability and will be twice as fit as amphimictic females; conversely, they will tend to decrease when they are common, since at some point their twofold advantage in the rate of reproduction will be annulled by a scarcity of males. More precisely, a little elementary algebra shows that the stable equilibrium frequency of parthenogenetic females is expected to be $1 - (\ln 2/b)^{\frac{1}{2}}$, or about 0.85 for the data analyzed by Moore. (As one would expect, this is close to the point of inflection of the graph of the rate of pregnancy of parthenogenetic females on the frequency of amphimictic females.) In the most southerly populations the observed frequency of parthenogenetic females is very close to expectation. In more northerly populations their frequency is much less, which Moore attributed to a lower primary fitness. Moore's study is the most successful attempt yet made to quantify the evolutionary dynamics of any system of parthenogenesis.

What we do not know is why parthenogenetic development should ever be dependent on sperm. The problem is especially acute in gynogenetic animals, where we cannot invoke the possibility of heterosis. Indeed, there may be no escape from the conclusion that gynogenesis is always a functionally suboptimal system which evolves through historical accident. It is conceivable that the benefit of sperm dependence is a very slight leakage of genetic material that will minimally diversify progeny; the fact that gynogenetic vertebrates have somewhat greater genic variance than autonomously parthenogenetic species might be held to support this opinion. It seems more likely to me that dependence on sperm entry is the price that gynogenetic females pay for an efficient mechanism of sperm destruction. Certainly, gynogenesis seems to be prevalent only among the parthenogenetic members of taxa with efficient internal fertilization through copulation, such as turbellarians and nematodes.

Double Fertilization. In flowering plants, the pollen fertilizes not only an ovule nucleus but also an endosperm nucleus. Nothing comparable to this curious phenomenon has been reported from any other group of organisms (though I am curious about the relationship between nurse cells and oligopyrene sperm). Generations of botanists have passed over the function of double fertilization in almost unbroken silence, but very recently a most ingenious suggestion has been advanced by E. L. Charnov (personal communication). Charnov reasons that the fate of a zygote depends critically on the stored reserves donated by the parent which is fertilized. It is in the interests of this parent — let us call it the female — to maximize its total reproductive success by trading off ovule number against ovule size, and the fact that producing larger ovules will mean producing fewer implies that optimal ovule size from the point of view of the female will generally be less than the ovule size which maximizes seed survival. But suppose that the male parent, donating pollen, fertilizes only one ovule of any given female. From his point of view, this ovule should be as large as possible, without regard to the fate of the other ovules; there will therefore be a conflict of interest between male and female. In general, this conflict will be decided in favour of the female, since she is able to control the size of her ovules. Charnov's suggestion is that endosperm fertilization is a male device to subvert this control by introducing genetic material into the endosperm capable of instructing the plant to direct more reserves into one particular fertilized ovule. If more than one ovule is fertilized by pollen from the same plant this argument is weakened, but it continues to hold up to the point where two adjacent plants mutually and exclusively cross-fertilize. A comparative test of Charnov's hypothesis seems scarcely possible, but the predicted correlation between pollen genotype and seed size should provide the basis for an experimental test.

Polar Bodies. In the formation of macrogametes, it is usual, though not invariable, for only one of the four haploid nuclei resulting from meiosis to be incorporated into a gamete, the others being extruded as polar bodies and degenerating. Williams (1976, p.114) speculates that this may be due to the mechanical difficulty of cleaving a large mass of yolky cytoplasm. I have also heard the opinion that polar-body formation is a randomizing device akin to reassortment. It seems to me more plausible to ascribe the extrusion of polar bodies to the fact that the oocyte is not homogeneous, but rather has a highly differentiated cytoplasm, with different parts having different epigenetic fates. Eggs which are experimentally subdivided often give rise to abnormal or incomplete embryos. It is therefore impracticable to form a tetrad of equally competent macrogametes, but very easy to expel a small quantity of chromatin in polar bodies.

Polyploidy. Polyploidy is very common among higher plants and very rare among amphimictic animals. Muller (1925) attributed this pattern to the much greater frequency of hermaphroditism in plants, reasoning that polyploidy would upset chromosomal sex determination in animals. White (1973), on the other hand, pointed out that even-numbered polyploidy will evolve rather easily in automicts and selfers, and attributed the prevalence of polyploidy in plants to the more frequent occurrence of self-fertilization rather than to hermaphroditism *per se*. White's comparative analysis (White, 1973, pp.454–60), which it would be superfluous to repeat here, leaves little room for doubt that his interpretation is broadly correct. Table 4.4 and the text of Chapter 3 can be consulted for the ploidies of automictic and self-fertilized animals.

5.2 Recombination

Paradox of Recombination. Imagine two diallelic loci occurring on the same chromosome in a sexual diplont. There are four types of chromosome – say, *A1B1*, *A1B2*, *A2B1* and *A2B2* – and we may assign any fitness we wish to these genotypes. We may choose to imagine that the two loci have completely independent effects on fitness, so that the performance of either allele at the *A* locus is unaffected by the state of the *B* locus, and vice versa. Any limit of this sort is artificial, however, and in general the fitness of an allele at either locus will depend to some extent on its genetic environment. For example, the epistatic interaction between the loci might result in the fitness of *A1B1* and

A2B2 chromosomes being much greater than that of *A2B1* and *A1B2* chromosomes. Now consider a third locus, situated somewhere on the same chromosome, whose sole effect is on the rate of crossing-over between the *A* and *B* loci. This recombination locus is fixed for an allele *M1*, until a few *M2* alleles arise by mutation. If the effect of *M2* is completely to suppress crossing-over between the two fitness loci, it is obvious that at equilibrium the population will include only the two chromosomes *A1B1M2* and *A2B2M2*, since any other type is less fit: those including *A1B2* or *A2B1* sequences are less fit because of the fitness scheme, whilst *A1B1M1* and *A2B2M1* are less fit because they may give rise through recombination to *A2B1M1* and *A1B2M1* chromosomes. The *M2* allele will thus become fixed quite rapidly if it abolishes recombination between the fitness loci; and by an extension of the argument, it follows that, in general, *M2* will tend to spread, eventually to fixation, if and only if it has the effect of reducing the rate of recombination between the two fitness loci, relative to the rate associated with the alternative *M1* allele. In this way, selection will always favour a reduction in the rate of recombination, provided that there is any degree of epistatic interaction between loci.

A verbal argument of this sort was first put forward by Fisher (1930), and his conclusion has since been amply verified by several more formal analyses: the major papers are those by Kimura (1956), Turner (1967), Nei (1967, 1969), Lewontin (1971), Feldman (1972), Karlin and McGregor (1974), Charlesworth and Charlesworth (1973) and Charlesworth (1976). Briefly, the argument given above is relevant only to the case in which there is a stable equilibrium point at which all four alleles are maintained. The conditions for such an equilibrium are not known for the general two-locus case, but they have been worked out for a few special selection schemes. Assuming them to be satisfied, the selection of a modifier affecting recombination will depend on linkage disequilibrium at the joint gene-frequency equilibrium. If this is zero (a result possible even with certain patterns of epistasis) then any modifier will be neutral, since the only effect of recombination is to move a nonzero value of linkage disequilibrium towards zero. If it is not zero, then a modifier will be selected if, and only if, its effect is to decrease recombination. It is still possible to dispute this result on technical grounds, the most important of which is that all the analyses deal with genetic variation maintained by heterosis in a constant and uniform environment. However, there is remarkable unanimity among theorists that the ineluctable suppression of recombination is not only a very general but also a very robust result.

This conclusion is manifestly false; that is the paradox of recombination. A greater rate of recombination does not bear a cost relative to a lesser rate in the same way that sex is costly relative to parthenogenesis; nevertheless, a lesser rate of recombination is almost invariably autoselected because it is more effective in freezing the fittest genotypes and maintaining linkage disequilibrium. We know that autoselection in a constant and uniform environment is not a sufficient theory of recombination, since most chromosomes in most sexual organisms show at least one cross-over during meiosis, and no eukaryote is known to have reached the end-point predicted by the theory, a single large chromosome unable to recombine. The resolution of the paradox, in general terms, is obvious enough: the model of the environment which leads to this conclusion is not an accurate portrait of nature and, in a realistic model, natural selection would oppose auto-selection and prevent the abolition of recombination. But in what way must the model be altered so as to procure an advantage for greater rates of recombination? Since the major consequence of recombination is the same as that of sex, in the broad sense that both diversify progeny, it has been assumed in the past that a theory which explains the occurrence of sex will be equally competent to explain patterns of recombination; thus, in my account of theories of sex, I included several hypotheses which address themselves specifically to recombination. I am not now satisfied that this proposition is true. The tangled bank offers an excellent explanation of patterns of sex and parthenogenesis among multicellular animals but, at least in its simpler versions, the principle of imperfect competition is not capable of generating selection for a nonrecessive allele whose effect is to increase the rate of recombination.

Something might be done with a model similar to that of Maynard Smith (1976b), in which the lottery principle were replaced by imperfect competition, but I am not inclined to be optimistic. In this section, my primary object will not be to identify the function of recombination; instead, I shall attempt to solve the easier problem of whether the correlates of recombination are the same as those of sex. If this hypothesis can be falsified, we must conclude that sex and recombination cannot both be explained by the same theory. I shall proceed in the same way as before: first by separating historical from functional theories; next, by separating balanced hypotheses of individual selection from unbalanced hypotheses of group selection; and finally, by using the correlates of recombination to decide between rival hypotheses.

5.2.1   The Historical Hypothesis

The fact that one can trace phylogenies through the organization of karyotypes is sufficient to demonstrate an historical component in the evolution of recombination. So far as I know, no-one has gone further than this to suggest that recombination is moulded by history to nearly the same extent that it can plausibly be maintained that the evolution of sexual systems is restrained by the lack of relevant genetic variation. Nevertheless, there is a major theory of recombination which is historical, to the extent that it denies that the genetic consequences of crossing-over are functional. It has been repeatedly suggested by cytologists that chiasmata normally serve a primarily mechanical function in the orderly pairing and separation of homologues. This is held to explain why, in karyotypes which comprise a large number of small chromosomes, each bivalent regularly exhibits a single chiasma. The major piece of evidence against the theory is that, in many taxa of higher animals, normal union and disjunction occur during meiosis in the absence of any crossing-over whatsoever (achiasmate meiosis, reviewed below); and if the mechanical indispensability of chiasmata is not invariable it is difficult to see why it should be general. It can also be argued that, if the genetic consequences of crossing-over are genetically unfavourable but chiasmata must be retained nonetheless, for mechanical reasons, chiasmata should occur only at the two ends of a chromosome, where they will hold the two homologues together with minimal genetic impact. However, such extreme terminal localization of chiasmata is very exceptional. The hypothesis must be rejected as a general theory of crossing-over; it remains conceivable that chiasmata have come to serve a subsidiary mechanical function in taxa with a long history of chiasmate meiosis, and this addiction might inhibit the evolution of achiasmy, even if a total suppression of crossing-over were adaptive.

5.2.2 The Balance Hypothesis

An accurate evaluation of the balance hypothesis is crucial for theories of sex, since it determines whether we look for an advantage to the group or to the individual. In the case of recombination the situation is much clearer: there is no doubt that there is considerable genetic variance for recombination rates, which can be changed in the short term by individual natural selection.

Response to Selection. About a dozen attempts have been made to alter recombination rates in *Drosophila melanogaster* by artificial selection. All have concerned crossing-over in the interval between specified marker genes on a single chromosome, and their results may therefore refer to shifts in chiasma localization rather than to genuine overall increases in chiasma frequency (compare, for example, the experiments of Parsons, 1958, and Chinnici, 1971). The earliest attempt, by Gowen (1919), was unsuccessful in altering recombination values between six loci on chromosome III, and later Acton (1961) failed to achieve a reduction of recombination between *cn* and *vg* on chromosome II. At the other extreme, Chinnici (1971) was able both to increase and to decrease recombination between the X-chromosome loci *sc* and *cv*. Kidwell (1972a) obtained an upward but not a downward response to selection on recombination between *gl* and *sb* on chromosome III, but later (Kidwell, 1972b) successfully selected for reduced recombination. Detlefson and Roberts (1921, X-chromosome), Moyer (1964, cited by Kidwell, 1972a; chromosome III) and Mukherjee (1961; two chromosome II loci in male *D. ananassae*) all achieved a reduction but not an increase of recombination under selection. Abdullah and Charlesworth (1974) were also successful in reducing recombination. These facts are most instructive. If, as the elementary theory would have us believe, natural selection almost invariably favours low rates of recombination, then artificial selection should be impotent to reduce them further. The experience of Chinnici (1971) is especially valuable, since he was able to procure substantial changes both up and down, while being confident that the chromosomes underwent no structural change. The simplest interpretation of these facts is that substantial genetic variation for the rate of recombination is maintained by balancing selection.

A few experiments have been done on other organisms. Shaw (1972) succeeded in both increasing and decreasing total chiasma frequency in *Schistocerca gregaria*. Allard (1963) was able to increase but not to decrease rates of recombination in *Phaseolus*.

Genetics of Recombination. The response to selection suggests that there is often considerable additive genetic variance for rates of recombination, and this has in general been borne out by direct studies of inheritance. Estimates of heritability are moderate or high: 0.12 and 0.78–0.98 for site-specific recombination rates in *Drosophila* (Kidwell, 1972a) and *Mus* (de Boer and van den Hoeven, 1977), respectively, and 0.42 and 0.53 for total chiasma frequency in *Corchorus* (Paria and Basak, 1973) and *Schistocerca* (Shaw, 1974). However, it would be superfluous here to review the quite voluminous literature on genetic elements affecting recombination, since my only purpose is to point out that such genes have often been described; instead, I shall point to the reviews by Bodmer and Parsons (1962; general), Catcheside (1977; microorganisms, including prokaryotes), Baker and Hall (1977; *Drosophila*) and Riley (1966; higher plants, see also Riley and Law, 1965).

Simchen and Stamberg (1969) suggested that recombination was under two sorts of genetic control: a coarse control, which by aborting the meiotic process produced asynapsis or desynapsis and a consequent abolition of crossing-over; and a fine control, which alters rates of recombination in specific chromosome segments much less drastically. However, it might be more useful to think of three levels of control. The coarse control, as before, abolishes recombination altogether throughout the genome; it is often determined by major genes, and these are usually recessive (e.g., the desynaptic mutants of *Crepis* described by Tease and Jones, 1976, or the nil-recombination genes of *Saccharomyces* studied by Fogel and Roth, 1974). The fine, or site-specific, control regulates recombination between given loci; in this case, alleles for low recombination tend to be dominant (e.g., Kidwell, 1972b, for *Drosophila melanogaster*; Smith, 1976, and Catcheside, 1968, 1975, for *Neurospora crassa*; Simchen, 1967, for *Schizophyllum commune*). It is a general but not a universal rule that the regulating loci are unlinked to the loci between which recombination is being regulated, and they are often on different chromosomes. The intermediate level of control is represented by genes which are responsible for non-specific fine control, altering recombination rates to a greater or lesser extent throughout the genome, but without abolishing recombination. The genetics of this situation is less well understood — except where it involves supernumerary chromosomes, which are discussed below — and may be less readily generalizable. Shaw (1974) described the polygenic inheritance of chiasma frequency in *Schistocerca gregaria*, and inferred that alleles for low recombination tend to be recessive. The chiasma frequency of inbred lines of rye shows overdominance (Rees and Thomson, (1956).

Recombination may also be affected by the structure of the karyotype. The effect of heterochromatin and inversions is too well-known to labour. A less specific effect may be exerted by supernumerary ('B') chromosomes, which are not homologous with the usual autosomal ('A') complement and may not behave regularly during nuclear division. B-chromosomes increase chiasma frequency in *Crepis* (Brown and Jones, 1976), *Festuca* (Malik and Tripathi, 1970), *Listera* (Vosa and Barlow, 1972), *Pennisetum* (Pantulu and Manga, 1972), *Pushkinia* (Barlow and Vosa, 1970), *Secale* (Zecevic and Paunovic, 1969), *Zea* (Ayonoadu and Rees, 1968), *Aster* (Koul and Wakhlu, 1976), *Gibasis* (Brandham and Bhatterai, 1977), *Melanoplus* (Abdel-Hameed *et al.*, 1970) and *Myrmeleotettix* (Barker, 1960; John and Hewitt, 1965a). They often also affect the distribution of chiasmata between bivalents. These effects are not universal, however: they were not found in *Brachycome* by Carter and Smith-White (1972) or in *Melanoplus* by Stephens and Bergmann (1972), and B-chromosomes were found to be associated with a decreased rate of crossing-over in *Lolium* by Cameron and Rees (1967) and in *Najas* by Viinika (1974).

It seems that most of the physical and chemical abuses to which organisms can be subjected may cause changes in rates of recombination (see review in Bodmer and Parsons, 1962). A few examples, more or less at random, are: temperature (Plough, 1917; Stern, 1926; White, 1934; Moffet, 1936; Mather, 1939; Elliott, 1955, 1958; Dowrick, 1957; Rifaat, 1959; Wilson, 1959; Hayman and Parsons, 1961; Church and Wimber, 1969; Church, 1974); humidity (Levine, 1955); x irradiation (Muller, 1925; Lawrence, 1961; Westerman, 1967; Church and Wimber, 1969); and various chemicals (Levine, 1955; Law, 1963). The extent to which these factors influence recombination in natural populations is not known; several workers on orthopteran cytogenetics have reported that the chiasma frequency in their material did not vary with the time of day when it was collected.

5.2.3 Correlates of Recombination

From the facts I have briefly presented above, I think it must be inferred that differences in the amount of recombination achieved by meiosis represent functional differences and are established in the short term by individual natural selection. This conclusion is entirely uncontroversial. I shall now examine the hypothesis that the difference between amphimixis and parthenogenesis is a matter of degree. If this is so, then the correlates of lesser and greater amounts of recombination among exclusively amphimictic organisms should be the same as the correlates of parthenogenesis and sex, respectively.

The Recombination Index. To identify these correlates, we must first find a way of measuring the amount of recombination. The best simple measure is the index of recombination R, defined to be the number of independently transmitted genetic elements. Imagine a diplont with N pairs of autosomes, each of which forms C chiasmata during meiosis; then the number of different possible gametic types is $2^{N(1+C)}$. It is conventional to define R as the $\log_2$ of this quantity:

$$R = N(1 + C).$$

The recombination index is thus equal to the sum of the haploid complement and the number of chiasmata per nucleus. In practice, C will be an average rather than a constant, and this simple version of the index ignores such awkward features of the real world as sex chromosomes, autosomal heterochromatin, chiasma localization and karyotype asymmetry. The index is intended only as a rough comparative measure of the amount of recombination achieved by meiosis.

The recombination index varies with chromosome number and with chiasma frequency; the former has been measured for many thousands of species, but the latter for only a few dozen. Our knowledge of recombination indices is therefore limited by measurements of chiasma frequency, and in Table 5.2 I have listed the results of a brief search through the literature; Table 5.2 includes only cases in which chiasma frequency has been measured explicitly. The smallest recombination indices are probably shown by animals with very few chromosomes and either male haploidy or achiasmy; these would take values, for instance, in arrhenotokous mites, of about three or four, just below the range of values in Table 5.2. An even more extreme situation concerns the permanent structural heterozygotes in the genus *Oenothera*, which simply reproduce the two gametic types from which they sprang. The largest indices are possessed by organisms with a great many small chromosomes, each of which is presumed to form a single chiasma; amongst animals, recombination indices of up to 500 may exist in some butterflies and decapods, far above the range of values in Table 5.2.

To elucidate the behaviour of the index as a function of its two components I have regressed the value of the index on the haploid number of autosomes (Figure 5.2). Since the two variables are autocorrelated, the result is naturally of no intrinsic interest, but it shows that the index does indeed rise steeply and more or less linearly with chromosome number. Over a wide range of chromosome numbers, therefore, the chiasma frequency per bivalent has only a small effect; this is simply because the range of chiasma frequencies (about fourfold) is much smaller than the range of chromosome numbers (over a hundredfold). In comparisons at a low phyletic level chromosome numbers are often rather uniform and the chiasma frequency is an appropriate measure of recombination. But at higher phyletic levels there is little systematic relationship between chiasma frequency and recombination index, while chromosome number becomes a good criterion of recombination. This has the important consequence that, provided comparative analysis is done at a sufficiently high phyletic level, or within a sufficiently variable taxon, the enormous literature on chromosome number can be quarried to investigate the correlates of recombination.

Correlates of Chiasma Frequency. The best work on the variation of chiasma frequency within species has been done on orthopterans. Barker (1960) found that individuals from two high-altitude northern populations of the British grasshopper *Myrmeleotettix maculatus* have fewer chiasmata than those from two low-altitude southern populations. This difference was retained in samples collected in the wild and reared together in a greenhouse. He later reported (Barker, 1966) that British populations of *M. maculatus* are polymorphic for a B-chromosome whose presence is associated with elevated chiasma frequencies and whose distribution is related to climate: populations living at lower altitude or experiencing higher mean July temperatures had a higher frequency of B-chromosomes and thus a higher average chiasma frequency. Hewitt and John (1967) extended Barker's work but, while confirming the existence of a north-south cline, noticed that some southern populations lacked B-chromosomes altogether. This observation led them to attempt a deeper analysis of the situation by collecting along a transect in Wales, in a region where altitude, rainfall and B-chromosome frequency were highly variable. They found a very significant negative relationship between chiasma frequency and altitude, but this was inevitably associated with an equally significant correlation of the same sign between chiasma frequency and rainfall, since this area of high relief lies in the path of the North Atlantic rain clouds. Since localities at comparable altitudes east of the watershed have much lower rainfall but high B-chromosome frequencies, they concluded that B-chromosome frequency is correlated with rainfall rather than with altitude *per se*; thus, when localities throughout the Welsh mountains are considered together, the significant correlation between B-chromosome frequency and rainfall is maintained, but that with altitude becomes insignificant. Over all British localities they found the relationship between B-chromosome frequency and rainfall to be nonlinear, being rather flat from 20–30″ per year but descending steeply for values of more than 40″ per year (Figure 5.3). In a remarkable study of a Welsh spoil heap, Hewitt and Ruscoe (1971) confirmed this pattern on a very small scale. They found a very steep cline in the frequency of B-chromosomes, which went from four to 37 per cent over only 320 m distance and 100 m altitude. The higher B-chromosome frequencies were associated with lower, warmer, drier and more sheltered sites. In short, British populations of *M. maculatus* have high chiasma frequencies in optimal (warm, dry) habitats, whilst low values characterize unfavourable (cool, damp) habitats; in conditions cooler or wetter than those sampled by Hewitt and John, the species does not occur.

Similar but less detailed results have been obtained for British populations of *Chorthippus parallelus* by Barker (1960) and Hewitt (1964).

The situation in the Australasian *Phaulacridium* is less clear. In mainland populations of *Ph. vittatum* there is a substantial but not formally significant positive correlation between chiasma frequency and altitude ($r = +0.322$, $n = 23$; data from Table I of Rowe and Westerman, 1974). In New Zealand populations of *Ph. marginale* the correlation with altitude has the same sign but is much smaller ($r = +0.063$, $n = 27$; data from Westerman, 1974; similar small positive correlation for different series of localities in Westerman, 1975), though

Table 5.2: Values of the Recombination Index in a Variety of Plants and Animals. Plants are listed first, followed by animals. Data columns are chiasmata per bivalent (XTA/BIV), haploid number of chromosomes (HAPLOID N) and recombination index (RECOMB. INDEX). Values given are the extreme range of means over species, sexes, localities, years and experimental treatments. Further information on chiasma localization is available from the authorities cited for: *Phaseolus, Atractomorpha, Aedes, Dendrocoelum, Lilium, Rhoeo, Mus, Triturus, Stethophyma, Trimerotropis, Cepaea, Triticum, Phyllodactylus, Allium* and *Galphimia.* For effect of heterochromatin: *Atractomorpha* and *Tulbaghia.* For B-chromosomes: *Zea, Gibasis, Crepis, Myrmeleotettix* and *Secale.* Tables includes only diplonts with chiasmate meiosis in both sexes.

TAXON	GENUS	XTA/BIV	HAPLOID N	RECOMB. INDEX	AUTHORITY
ALLIACEAE	*Allium* (seven spp.)	1.53–2.66	7–8	20.3–29.3	Ved Brat (1966) Ved Brat and Djingia (1973)
	Tulbaghia (four spp.)	1.85–2.63	6	17.1–20.4	Vosa (1972)
COMMELINACEAE	*Gibasis* (one sp.)	1.83–2.65	6	17.0–21.9	Brandham and Bhattarai (1977)
	Rhoeo (one sp.)	2.21–2.28			Carneil (1960)
	Tradescantia (two spp.)	1.7–2.2	6	16.2–19.2	Sax (1935)
COMPOSITAE	*Crepis* (one sp.)	1.07–1.83	3	6.2–8.5	Brown and Jones (1976) Tease and Jones (1976)
	Picris (1 sp.)	1.12	5	10.6	El-Bayoumi (1973)
	Senecio (1 sp.)	0.92	9	17.3	El-Bayoumi (1973)
GRAMINEAE	*Briza* (five spp.)	1.87–2.36	5–7	16.8–23.1	Murray (1976)
	Hordeum (five spp.)	1.94–2.23	7	20.6–22.6	Gale and Rees (1970)
	Lolium (one sp.)	1.30–2.14	7	16.1–22.0	Law (1963)
	Secale (one sp.)	1.77–3.07	6	16.7–24.4	Darlington (1933) Sax (1935) Zecevic and Paunovic (1969) Davies and Jones (1974)

LEGUMINOSAE	*Triticum* (two spp.)	1.90–1.93	7	20.3–20.5	Zarchi *et al.* (1972)
	Zea (one sp.)	1.89–2.03	10	28.9–30.3	Ayonoadu and Rees (1968)
	Phaseolus (16 spp.)	1.77–2.03	11	30.5–33.3	Sinha and Roy (1979)
	Vicia (one sp.)	3.1–4.3	6	24.6–31.8	Maeda (1930); Sax (1935); Rowlands (1958)
LILIACEAE	*Endymion* (one sp.)	2.11–2.48	8	24.9–27.8	Elliott (1955, 1958); Wilson (1960)
	Fritillaria (one sp.)	2.07–3.15	12	36.3–49.8	Fogwill (1958)
	Lilium (six spp.)	2.32–3.70	12	39.8–53.2	Fogwill (1958)
	Tulipa (four spp.)	1.3–1.6	12	27.6–30.9	Couzin and Fox (1974)
LINACEAE	*Linum* (seven spp.)	1.47–1.56	15	37.0–38.4	Seetharam and Srinivasachar (1972)
MALPIGHIACEAE	*Galphimia* (one sp.)	1.76	12	33.1	Zaman *et al.* (1977)
PLANTAGINAECEAE	*Plantago* (two spp.)	1.0–1.1	5	10.0–10.5	El-Bayoumi (1973)
SCROPHULARIACEAE	*Collinsia* (eight spp.)	1.20–1.90	7	15.4–20.3	Garber (1956)
TURBELLARIA	*Dendrocoelum* (one sp.)	1.69–2.91	7	18.8–27.4	Borragan and Callan (1952)
MOLLUSCA	*Cepaea* (one sp.)	1.01	22	44.5	Price (1974)
ORTHOPTERA	*Atractomorpha* (four spp.)	1.06–2.08	10	18.5–27.7	Nankivell (1976)
	Chorthippus (three spp.)	1.5–1.9	9	21.5–24.8	Barker (1960); Hewitt (1964, 1965); Hewitt and John (1968)
	Locusta (one sp.)	1.3–1.6	11	25.3–28.4	Elliott (1955); Nolte (1964)
	Locustana (one sp.)	1.0–1.3	11	21.9–25.7	Nolte (1964)

Table 5.2 (*cont.*)

TAXON	GENUS	XTA/BIV	HAPLOID N	RECOMB. INDEX	AUTHORITY
	Myrmeleotettix (one sp.)	1.6–1.8	9	22.1–23.9	Barker (1960) John and Hewitt (1965a, b) Hewitt (1965) Hewitt and John (1967)
	Omocestus (one sp.)	2	9	27	Hewitt (1964, 1965)
	Phaulacridium (two spp.)	1.07–1.48	12	23.8–28.8	Rowe and Westerman (1974) Westerman (1974, 1975)
	Schistocerca (one sp.)	1.73–1.89	11	30.0–31.8	Henderson (1963) Nolte (1964)
	Strenobothrus (one sp.)	1.74	9	23.7	Hewitt (1964)
	Stethophyma (one sp.)	0.94–1.25	12	22.3–27.0	Perry and Jones (1974)
	Tolgadia (three spp.)	1.15–1.37	10–12	22.7–26.6	John and Freeman (1976)
	Trimerotropis (one sp.)	1.13–1.44	12	24.5–28.3	Weissman (1976)
DIPTERA	*Aedes* (one sp.)	1.43	3	7.3	Ved Brat and Rai (1974)
HOMOPTERA	60 sp in six families	1–1.1	6–15	12–30	Halkka (1964)
VERTEBRATA	*Bos* (one sp.)	1.23	30	66.9	Jagiello *et al.* (1976)
	Cavia (one sp.)	1.04	32	65.3	Jagiello *et al.* (1976)
	Homo (one sp.)	1.89	23	66.5	Jagiello *et al.* (1976)
	Macaca (one sp.)	1.62	21	55.0	Jagiello *et al.* (1976)
	Ovis (one sp.)	1.39	27	64.5	Jagiello *et al.* (1976)
	Papio (one sp.)	2.00	21	63.0	Jagiello *et al.* (1976)
	Mus (one sp.)	1.17–1.97	20	43.3–59.4	Crew and Koller (1932)
	Saguinus (one sp.)	2.97	21	62.4	Jagiello *et al.* (1976)
	Triturus (two spp.)	1.83–3.40	12	34.0–52.8	Watson and Callan (1963)

Figure 5.2: Relationship Between Recombination Index and Chromosome Number. Data included are midpoint of range for primary spermatocytes or pollen mother cells of species referred to in Table 5.2. Key: ●, plants; ○, mammals; X, orthopterans; Δ, other animals; line encloses Halkka's data for homopterans.

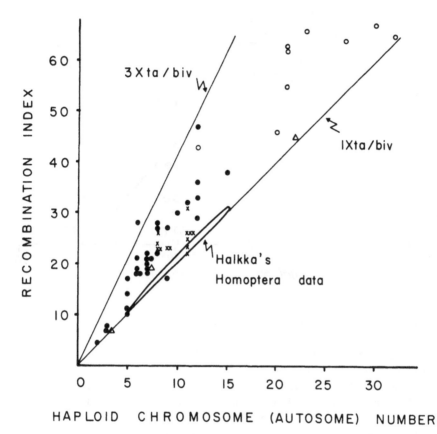

it is very clear that chiasma frequency increases from the coast to the interior of the South Island (cf. Figure I and Table I of Westerman, 1974). Climatic variables are more highly correlated with chiasma frequency, higher frequencies being associated with cooler, drier habitats (Westerman, 1974, Table 6).

Weissman (1976) studied populations of *Trimerotropis pseudofasciata* on the Pacific coast of North America. He found that mainland populations had more chiasmata than island populations, but that they had fewer interstitial chiasmata and thus probably achieved less recombination per meiosis. When different island populations were compared, the highest chiasma frequencies were found on the island with the lowest population density; moreover, although the data were scanty, this island probably also had the highest frequency of interstitial

Figure 5.3: Chiasma Frequency and Climate in *Myrmeleotettix maculatus*. The occurrence of B-chromosomes is associated with an increase in chiasma frequency.

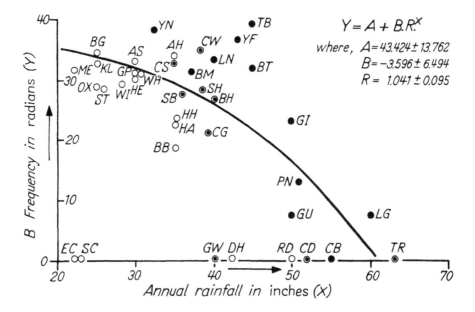

Source: Hewitt and John (1967), p. 157, Figure 11.

chiasmata. Schroeter (1968) found that peripheral populations of *T. helferi* had more chiasmata than central populations.

In the Australian *Tolgadia*, a generally hygrophilous acridid, two species were found to have about 13.5-14.0 chiasmata per nucleus, whilst a third species, *T. infirma*, had substantially more (about 14.5 or about 15.5, depending on whether or not supernumerary segments were present). *T. infirma* is brachypterous and has a wider area of distribution than the other two species, extending into much drier habitats and even into sclerophyll forest (John and Freeman, 1976). This association of brachyptery with a greater degree of recombination is also found in *Chorthippus*, where *C. parallelus* forms more chiasmata than *C. brunneus* (Hewitt, 1964).

In *Schistocerca gregaria* an increase of crossing-over at high population density has been known for some time (see Nolte, 1964).

In the snail *Cepaea hortensis* there is a negative correlation between overall chiasma frequency and altitude, compounded of a strong negative correlation with interstitial chiasma frequency and a weak positive correlation with terminal chiasma frequency (Price, 1974). However, according to Price, climate in the study area was more variable at lower altitude, so that populations with more recombination were experiencing a more variable physical environment. The correlation between chiasma frequency and population density was positive but not formally significant.

In wild populations of wheat, *Triticum*, Zarchi *et al.* (1972) found the highest chiasma frequencies in marginal and disturbed environments.

Achiasmy. The instances cited above seem sufficient to demonstrate that the occurrence of parthenogenesis in disturbed habitats is not reflected by a general correlation between disturbance and low rates of crossing-over in amphimicts. Indeed, it is difficult to propose any general rule that would encompass these results. We must therefore hedge. Many of the models which generate an advantage for recombination contrast an allele causing some degree of recombination with an alternative allele which completely suppresses crossing-over. It is often not clear whether any higher rate of recombination would be favoured over a lower but still substantial rate, and in some cases it is clear that this would not be the case. The correlates of recombination should then be most clearly displayed when we compare organisms having a normal chiasmate meiosis with those in which a meiosis occurs without any crossing-over.

I have listed the known instances of this curious phenomenon of achiasmy in Table 5.3. The most obvious interpretation of achiasmy is that it represents a brake on recombination analogous with, though not as severe as, parthenogenesis. Indeed, so distinguished a cytologist as M. J. D. White somewhere speaks of those enchytraeids in which both spermatogenesis and oogenesis are achiasmate as being 'halfway to parthenogenesis'. The evidence for this point of view is that parthenogenesis and achiasmy often occur in the same taxa. In enchytraeids, mantids and some families of Diptera, both parthenogenesis and achiasmy are widespread; in the scorpions, parthenogenesis is known only from Buthidae, which also includes the sole examples of achiasmy; and in the enchytraeid *Lumbricillus*, the triclad *Dugesia* and the isopod *Trichoniscus*, we find an achiasmate automixis. Consequently, the first piece of evidence against the theory is that in the Lepidoptera, Trichoptera and Diptera, together with isolated species of molluscs, water-mites, copepods, grasshoppers and alder-flies, achiasmy occurs in taxa where parthenogenesis is rare or unknown.

A second and even more potent objection to the hypothesis is that the occurrence of achiasmy follows two simple rules, neither of which is predicted by the theory. The first is that, where achiasmy occurs in gonochores (all cases except enchytraeids and *Dugesia*), achiasmy is restricted to one sex. The second is that this sex is always the heterogametic sex: the male in most instances, but the female in copepods, Lepidoptera and Trichoptera. Not only are these rules not explained by simply drawing a parallel between parthenogenesis and achiasmy; it is very difficult to think of any way in which they can be explained.

The heterogametic sex differs from the homogametic in that its karyotype includes two different sex chromosomes (naturally, there are many other systems of sex determination, but this is going to be a very simple argument); moreover,

Table 5.3: Synopsis of Achiasmate Meiosis in Animals. Table lists major taxon, genus, gamete affected by achiasmy, mode of reproduction (REP), the heterogametic sex in gonochores (HET. SEX), the haploid number of chromosomes (HAP. N), ploidy and authority. The only plant known to have an achiasmate meiosis is *Fritillaria* (Liliaceae) (Noda, 1968).

TAXON	GENUS	ACHIASMY	REP	HET. SEX	HAP. N	PLOIDY	AUTHORITY
TURBELLARIA	*Dugesia*	Oogenesis	Pseudogamy	(Herm.)	8	$4n$	Benazzi-Lentati (1961, 1962, 1970) Benazzi-Lentati and Bertini (1961) Benazzi and Pulcinella (1961)
OLIGOCHAETA	Enchytraeidae; seven genera, 22 spp.	Sperm and oogenesis	Amphimixis and thelytoky	(Herm.)	16–24	Mostly polyploid	Christiansen (1961)
	Lumbricillus	Asynaptic oogenesis	Thelytoky	(Herm.)	13	$3n$	Christiansen (1960, 1961)
PROSOBRANCHIA		Oligopyrene sperm undergo asynaptic meiosis					
PELECYPODA	*Sphaerium*	Sperm	Amphimixis	Male		$2n$	Keyl (1956)
CRUSTACEA	*Ectocyclops*	Oogenesis	Amphimixis	Female		$2n$	Beermann (1954)
	Cyclops	Oogenesis	Amphimixis	Female	4–14	$2n$	Beermann (1959)
	Tigriopus	Oogenesis	Amphimixis	Female		$2n$	Ar-Rushdi (1963)
ISOPODA	*Trichoniscus*	Asynaptic oogenesis	Thelytoky		8	$3n$	Vandel (1934)
ACARI	*Eylais*	Sperm	Amphimixis	Male		$2n$	Keyl (1957)
	Hydrodroma	Sperm	Amphimixis	Male		$2n$	Keyl (1957)
SCORPIONIDA	*Tityus*	Sperm	Amphimixis	Male	3	$2n$	Piza (1943)
	Isometrus	Sperm	Amphimixis	Male	6	$2n$	Piza (1947)

MANTODEA	14 genera	Sperm	Amphimixis	Male	2n	Hughes-Schrader (1943, 1948, 1950, 1953) White (1938, 1965) Gupta (1966)
ORTHOPTERA	*Thericles*	Sperm	Amphimixis	Male	2n	White (1965)
MECOPTERA	*Panorpa*	Sperm	Amphimixis	Male	c. 21	Ullerich (1961)
LEPIDOPTERA	? All genera	Oogenesis	Amphimixis	Female	2n	Smith (1938) Suomaleinen (1953, 1965b)
TRICHOPTERA	? All genera	Oogenesis	Amphimixis	Female	2n	Suomaleinen (1966b)
HOMOPTERA						Oligopyrene sperm produced in one lobe of testis through asynaptic meiosis in Pentatomida: Discocephalini
HOMOPTERA: COCCOIDEA						Variety of complicated meiotic mechanisms, many of which involve asynaptic or achiasmate spermatogenesis. However, males are functionally haploid because paternal chromosomes are made heterochromatic and are not transmitted to sperm. *Puto* has normal chiasmate meiosis: the male is the heterogametic sex. See Brown (1965)
DIPTERA: NEMATOCERA						Achiasmate spermatogenesis in Phryneidae (Bauer, 1946; Wolf, 1946, 1950), Bibionidae (Wolf, 1941), Scatopsidae (Wolf, 1941), Thaumalidae (Wolf, 1941), Blepharoceridae (Wolf, 1946), Asilidae (Metz and Nonidez, 1921, 1923), some Tipulidae (Bauer and Beerman, 1952) SCIARIDAE. Two 'limited' (L) chromosomes present in germ line but not in somatic cells. Oogenesis is synaptic and chiasmate for other chromosomes; behaviour of L chromosomes unknown. In spermatogenesis, first meiotic division is asynaptic and paternal chromosomes are expelled; one of two products of second division also expelled; single spermatocyte produced by meiosis has two L chromosomes, two X-chromosomes and a haploid complement of autosomes. Males are somatically XO. CECIDOMYIDAE. Complex cyclical parthenogenesis in some species, see Chapter 3. Most species amphimictic with asynaptic spermatogenesis. Female is diploid and male haploid for two nonhomologous sex chromosomes. Like sciarids there are genetic elements which are eliminated during early cleavage from somatic-line nuclei. CHIRONOMIDAE: ORTHOCLADIINAE. In some spp. a set of chromosomes is transmitted through the germ line but eliminated from somatic nuclei: these chromosomes usually form achiasmate bivalents during spermatogenesis

Table 5.3 (*cont.*)

TAXON	GENUS	ACHIASMY	REP	HET. SEX	HAP. N	PLOIDY	AUTHORITY
DIPTERA: BRACHYCERA			Achiasmate spermatogenesis universal. Almost all Diptera are amphimictic diplonts with heterogametic males				
ANOPLURA							
	Pediculus	Sperm	Amphimixis	? Male	6	2n	Doncaster and Cannon (1919)
	Haematopinus	Sperm	Amphimixis	? Male		2n	Cannon (1922) Hindle and Pontecorvo (1942) Bayreuther (1955a)
MALLOPHAGA							
	Genoides	Sperm	Amphimixis		2	2n	Perrot (1934)
	Gyropus	Sperm	Amphimixis			2n	Scholl (1955)

this is the only general cytogenetic difference between the sexes. It is tempting to infer that the function of achiasmy is to prevent crossing-over between the sex chromosomes. This must be false, however, not only because the Y-chromosome is heterochromatic but also because in mantids with an achiasmate spermatogenesis the males are XO. The function of achiasmy must therefore be to prevent crossing-over between autosomes; but its restriction to the heterogametic sex is then inexplicable, unless one member of an autosomal bivalent regularly segregates with one or other of the sex chromosomes. Suppose we have two alleles at an autosomal locus: A_m has the effect of elevating fitness in males and A_f a similar effect in females. At a second locus, chromosomes bearing the allele SD^y tend to segregate with the Y-chromosome and those bearing SD^x with the X-chromosome, while those bearing SD^0 segregate randomly. If the male is the heterogametic sex, the chromosomes $A_m SD^y$ and $A_f SD^x$ are optimal, but they will tend to be broken up by crossing-over between the A and SD loci. Thus, if there is a third locus at which R^+ determines chiasmy and R^0 achiasmy, the fittest chromosomes will be $A_m SD^y R^0$ and $A_f SD^x R^0$, and the allele for achiasmy will spread to fixation.

This is little more than an extension of the elementary proposition that nil-recombination alleles will be favoured by epistasis. The kindest thing that can be said about it is that the sex-ratio distortion which is to be expected if either SD^y and SD^x happen to be absent has indeed been observed in some fleas, butterflies, sciarids and isopods (see next section). But the failure of half a century of research into the genetics of the achiasmate males of *Drosophila* to detect the proposed phenomenon must be regarded as fatal. Achiasmy thus presents us with the dismal spectacle of a phenomenon for which the crucial generalization is already known, but for which no explanation is available.

There is worse to come, however, since several workers have found that different numbers of chiasmata occur during the formation of male and female gametes. According to the data shown in Figure 5.4, female gametes experience more crossing-over among hermaphroditic plants (and perhaps animals), but this is not invariably the case among gonochoric animals. It is not clear whether or not this is a general rule; certainly it has never received any explanation; and no rule seems consistent with general hypotheses of recombination.

By invoking achiasmy to rescue the theory from its failure to predict rates of crossing-over in chiasmate meioses, we have instead landed it in even deeper trouble. This initial failure threw doubt on the existence of a monolithic theory of sex and recombination; the failure to predict the correlates of achiasmy throws doubts on the existence of a monolithic theory of recombination. It suggests that we may need different theories to explain different levels of recombination, and that achiasmy is not merely a limiting low rate of recombination, but rather a phenomenon qualitatively different from intermediate rates of recombination.

Purely as a speculation, I would suggest that the crucial mistake may have been the assumption that recombination diversifies zygotes. Naturally, it also

Figure 5.4: Chiasma Frequency in Oogenesis and Spermatogenesis. Includes only those forms with a chiasmate meiosis in both sexes. Authorities given in Table 5.2.

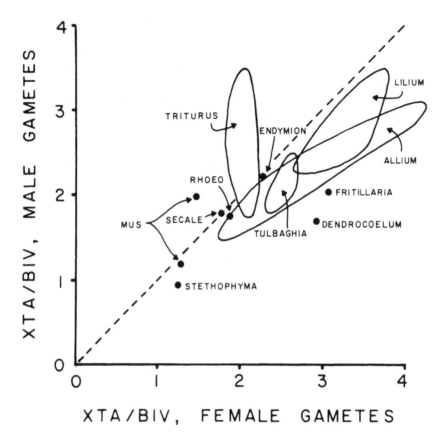

diversifies gametes; and the restriction of achiasmy to the heterogametic sex, together with its implication in the production of oligopyrene sperm, suggests irresistibly to me that we should turn our attention to the comparative biology of the gamete. I have no more concrete suggestion to offer, though the relationship between gamete competition and the site of meiosis is discussed in Section 5.4.

5.2.4   Chromosome Number

Suitable comparative series of chiasma frequency are very scarce and are confined largely to grasshoppers but, by turning to the other component of the recombination index, we can broaden the analysis by making use of the enormous accumulation of facts about chromosome number.

If the basic hypothesis is true, then the correlates of obligate sexuality should be mirrored by similar trends in chromosome number. For the most striking of these correlations, the prevalence of parthenogenesis in fresh water and amphimixis in the ocean, the truth of this proposition is far from certain. The best single comparison is perhaps that between freshwater and marine prosobranch molluscs, where there is a considerable range in chromosome number which is not complicated by any suggestion of polyploidy. Scanning the lists in Makino (1965), there is indeed a tendency for freshwater species ($\bar{x} = 12.6, N = 16$) to have fewer chromosomes than marine species ($\bar{x} = 17.4, N = 19$), and applying the Kolmogorov-Smirnov two-sample test shows that this result is unlikely to be due to chance ($\chi^2 = 9.74$, d$f = 2, P \sim 0.01$). However, if we go on to make the apparently appropriate comparison between the freshwater order Basommatophora and the marine subclass Opisthobranchia, we get the reverse result: pond-snails have a rather well-marked mode at 18, whilst opisthobranchs usually have twelve or 16 chromosomes. Freshwater cyclopoid and harpacticoid copepods often have seven or fewer chromosomes, whereas marine species have eight or more; but the data are heavily weighted by the single genus *Cyclops*. Among calanoid copepods both freshwater and marine species usually have 16 or 17 chromosomes. There is a slight tendency amongst amphipods and a strong tendency amongst isopods for the freshwater representatives to have fewer chromosomes than marine species, but in both cases the data are scanty, and the equally scanty data for decapods suggest the reverse correlation. The freshwater triclad turbellarians have fewer chromosomes than the marine polyclads, but the difference is small and statistically insignificant.

Among higher plants, it has been repeatedly suggested that the suppression of recombination by various means is associated with the occupancy of novel, disturbed or marginal environments. This is not an area that I shall review in any detail, since it is already familiar from the work of G. L. Stebbins, Verne Grant and others (e.g., Stebbins, 1958; Grant, 1958). The basic fact which they have established is that weediness is correlated both with an annual or herbaceous habit, or both, and with features of the meiotic and mating systems which will have the effect of reducing recombination − notably selfing and low chromosome number but also, in some forms, an increased occurrence of heterochromatic segments and of karyotypic asymmetry.

What seems to be lacking in the literature is a rigorous statistical description of the correlation between chromosome number (excluding polyploidy), on the one hand, and ecological and morphological variables, on the other. Considering the enormous effort that has gone into measuring chromosome numbers, it is astonishing, and even shameful, that we still have no detailed picture of their correlations, let alone a satisfactory interpretation of these correlations. This will doubtless be remedied by O. Solbrig's study of chromosome number in the Compositae, but his preliminary report (Solbrig, 1976) is confined to historical arguments and to establishing the simple correlation, already known from other taxa, between low chromosome number and the annual habit. I cannot supply the deficiency, but I can at least take a first step by establishing the correlates of chromosome number within a single small taxon.

The Aneuploid Series in Carex. The material that I have chosen for this survey comprises the British and North American species of *Carex* (Cyperaceae), a genus of sedges. My reasons for making this choice were the great number of species in the genus, which made it possible to use 252 species (181 from North America and 71 from Britain) in the analysis; the existence of a very long series in chromosome number, from a minimum of 18 to a maximum of 112 in the species I have used; and the availability of an adequate monographic treatment of the North American species. All species of *Carex* are perennial and almost all are monoecious. The basis for my analysis is the assumption that variation in chromosome number is genuinely aneuploid, and that differences in chromosome number therefore represent, and are rather precisely proportional to, differences in the recombination index. Granted this premise, I shall propose the hypothesis that recombination has the same function as sex, and that the correlates of high chromosome number in *Carex* will therefore mirror those of amphimixis in higher animals. To test this prediction, chromosome numbers were taken from Heilborn (1928, 1932), Davies (1956) and other authors whose counts are listed by Fedorov (1974). All British and North American species for which a count was available were included in the analysis, provided that there were no nomenclatural difficulties. Ecological and morphological data for the North American species were abstracted from the monograph by Mackenzie (1935) and for the British species from the much less complete account given by Butcher (1961), supplemented by distributional data from the *Atlas of the British Flora*. The hypothesis was tested with regard to the North American species, with the British material serving as a check on the generality of the correlations.

It is essential to the analysis that the series in chromosome number be genuinely aneuploid, rather than polyploid. This is almost certainly not wholly true: in the section Hirtae we find *C. lasiocarpa* with 56 and *C. hirta* with 112 chromosomes; in the Vesicariae *C. milliaris* has 40, *C. physocarpa* 60 and *C. saxatilis* 80 chromosomes; and according to Heilborn, there is autopolyploidy within *C. siderosticta*. Nevertheless, the prevalence of aneuploidy is very strongly indicated by four lines of evidence. Firstly, I have attempted an autocorrelation analysis of the frequency distribution of chromosome numbers in the world flora and in the North American flora, and have found no indication of the strongly periodic structure expected in a polyploid series. Second, chromosomes of *Carex* have diffuse centromeric activity (Davies, 1956). This implies that even small fragments will segregate regularly, and provides a cytological rationale for aneuploidy. Third, although I have found no extensive data on chromosome morphology, what there is suggests that chromosome number and chromosome size vary inversely. Thus, most species in Montanae and Digitatae have few, large chromosomes, while those in Acutae have many more much smaller chromosomes (Davies, 1956). Finally, several workers (reviewed in Davies, 1956) have found short aneuploid series within species. It seems very probable, therefore, that an increase in chromosome number will be associated with a loosening of linkage.

The obvious place to start is with the simple direct correlations between chromosome number and habitat. The results of the analysis are given in Table 5.4. For our present purposes, the hypothesis under test predicts that low chromosome numbers will be characteristic of open, xeric, montane, novel and northerly habitats. This is supported by the positive correlation between chromosome number and the ranking of habitat descriptions on a seven-point scale from most xeric to most mesic: species inhabiting more mesic situations tend to have more chromosomes. This correlation is quite strong; it applies to both the North American and the British material; and it was sustained when the ranking was done by a class of undergraduate students working in ignorance of the hypothesis, rather than by myself. It is not clear how this ranking relates to environmental uncertainty: a waterside habitat, where plants are liable to be uprooted by flooding, might actually be less equable and less predictable than a moorland or woodland site. Moreover, the equally strong correlation between chromosome number and the openness of the environment goes against the hypothesis, with low chromosome numbers characteristic in the relatively sheltered, shaded woodland habitat. Further, there is no hint of a correlation, either in the North American or in the British material, between chromosome number and a montane or a northerly distribution. By combining these four habitat descriptors we can fabricate an index of 'exposure', with one extreme representing the most open, xeric, montane and northerly habitats whilst the other extreme represents the most sheltered, mesic, low-lying, southerly habitats. The value of this index is not correlated with chromosome number. Introduced species do not differ in chromosome number from established native species, nor do they have unusually low chromosome numbers for their European source area.

There are other correlations between chromosome number and ecology which do not follow straightforwardly from the hypothesis. There is a rather weak but significant and satisfactorily quantified correlation between chromosome number and geographical range within Britain (this relationship was sought but not found by Haskell, 1952, who used a much coarser data set based on distribution by vice-counties rather than by the 10-km squares of the National Grid). Maritime species tend to have somewhat high chromosome numbers; more unexpectedly, species characteristic of either acidic or basic environments tend to have fewer chromosomes than species with no pronounced preference. Indeed, whilst analyzing the habitat data I noticed a slight but apparently consistent tendency for species inhabiting unusual habitat types to have fewer chromosomes than expected. To quantify this impression I counted the number of species inhabiting different combinations of the variable HMES, WOOD, MONT and ACID. Thus, one species might inhabit mesic, lowland, basic, woods, another montane, xeric, acid grassland, and so forth. Since HMES, WOOD and MONT were scored at two levels and ACID at three, there were in all 24 combinations, of which 18 were actually occupied by one or more species. Three of these combinations supported 22, 26 and 29 species, respectively; the

Table 5.4: Search for Direct Correlations Between Chromosome Number and Habitat Description in *Carex*; see text. Column headings are self-explanatory, except that AGREE records whether the result agrees (+) or disagrees (−) with the direction of correlation expected under the hypothesis, or shows essentially no correlation (0).

VARIABLE	HABITAT CLASSIFICATION	REGION	TEST	STATISTIC	AGREE	SIG
1. HMES	From most xeric to most mesic on a seven-point scale, roughly as follows. *1* sand-dunes and dry soil; *2* dry grassland; *3* meadows and dry woodland; *4* woodland and damp grassland; *5* wet grassland or woodland; *6* by water-side; *7* in standing water. For dichotomous tests, categories 1–4 and 5–7 pooled	N. America	Chi-square with two habitat categories and six chromosome no. categories	$x_5^2 = 21.6$	? +	< 0.001
		Britain	Kendall rank correlation	$T = 0.302$	? +	< 0.001
2. WOOD	*1* Habitat description mentions woodland; *2* woodland not mentioned	N. America	Chi-square with five chromosome no. categories	$x_4^2 = 32.9$	−	< 0.001
3. MARI	*1* Habitat description mentions coastal distribution; *2* otherwise	N. America	Median test	$x_1^2 = 6.14$	? +	~0.015
	Proportion of grid squares from which species recorded which fall on coastline	Britain	Kendall rank correlation	$T = 0.179$	? +	0.014
4. MONT	*1* Habitat description mentions mountains, summits, etc., other than in most southerly part of range; *2* otherwise	N. America	Median test	$x_1^2 = 0.37$	0	~0.55
5. NORT	*1* Distribution predominantly Canada, Alaska, Greenland; *2* distribution predominantly contiguous states of USA	N. America	Median test	$x_1^2 = 0.40$	0	~0.55
	Proportion of grid squares in which species occurs falling in northern half of island	Britain	Kendall rank correlation	$T = 0.005$	0	0.48
6. ACID	*1* Favours acid localities; *2* no stated preference; *3* favours basic localities	N. America	Median test	$x_2^2 = 8.38$	?	~0.015
7. INTR	*1* Exotic; *2* native	N. America	Median test	$x_1^2 = 0.10$	0	~0.75
8. RANG.	Number of 10-km grid squares occupied	Britain	Kendall rank correlation	$T = 0.219$	? −	0.003
9. HEXP	Compounded from HMES, WOOD, MONT and NORT; in four categories from xeric, open, montane, northerly habitats to mesic, closed, lowland southerly habitats	N. America	Median test	$x_3^2 = 0.65$	0	~0.90

other 15 each supported only twelve or fewer species. A median test on these two groups gave $\chi^2 = 6.1$ ($0.02 > P > 0.01$), with a weak tendency for species in the more overstocked habitat combinations to have more chromosomes.

These results are at best highly equivocal. It is reassuring to find that chromosome number is indeed correlated with habitat and, given the crude descriptions of habitat provided in taxonomic monographs, some of the correlations are surprisingly strong. However, they are about equally divided between being favourable, neutral, unrelated and hostile to the hypothesis. Given the limitations of the habitat descriptions, this sort of analysis can be pushed no further; instead, I shall turn to the much more extensive and precise morphological data, and use this to test the hypothesis through the device of setting up a secondary hypothesis. This secondary hypothesis is a commonplace of life-history theory: it states that a high rate of production of small propagules will be characteristic of species which occupy disturbed habitats, while the species of more stable habitats will reproduce more slowly by means of large propagules. Linking this with our main hypothesis, via the very firmly established generalization that parthenogenesis is characteristic of disturbed habitats, we are led to predict that the morphological features associated with a rapid rate of production of small propagules will be correlated with low chromosome number.

To test this prediction, eighteen readily quantifiable morphological variables were identified (see caption to Table 5.5), and estimates of as many as possible of these eighteen were obtained from Mackenzie's monograph for each of the 181 North American species whose chromosome number was known. Eight of these variables could also be estimated for 71 British species from Butcher's text, and these were used as before to check the generality of the North American results. The resulting matrices were then analyzed for the linear effects of morphological variation on chromosome number. The simple correlation coefficients are given in Table 5.5. These correlation coefficients have about the same value whether the analysis is conducted at the level of species or of sections, whether this analysis uses parametric or nonparametric correlation techniques, and whether it uses the British or the North American material. The results are therefore both general and robust (although a glance at the scatter diagrams confirms that the generally higher values of nonparametric correlation coefficients are caused by a certain degree of nonlinearity).

The leading fact established by the simple correlation analysis is that there is a strong correlation between chromosome number and total reproductive output: species with many chromosomes tend to be larger, with a greater area of leaves, a greater number of larger spikes and a greater number of larger perigynia. Moreover, there is a significant positive correlation between chromosome number and an 'index of allocation to reproduction', RALL. This suggests that high chromosome number is not merely correlated with large size and its allometric consequences, but with heavy investment in reproduction at the expense of somatic function.

More crucial to the test is the way in which different components of

Table 5.5: Chromosome Number and Morphology in *Carex*. Statistics given are: sample size (N); Pearson correlation coefficient (r); Spearman rank correlation coefficient (r_S); Kendall's rank correlation coefficient (T); and probability of chance departure from zero correlation (P). The dependent variable is in all cases the diploid chromosome number, **CHRN**. Where more than one count is available, the following criteria were used: (1) the work of Davies (1956) is preferred; (2) the count supported by the greatest number of authorities is preferred; (3) the count supported by the latest authorities is preferred. In a few cases where the sole authority gives two similar counts the midpoint of these is used. The independent variables are as follows:

(1) *PHGT* (cm). Culm height. Calculated as midpoint of maximum and minimum values.

(2) *BASE*. Nature of plant base. Three indices were calculated; caespitose, from one (very densely caespitose) to five (not caespitose); rhizomatous, from one (not rhizomatous) to seven (long creeping rhizome or rootstock); and stoloniferous, from one (not stoloniferous) to four (long stolons, freely stoloniferous). A compound index **BASE** was then calculated as the sum of these three simple indices. **BASE** is intended to measure the degree to which an individual plant is spread out in space; it ranges between three (highly aggregated, little potential for vegetative propagation) and 16 (highly dispersed, great potential for vegetative propagation).

(3) *NLLC*. Number of leaves with well-developed blades on fertile culms. Value used is midpoint of range.

(4) *LARE* (cm²). Area of leaf on fertile culm. Calculated as product of midpoint of leaf length and midpoint of leaf width, divided by two.

(5) *NARE* (cm²). Total area of leaves on fertile culm. NARE = NLLC × LARE.

(6) *NSPK*. Total number of spikes: sum of midpoint numbers of male, female and united spikes. Entirely male spikes, if present, are usually solitary.

(7) *SSPP* (cm²). Size of pistillate spikes. Calculated as product of midpoint of length and midpoint of breadth.

(8) *SSPS* (cm²). Size of staminate spikes; cf. **SSPP**.

(9) *SSPU* (cm²). Size of united (bisexual) spikes; cf. **SSPP**.

(10) *NPGA*. Number of perigynia per spike. Value used is midpoint of range.

(11) *TPGA*. Total number of perigynia per fertile culm. TPGA = NSPK × NPGA, omitting male spikes.

(12) *SPGA* (cm²). Size of perigynium. Calculated as product of midpoint of length and breadth.

(13) *LACH* (cm). Length of achene. Value used is midpoint of range.

(14) *RALL* (cm²/cm²). An attempt to measure allocation to reproduction in relation to vegetative function. The value used is the total area of all spikes divided by the total area of all leaves on a fertile culm.

(15) *BSYS*. An attempt to measure the probability that pollen will fertilize ovules on the same culm because of the arrangement of floral parts, assuming self-compatibility. Plants scored one for monoecy, two for dioecy; one for united spikes, three for separate spikes and two for an intermediate condition; one for androgyny and two for gynaecandry. These scores were summed to give an index ranging from three (monoecious culms with androgynous spikes) to seven (dioecious, or monoecious with separate spikes, the female above the males); invariable dioecy would receive a score of eight.

(16) *INBR*. An attempt to measure the probability that pollen will fertilize ovules on the same plant, but not necessarily on the same culm, because of the arrangement of floral parts on the culm and the aggregation of the culms in space: INBR = (9 − BSYS) × BASE.

Note: it seems likely that most *Carex* spp. are self-incompatible, in which case **BSYS** and **INBR** are meaningless.

(17) *MFUN* (cm²/cm²). An attempt to measure allocation to male function. The value used is the total area of male spikes per culm divided by the total area of spikes per culm. Calculated for species with separate (unisexual) spikes only.

In all cases the value of an entry for a section is the arithmetic mean of all entries of that variable for species in the section. Sources of data cited in text.

VARIABLE	N. America (species)					N. America (sections)			Britain (species)		
	N	r	P	r_s	P	N	r_s	P	N	T	P
1. PHGT	175	+0.372	<0.001	+0.369	<0.001	51	+0.173	0.226	67	+0.333	<0.001
2. BASE	174	...		−0.005	0.946	49	+0.143	0.327	71	+0.052	0.260
3. NLLC	153	−0.038	0.639	−0.022	0.791	48	−0.011	0.940	...	......	......
4. LARE	164	+0.220	0.005	+0.291	<0.001	49	+0.082	0.575	71	+0.207	0.005
5. NARE	148	+0.165	0.045	+0.206	0.012	46	+0.180	0.232	...	......	......
6. NSPK	170	+0.330	<0.001	+0.370	<0.001	46	+0.344	0.019	62	+0.283	0.001
7. SSPP	103	+0.420	<0.001	+0.592	<0.001	36	+0.326	0.052	56	+0.169	0.049
8. SSPS	94	+0.432	<0.001	+0.577	<0.001	35	+0.603	<0.001	...	......	......

Table 5.5 (*cont.*)

VARIABLE	N	N. America (species)				N. America (sections)			Britain (species)		
		r	P	r_s	P	N	r_s	P	N	T	P
9. SSPU	58	+0.110	0.410	+0.025	0.852	18	+0.110	0.664	...	...	
10. NPGA	143	+0.388	<0.001	+0.429	<0.001	43	+0.458	0.002	...	...	
11. TPGA	140	+0.451	<0.001	+0.520	<0.001	40	+0.528	<0.001	...	...	
12. SPGA	177	+0.279	<0.001	+0.218	0.004	51	+0.051	0.721	71	−0.006	0.468
13. LACH	176	−0.002	0.983	−0.053	0.489	51	−0.155	0.276	71	−0.035	0.334
14. RALL	115	...		+0.184	0.049	37	+0.282	0.091		...	
15. BSYS	176	...		+0.043	0.570	50	−0.050	0.730	71	−0.028	0.364
16. INBR	172	...		+0.040	0.605	48	+0.087	0.556		...	
17. MFUN	87	...		+0.039	0.717	33	+0.158	0.379		...	

Source: As cited in text.

reproduction are traded off against one another. This was investigated using multiple linear regression techniques, and a sketch of the correlation structure of the data matrix for North American species is shown in Figure 5.5. The major effect in the centre of this diagram is reassuringly self-evident: as the size of the pistillate spikes increases, both the number and the size of the perigynia increase but, when other factors are held constant, a greater number of perigynia is associated with a reduction in their size, whilst an increase in the size of the perigynia is associated with a reduction in their number. The most important result of this analysis is that it is the number, and not the size, of the perigynia which is associated with variation in chromosome number. The number of perigynia and of spikes is in turn dependent on strong independent effects acting through plant size and leaf area, but neither of these variables has any substantial independent effect on chromosome number. Finally, a compound variable BASE, which expresses the dispersion of an individual plant in space and which was designed to measure the proclivity for vegetative reproduction, was only loosely connected to other morphological variables and had no appreciable effect on chromosome number.

The picture which emerges from this analysis is quite clear. There are very strong correlations between morphology and chromosome number: depending on the data set and the statistical model used, between 30 and 50 per cent of variation in chromosome number can be accounted for by variation in morphology. When one considers that the analysis is restricted to the linear effects of midpoint estimates of morphological variables abstracted from the secondary literature by an amateur, this result is most encouraging: no doubt a determined effort by specialists in other groups would quickly reap the harvest of the last 70 years of chromosome-counting. In *Carex*, there is a strong positive correlation between chromosome number and any measure of total reproductive output. There is a significant positive correlation between chromosome number and a measure of relative reproductive output. And it is the number, and not the size, of propagules which is the component of reproduction associated with variation in chromosome number. In short, high chromosome numbers are associated with a high rate of production of many small propagules and thus, by inference, with the occupation of disturbed habitats. The hypothesis originally proposed is decisively falsified.

Since this falsification has important consequences, it seems worthwhile to retrace the chain of reasoning which led up to it. The fundamental concept under scrutiny is that the correct theory of sex is also the correct theory of recombination, and the argument proceeds as follows.

(1) Amphimixis is to parthenogenesis as high rates of recombination are to low; the correlates of low levels of recombination will therefore be the same as the correlates of parthenogenesis.

(2) The recombination index is an adequate measure of the amount of recombination achieved by meiosis.

Figure 5.5: Correlation Structure of Chromosome Number and Leading Morphological Variables in North American Species of *Carex*. Each variable was in turn used as the dependent variable, the others (except CHRN) serving as independent variables. The multiple linear regression equation was then calculated for all models from two to six independent variables (using an r^2-maximization technique with procedure STEPWISE on SAS). All independent variables showing a significant correlation with the dependent variable for any model were retained, and are plotted in the figure. A solid line indicates positive correlation, a broken line negative correlation. A thin line, broken or solid, represents correlations with $P < 0.05$; a thick line correlations with $P \ll 0.01$.

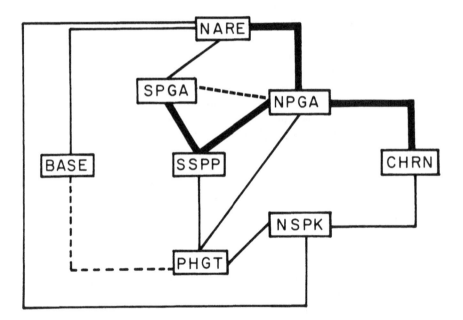

(3) The effect of chromosome number on the recombination index, given a sufficiently extensive range of chromosome numbers, is strong and nearly linear; it dominates any effect of chiasma frequency.

(4) The chromosome series in *Carex* is genuinely aneuploid.

(5) Parthenogenesis tends to occur in disturbed environments.

(6) In disturbed environments, selection will favour intense reproduction through the dissemination of many small propagules; in more equable environments, less intense reproduction through smaller numbers of larger propagules will be the rule.

(7) There will be a correlation between chromosome number and morphology, such that plants with low chromosome number will have morphological features associated with intense reproduction via small propagules.

(8) All correlations observed in *Carex* are the reverse of those expected: intense reproduction through many small propagules is associated with high rather than low chromosome number. These correlations are moderately large and often very highly significant; they account for a substantial proportion of the total variance in chromosome number; they apply to a range of characters, and to all characters examined; where it is possible to check, they apply both to British and to North American material.

(9) Further, there is no overall direct correlation between chromosome number and environment in the expected direction; and there is no correlation between chromosome number and proclivity for vegetative reproduction.

(10) This evidence is sufficient to falsify the hypothesis.

Points (1) and (7) are logical operators which will not be disputed. Points (3), (5) and (8) are factual statements based on overwhelming evidence; point (9) is not as securely based but, even if false or wrongly interpreted, its failure would not prejudice the integrity of the reasoning as a whole. Point (2) seems safe enough, since the argument fails only if heterochromatin, chiasma localization and so forth increase disproportionately with chromosome number. Point (4) is vulnerable, but the evidence in its favour, given above, seems adequate. The weakest link in the chain is undoubtedly point (6), an hypothesis which has not, to my knowledge, been tested in *Carex*. The nearest plant (in growth form and habitat) for which an adequate comparative study has been published is *Typha* (Juncaceae), where McNaughton (1975) showed that northern populations lived in disturbed, density-independent conditions whilst southern populations lived under a more stable, density-dependent regime. The northern populations produced both a greater number of smaller fruits and a greater number of smaller rhizomes, so that both sexual and vegetative reproduction followed the pattern suggested here. Unless a careful direct test falsifies point (6), I think that point (10) must be accepted.

In retrospect, the evolution of sex begins to seem a relatively simple problem, insofar as the correlates of parthenogenesis are rather clear-cut and its occurrence can be explained by a single general principle – even to the extent that we can largely ignore the fundamental cytogenetic difference between the two principal modes of thelytoky. The prospects for such a simple, unified theory of recombination seem to me very dim. The fact that neither of the two major determinants of the amount of recombination, the chiasma frequency and the chromosome number, have the same correlates as amphimixis makes it impossible to sustain the proposition that sex and recombination are merely different aspects of the same phenomenon, to be explained by any single rival hypothesis. To take one example at random, the Red Queen seems to offer an elegant interpretation of recombination. If we wished to advocate it, we might point out that the positive correlation between chromosome number and the number of species inhabiting a particular combination of habitat-descriptors is to be expected, since interspecific competition will be more intense when more species occur together.

With a little ingenuity, such correlations as those between chromosome number and geographical range in *Carex* and between habitat and chiasma frequency in *Myrmeleotettix* could also be advanced in support of the theory, and we could go on to quarry the literature on habitat and breeding system in higher plants, as Levin (1975) has done. Of course we would have to shut our eyes to such awkward facts as the morphological correlates of chromosome number in *Carex*, the occurrence of high chiasma frequencies in disturbed habitats in organisms such as *Triticum*, and the restriction of achiasmy to the heterogametic sex of gonochores. If the analysis were pursued further, we would recall that cave-dwelling turbellarians do not display the expected reduction in chromosome number, and that parasites do not have unusually high chromosome numbers, but rather the reverse. And if we had chosen any of the other permissible general theories of recombination, such as the hitch-hiker or certain versions of the best man, we would have come across exactly the same problem. A determined advocate will have no difficulty in selecting a body of facts which support his hypothesis, but a more disinterested search will quickly reveal a larger body of facts which contradict the hypothesis. I suspect that the complete suppression of recombination throughout the genome, the regulation of crossing-over between particular loci and the fine-scale adjustment of recombination rates throughout genomes or substantial fractions of genomes may be distinct phenomena, controlled by different genetic entities and requiring different sorts of explanation. I also suspect that too much attention has been paid to zygotes and too little to gametes; certainly, it would be very satisfying to discover that the correlates of recombination differ from those of sexuality in part because the function of recombination is to diversify the gametes which thelytokous organisms do not possess. Above all, I suspect that we must take a much closer look at the biology of recombination if we are to make any progress towards its explanation.

5.2.5 *Mutation*

The recent history of theories of mutation has been similar to that of theories of sex and recombination: first, a period in which it was assumed that mutation fuelled long-term evolution, to the benefit of the species, followed by the eclipse of this group-selectionist position by a theory of individual selection, and culminating recently in a theory of genic autoselection.

The idea that mutation rates are optimized by a balance between short-term disadvantage and long-term advantage, being a sort of disagreeable necessity for continued response to selection, was a commonplace until 1966, when, along with many other group-selectionist interpretations of genetic systems, it was swept away by Williams (1966, pp. 139–41). Williams argued that mutation is merely a copying mistake, almost invariably deleterious to individual fitness, whose incidence will always tend to be reduced towards zero by individual natural selection. It was subsequently shown mathematically that the optimal mutation rate, at least for certain simple selection schemes, is indeed zero (Kimura, 1967; Leigh, 1970). The only serious opposition to this opinion was mounted by Levins (1967, 1968). He argued that in any environment there will be an equilibrium gene frequency at any given locus; if sampling error or environmental fluctuation shift the frequency of a gene above this equilibrium point, mutation will act so as to bring it back and will, therefore, be favoured. In a model of this sort, a mutator gene linked to a fitness locus is favoured because it increases the genotype-environment correlation at the fitness locus. There seems to be no empirical evidence for this view; on the other hand, Williams argued that the existence of a zero optimum is supported by the very small magnitude of mutation rates, by the fact that they are less in haplonts than in diplonts, and by the fact that they are less variable between species when measured on a scale of generations than when measured in absolute time.

Some doubt has been thrown on this simple hypothesis by the discovery that genetic elements causing either an increase or a decrease in the mutation rate can be selected in bacterial populations maintained in chemostats (for a selection of papers, see vol. 73 of *Genetics*). Obviously, if the mutation rate were under constant downward selection pressure, it should not be easy or even possible to select in either direction. What seems to happen is that mutator genes hitch a ride on the favourable chromosomes they create, as suggested by Leigh (1970, 1973). A chemostat is a novel environment, whose effect will be to favour a number of genetic changes. Some of those genes which induce mutations at linked fitness loci will procure the spread and fixation of the newly favoured alleles at these loci, and in the process their own frequency will automatically increase. Once equilibrium has been reached with the fixation of all the favourable mutants, selection will revert to favouring mutation rates as close as possible to zero.

The results obtained during an experiment by Cox and Gibson (1974) make it possible to test this interpretation quantitatively. Since the initial linkage disequilibrium between mutator and fitness loci is likely to arise only if the population is fairly small, the fitness of the mutator alleles should be greater when population size is smaller or when it fluctuates with greater magnitude. Since the effect will decay in time as the population approaches equilibrium under selection, the fitness of the mutator alleles should be greater when it is measured after a shorter period of time. These predictions are tested in Table 5.6, which proves that there is a highly significant positive partial correlation between mutator fitness and the amplitude of population fluctuations (Cox and Gibson had previously obtained a significant simple correlation), and a weakly significant negative partial correlation between mutator fitness and the duration of the experiment. The weak positive correlation with mean population density is unwelcome, especially since Cox and Gibson state that population sizes were large enough to generate most mutations in every generation; however, given that recombination will be infrequent, it does not seem implausible that mutator genes will be brought into coupling with new favourable mutants during periods of low population density and can then hang on to these genes for appreciable periods of time. All in all, the experiment provides excellent confirmation of a hitch-hiker theory of prokaryote mutation rates.

Table 5.6: Analysis of Cox and Gibson's Experiment Measuring the Fitness of Mutator Genes in Chemostat Populations of *E. coli*. First table gives data from Table 2 of Cox and Gibson (1974). Column headings are: W, rate of increase of mutator genotypes over period of experiment, relative to that of non-mutator genotypes; N, mean cell density during experiment ($\times 10^{-7}$); $CV(N)$, coefficient of variation of N; T, duration of experiment in generations. Second table is multiple linear correlation analysis: b is regression coefficient, with standard error (both $\times 10^3$); t is Student's t with associated probability P; *part r* is the partial correlation coefficient.

W	N	$CV(N)$	T
1.063	120	67	100
1.040	16.0	94	165
1.022	1.10	33	229
1.006	0.95	35	518
1.006	1.40	48	715
1.012	1.20	29	574
1.012	1.30	54	368
1.003	0.57	39	837
1.035	0.73	62	378
1.002	0.72	16	780
1.053	0.44	77	314
0.996	0.64	9	437
1.007	0.79	47	636
1.066	16.6	86	145

	$b + se$	t	P	*part r*
N	$+0.18 \pm 0.11$	$+1.71$	0.117	$+0.476$
$CV(N)$	$+0.51 \pm 0.15$	$+3.45$	0.006	$+0.737$
T	-0.03 ± 0.02	-1.99	0.074	-0.534

Multiple correlation coefficient = 0.924

$F = 19.56$ with $P = 0.00017$

Transformations: Fit is slightly loosened if N is logged, but signs and approximate magnitudes of regression and correlation coefficients are conserved

5.2.6　Mendel's First Law

The evolutionary significance of Mendel's first law has aroused much less interest than that of his second law. More precisely, the principle of random segregation raises two problems for evolutionary biologists. The first concerns phenotype ratios and is the subject of the theory of dominance, a well-worn topic to which I have nothing new to add. The second and more fundamental problem concerns genotype ratios. It is very generally true that heterozygotes produce the two gametic types in equal numbers, but it is not easy to see why this should be so. Imagine a locus fixed for a particular allele. A mutant allele arises which has no effect on the viability, fertility or fecundity of the individual which bears it, but which in heterozygotes is incorporated into gametes more frequently than the wild-type allele. Clearly, this mutant allele will spread through the population by genic autoselection, and at no point during its spread will Mendel's first law be obeyed. This argument is readily extended to the case of two linked loci. At one locus two alleles are maintained in equilibrium by balancing selection. At the second locus a mutation arises whose effect is to distort the segregation ratio at the fitness locus, so that one or other of the alleles at this locus is incorporated more frequently into gametes If the segregation-distorting mutation arises in coupling with the fitness allele whose transmission it favours, it will clearly increase in frequency through a hitch-hiking effect, and will become fixed unless its spread is eventually checked by countervailing zygotic selection at the fitness locus. The more rigorous and general arguments developed by Prout *et al.* (1973), Thomson and Feldman (1974, 1976), Karlin and MacGregor (1974) and Liberman (1976) strongly suggest that random Mendelian segregation should seldom be evolutionarily stable. And yet it is certainly the general case; so we must recognize another major paradox in the evolution of genetic systems.

Quite a lot is now known about particular non-Mendelian systems, such as the *t*-alleles of mice, the B-chromosomes of mites, the driving Y-chromosomes of mosquitoes and the *SD*-chromosomes of *Drosophila*. In some cases we may even be quite close to understanding how the system evolves (e.g., Charlesworth and Hartl, 1978). What we lack is a general principle telling us why such cases are the exception rather than the rule. Liberman and Feldman (1980) have made a brave attempt to show that the stability of single-locus polymorphism maintained by overdominance is greatest when segregation is Mendelian, but this is true only for the symmetrical case in which the two homozygotes are of equal fitness. Their models demonstrate only a weak and debatable form of group selection for Mendelian segregation, and that only in a special case; and yet if we attempt to go beyond them we step into a morass of confusion and uncertainty. Possibly Liberman and Feldman's results also apply to individual selection, and given heterosis and a symmetrical selection scheme a line of descent with Mendelian segregation will have a similar variance of fitness and thus a greater long-term geometric mean fitness than any non-Mendelian line. Possibly we should attempt to resuscitate the old idea that Mendelian segregation maximizes the genotypic variance of sexual progeny, and thereby amplifies the diversifying effect of sex. I have even entertained the notion that the elementary arguments are correct; that alleles which distort segregation ratios will almost always increase at the expense of those which do not; that as a consequence most genomes are saturated with such alleles; and that the Mendelian ratios usually observed do not result from a virtual absence of segregation distortion, but from the fact that all chromosomes are very nearly balanced for alleles which, relative to the aboriginal wild-type, cause extreme distortion.

(This last fantasy is part of an undercurrent in recent thinking on genetic systems, that conventional notions of function at the gene level may have to be abandoned in favour of the concept of an almost unrestricted civil war within the genome. Take the recent discovery that the genomes of higher eukaryotes include long stretches of highly repetitious DNA. The conventional interpretation of its occurrence is that repetitious DNA has some function related to the performance of the individual organism; for example, it might facilitate the pairing of homologous chromosomes. However, repetitious DNA, like B-chromosomes, is known to affect recombination rates elsewhere in the genome. It is not clear whether this effect is incidental and unselected; or a primary function which increases individual fitness; or represents the attempt of the rest of the genome to counteract the presence of parasitic, dysfunctional chromatin; or represents the attempt of the repeated sequences themselves to nullify the counteradaptation of the rest of the genome. At all events, I wonder whether recent research into the fine structure of the genome might not eventually revolutionize our thinking about genetic systems.)

An interesting and potentially crucial special case of segregation is the sex ratio of progeny in gonochores. The confusion surrounding this topic has only quite recently been dispelled by G. Williams; it is a confusion between the population equilibrium and the individual optimum. Briefly, a population in which there is unrestricted competition for mates will evolve towards an equal frequency of the two sexes (Fisher, 1930; for the ramifications of this argument see Shaw, 1958; Bodmer and Edwards, 1960; MacArthur, 1965; Eshel, 1975; Charnov, 1975; it breaks down when there is inbreeding or local competition for mates, a line of argument descending from the classic paper by Hamilton, 1967). The reasoning is well-known: individuals of the minority sex will always have an advantage through leaving the greater number of grandchildren, and the equilibrium sex ratio is therefore stabilized by frequency-dependent selection. However, while this argument identifies the equilibrium sex ratio of the population, it tells us nothing about the variance of the sex ratio between progenies. An equal population sex ratio will follow if every female produces equal numbers of sons and daughters, but it will also follow if half the females produce only sons and the other half only daughters. The fact that progeny sex ratios are usually near equality emphatically does not follow from the fact that population sex ratios are usually near equality, and it is very difficult to understand what it does follow from. The relevant theory is admirably reviewed by Williams (1979), but the only general principle which has been discovered is that suggested by Verner (1965). He argued that if the population sex ratio fluctuated stochastically around one-half, a line of descent which has an equal progeny sex ratio will have a smaller variance of fitness and thus a greater long-term geometric mean fitness than any other line of descent. Williams cites some unpublished analytical work which supports this conclusion. It is an extension of Liberman and Feldman's argument to individual selection in the special case of an equilibrium frequency of one-half maintained by frequency-dependent selection. Williams argues that if it were true, selection should constrain progeny sex ratios to be very nearly one-half with much less variance than would be produced by random segregation. He cites data from vertebrates which show that progeny sex ratios are distributed around one-half with binomial variance, and on these grounds rejects Verner's hypothesis.

Monogeny. I thought it might be instructive to investigate an extreme special case, comprising those gonochores known to produce unisexual broods of offspring. This condition is known as monogeny (or merogony), in contrast to the usual condition of amphogeny, in which females produce nearly equal numbers of daughters and sons. Females which produce only daughters are said to be thelygenous; those which produce only sons are said to be arrhenogenous. 'True monogeny' is defined to be restricted to those cases in which both thelygenous and arrhenogenous females occur, whether or not amphogenous females also occur; it therefore excludes cases in which there are only thelygenous and amphogenous females, or only arrhenogenous and amphogenous females, such as the 'sex-ratio' condition of *Drosophila* (Gershenson, 1928; Sturtevant and Dobzhansky, 1936; Novitski *et al.*, 1965; and many others), scolytid beetles (Lanier and Oliver, 1966) and the butterfly *Acraea encedon* (Chanter and Owen, 1972). The relevant theory is easy to obtain, and to avoid a long stretch of simple algebra I shall merely give its conclusions. If monogeny is determined exclusively by a cytoplasmic factor, then it is easy to see that the factor will either die out (if it causes arrhenogeny) or will become fixed and thereby sterilize the population (if it causes thelygeny). It is conceivable that a cytoplasmic factor causing thelygeny might persist as the result of a balance between individual selection and group selection (see Heuch, 1978). On the other hand, suppose that monogeny is determined by nuclear genes, or by an interaction between nucleus and cytoplasm. Selection is then driven by the population sex ratio, and ceases when this has reached equilibrium. There are three relevant genetic equilibria. The first is the population sex ratio itself, which in simple models always has a stable equilibrium point of one-half. The second is the equilibrium between arrhenogeny and thelygeny, when both are present. This equilibrium is also stable; its position varies according to the mode of inheritance. The third equilibrium is that between monogeny and amphogeny. If only a single monogenous type (arrhenogenous or thelygenous) is present, this type is always eliminated by competition with an amphogenous type. The reason is obvious enough: an arrhenogenous mutant, for example, will either fail to spread or, if it does spread, will come to bias the population sex ratio towards males and thus incur a selective disadvantage. (Naturally, this conclusion does not hold in the case of a driving sex chromosome). If both monogenous types are present, the equilibrium between monogeny and amphogeny is only neutrally stable, and its position depends on the initial conditions. Thus, a population which includes arrhenogenous, thelygenous and amphogenous females has a stable population sex ratio, a stable ratio of arrhenogenous to thelygenous females and a neutral ratio of monogenous to amphogenous females. An amphogenous population is protected against invasion by either monogenous type, but it is not protected against simultaneous invasion by both types. However, I have confirmed by simulation that, in finite populations initially comprising a mixture of equal proportions of monogenous and amphogenous females, and with both the ratio of arrhenogenous to thelygenous females and the population sex ratio at their stable equilibrium values, amphogeny eventually becomes fixed much more often than monogeny, presumably because it is associated with a smaller variance of fitness.

I have not presented these arguments in detail, partly because they are entirely compatible with conventional sex-ratio theory, but also because they seem to be irrelevant to the distribution of monogeny in nature. It seems that true monogeny is restricted to three categories of animals. The first is a group of taxa with cyclical parthenogenesis – cladocerans, plankton rotifers and cynipid wasps (see Chapter 3). The second comprises a few ectoparasitic arthropods, such as the louse *Pediculus* (Hindle, 1919). The third includes a mixed bag of arthropods: sciarids (White, 1950), cecidomyids (Painter, 1930; Barnes, 1935, 1944; these also have a peculiar cyclical parthenogenesis); isopods (large literature, chiefly in French: see Vandel, 1938, 1941; Howard, 1942; Johnson, 1961; Legrand and Juchault, 1969; Juchault and Legrand, 1970; Martin *et al.*, 1973; Legrand and Legrand-Hamelin, 1975); amphipods (Charniaux-Cotton, 1957; Ginsburger-Vogel, 1973a, b); and centipedes (Rantala, 1974). Apart from a naive group-selectionist argument by Wildish (1971), who interprets monogeny in the amphipod *Orchestia* as a device to stabilize population number, the only extant theory of monogeny other than that developed above is that it reduces close inbreeding by preventing mating between sibs from the same brood (Howard, 1940; Ghiselin, 1974a). The distribution of monogeny seems to bear this out rather well. The nearest neighbours of any given sexual individual of a species reproducing by cyclical parthenogenesis will often be members of the same clone, and mating with them would be equivalent to full self-fertilization. This problem will be especially acute in animals such as monogonont rotifers, where the males are extremely precocious. Ectoparasites inevitably have a highly clumped distribution, which would lead to close inbreeding unless this was prevented by a device such as monogeny. The remaining cases are less obvious, but also seem to involve spatial aggregation. Isopods and centipedes characteristically inhabit island-like patches of suitably damp microhabitats, and cecidomyids and amphipods are often clumped around food sources. Monogeny, therefore, seems to occur in situations where inbreeding might otherwise be a serious problem. The theory is badly in need of a rigorous investigation in the field, but it certainly seems much more plausible than any general theory of the sex ratio. As with achiasmy, monogeny is an extreme special case governed by a rule which seems to tell us little about the more general phenomenon.

5.3 Alternation of Generations

The life cycle of sexual organisms includes a meiosis, which halves nuclear ploidy, and a syngamy, which redoubles it. An alternation of nuclear phases is therefore a necessary attribute of sexuality. It is usually accompanied by an alternation of morphologically distinct haploid and diploid individuals. In the higher multicellular plants and animals the vegetative individual is usually diploid, and produces transient haploid gametes or morphologically reduced gametophytes; in unicellular protists the vegetative individual is often haploid, and a transient zygote is the only diploid phase in the life cycle. Finally, amongst the algae both haploid and diploid phases may be relatively large and morphologically complex, and may or may not be morphologically different. Four common types of life cycle are illustrated diagrammatically in Figure 5.6, where some of the terminology used below is defined.

This regular succession of haploid and diploid individuals during the life cycle is called 'the alternation of generations'. The term was originally applied to the alternation of polyp and medusa in the life cycle of coelenterates by Steenstrup in 1842 (*'Generationswechsel'* – literally a 'change of generations'). It was applied to the spore-producing and gamete-producing generations of conifers in 1851 by Hofmeister, but it was not until 1894 that Strasburger inferred the occurrence of an underlying alternation of nuclear ploidies. The term was thus originally applied to an alternation of morphological states; but priority of usage is meaningless in a term which has undergone so many changes of meaning (Svedalius, 1927), and I shall follow Hyman (1940) and others in using 'the alternation of generations' to mean the regular succession of nuclear ploidies in sexual organisms.

The alternation of generations is not only a necessary consequence of sexuality; it is the *only* necessary consequence. The consequences of mixis in terms of diversity, heterozygosity and economy vary almost indefinitely, to the point at which mixis and amixis are functionally equivalent; but, in all mictic systems whatsoever, there is both a doubling and a compensatory reduction in the amount of chromatin. The alternation of generations therefore poses a problem which is second in importance only to the fundamental problem of mixis itself. But the phenomenon has been viewed by the great majority of biologists as merely an aid to taxonomy; and the casualness of the few attempts to provide a functional account of haploidy and diploidy constitutes a major scandal. Some early theories were reviewed by Svedalius (1927), but all are pre-Darwinian, in content if not in date. The attempts to provide an explanation during the fifty years following Svedalius' article have recently been summarized by Maynard Smith (1978) as follows: 'We do not know. Two possibilities are that diploidy protects against the deteriorative effects of somatic mutation, and that it in some way assists complex processes of cellular differentiation, but these are unsupported speculations'. I think the hollowness of these hypotheses needs no further comment.

Multicellularity. For convenience, let us begin by comparing two extreme types of life cycle: that in which the vegetative phase is haploid and the only diploid phase is a transient zygote (Figure 5.6b) with that in which the vegetative phase is diploid and the only haploid phase is a transient gamete (Figure 5.6c). These can be derived from a cycle in which both haploid and diploid phases are unicellular (Figure 5.6a) by the evolution of multicellularity in either the haploid

Figure 5.6: Life Cycles. A diagrammatic representation of the four principal sorts of life cycle. (a) Unicellular in both haploid and diploid generations, with a zygotic meiosis. (b) Multicellular in the haploid generation only, with a zygotic meiosis (haplont). (c) Multicellular in the diploid generation only, with a gametic meiosis (diplont). (d) Multicellular in both haploid and diploid generations, with a sporic meiosis (haplodiplont).

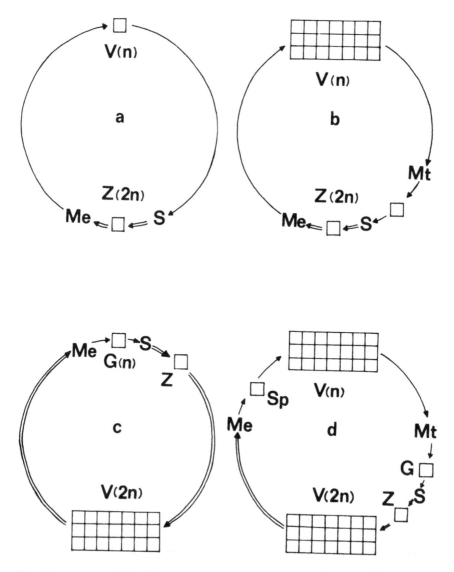

Key: V, vegetative phase; Z, zygote; G, gamete; Mt, mitosis; Me, meiosis; S, syngamy; Sp, spore; n, haploid; 2n, diploid. The unicellular condition is indicated by a single square, the multicellular by a rectangle divided into squares

or the diploid phase. Their evolution therefore raises two rather different questions: first, why should multicellularity evolve; and secondly, should it be the haploid or the diploid phase which becomes multicellular?

The evolution of multicellularity can be handled rather easily by the modern theory of life histories. A postponement of reproduction will be favourably selected, provided that it is associated with growth to a greater size, and therefore a greater fecundity, which more than compensates for any additional mortality. The advantage of the multicellular condition is created by a division of labour, which, by increasing the rate at which energy and nutrients are captured or the efficiency with which they are used, enables a group of cells to produce more reproductive propagules than the same number of cells in isolation. The primary division of labour in multicellular organisms is therefore between somatic and reproductive cells; and since there must exist some pathway for the transfer of energy and nutrients from somatic to reproductive cells, the most primitive multicellular organisms must have been coenobia. As evolution proceeds and the soma becomes larger, it will undergo a further division of labour into − for example − locomotory, attachment and photosynthetic tissue. This sequence is nicely exemplified by the chlorophyte Volvocales, which includes unicells, noncoenobial colonies, small primitive coenobia with no apparent division of labour, coenobia with distinct somatic and reproductive cells, and large advanced coenobial forms, with a further division of labour among the somatic cells. This does not necessarily imply that the common ancestors of multicellular plants and animals resembled the modern coenobial green algae, but it does illustrate how simple Darwinian arguments might be used to elucidate fundamental phylogenetic problems.

Haploidy and Diploidy. Let us imagine that, once multicellularity has evolved, it is restricted either to the haploid or to the diploid phase of the life cycle. Which of these two alternatives is selected must depend on some major consequence of vegetative ploidy. An organism which is vegetatively diploid produces diverse gametes through meiosis, which through syngamy return diverse diploid vegetative offspring; an organism which is vegetatively haploid produces identical gametes through mitosis, which fuse to form diverse diploid zygotes, and these, through meiosis, return diverse haploid vegetative offspring. I think that there are only two possible theories.

The first theory states that the vegetative progeny of an organism which is vegetatively haploid are less diverse than those of a comparable organism which is vegetatively diploid. This is because the zygotes which descend from a vegetatively haploid individual will tend to resemble one another, since all have at least one haploid genome in common. Meiosis will recombine and reassort this haploid genome, but its ability to do so is inevitably restricted by the fact that all the zygotes inherit the same set of chromosomes from the parent in question. By contrast, a diploid parent transmits a different set of chromosomes to each zygote. When we compare organisms with a zygotic and with a gametic meiosis, therefore, we expect to find less genetic variance within broods of progeny when the zygote is the site of meiosis; this will be balanced by the greater genetic variance between broods, so that the population as a whole will exhibit the same level of genetic variation whether meiosis is zygotic or gametic. Strictly speaking, this argument applies only to neutral genes, and it will be sensitive to the way in which genetic variation is actually maintained; for example, it is obvious that, if the bulk of genetic variation is maintained by heterosis, the overall level of variation will be much lower in vegetatively haploid than in vegetatively diploid organisms. I propose to deal with such objections by dismissing them with a broad wave of the arm, asserting that, for some set of population models, and especially for those in which sampling error and frequency-dependent selection are the dominant forces acting on gene frequencies, a gametic meiosis will be more effective in diversifying offspring genotypes than a zygotic meiosis. I have unearthed a theory of this sort from several texts of botany and phycology, but it appears not to have received any further development, nor even to have been plainly restated, since the classic paper by Svedalius (1927).

The second theory, which is to the best of my knowledge original, states that the function of a gametic meiosis is to diversify the gametes, without respect to any effect on the diversity of vegetative progeny. I assume that the genome will include a certain number of loci that are segregating for alleles which affect gamete function. Such genes can be maintained at stable equilibria if the allele which contributes less to gametic function contributes more to vegetative function, or if alternative alleles with different effects on gamete function experience heterosis or frequency-dependent selection in the vegetative phase. The gametes produced by a haploid individual will all be alike, and will all bear the same number of those alleles which elevate the competence of gametes to achieve syngamy. The gametes produced by a diploid individual will be diverse; genetic recombination during meiosis will create genetic variance for gamete function, so that some gametes will include very few superior alleles and will therefore suffer a loss of function, whilst others will bear very many superior alleles and will therefore be extremely competent as gametes.

The effect of the greater genetic variance of gamete function created by a gametic meiosis will depend on the intensity of competition between the gametes for fusion partners. Two extreme cases are illustrated in Figure 5.7. In the first place, suppose that there is very little competition: virtually all gametes will succeed in forming zygotes, provided that they have more than some minimal degree of gamete function (Figure 5.7a). In this situation, vegetatively diploid individuals will stand at a disadvantage, since a considerable proportion of their gametes may lack this minimal degree of function. On the other hand, suppose that there is intense competition between the gametes, so that only those with more than some high degree of function will succeed in forming zygotes (Figure 5.7b). In this case, vegetatively diploid individuals will stand at an advantage, since virtually all of the successful gametes will be those in which large numbers of alleles contributing to gametic function have been combined by meiosis. In

Figure 5.7: Gamete Competition and the Site of Meiosis. The curves are the frequency distributions of fitness dosage for gamete function when gametes are produced by mitosis (solid line) or meiosis (broken line). (a) If only a small proportion of very inferior sperm are incompetent to achieve fertilization, this proportion will be greater among meiotically diversified than among mitotically replicated sperm. (b) On the other hand, if gamete competition is so intense that almost all sperm have a vanishingly small probability of achieving fertilization, meiosis is the only way of producing a substantial number of highly competent gametes. The hatching is a reminder that the cut-off separating incompetent (NF) from competent (F) sperm will not be perfectly sharp.

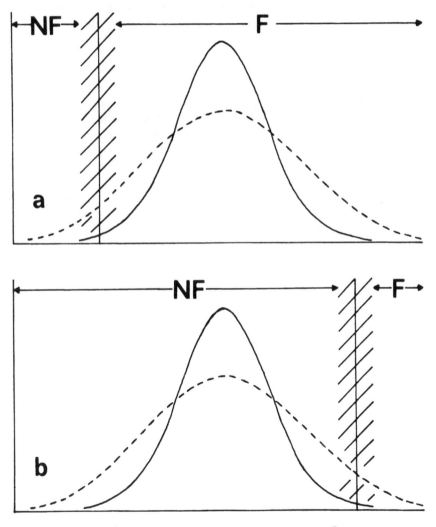

Fitness dosage for gamete function

this way, selection may favour vegetative haploidy or vegetative diploidy, according to the intensity of competition between gametes. If the hypothesis is correct, the trend towards the dominance of the diploid generation should be associated with a trend in gamete competition.

We must now attempt to falsify one or both of these rival hypotheses by comparative analysis. Unfortunately, multicellular animals are not suitable material, since all sexual taxa have a gametic meiosis; suitable comparative material is provided only by algae and protozoans, and to a lesser extent by fungi and primitive vascular plants. Rather than review the life cycles of these organisms at length, I have chosen, for reasons of time and space, to summarize them in Table 5.7, which includes rather more than 300 genera from all the major taxa of protozoans and algae, excluding the red algae. The use of a table such as this, drawn up from secondary sources by an amateur, is naturally open to cavil; but I found it impossible to proceed without some means of making objective and at least semiquantitative assessments of the predictions of the rival hypotheses. I have tried to avoid the more obvious pitfalls of working with such a diverse array of organisms by performing each test, wherever possible, at each of three phyletic levels: the genus, the family within the order Chlorophyceae and the order. The weakness of tests at generic level is that I am not satisfied that the number of genera listed for any given taxon is truly representative of the importance of that taxon; nor do I know what criteria should be applied in order to determine what number would be representative. The weakness of tests at familial or ordinal level is that I have been forced to assess 'typical' states of characters for these higher taxa, a procedure which is not always realistic. The ground rules for the analyses are as follows. Entries in tables are taken directly from Table 5.7 without modification; rare, occasional and aberrant values (given in parentheses in Table 5.7) are excluded; doubtful values (given with a question mark in Table 5.7) are included; where two values are given in Table 5.7, both are included, so that marginal totals in subsequent tables may slightly exceed the number of taxa available for comparison.

I shall distinguish the rival hypotheses as the 'vegetative' and 'gametic' hypotheses, and deal first with the vegetative hypotheses. The vegetative hypothesis states that gametic meiosis has the effect of increasing the genotypic diversity of vegetative progeny; from this I shall infer two general rules, each of which leads to specific predictions which can be tested with the data of Table 5.7.

The first rule is that the correlates of zygotic meiosis among algae and protozoans will be similar to the correlates of thelytoky among multicellular animals, since both have the effect of suppressing recombinational diversity. This rule leads to two specific predictions.

First, zygotic meiosis will prevail among vegetatively smaller, simpler types. The basis for the prediction can be confirmed by glancing through Table 5.7, which shows clearly that the taxa in which sexuality is rare or unknown are predominantly unicellular. In this test, and in subsequent tests unless otherwise noted, forms with a sporic meiosis should be intermediate between the extremes represented by zygotic and gametic meioses. In constructing a quantitative test, I have assessed vegetative development by calculating a 'vegetative index' as the product of entries in SIZ, DIF and LAB columns of Table 5.7. There is no particular reason for using this index, other than that it seems to correspond more or less to my intuitive assessment of the vegetative complexity of different forms; and in every analysis in which the vegetative index is used I have repeated the analysis using different indices (obtained by adding SIZ, DIF and LAB instead of multiplying them, or by omitting the LAB column, and so forth) without in any case obtaining a substantially different result. The test is shown in Table 5.8. It is indecisive. The highest degree of vegetative development is shown by taxa (largely of brown algae) with a sporic rather than a gametic meiosis; the Fucales is the only taxon with a gametic meiosis which clearly exceeds in vegetative development any taxon with a zygotic meiosis; and there are many unicellular taxa in which meiosis is gametic. However, the prediction is supported at the familial (but not the generic) level among Chlorophyceae.

Secondly, a zygotic meiosis will predominate among freshwater and a gametic meiosis among marine forms. Table 5.9 shows that this prediction is accepted at all levels of analysis. Freshwater and soil algae almost invariably have a zygotic meiosis, marine algae almost invariably a sporic or a gametic meiosis. The same result is obtained if we analyze chlorophyte and non-chlorophyte algae separately, or if we analyze unicells and multicells separately. The possibility that the true causal relationship lies between the marine habitat and the degree of vegetative development is excluded by Table 5.8. Protozoans are not included in Table 5.9 because they are so frequently symbiotic; their inclusion does not disturb the result, though the Heliozoa include two aberrant freshwater genera with gametic meiosis.

The second general rule is that a zygotic meiosis will be part of an adaptive complex of characters, all of which function to reduce the genotypic diversity of progeny; there will therefore be a correlation between the site of meiosis and the state of other characters which affect progeny diversity.

The first specific prediction concerns the number of products of gametogenesis. When meiosis is gametic, up to but not more than four unique haploid gametes may be produced from a single nucleus by a single meiosis. If the function of gamete diversification is to diversify the vegetative phase, then the mitotic replication of meiotically derived gametes would be dysfunctional. But when meiosis is zygotic all gametes are produced by mitosis, and there is no necessary restraint on the number of gametes which should descend from a single original nucleus. We are led to expect that the number of gametes elaborated from each germ-cell nucleus will be smaller when meiosis is gametic and larger when it is zygotic. I have attempted this analysis for isogametogenesis and microgametogenesis, but in both cases the data are too few and variable for the result to be decisive. Amongst isogametic taxa the number of gametes formed is small or moderate in genera with a zygotic meiosis (one in Polyblepharidaceae and most Polymastigina; two to 32 or 64 in other Volvocales and Tetrasporales)

Table 5.7: A Digest of Modes of Reproduction in some Lower Eukaryotes. The purpose of the table is to provide a body of comparative material adequate to test theories of the alternation of generations and of anisogamy. A glossary is given separately. In using the table, two caveats should be borne in mind. First, the attempt to force each character state into the straightjacket of a single symbol inevitably results in the loss of a considerable amount of detail and in the occurrence of some degree of distortion. Secondly, the table was composed entirely from secondary sources by someone who has no first-hand experience with any of the taxa; there may be howlers.

TEXTNAME: Tax (R)P: 01   TEXTNAME: Belitbl (R)P: 01

Tax	LCY	MEI	GAM	ISO	MIC	MAC	FUS	GER	STA	COM	ASX	MOT	IMM	MFN	DIV	HAP	SIZ	DIF	LAB	HAB
C 1 CHLOROPHYCEAE	v	v	v	v	v	i		v	v	v	v	v	v	v	v	v	v	v	v	Cos
O 1 VOLVOCALES	H	Z	v	v	v	1	Get	v	v	v	3	F,As		Cy,As	v		v	v	1	Fu
F 1 POLYBLEPHARIDACEAE	H	Z	1	1		1	Get	?			?	F		Cy	?		1	1	1	
G 1 Pedinomonas							Get					F			?		1	1	1	So,FuP
G 2 Dunaliella	H	Z	1H	1			Get					F			?		1	1	1	SalP
G 3 Pyramimonas							Get				Excl	F		Cy	?	'.	1	1	1	Fw,Mr,Mar
G 4 Polytomella			1H	1			Get					F			?					
G 5 Raciborskiella			1H	1			Get					F			?		2	2	1	
G 6 Dangeardinella			PA														2	2	1	BrP
F 2 CHLAMYDOMONADACEAE	H	Z	v	12	12	1	Get	4	C		3	F		Ak	?	v	1	1	1	Cos
G 7 Chlamydomonas	H	Z	1-0	16	16-64	1	Get	4(N)	C,1			F,As		Ak	?-N	N-3N	1	1	1	Cos
G 8 Carteria	H	Z	1,0						C,1						?-N		1	1	1	Fu
G 9 Chlorogonium	H	Z	1,0	12-64		1	Get					F		Ak	2-4	?0-30	1	1	1	So,Fw
G 10 Haematococcus			1									F			?		1	1	1	Fu
G 11 Polytoma																				
F 3 PHACOTACEAE											1	F			4		1	1	1	Fu
G 12 Phacotus											Excl	F			2-N		1	1	1	Fu
G 13 Pteromonas												F			2-4		1	1	1	Fu
G 14 Dysmorphococcus												F			4		1	1	1	Fu
F 4 ASTREPHOMENACEAE			1				Gol	1	D		3	Ac					2	3	3	Fu
G 15 Astrephomene			1	-			Gol	1	D			Ac				4-N	?	3	3	Fu
F 5 SPONDYLOMORACEAE			1								3	Ac					2	3	1	FuP
G 16 Pyrobotrys			1									Ac					2	3	1	FuP
F 6 VOLVOCACEAE	H	Z	v	16	many	1	v	2	v	1	4	Ac	(Ag)		2-3	2-3	2-3	3	1-2	FuP

TEXTNAME: Tax (R)P: 02   TEXTNAME: Belltbl (R)P: 02

Table cont'd

Tax		LCY	MEI	GAM	ISO	MIC	MAC	FUS	GER	STA	COM	ASX	MOT	IMM	RES	DIV	HAP	SIZ	DIF	LAB	HAB
G 17	Gonium	H	Z	I,PA				Get	4		I		Ac				10,17	2	3	1	PvP
G 18	Pandorina	H		I,PA									Ac				10	2	3	1	PvP
G 19	Volvulina	H	Z	I					1	D	I		Ac				7	2	3	1	PvP
G 20	Eudorina	H	Z	A		64	1	SP	1	D			Ac	(Ag)			10-28	2	3	1	PvP
G 21	Platydorina			O						D			Ac					3	3	2	PvP
G 22	Pleodorina	H	Z	A				SP	1	M,D			Ac	(Ag)			15	3	3	2	PvP
G 23	Volvox	?H	?Z	O	4-32	16-512	1	SP	4	M	I(c)		Ac	(Ag)			5-15	2	3	1	PvP
G 24	Stephanosphaera			I																	PvP
O 2	TETRASPORALES	?H	?Z	I	4			Get	4	D		2	Zo	(Ap)	Cy	v		2	2	1	Fw
F 7	PALMELLACEAE	?H		I				Get				2	Zo	(Ap)	Cy			2	2	1	Fw,Sub
G 25	Gloeococcus												Zo					5	2	1	FwB
G 26	Palmella			I				Get										2	2	1	Sub
G 27	Gloeocystis																	2	2	1	Sub,FwB
G 28	Asterococcus												Zo	Ap				2	2		
G 29	Hormotila												Zo					2	2	1	Sub,FwB
F 8	CHLORANGIACEAE			I				Get				2	Zo		Cy	v		2	v	1	Fw,Br,Mar
G 30	Prasinocladus												Zo		Cy			3	5	1	Br,MarB
G 31	Chlorangium												Sw			1		2	2	1	PvP,Epi
G 32	Malleochloris			I				Get					Sw				2-4	1	1	1	Epi
F 9	TETRASPORACEAE	?H	?Z	I	4			Get	4			2	Zo		Cy	4		3	3	1	Fw
G 33	Tetraspora	?H	?Z	I	2-9			Get	4-R	D			Zo					4	5	1	FwB
G 34	Apiocystis			I				Get		D			Zo					3	3	1	FwB
G 35	Paulschulzia												Zo					2	2	1	PvP
G 36	Schizochlamys												Zo			2-R		4	2		FwB
O 3	CHLOROCOCCALES	H	Z	I	8			GetE		M	v	4	Zo	Ap,Ac	Hy	8		v	v	1	Fw
F 10	CHLOROCOCCACEAE			I	8			Get				2	Zo	Ap		8		1	1	1	So,Fw
G 37	Chlorococcum			I	8+			Get					Zo	Ap		8+		1	1	1	So
G 38	Characium			I,A				Get					Zo					1	1	1	So,FwEpi

TEXTNAME: Tax (R)P: 03   TEXTNAME: Belltbl (R)P: 03

Table cont'd

Tax	LCT	MEI	GAM	ISO	MIC	MAC	FUS	GER	STA	COM	ASX	MOT	IMM	RES	DIV	HAP	SIZ	DIF	LAB	HAB
F 11 PROTOSIPHONACEAE	?H	?Z	I				GetE			C	4	Zo	Ap,Ag	Hy			3	4	1	Sub
G 39 Protosiphon	?H	?Z	I				GetE			C		Zo	Ap,Ag	Hy			3	4	1	Sub
F 12 CHARACIOSIPHONACEAE	H	Z	I				GetE			I	4	Zo	Ap				4	4	1	FwB
G 40 Characiosiphon	H	Z	I				GetE			I		Zo	Ap				4	4	1	FwB
F 13 HYDRODICTYACEAE	H	Z	I				GetE		M	C	5		Ac	Ap			3	4	1	Fw
G 41 Pediastrum			I										Ac				2	3	1	FwP
G 42 Sorastrum													Ac	(Hv)			2	3	1	Fw
G 43 Hydrodictyon	H	Z	I				GetE	4	M	C	(Unk)		Ac,Az	Ap			5	5	1	Fw
O 4 CHLOROSARCINALES											2	Zo	Ap				2	2	1	So
F 14 CHLOROSARCINACEAE			I								2	Zo	Ap				2	2	1	So
G 44 Chlorosarcinopsis			I									Zo	?Ak				2	2	1	So
G 45 Tetracystis			I									Zo					2	2	1	So
G 46 Pseudotetracystis			I									Zo					2	2	1	So
G 47 Planophila													Ap				2	2	1	So
O 5 CHLORELLALES	?H	?Z	O(I)		16	1	GetE				2	?Zo	Au		4		v	v		Fw
F 15 CHLORELLACEAE	?H	?Z	O		16	1	GetE				2	?Zo	Au		4		1	1		Cos
G 48 Chlorella											Excl		Au		4-8		1	1	1	Cos
G 49 Prototheca											?Excl		Au				1	1	1	
G 50 Golenkinia			O		8-16	1						?Zo	Au		2-8		1	1	1	FwP
G 51 Oocystis													Au		4-8		1-2	1-2		So,FwP
G 52 Eremosphaera			O		16-84	1					Excl		Au		2-4		1-2	1-2		
G 53 Ankistrodesmus													Au		2-16		1-2	1-2		So,FwP
F 16 SCENEDESMACEAE			v				GetE				3	?Zo	Au				2	1	1	Fw
G 54 Scenedesmus		I					GetE						Au				2	3	1	So,FwP
G 55 Coelastrum		I											Au				2	3	1	So,FwP
G 56 Dictyosphaerium			O									?Zo	Au				2	2	1	FwP

TEXTNAME: Tax (R)P: 04   TEXTNAME: Belltbl (R)P: 04

Table cont'd

Tax	LCY	MEI	GAM	ISO	MIC	MAC	PUS	GER	STA	COM	ASX	MOT	IMM	RES	DIV	HAP	SIZ	DIF	LAB	HAR
O 6 ULOTRICHALES	H	Z	V	2+			Get	4			3	Zo	Ap,Fr	Hy	4		4	4	2	Fw
F 17 ULOTRICHACEAE	H	Z	1				GetE	4			2	Zo	Fr,Ap	Hy	4		4	4	1	Cos
G 57 Ulothrix	H	Z	1				GetE	4+				Zo	Ap	Hy	1-8	5-14	4	4	2	Fw,MarR
G 58 Klebsormidium			1								?Excl	Zo	Ap,Fr				4	4	1	Sv/Fw
G 59 Geminella											?Excl		Fr				4	4	1	
G 60 Radiofilum											Excl		Fr				3	4	1	Fw
G 61 Stichococcus													Fr				3	4	1	So,Fw,Rr
F 18 MICROSPORACEAE			1	2-16			GetE				3	Zo	Ap,Fr	Hy,Ak			4	4	2	Fw
G 62 Microspora			1	2-16			GetE					Zo	Ap,Fr	Hy,Ak		8-32	4	4	2	Fw
F 19 CYLINDROCAPSACEAE			0	2	1		GetI				4	Zo	Ap,Fr		2		4	4	2	Fw
G 63 Cylindrocapsa		-	0	2	1		GetI					Zo	Ap,Fr		2-4	16	4	6	2	Fw
O 7 CHAETOPHORALES	V	V	V		1,2	1	Get				3	Zo	Ap	Hy,Ak	1		4	6	?	Fw
F 20 CHAETOPHORACEAE	V	V	1				Get				3	Zo		Hy,Ak			4	6	?	Fw
G 64 Chaetophora			1				Get					Zo		Ak		11,12	4	6	2	FwB
G 65 Stigeoclonium		?Z										Zo		Hy		8-20	4	6	2	FwB
G 66 Draparnaldia		S	1									Zo				13,14	4	6	3	FwB
G 67 Fritschiella	?H		1									Zo					3	5	2	Sv,SiA
G 68 Microthamnion	Iso										?Excl	Zo					3	5	2	Fw,So
G 69 Gongrosira							Get				?Excl	Zo		Ak		8	4	6	2	Fw/Cos
F 21 APHANOCHAETACEAE			A		2	1					2	Zo					4	5	2	Fw
G 70 Aphanochaete			A		2	1						Zo	Ap				4	5	2	FwB,Epi
F 22 COLEOCHAETACEAE	H	Z	0		1	1	GetI	15			3	Zo	Ap		1		4	5	3	Fw
G 71 Coleochaete	H	Z	0		1	1	GetI	8-32	M(D)			Zo	Ap		1	36,42	4	5	3	FwB,Epi
O 8 OEDOGONIALES	H	Z	0		2	1	GetI,Dm	4	M,D		4	Zo	Fr,Az	Ak	1		4	5	3	Fw

TEXTNAME: Tax (R)P: 05   TEXTNAME: Belltbl (R)P: 05

Table cont'd

Tax	LCY	MEI	GAM	ISO	MIC	MAC	FUS	GER	STA	COM	ASX	MOT	IMM	RES	DIV	HAP	SIZ	DIF	LAB	HAB
F 23 OEDOGONIACEAE	H	Z	O		2	1	Get,I,Dm	4	M,D		4	Zo	Fr,Az	Ak	1		4	5	3	Fv
G 72 Oedogonium	H	Z	O		2-4	1	Get,I,Dm	4	M,D			Zo	Fr,Az	?Ak	1	13-32	4	4	3	FvB,Epi
G 73 Bulbochaete	?H	?Z	O		2	1	Get,I,Dm		M,D			Zo		Ak	1		4	5	3	FvB,Epi
G 74 Oedocladium					2	1	Get,I,Dm		M,D			Zo					4	5	3	Fv,So
O 9 ULVALES	V	V	V				GetE		V		3	Zo	Fr,Ap	Ak	4		4	5	3	Cos
F 24 PERCURSARIACEAE	Iso	S	A						D		3	Zo	Ag				4	6	3	MarB
G 75 Percursaria	Iso	S	A						D			Zo	Ag				4	6	3	MarB
F 25 MONOSTROMATACEAE	Het	S	I				GetE		M,D		3	Zo	Fr,Ap	Ak	4		5	5	3	MarB
G 76 Monostroma[2]	Het	S	I,A				GetE		M,D			Zo	Fr,Ap	Ak	4-8		5	5	3	MarB,Epi
F 26 ULVACEAE	Iso	S	A				GetE		D		3	Zo,Ag	Fr	Ak	4		5	5	3	MarB
G 77 Enteromorpha[3]	Iso	S	A				GetE		D			Zo,Ag	Fr		4-8	9-16	4	5	3	MarB
G 78 Ulva[3]	Iso	S	A				GetE		D			Zo,Ag	Fr			10	5	5	3	MarB,Br
G 79 Ulvaria[3]	Iso	S	A				GetE		M			Zo,Ag					5	5	3	MarB
F 27 SCHIZOMERIDACEAE			A								2	Zo	Fr		1		4	5	1	Fv
G 80 Trichosarcina											?Excl	Zo					4	5	1	Fv,So
G 81 Schizomeris			A									Zo	Fr		1		4	5	1	Fv
F 28 PRASIOLACEAE	D	G	O				GetE		M		2		Fr,Ap	Ak			4	5	3	Cos
G 82 Prasiola	D	G	O				GetE		M				Fr,Ap	Ak		3-8	4	5	3	Fv,Mar,Sub
O 10 CLADOPHORALES	V	V	V				Get		D		2	Zo	Fr	Ak			4	4	2	Fv
F 29 CLADOPHORACEAE	Iso	S	I						D		2	Zo,Ag	Fr	Ak			4	4	3	Fv,Mar

TEXTNAME: Tax (R)P: 06   TEXTNAME: Bel1tb1 (R)P: 06

Table cont'd

Tax	LCY	MEI	GAM	ISO	MIC	MAC	FUS	GER	STA	COM	ASX	MOT	IMM	RES	DIV	HAP	SIZ	DIF	LAB	HAB
G 83 Rhizoclonium[4]	Iso	S	I				Get		D			Zo	Fr				11-18	4	4	2	Fv,Br,Mar
G 84 Chaetomorpha	Iso	S	I				Get					Zo,Ag	Fr				10-18	4	4	3	Fv,Br,Mar
G 85 Cladophora[5]	Iso	S	I				Get		D			Zo	Fr	Ak			12-18	4	5	3	Fv,Br,Mar
G 88 Pithophora	Iso						Get				Excl			Ak			12	4	5	3	Fv
F 30 SPHAEROPLEACEAE	?H	?Z	A			1	GetI,E	4			3		Fr					4	4	2	Fv
G 89 Sphaeroplea	?H	?Z	A,O			1	GetI,E	4					Fr				10-16	4	4	2	Fv
O 11 ACROSIPHONIALES	HetH	S	A				GetE		D		2	Zo		Ak				4	5	2	MarB
F 31 ACROSIPHONIACEAE	HetH	S	A				GetE		D		2	Zo		Ak				4	5	2	MarB
G 90 Acrosiphonia[6]	Alt	S	A				GetE		D		(Excl)	Zo				6		4	5	2	MarB
G 91 Urospora	?HetH	S	A						D		(Excl)	Zo,Ag		Ak				4	5	2	MarB
G 92 Spongomorpha	HetH	S	I						D							6		4	5	2	MarB
O 12 CAULERPALES	D	G	A				Get		v		2	Zo	Fr,Ap	Ak				5	6	3	MarB
F 32 CODIACEAE	D	G	A			1	GetE		D		2		Ag					5	6	3	MarB
G 93 Codium	D	G	A			1	GetE		D(M)				Ag				10,20	5	6	3	MarB
F 33 UDOTACEAE	?D	?G	A				GetE	1	D		2							5	6	3	MarB
G 94 Halimeda	?D	?G	A				GetE											5	6	3	MarB
G 95 Udotea		?G	A				GetE	1	D									5	6	2	MarB
F 34 CAULERPACEAE	D	G	A				GetE	1	?D		2		Fr					5	7	2	MarB
G 96 Caulerpa	D	G	A				GetE	1	?D				Fr					5	7	2	MarB
F 35 BRYOPSIDACEAE	v	v	A				GetE	1	v		4		Fr					5	7	3	MarB
G 97 Bryopsis[7]		G,S	A				GetE	1	D				Fr					5	7	3	MarB
G 98 Pseudobryopsis		G,S	A				GetE	1	M									5	7	3	MarB
F 36 DERBESIACEAE	HetD	S	A				Get		D		4	Zo						5	6	3	MarB

TEXTNAME: Tax (R)P: 07   TEXTNAME: Belltbl (R)P: 07

Table cont'd

Tax	LCY	MEI	GAM	ISO	MIC	MAC	FUS	GER	STA	COM	ASX	MOT	IMM	RES	DIV	HAP	SIZ	DIF	LAB	HAB
G 99 Derbesia	Iso	S	A				Get		D		4	Zo					5	6	3	MarB
F 37 PHYLLOSIPHONACEAE											1		Ap				3	4	1	Par
G 100 Phyllosiphon											Excl		Ap				3	4	1	Par
F 38 DICHOTOMOSIPHONACEAE			0				GetI		M		4			Ak			4	5	3	Fw
G 101 Dichotomosiphon			0				GetI		M					Ak			4	5	3	Fw
O 13 SIPHONOCLADALES	v	v	I						?M		2	Zo		Cy			5	5	2	MarR
F 39 SIPHONOCLADACEAE											2	Sw		Cy			5	5	2	MarR
G 102 Siphonocladus												Sw		Cy		8	5	5	2	MarR
F 40 VALONIACEAE	?D	?G	I				Get				2	Sw					5	5	2	MarR
G 103 Valonia	?D	?G	I				Get					Sw				14,16	5	5	2	MarR
F 41 ANADYOMENACEAE	Iso	S	I				Get		M		3	Zo					5	5	3	MarR
G 104 Anadyomene	Iso	S	I				Get		M			Zo					5	5	3	MarR
O 14 DASYCLADALES	?D	?G	I				GetE	1	D	I	4			Cy			5	5	3	MarR
F 42 DASYCLADACEAE			I				GetE	1	D	I	4			Cy			5	5	3	MarR
G 105 Dasycladus			I				GetE	1	D	I				Cy			5	5	3	MarR
F 43 ACETABULARIACEAE	?D	?G	I				GetE		D	?I	4			Cy			5	5	3	MarR
G 106 Acetabularia	?D	?G	I				GetE		D	?I				Cy		4-10	5	5	3	MarB
O 15 ZYGNEMATALES	H	Z	I	1	1		Gia	v	V		3		F,Fr		2,4		1,4	1,4	1	Fw
F 44 ZGNEMATACEAE	H	Z	I	1	1		Gia	1		C	4		Fr,Ak				4	4	1	Fw

Table cont'd

Tax	LCY	MEI	GAM	ISO	MIC	MAC	FUS	GER	STA	COM	ASX	MOT	IMM	RES	DIV	HAP	SIZ	DIF	LAB	HAB
G 107 Spirogyra	H	Z	I	1			Gia	1		C,I			Fr			4-52	4	4	1	Fv
G 108 Sirogonium	?H	?Z	A			1	Gia	1		C							4	4	1	Fv
G 109 Zygnema	?H	?Z	I	1	1		Gia			C			Ak			16-82	4	4	1	Fv
G 110 Mougeotia	?H	?Z	I	1		1	Gia			C						35-94	4	4	1	Fv
F 45 MESOTAENIACEAE	H	Z	I	1			Gia	4			4		F		2		1	1	1	FwP
G 111 Mesotaenium	H		I	1			Gia	2-4					F		2		1	1	1	FwP
G 112 Spirotaenia		Z	I	1			Gia	2-4					F		2		1	1	1	FwP
G 113 Cylindrocystis	?H	?Z	I	1			Gia	4								20,80	1	1	1	FwP
G 114 Netrium	H	Z	I	1			Gia	4	M,D				F		2	122-592	1	1	1	FwP
F 46 DESMIDIACEAE	H	Z	I	1			Gia	2	V		2		F		4		1	1	1	FwP
G 115 Closterium	H	Z	I	1(2)			Gia	2	M,D				F	(Ak)	4	60-220	1	1	1	FwP
G 116 Cosmarium	H	Z	I	1			Gia	2	D,?M				F		4	10-94	1	1	1	FwP
G 117 Micrasterias			I	1			Gia	2	M				F		4	34-200	1	1	1	FwP
G 118 Staurastrum			I	1			Gia	1-4					Fr			14-103	1	1	1	FwP
G 119 Hyalotheca	H	Z	I	1			Gia	2					Fr			12-112	4	4	1	Fv
G 120 Desmidium			I				Gia						Fr			28,33	4	4	1	Fv
C 2 CHAROPHYCEAE	?H	?Z	0		c150	1	GetI	1	V		5		Fr				5	7	4	FwB
O 16 CHARALES	?H	?Z	0		c150	1	GetI	1	V		5		Fr	Bu			5	7	4	FwB
G 121 Chara	?H	?Z	0		c150	1	GetI	1	M,D				Fr,Az				5	7	4	FwB
G 122 Nitella	?H	?Z	0		c150	1	GetI	1					Fr				5	7	4	FwB
G 123 Tolypella			0				GetI						Fr				5	7	4	FwB
C 3 EUGLENOPHYCEAE											1	F		Cy	2		1	1	1	Fv,Mar
O 17 EUTREPTIALES											1	F			2		1	1	1	Mar,Br
G 124 Eutreptia											Excl	F		Cy	2	44-90	1	1	1	Mar,Br
O 18 EUGLENALES		?I									1	F			2		1	1	1	Fv,Mar

TEXTNAME: Tax (R)P: 09   TEXTNAME: Belltbl (R)P: 09

Table cont'd

Tax	LCY	MEI	GAM	ISO	MIC	MAC	FUS	GER	STA	COM	ASX	MOT	IMM	RES	DIV	HAP	SIZ	DIF	LAB	HAB
G 125 Euglena			??I					?? 4			?Excl	F		Cy	2	42-86	1	1	1	Fw
G 126 Astasia											Excl	F		Cy	2	18	1	1	1	Fw,Mar
G 127 Trachelomonas											Excl	F		Cy	2	25-601	1	1	1	Fw,Epi
G 128 Colacium										?Aut	Excl	F		Cy	2	35	1	1	1	Fw,Epi
G 129 Phacus			?I				?GN				?Excl	F			2	4-2	1	1	1	Fw(Mar)
G 130 Hyalophacus											?Excl	F				92	1	1	1	Fw(Mar)
O 19 HETERONEMATALES											1	F			2		1	1	1	
G 131 Peranema											Excl	F			2	177	1	1	1	
C 4 PHAEOPHYCEA	Alt	S	V			V		V	V		4	Zo	Ag,Fr		V		5	6	4	MarB
O 20 ECTOCARPALES	Iso	S	I				Get		D		4	Zo,Ag					4	5	3	MarB
G 132 Ectocarpus	Iso	S	I,A				GetE		D			Zo,Ag					4	5	3	MarB,Epi
G 133 Pilayella	Iso	S	I				GetE					Zo,Ag					4	5	3	MarB
G 134 Sorocarpus												Zo								MarB
O 21 RALFSIALES	?Iso										3	Zo					5	5	3	MarB
G 135 Ralfsia	?Iso											Zo					5	5	2	MarEpi
G 136 Analipus	?Iso											Zo					5	6	3	MarB
O 22 Chordariales	Alt	S	I				GetE				3	Zo					5	5	3	Mar
G 137 Lithoderma	Iso	S	I				GetE					Zo					4	5	3	MarEpi
G 138 Nemoderma	Iso	S	A				GetE					Zo					4	5	3	MarEpi
G 139 Myrionema	HetD	S	I				GetE					Zo					4	5	3	MarEpi
G 140 Elachista	HetD	S									?Excl	Zo					5	6	3	MarFpi
G 141 Leathesia	HetD	S															5	5	3	MarB
G 142 Chordaria	HetD	S	?A								?Excl	Zo,Ag					5	6	4	MarB
G 143 Heterochordaria	Iso	S	I														5	6	4	MarB
G 144 Mesogloea	HetD	S	I														5	6	4	MarB
O 23 SPOROCHNALES	HetD	S	O				GetI		V		3	Zo	Ag				5	6	5	MarB

TEXTNAME: Tax (R)P: 10   TEXTNAME: Belltbl (R)P: 10

Table cont'd

Tax	LCY	MEI	GAM	ISO	MIC	MAC	FUS	GER	STA	COM	ASX	MOT	IMM	RES	DIV	HAP	SIZ	DIF	LAB	HAR
G 145 Sporochnus	HetD	S	O				GetI		M,D			Zo	Ag			20	5	6	5	MarB
G 146 Carpomitra	HetD	S	O				GetI		M			Zo	Ag				5	6	5	MarB
G 147 Nereia	HetD	S	O				GetI		D			Zo					5	6	5	MarB
O 24 DESMARESTIALES	HetD	S	O		1	1	Get(1)		V		5	Zo					5	6	5	MarB
G 148 Desmarestia	HetD	S	O		1	1	Get(1)		M,D	??C		Zo					5	6	5	MarB
O 25 CUTLERIALES	Alt	S	A				GetE	8	V		5	Zo					5	6	5	MarB
G 149 Cutleria	HetH	S	A				GetE	8+	D(M)			Zo	Ag				5	6	5	MarB
G 150 Zanardinia	Iso	S	A				GetE	4	M			Zo	?Ag				5	6	5	MarB
O 26 SPHACELARIALES	Iso	S	A				GetE		D		4	Zo	Fr				5	6	4	Mar
G 151 Sphacelaria	Het(D)	S	A				GetE	1	D(M)			Zo,Ag	Fr				5	6	4	MarEpi
G 152 Cladostephus	Iso	S	I				GetE	1				Zo	Fr				5	6	4	Mar
G 153 Halopteris	Iso	S	I,?O				GetE	1				Zo	Fr				5	6	4	Mar
O 27 TILOPTERIDALES	Iso	S	O				GetE	1			3		Msp,Fr				4	5	3	Mar
G 154 Haplospora	Iso	S	O				GetE	1					Msp,Fr				4	5	3	Mar
O 28 DICTYOTALES	Iso	S	O		c1500		GetE	1	D		5		Te		4		5	6	4	MarB
G 155 Dictyota	Iso	S	O		c1500	1	GetE	1	D				Ag,Te		4	14–32	5	6	4	MarB
G 156 Dilophus	Iso	S	O				GetE						Te		4		5	6	4	MarB
G 157 Padina	Iso	S	O				GetE		D(M)				Te,Ag		4	32	5	6	4	MarB
G 158 Zonaria	Iso	S	O				GetE						Te		8		5	6	4	MarB
O 29 DICTYOSIPHONALES	HetD	S	I				GetE				4	Zo					5	6	3	MarB
G 159 Myriotrichia	HetD	S	I				GetE					Zo					4	5	3	MarEpi
G 160 Stictyosiphon	HetD	S	I				GetE					Zo				26	4	5	3	MarB
G 161 Punctaria	?Het	?S	?I				GetE				?Excl	Ag,Zo					5	6	3	MarEpi
G 162 Soranthera	HetD	S	I,A				GetE					Zo					5	6	3	MarEpi
G 163 Phloeospora	HetD	S	I				GetE					Zo					5	6	3	Mar
G 164 Dictyosiphon	HetD	S	I				GetE					Zo				18	5	6	3	MarB

TEXTNAME: Tax (R)P: 11   TEXTNAME: Belltbl (R)P: 11

Table cont'd

Tax	LCY	MEI	GAM	ISO	MIC	MAC	FUS	GER	STA	COM	ASX	MOT	IMM	RES	DIV	HAP	SIZ	DIF	LAB	HAB
O 30 SCYTOSIPHONALES	HetD	S	1				GetE		D		3	Zo					4	5	3	MarB
G 165 Petalonia	Het	S	A				?GetE				?Excl	Zo,AR					4	5	3	MarEpi
G 166 Scytosiphon	HetD	S	I				GetE		D			Zo,AR					4	4	3	MarB
G 167 Aspercoccus	HetD	S					GetE					Zo				R	4	5	3	MarB
O 31 LAMINARIALES	HetD	S	O				Get(I)		D		4	Zo			32		6	7	5	MarB
G 168 Chorda	HetD	S	O				Get(1)		D,M			Zo			16	28,30	6	6	5	MarB
G 169 Laminaria	HetD	S	O				Get(1)		D			Zo			32	8-31	6	7	5	MarB
G 170 Hedophyllum	HetD	S	O				Get(1)					Zo					6	7	5	MarB
G 171 Saccorhiza	HetD	S	O				Get(1)		D			Zo			128	31	6	7	5	MarR
G 172 Nereocystis	HetD	S	O				Get?I					Zo				31	6	7	5	MarR
G 173 Macrocystis	HetD	S	O				Get(1)					Zo				16	6	7	5	MarR
G 174 Alaria	HetD	S	O				Get(1)					Zo			32	22-28	6	7	5	MarR
G 175 Eisenia	HetD	S	O				Get(1)					Zo			32(44)		6	7	5	MarR
O 32 FUCALES	D	G	O		64	V	GetE	1	V		5		Fr				5	7	5	MarB
G 176 Fucus	D	G	O		64	8	GetE	1	D,M							32	5	7	5	MarB
G 177 Ascophyllum	D	G	O			4	GetE	1	D							32	5	7	5	MarB
G 178 Pelvetia	D	G	O			2	GetE	1	D,M				Fr			22,32	5	7	5	MarB
G 179 Sargassum	D	G	O			1	GetE	1	D,M		(Excl)					32	6	7	5	Mar
G 180 Cystoseira	D	G	O			1	GetE	1	M				Fr				5	7	5	MarR
C 5 CHRYSOPHYCEAE	V	v	I					4			1	F,Zo		Cy	2		1	1	1	Fw,Mar
O 33 OCHROMONADALES	?H	?Z	I				Get				2	F		Cy,St	2		1	1	1	FwP
G 181 Ochromonas	?H	?Z	I				Get			(Aut)		F		Cy	2		1	1	1	FwP(Mar)
G 182 Mallomonas	?H	?Z	IH				Get					F		Cy	2		1	1	1	FwP
G 183 Synura			IH														2	2	1	Fw
G 184 Dinobryon			I									F		Cy,St	2-4		2	2	1	Fw
O 34 PHAEOTHAMNIALES											1	Fo		Cy	4		2	2	1	Fw
G 185 Phaeothamnion											?Excl	Zo		Cy	4-8		2	2	1	Fw,Epi

TEXTNAME: Tax (R)P: 12   TEXTNAME: Belltbl (R)P: 12

Table cont'd

Tax	LCY	MEI	GAM	ISO	MIC	MAC	FUS	GER	STA	COM	ASX	MOT	IMM	RES	DIV	HAP	SIZ	DIF	LAB	HAB
O 35 CHROMULINALES											1	F		Cy	2		1	1	1	Fw,Mar
G 186 Chromulina												F		Cy	2		1	1	1	Fw,Mar
O 36 DICTYOCHALES											1	F		Cy	2		1	1	1	MarP
G 187 Distephanum											Excl	F		Cy	2		1	1	1	MarP
O 37 CHRYSOSPHAERALES											1	F,Zo	Au		2		1	1		
G 188 Chrysosphaera											Excl	F,Zo	Au		2		1	1		
O 38 PRYMNESIALES	Het	S	I				Get	4			2	F,Zo			2		1	1	1	Fw,Mar
G 189 Chrysochromulina											Excl	F			2		1	1	1	Fw,Mar
G 190 Prymnesium											Excl	F			2		1	1	1	Fw,Mar
G 191 Phaeocystis											?Excl	F,Zo			2		2	2	1	Mar
G 192 Ochrosphaera	H	Z	I				Get	4				Zo					2	1	1	Fw
G 193 Hymenomonas	Het	S	I				Get	4				F,Zo			2		1	1	1	
C 6 XANTHOPHYCEAE	H	Z	V			I	GetI	1	M	C	2	F,Zo	v	v	v		v	v	v	Cos
O 39 HETEROCHLORIDALES											1	F		Cy	2		1	1	1	BrP
G 194 Olisthodiscus											Excl	F			2		1	1	1	Br,FwP
G 195 Chloramoeba											Excl	F		Cy	2		1	1	1	BrP
G 196 Heterochloris											Excl	F		Cy	2		1	1	1	BrP
O 40 HETEROGLOEALES											1	Zo	F,Ap	Cy	2		2	2	1	Fw
G 197 Chlorosaccus											Excl	Zo	F	Cy	4		2	2	1	Fw
G 198 Botryococcus											Excl	?Zo	F,Ap		2		2	2	1	Fw
O 41 RHIZOCHLORIDALES											1	Zo					1	1	1	FwEpi
G 199 Stipitococcus											Excl	Zo					1	1	1	FwEpi
O 42 MISCHOCOCCALES											1	Zo	Ap,Au	Hy	v		1	1	1	Fw,MarP

TEXTNAME: Tax (R)P: 13   TEXTNAME: Belltbl (R)P: 13

Table cont'd

Tax	LCY	MEI	GAM	ISO	MIC	MAC	FUS	GER	STA	COM	ASX	MOT	IMM	RES	DIV	HAP	SIZ	DIF	LAB	HAB
G 200 Mischococcus											Excl	Zo	Ap	Hy	1-2		3	5	1	FwP,So
G 201 Botrydiopsis											Excl	Zo	Au		?32		1	1	1	FwB
G 202 Characiopsis											?Excl	Zo	Ap,Au	Cy	16-32		1	1	1	FwP
G 203 Ophiocytium											Excl	Zo	Ap	Hy	4-8		1	1	1	MarP
G 204 Halosphaera											Excl	Zo	Ap		8		1	1	1	FwP
G 205 Tetraedriella															4					
O 43 TRIBONEMATALES			I								2	Zo	Ap,Fr	St,Ak	1		4	4	1	Fw
G 206 Tribonema												Zo	Fr,Ap	St	1-2		4	4	1	Fw
G 207 Monocilia			I									Zo	Ap	Ak			2	2	1	So
O 44 VAUCHERIALES	H	Z	V			1			M	C	3	Zo	Ap	Cy	1		4	5	3	Sub
G 208 Botrydium	H	Z	I,A						M	C		Zo	Ap	Cy			4	6	2	Sub
G 209 Vaucheria			O			1	GetI	1	M(D)	?C		Zo	Fr,Ap		1		5	5	3	Sub
C 7 CHLOROMONADOPHYCEAE		-									1	F			2		4	1	1	Fw
O 45 CHLOROMONADALES											1	F			2		1	1	1	Fw
G 210 Gonyostomum											Excl	F			2		1	1	1	Fw
C 8 BACILLARIOPHYCEAE	D	G	V		2,4	1,2	V	1	M	C	4		F		2		1	1	1	Fw,MarP
O 46 CENTRALES	D	G	O		4	1	Get	1	M	C	4		F		2		1	1	1	Fw,MarP
G 211 Cyclotella	D	G	O		4	1	GN	1	M	Aut			F		2		1,2	1,2	1,2	Fw
G 212 Melosira	D	G	O		4	1	GN,GetI	1	M	C,Aut			F		2		2	4	1	Mar,Fw
G 213 Stephanopyxis	D	G	O		4	1	GetI	1	M	C			F		2		1,2	1	4	MarP
G 214 Biddulphia	D	G	O		4	2	Get(I)	1	M	C			F		2		1	1	1	MarP
G 215 Lithodesmium	D	G	O		4	2	GetE	1	M	C			F		2		1	1	1	MarP
G 216 Streptotheca	D	G	O		4	1	GetE	1	D,M	C			F		2		2	4	1	MarP
G 217 Chaetoceros	D	G	O		4		GetI	1	M	C			F	Cy	2		2	4	1	MarP
O 47 PENNALES	D	G	I		2	1	Gia	1	M	C	4		F		2		1	1	1	MarP

TEXTNAME: Tax (R)P: 14   TEXTNAME: Belltbl (R)P: 14

Table cont'd

Tax	LCY	MEI	GAM	ISO	MIC	MAC	FUS	GER	STA	COM	ASX	MOT	IMM	RES	DIV	HAP	SIZ	DIF	LAB	HAB
G 218 Rhabdonema	D	G	A		2	1	GetI	1	D	C			F		2		2	4	1	MarP
G 219 Gomphonema	D	G	1				Gia	1	M	C			F		2		1	1	1	MarP
G 220 Navicula	D	G	1				Gia	1	M	C			F		2		1	1	1	MarP
G 221 Eunotia	D	G	1				Gia	1	M				F		2		2	4	1	MarP
G 222 Amphora	D	G	1				Gia,(GN)	1	M	C(Aut)			F		2		1,2	1	1	MarP
G 223 Cymbella	D	G	1				Gia(GN)	1	M	C(Aut)			F,Ag		2		1	1	1	MarP
G 224 Cocconeis	D	G					Gia	1	M	C			F,Ag		2		2	4	1	MarP
C 9 DINOPHYCEAE	H	Z	1				Get	V			2	F,Zo	F,Au,Scy	Cy	V		1	1	1	Mar
O 48 BLASTODINIALES											1	Zo	Scy		c400		1	1	1	Mar,Par
G 225 Blastodinium											?Excl	Zo	Scy		c400		1	1	1	Mar,Par
O 49 DINAMOEBALES											1	Zo		Cy	4		1	1	1	Mar
G 226 Dinamoebidium												Zo		Cy	4-8		1	1	1	Mar
O 50 DINOPHYSIALES											1		F		2		1	1	1	MarP
G 227 Dinopysis											?Excl		F		2		1	1	1	MarP
G 228 Ornithocercus											?Excl		F		2		1	1	1	MarP
O 51 DINOTRICHALES											1	Zo	F		2		2	2	1	Mar
G 229 Dinothrix												Zo	F		2		2	2	1	Mar
G 230 Dinoclonium												Zo			1		2	6	1	Mar,Epi
O 52 GLOEODINIALES											1	Zo					2	2	1	Fw
G 231 Gloeodinium												Zo					2	2	1	Fw
O 53 GYMNODINIALES	H	Z	1				GetE	4			3	F			2		1	1	1	Fw,Br,Mar
G 232 Gymnodinium	?G	?Z	1				GetE	4				F			2	44-64	1	1	1	Fw,Br,Mar
G 233 Woloszynskia	?H	?Z	1				GetE	4				F			2		1	1	1	
G 234 Amphidinium	H	Z	1				GetE	4				F			2	24-34	1(2)	1	1	Fw,Br,Mar

TEXTNAME: Tax (R)P: 15   TEXTNAME: Belltbl (R)P: 15

Table cont'd

Tax	LCY	MEI	GAM	ISO	MIC	MAC	FUS	GER	STA	COM	ASX	MOT	IMM	RES	DIV	HAP	SIZ	DIF	LAB	HAB
O 54 NOCTILUCALES	?D	?G	I	c1000			GetE	1	M	C	2	F			2		1	1	1	MarP
G 235 Noctiluca	?D	?G	I	c1000			GetE	1	M	C		F			2		1	1	1	MarP
O 55 PERIDINIALES	H	Z	I		1	1	V	1,2			3	Zo	F	Cy	2		1	1	1	Mar,Fw
G 236 Peridinium	H	Z	I		1	1	Get	1				Zo	F	Cy	2	44	1	1	1	Mar,Fw
G 237 Ceratium	H	Z	A				Gia	2					F		2	200;274	1(2)	1	1	Mar,Fw
O 56 PHYTODINIALES	?H	?Z	?I				?Get	?4			2	Zo	F,Au		2		1	1	1	So
G 238 Cystodinium	?H										?Excl	Zo	Au		2-4		1	1	1	So
G 239 Hypnodinium	?H	?Z	?I				?Get	?4				Zo	F		2		1	1	1	So
O 57 PYROCYSTALES											1	Zo	Au		4		1	1	1	MarP
G 240 Pyrocystis											?Excl	Zo	Au				1	1	1	MarP
G 241 Dissodinium											?Excl	Zo	Au		4-8		1	1	1	MarP
C 10 DESMOPHYCEAE											1	F,Zo	F	Cy	2		1	1	1	Fw,Br,Mar
O 58 DESMOCAPSALES											1	Zo	F	Cy			2	2	1	MarEpi
G 242 Desmocapsa											?Excl	Zo	F	Cy			2	2	1	MarEpi
O 59 PROROCENTRALES											1	F			2		1	1	1	Fw,Br,Mar
G 243 Prorocentrum											?Excl	F			2(4)	24,68	1	1	1	Fw,Br,Mar
C 11 CRYPTOPHYCEAE											1	F			2		1	1	1	Fw,Mar
O 60 CRYPTOMONADALES											1	F			2		1	1	1	Fw,Mar
G 244 Cryptomonas											?Excl	F			2	42-209	1	1	1	Fw,Mar
C 12 FLAGELLATA (Protozoa)	V	V	V	1	1	1	GetE	1			3	F	F	Cy	2		1	1	1	Sym

Table cont'd

Tax	LCY	MEI	GAM	ISO	MIC	MAC	FUS	GER	STA	COM	ASX	MOT	IMM	RES	DIV	MAP	SIZ	DIF	LAB	HAB
O 61 PROTOMONADINA																				
G 245 Leishmania											Excl	F		Cy	2		1	1	1	Par
G 246 Leptomonas											Excl	F		Cy	2(4)		1	1	1	Par
G 247 Crithidia											Excl	F		Cy	2		1	1	1	Par
G 248 Trypanosoma											Excl	F			2(4)	3	1	1	1	Par
G 249 Bodo											Excl	F		Cy	2		1	1	1	Par
G 250 Pleuromonas											Excl	F		Cy	4-8		1	1	1	Par
O 62 POLYMASTIGINA	V	v	v	1	1	1	GetE				4		F		2		1	1	1	Sym
G 251 Oxymonas	H	Z*	I	1			GetE			(Aut)			F		2		1	1	1	Sym
G 252 Saccinobaculus	H	Z*	I	1			GetE			(Aut)			F		2		1	1	1	Sym
G 253 Notila	D	G*	I*				Gon						F		2	14	1	1	1	Sym
G 254 Euconomympha	H	Z	A(PA)		1		GetE								2	22-26	1	1	1	Sym
G 255 Trichonympha	H	Z	A(PA)	1	1		GetE						F		2	10	1	1	1	Sym
G 256 Leptospironympha	H	Z	I		1		GetE			(Aut)			F		2	26	1	1	1	Sym
G 257 Barbulanympha	H	Z	A(PA)				GN			Aut			F		2	19	1	1	1	Sym
G 258 Rhynchonympha	D	G					GN			Aut			F		2	8	1	1	1	Sym
G 259 Urinympha	D	G*			1		GetE						F		2		1	1	1	Sym
G 260 Macrospironympha	D -	G	A			1	GetE								2		1	1	1	Sym
O 63 OPALININA	D	G	A				GetE	1			3		F		2		1	1	1	Par
G 261 Opalina	D	G	A				GetE	1					F		2		1	1	1	Par
C 13 RHIZOPODA	V	I	I	V			V	1		V	2	F,Sw	F		2		1	1	1	Cos
O 64 AMOEBINA	V	I	I					1			1	F			2		1	1	1	Fw,So,Mar
G 262 Amoeba	Dk	IH	IH					1			Excl	F			2	c500	1	1	1	Fw,So
G 263 Sappinia												F			2		1	1	1	Fw
G 264 Thecamoeba											Excl	F			many		1	1	1	Fw,Mar
G 265 Entamoeba											Excl	F			24	4-12	1	1	1	Par
G 266 Pelomyxa											Excl	F								
O 65 TESTACEA		I									1		F		2		1	1	1	Fw

TEXTNAME: Tax (R)P: 17   TEXTNAME: Bellcb1 (R)P: 17

Table cont'd

Tax	LCY	MEI	GAM	ISO	MIC	MAC	FUS	GER	STA	COM	ASX	MOT	IMM	RES	DIV	HAP	SIZ	DIF	LAB	HAB
G 267 Euglypha											Excl		F		2		1	1	1	Fw
G 268 Arcella											Excl		F		2		1	1	1	Fw
G 269 Cromia			I										F		2		1	1	1	Fw
O 66 FORAMINIFERA	Het	S	I	v			v		D	v	4	Az					1	1	1	MarP
G 270 Alloxromia	Het	S	I				GetI			Aut		Az					1	1	1	MarP
G 271 Iridia	Het	S	I	c10¹⁷			GetF					Az				10	1	1	1	MarP
G 272 Rubratella	Het	S	PA				GonPA			I		Az					1	1	1	MarP
G 273 Patellina	Het	S	I	c10			Gon		D	I		Az			16	8	1	1	1	MarP
G 274 Rotaliella	Het	S	I				GetI			Aut		Az			c3	24	1	1	1	MarP
G 275 Metarotaliella	Het	S	I	c250			GonA					Az				9,18	1	1	1	MarP
G 276 Dicorbis	Het	S	I				Gon,Get			?I							1	1	1	MarP
G 277 Glabratella	Het	S	I				GonI					Az				6	1	1	1	MarP
G 278 Myxotheca	Het	S	I				GetE			C		Az				9	1	1	1	MarP
G 279 Tretomphalus	Het	S	I				GetE					Az					1	1	1	MarP
O 67 HELIOZOA	D	G	I				GetI	1		Aut	2	Sw	F		2		1	1	1	FwP
G 280 Actinosphaerium	D	G	I				GetI	1		Aut		Sw	F		2		1	1	1	FwP
G 281 Actinophrys	D	G	I				GetI(Gon)	1		Aut						22	1	1	1	FwP
G 282 Clathrulina												Sw	F,B		2,20		1	1	1	FwP
G 283 Acanthocystis																	1	1	1	FwP
O 53 RADIOLARIA											1	Sw	F		2		1	1	1	MarP
G 284 Thalassicolla											?Excl	Sw	F		2	4	1	1	1	MarP
G 295 Aulacantha											?Excl	Sw				4	1	1	1	MarP
O 54 ACANTHARIA			I					1			2						1	1	1	MarP
G 286 Acanthometra			I					1									1	1	1	MarP
G 287 Acanthostaurus			I					1									1	1	1	MarP
C 14 SPOROZOA	H	Z	v		v	v	v	v			4						1	1	1	Par
O 70 GREGARINIDA	H	Z	v				GonI	R			4						1	1	1	Par

TEXTNAME: Tax (R)P: 18   TEXTNAME: Belltbl (R)P: 18

Table cont'd

Tax	LCY	MEI	GAM	ISO	MIC	MAC	FUS	GER	STA	COM	ASX	MOT	IMM	RES	DIV	HAP	SIZ	DIF	LAB	HAB
G 288 Monocystis	H		I				GonPA	8									1	1	1	Par
G 289 Gregarina		Z	A				GonI	8								3	1	1	1	Par
G 290 Stylocephalus	?H	?Z	O⁸				GonI	8								4	1	1	1	Par
G 291 Lipocystis	H	Z	I	2			GonI						Sch				1	1	1	Par
G 292 Mattesia			A										Sch				1	1	1	Par
G 293 Schizocystis			I				GonI	8					Sch				1	1	1	Par
G 294 Ophryocystis				1									Sch				1	1	1	Par
O 71 COCCIDIA	H	Z*	O		v	1	v	v			4	v					1	1	1	Par
G 295 Eococcidium	H	Z*	0		12-32	1	GetE						Az				1	1	1	Par
G 296 Eimeria	H	Z*	0		many	1	GetE	8					Sch				1	1	1	Par
G 297 Aggregata	H	Z*	0		4-8	1	GetE						Sch			6	1	1	1	Par
G 298 Haemoproteus	H-	Z*	0		4-8	1	GetE	many					Sch				1	1	1	Par
G 299 Plasmodium	H	Z*	0		4	1	GetE	many					Sch				1	1	1	Par
G 300 Klossia	H	Z*	0		2	1	GonA	many								4	1	1	1	Par
G 301 Karyolysus	H	Z*	0		2-4	1	GonA	12-15									1	1	1	Par
G 302 Adelina	H	Z*	0			1	GonA									10	1	1	1	Par
C 15 CILIATA	D	G	I	1,2			Gon	1			4	F,Sw	F,Az		v		1	1	1	Cos
O 72 Holotricha	D	G	I	2			GonI				4	F	F,Az		2		1	1	1	Fw,Par
G 303 Didinium	D	G	I	2			GonI	1				F			2	8	1	1	1	Fw
G 304 Tetrahymena	D	G	I	2			GonI	1		(Aut)		F			2	5	1	1	1	Fw
G 305 Colpidium	D	G	I	2			GonI	1				F			2		1	1	1	Fw
G 306 Paramecium	D	G	I	2			GonI	1		(Aut)		F			c1000	30-40	1	1	1	Fw
G 307 Ichthyophthirius	D	G	I	2			GonI	1				Sw			2		1	1	1	Par
G 308 Gymnodinioides	D	G	I	2			GonI	1				F			128		1	1	1	Par
G 309 Ophryoglena	D	G	I	2			GonI	1				Sw					1	1	1	Par
G 310 Radiophrya												B					1	1	1	Par
O 73 PERITRICHA	D	G	I				GonA	1			3	Sw	F,Az		2		1	1	1	FwEpi
G 311 Zoothamnion	D	G	I	1			GonA	1				Sw	F,Az		2		2	2	1	FwEpi
G 312 Vorticella	D	G	I	2			GonA	1				Sw	F,Az		2	4	1	1	1	FwEpi
G 313 Opercularia	D	G	I				GonA	1				Sw	F,Az		2		1	1	1	FwEpi

TEXTNAME: Tax (R)P: 19   TEXTNAME: Belltbl (R)P: 19

Table cont'd

Tax	LCY	MEI	GAM	ISO	MIC	MAC	FUS	GER	STA	COM	ASX	MOT	IMM	RES	DIV	HAP	SIZ	DIF	LAB	HAB
O 74 SPIROTRICHA	D	G	1	1			GonI	1			3		F		2		1	1	1	Fw
G 314 Metopus	D	G	1	1			GonI	1					F		2		1	1	1	Fw
O 75 CHONOTRICHA	D	G	1	1			GonA				3	BSw					1	1	1	Fw
G 315 Spirochona	D	G	1	1			GonA					BSw					1	1	1	FwEpi
O 76 SUCTORIA	D	G	1	1			GonA				4	BSw			4		1	1	1	FwEpi
G 316 Tokophrya	D	G	1	1			GonA					BSw					1	1	1	FwEpi
G 317 Stylocometes	D	G	1	1			GonI					BSw					1	1	1	FwEpi
G 318 Ephelota	D	G	1	1			GonA					BSw			4		1	1	1	FwEpi

Glossary

TAX, taxon. C, class; O, order; F, family; G, genus. Each rank is numbered consecutively. All major taxa of eukaryotic algae (except red algae) and protozoans are included. Familial rank is given for Chlorophyceae only. Algal genera included are those for which useful information is given by Bold and Wynne or by Fritsch. The table is biased in favour of genera which reproduce sexually, and in favour of forms with unusual modes of reproduction; otherwise some attempt has been made to list a number of genera proportional to the diversity of the taxon. Classification follows Bold and Wynne (algae) and Grell (protozoans), with the flagellates split

LCY, life cycle. H, vegetatively haploid; D, vegetatively diploid; Iso, isomorphic alternation of generations; Het, heteromorphic alternation; HetH, heteromorphic with haploid phase dominant: HetD, heteromorphic with diploid phase dominant; Alt, alternation, nature unknown; Dk, dikaryon

MEI, site of meiosis. Z, zygotic; S, sporic; G, gametic; asterisk* indicates a one-division meiosis

GAM, gametic dimorphism. I, isogametic; IH, isogametic and hologametic; PA, pseudo-anisogametic; A anisogametic; O, oogametic; asterisk* indicates diploid gametes

ISO, isogametogenesis

MIC, microgametogenesis

MAC, macrogametogenesis

These three columns give the number of gametes derived from a single initial germ-line cell

FUS, fusion; the nature of entities undergoing sexual fusion. GN, gametic pronuclei only; Get, gametes only; GetE, gametes liberated, fusion external; GetI, macrogamete retained, fusion internal; Get(I), macrogamete released from oogonium but remains adherent to its surface and is fertilized in this position. In all other cases some other entity in addition to gametes or gametic nuclei undergoes fusion. Col, colonies; SP, sperm packets; Gia, gametangia (algae); Gon, gamonts (protozoans); GonI, isogamonty; GonPA, pseudoanisogamonty; GonA, anisogamonty; Dm, dwarf male filaments

GER, germination. The number of products of germination of the cell resulting from syngamy (zygote)

STA, gender state. M, monoecious; D, dioecious

COM, compatibility. C, self-compatible; Aut, autogamous; I, self-incompatible

ASX, asexuality. A guess at the frequency of asexual reproduction. For genera: Excl, exclusively asexual; Unk, asexual reproduction unknown. For higher taxa the guess is scaled from one (wholly asexual or nearly so) to five (wholly sexual or nearly so)

MOT, motile propagules, other than gametes. F, produced by fission; Ag, agametes; Ac, autocolonies; Zo, zoospores; Sw, swarmers; B, motile buds; BSw, budded swarmers

IMM, immotile propagules, other than macrogametes and zygotes. Ag, agametes; Az, azygotes; Ap, aplanospores; Ac, autocolonies; Ak, akinetes; Au, autospores; Fr, fragmentation; F, fission; Msp, monospores; Te, tetraspores; B, buds; Sch, schizonts; Scy, sporocytes

RES, resistant propagules, other than zygotes. Cy, cysts; Ak, akinetes; Hy, hypnospores; Ap, aplanospores; Bu, bulbils; St, statospores

DIV, division. The number of products of division (classified under MOT, IMM or RES) during sporogenesis. Note that in haplonts (entry H in column LCY) the spores are produced by mitosis, but where there is an alternation of generations the spores are produced by meiosis

HAP, haploid complement. The haploid number of chromosomes; for asexual taxa, the somatic number of chromosomes

SIZ, size. Size of the dominant vegetative phase, scaled as follows; 1, unicellular; 2, 2–10^2 cells in colony; 3, 10^2–10^4 cells in colony, microscopic coenocytes; 4, macroscopic colonies, to 1 cm, several cm of filament, macroscopic coenocytes; 5, macroscopic, greater than 1 cm, several cells thick

Table 5.7 (cont.)

Glossary

DIF, differentiation. Form of the dominant vegetative phase, scaled as follows: 1, unicellular, motile or coccoid; 2, colonial, noncoenobial, palmelloid; 3, colonial, coenobial; 4, simple filaments; 5, branched filaments, parenchymatous thallus; 6, heterotrichous; 7, advanced heterotrichous forms

LAB, division of labour. A measure of the extent to which different functions are allocated to separate cells or groups of cells. A minimal score of one indicates that each cell performs all vital functions; for each function (e.g., sexual reproduction, zoosporogenesis, attachment, meristematic growth) performed by a special subset of cells only, one point is added, to a maximum of five

HAB, habitat or way of life. Fw, freshwater; Br, brackish water; Mar. marine; P, planktonic; B. benthic or attached to substrate; Epi, epiphytic or epizooic; So, soil; Sub, subaerial; Cos, cosmopolitan; Par, parasitic; Sym, symbiotic

General. A dash indicates a range of values. A comma separates two values both of which occur. An absence of punctuation denotes qualification (e.g., FwP, freshwater plankton). Parentheses enclose values which are true only occasionally or for a few species within a genus. A question mark indicates doubt concerning the truth of an entry. V (for taxa higher in rank than genera) indicates that a character varies too widely for any simple entry to be made. Note that for taxa higher than genera the entries attempt to express the usual or typical character states, no doubt with a very uneven degree of success.

Notes: 1. Marine species of *Ulothrix* may be exclusively asexual, may be vegetatively haploid or may have either a heteromorphic or an isomorphic alternation of generations. 2. Very variable life histories. *Monostroma angicava* is dioecious and anisogametic, *M. groenlandicum* monoecious and isogametic; both have heteromorphic life histories with the haploid phase being dominant. In *M. zostericola* the diploid phase is dominant. Two or three species may be obligately asexual. 3. Some species of *Enteromorpha*, *Ulva* and *Ulvaria* appear to be exclusively asexual. 4. Freshwater species of *Rhizoclonium* reproduce mainly by fragmentation, and sexuality has not been described. 5. Some freshwater species of *Cladophora* may be exclusively asexual. *C. glomerata* may be a diplont with a gametic meiosis. 6. *Acrosiphonia* has very variable life cycles: some species have a heteromorphic and others an isomorphic alternation, whilst others are exclusively asexual. 7. Life cycles may differ even between conspecific populations of *Bryopsis*. 8. In *Stylocephalus* the motile gamete is the larger.

Source: Principally, Bold and Wynne (1978); Fritsch (1935, 1945); Grell (1967, 1973); Dogiel (1965). Chromosome numbers for algae are from Godward (1966); those for protozoa are taken from lists in Dogiel (1965), Grell (1967) and Kudo (1966).

Table 5.8: Relationship Between Vegetative Development and the Site of Meiosis. Calculation of vegetative index discussed in text: index = *SIZ* × *DIF* × *LAB*. Site of meiosis from *MEI* column of Table 5.7. Analysis by genera (G), families within Chlorophyceae (F) and orders (O).

| Vegetative index | MEI | | | | | | | | |
| | Z | | | S | | | G | | |
	G	F	O	G	F	O	G	F	O
1	35	4	7	11	0	2	31	0	10
2–10	5	1	1	0	0	0	9	0	0
11–50	11	5	2	8	2	1	1	1	0
51–100	5	2	2	25	5	6	5	5	2
100–50	2	0	1	17	0	5	1	0	0
150+	0	0	0	8	0	1	5	0	1

Table 5.9: Relationship Between Habitat and Site of Meiosis. Habitat from HAB column of Table 5.7. Site of meiosis from MEI column of Table 5.7. Analysis by genera (G), families within Chlorophyceae (F) and orders (O).

| HAB | MEI | | | | | | | | |
| | Z | | | S | | | G | | |
	G	F	O	G	F	O	G	F	O
1. So, Sub	3	1	2	1	0	0	0	0	0
2. Fw	29	10	8	1	0	0	1	0	0
3. Br; Fw, Mar	5	0	2	2	1	1	1	0	1
4. Mar	0	0	0	42	17	14	23	5	5
5. Cos	1	4	0	0	0	0	1	1	0

but large (about 1000) in *Noctiluca*, the only isogametic genus with a gametic meiosis for which I have data. (This applies only to taxa with exclusively gametic fusion; where fusion is gamentangial or gamontic the number of gametes formed is always less, a phenomenon which I discuss in the next section.) Amongst oogametic forms with a zygotic meiosis the number of microgametes produced may be small (one in *Coleochaete*, Peridiniales, Zygnemataceae and some Polymastigina), moderate (other taxa) or quite large (up to 64 in *Chlamydomonas*, *Eudorina* and *Eremosphaera*; about 150 in Charales; as many as 512 in *Volvox*); when meiosis is gametic the number may be small (four in Centrales) or moderately large (64 in Fucales). The absence of any real trend is illustrated by the dramatic contrast between two orders of oogametic brown algae, the Desmarestiales producing a single microgamete and the Dictyotales about 1500, whilst both have a sporic meiosis. It would also appear that the gametic hypothesis makes the same prediction; so that even if there were enough data to distinguish a consistent trend, the test could not support either hypothesis at the expense of the other.

The second specific prediction concerns the number of products of germination of the zygote. When meiosis is gametic there should be no mitotic replication of diploid individuals, and only a single germling or swarmer should emerge from the zygote. But when meiosis is zygotic it is asserted that genotypic diversity is not necessarily maximized, so that there should be cases in which more than four individuals germinate from a single zygote, any individuals in excess of four being necessarily produced by a mitosis. The data, set out in Table 5.10, give tentative support to the prediction. Forms in which the mitotic replication of zygote nuclei leads immediately to the production of new individuals are *Coleochaete*, three genera of Gregarinida and five genera of Coccidia and perhaps also *Tetraspora* and *Ulothrix*. All have a zygotic meiosis. Four plants emerge from the zygotes of *Zanardinia* and *Hymenomonas*, which have an essentially isomorphic alternation of generations, and eight from those of *Cutleria*, in which the haploid generation is dominant. All other genera with a sporic meiosis show only a single product of zygote germination and are either isomorphic or (in *Bryopsis* and *Sphacelaria*) have a dominant diploid generation. The number of genera in this category would doubtless be greatly increased if more data were available for brown algae.

Table 5.10: Relationship Between the Site of Meiosis and the Number of Products of Germination of the Zygote. Analysis by genera only

Site of meiosis	Productions of germination:		
	all unique	may or may not all be unique	not all unique
Zygote	28	2	3
Spore	4	0	3
Gamete	38	0	0

A third prediction I have examined states that the frequency and diversity of specialized immotile and resistant asexual propagules, relative to that of motile asexual propagules, will be less in forms with a gametic meiosis, where their place is supplied by the zygote. This prediction is fully supported by the data, but this pattern is accounted for by two other correlations: the tendency for marine forms to have a gametic meiosis (Table 5.9), and the well-known and easily understood prevalence of dormant and resistant propagules, both sexual and asexual, among freshwater taxa. The proposed prediction adds no further explanatory power.

Finally, the vegetative hypothesis predicts that meiosis will be zygotic in those taxa whose mode of reproduction has the effect of generating low levels of genetic variation. For instance, we expect to find a zygotic meiosis in the sexual members of taxa in which sexuality is rare or infrequent (ASX score for families and orders in Table 5.7). This test is unsatisfactory because the ASX score is highly subjective and of questionable accuracy. With this in mind, the hypothesis is rejected (Table 5.11). There is no pattern at the ordinal level. At familial rank within Chlorophyceae the trend is the reverse of that expected: five of six families with a gametic meiosis appear to have only occasional sexuality (Prasiolaceae,

Table 5.11: Relationship Between Prevalence of Asexual Reproduction and Typical Site of Meiosis in Major Taxa. First column is ASX score from Table 5.7. The typical site of meiosis (Z, zygotic; S, sporic; G, gametic) is entered for families within Chlorophyceae (F) and for orders (O).

ASX	RANK					
	F			O		
	Z	S	G	Z	S	G
2	5	2	5	4	2	3
3	3	4	0	5	5	4
4	6	1	1	5	4	5
5	1	0	0	1	4	1

Codiaceae, Udotaceae, Caulerpaceae, Valoniaceae), whilst in six of fifteen families with a zygotic meiosis sex is a regular and frequent feature of the life history (Protosiphonaceae, Characiosiphonaceae, Hydrodictyaceae, Oedogoniaceae, Zygnemataceae, Mesiotaeniaceae). A related prediction is that, if the hypothesis were correct, we would expect to find among forms with a zygotic meiosis an excess of: monoecious taxa; self-compatible monoecious taxa; and regularly autogamous self-compatible moneocious taxa. Indeed, this prediction does appear to be true, so long as we restrict the analysis to Chlorophyceae, Phaeophyceae and Xanthophyceae. However, the correlation is entirely abolished by the inclusion of the monoecious, self-compatible diatoms, and autogamy is found in diatoms, Polymastigina, Heliozoa and Holotricha, all of which have a gametic meiosis. A third prediction of the same general type that I have examined is that taxa with a zygotic meiosis should have fewer chromosomes. I have found no such pattern and do not believe that any exists; but the comparison is spoiled by the enormous range in chromosome number, the upper end of which, especially in Zygnematales and *Peridinium*, seems to reflect very high ploidy.

The predictions made by the rival gametic hypothesis are quite different. According to this hypothesis the function of gametic meiosis is to provide a few extremely competent gametes which will alone stand any appreciable chance of syngamy when there is intense competition between gametes. This competition will be generated by a greater production of gametes of one gender or matingtype than of the other since, even if all possible fusions occur, there will still remain an unfused residue comprising those gametes of the type initially present in excess which failed to encounter an appropriate fusion partner. Gamete competition may occur among isogametic organisms simply because there will often be an excess of one mating type in a given place at a given time by chance; but, since we expect isogamous mating types to be on average equally frequent (as I shall show in the next section), this competition will not be very intense. At the other extreme, in oogametic organisms there is almost always a vast overproduction of microgametes, only a minute fraction of which will achieve syngamy. The primary determinant of gamete competition is thus gamete dimorphism, with competition being least severe amongst isogametic and most severe amongst oogametic taxa. Suppose that we recognize three levels of gamete dimorphism (I; A or PA; and O in column GAM of Table 5.7) and five types of life history (using columns LCY and MEI of Table 5.7 to give: H with a zygotic meiosis; HetH, Iso and HetD, with a sporic meiosis; and D with a gametic meiosis), so that there are fifteen possible combinations of gamete dimorphism and life cycle. The gametic hypothesis then makes the following specific predictions: first, vegetatively haploid organisms will be isogametic; secondly, organisms with a heteromorphic life history in which the haploid phase is dominant will not be oogametic; thirdly, organisms with an isomorphic alternation of generations will be anisogametic; fourthly, organisms with a heteromorphic alternation in which the diploid generation is dominant will not be isogametic; finally, vegetatively diploid organisms will be oogametic. The effect of these predictions is to divide a table of fifteen cells, representing the fifteen possible combinations of gamete dimorphism and life cycle, into two roughly equal sets, seven of the cells being confirming instances and eight falsifying instances. If the gametic hypothesis is correct, we expect the confirming greatly to outnumber the falsifying instances at all phyletic levels. This prediction is not accepted at any phyletic level: at the generic level there are 76 confirming and 101 falsifying instances; at the familial level, 15 confirming and 13 falsifying instances; at the ordinal level, 13 confirming and 20 falsifying instances (see Table 5.12).

Table 5.12: Relationship Between Life Cycle and Gamete Dimorphism in Algae and Protozoans. The table has fifteen cells, each of which has nine entries arranged in three rows and three columns. The rows correspond to different ranks: genus (G), family within Chlorophyceae (F) and order (O). The three columns correspond to different states of the FUS variable of Table 5.7. The first column includes Get, GetE and GetI,E; the second includes GetI,E (note repetition) Get(I) and GetI; the third includes Gon, Gia and Dm.

GAM				LCY												
			H			HetH			Iso			HetD			D	
	G	23	0	11	1	0	0	14	1	4	9	0	0	3	2	23
I	F	7	0	3	0	0	0	2	0	0	0	0	0	2	0	0
	O	5	0	1	0	0	0	1	0	0	2	0	0	2	1	6
	G	7	2	0	3	0	0	8	0	1	6	0	0	6	1	0
A, PA	F	1	1	0	1	0	0	2	0	0	1	0	0	3	0	0
	O	0	0	0	1	0	0	1	0	0	0	0	0	2	0	0
	G	9	9	6	0	0	0	5	0	0	0	12	0	8	4	0
O	F	1	2	1	0	0	0	0	0	0	0	0	0	1	0	0
	O	1	2	1	0	0	0	2	0	0	0	3	0	2	0	0

This amounts to an apparently decisive rejection of the hypothesis, and ordinarily one would enquire no further. However, as I was compiling the data I was struck by the difference between organisms with exclusively gametic fusion and those with gametangial or gamontic fusion. For this reason I constructed

Table 5.12, in which each of the fifteen cells is subdivided according to the nature of the entities undergoing sexual fusion (column FUS of Table 5.7). This more complex table was then analyzed as follows. First, four categories of support for the hypothesis were recognized: confirming instances, as already detailed (cells H/I, HetH/I, HetH/A, Iso/A, Het'D/A, HetD/O and D/O); neutral instances, which I would not regard as sufficient to falsify the hypothesis (cells H/A, Iso/I, Iso/O and D/A); falsifying instances, in which the general rule proposed by the hypothesis is disobeyed (cells HetH/O and HetD/I); and strongly falsifying instances, in which the general rule is reversed (cells H/O and D/I). Next, the expected number of instances for each phyletic level and for each category of fusion was calculated, using the null hypothesis that instances should be distributed at random between cells. The ratio (observed−expected)/(expected) was then calculated for each case, and the result entered in Table 5.13. The gametic hypothesis predicts that this ratio should be positive for confirming instances; zero (or perhaps weakly positive) for neutral instances; negative for falsifying instances; and more highly negative for strongly falsifying instances. The result of this analysis (Table 5.13) is remarkable. The gametic hypothesis is accepted at all phyletic levels for those taxa in which gametes are released and fuse in the external medium − for those taxa, in other words, for which the hypothesis was implicitly designed. But when any restraint is placed on gamete fusion, by the retention of the macrogamete or by the prior fusion of gametangia or gamonts, the reverse of what is expected is observed. I conclude that the original hypothesis has not been falsified, but rather that its explanatory power must be enhanced by taking into account the nature and behaviour of fusing entities as well as the original crude criterion of gamete dimorphism. In particular, the hypothesis is to be accepted only if it can be shown that, in the strongly falsifying instances, sexual fusion somehow modulates gamete dimorphism so as to bring them into conformity with the hypothesis.

Consider first the oogametic haplonts. Since the zygote is the site of meiosis, the hypothesis predicts a low level of gamete competition despite oogamety. This can be true only if there is some special restriction on gamete fusion. One such restriction is autogamy, since individual reproductive success will be limited by the number of macrogametes which can be produced rather than by the competence of the microgametes, provided that they are competent in some minimal degree. Another is an extreme spatial restriction in outcrossing organisms. This is obviously true if we consider an isolated pair of self-incompatible monoecious plants, since the number of successful microgametes from either plant is limited by the number of macrogametes produced by the other. The circumstances in which we might expect the vegetative phase to be haploid, even in oogametic organisms, thus involve a spatial or a genetic constraint on syngamy.

Eudorina, *Platydorina* and *Volvox* liberate packets of sperm, which disintegrate only when they have penetrated a female colony. Since there are more than enough sperm in each packet to fertilize all the oogonia of the female colony it seems likely that all the oogonia will be fertilized by sperm from a

Table 5.13: Analysis of Table 5.12. Entries are values of (observed − expected)/ (expected), as explained in text. The rows for each category of FUS correspond to confirming (+), neutral (0), falsifying (−) and strongly falsifying (−−) instances. A value of (−1) indicates that no instances are known, some small (usually fractional) number of instances being expected under the null hypothesis. Abbreviation: nd, insufficient data.

| FUS | | RANK | | |
		G	F	O
GetE, etc.	+	+0.247	+0.481	+0.447
	0	+0.073	+0.154	+0.892
	−	−0.308	(−1)	+0.429
	−−	−0.372	−0.250	−0.375
GetI, etc.	+	−0.500	nd	−0.167
	0	+0.481	0	nd
	−	(−1)	nd	(−1)
	−−	+0.146	0	+0.579
Gon, etc.	+	−0.185	nd	−0.600
	0	−0.310	nd	nd
	−	nd	nd	nd
	−−	+0.346	nd	+0.273

single successful packet, and the meiotic diversification of these sperm would be pointless. Some monoecious species of *Volvox* are autogamous. In *Coleochaete* the antheridia are borne on short branches of the main system, often close to oogonia. However, some species are dioecious. In Charales the oogonia and antheridia are usually borne close together, but some species are protandrous and a few dioecious. According to Proctor (1976, p. 210), 'all monoecious charophytes yet examined have proven fully autogamous', but it is not clear whether this implies habitual self-fertilization in natural populations. It may be significant that monoecious species are often polyploid while dioecious species are not. Most species of *Vaucheria* are monoecious, with antheridia and oogonia arising close together on the main filament or on small side branches (but in the related *Dichotomosiphon* they are borne on separate side branches). Fritsch (1935, p. 437) remarks, apparently without empirical support, that 'in many cases probably the adjacent oogonium is fertilized'; certainly in the drawing of *V. debaryana* reproduced by Fritsch (1935, Figure 146C) the terminal pore of the antheridium appears to be directed towards the receptive area of the adjacent oogonial wall.

The situation in Oedogoniales is more complicated. The filaments may be unisexual or bisexual. In some species with unisexual filaments, the male filaments are minute (nannandrous, as opposed to macrandrous, species). These dwarf male filaments arise from androspores, which are formed singly in androsporangia borne on the same filaments as oogonia (gynandrospory) or on separate filaments (idioandrospory). The oogonium contains a single ovum and its wall bears a pore or fissure through which the sperm enters. In nannandrous species the androspore is attracted to the oogonium, where it attaches and grows into a dwarf filament bearing an antheridium. Competition therefore seems to occur between andros-pores for settling sites on the oogonial wall, rather than between sperms for receptive ova.

The bryophytes are the only taxon of vascular plants to combine oogamety with vegetative haploidy. Although terrestrial in habit, bryophytes have not liberated their reproduction from dependence on the water, and fertilization depends on a flagellated sperm swimming in the water film from one plant to another. This places a severe spatial restriction on gamete competition, and leads to high levels of self-fertilization. Anderson and Lemmon (1974) estimated that the maximum distance travelled by the sperm of a moss was 40 mm, and the median distance only 5 mm.

Most oogametic haplonts, therefore, can be shoehorned into conformity with the general rule that they appear to contradict, though I am not enough of a phycologist to know how much violence is being done to the facts. Three of the remaining cases, whilst they cannot be explained, might be shrugged off; there are a few oogametic species in the predominantly isogametic genera *Chlamydo-monas* and *Carteria*, and we might choose to imagine that evolutionary inertia has prevented the acquisition of oogamety from being accompanied by a shift in the site of meiosis; meiosis is said to be zygotic in *Golenkinia*, but the evidence is inferential rather than direct, and at a pinch could be suppressed. This leaves the peculiar *Sphaeroplea* as the only utterly indigestible counterexample. In this coenocytic alga, cell masses round off without nuclear division to form uni-nucleate macrogametes; the microgametes are formed in a more conventional manner, involving repeated nuclear division. The several to many ova within an oogonium may or may not be motile – perhaps usually not – and may be fertilized either within or outside the oogonium. In at least one species both the macrogametes and the microgametes are released into the external medium, so that fusion is wholly external. In no case does there seem to be any hint of the special restriction on fusion demanded by the hypothesis.

Consider next the other strongly falsifying instances, involving isogametic diplonts. If the gametes are the only entities which fuse, these instances cannot be reconciled with the hypothesis. But suppose that fusion occurs between gametangia or gamonts, with gametic fusion between gametes or gamete nuclei occurring subsequently within the space formed by the walls of these structures. There will then be little competition between gametes; instead, it is the gamete-bearing structures which will compete for fusion partners, and it is therefore these structures rather than the gametes which should be diversified by meiosis. A first prediction, therefore, is that fusion between entities other than gametes will predominate amongst isogametic diplonts. This prediction must be accepted: only six of 25 genera of oogametic haplonts but 23 of 28 genera of isogametic diplonts are gamontogamous (six genera of Pennales; *Notila*; 16 genera of ciliates in five orders. The exceptions are *Valonia*, *Acetabularia* and *Noctiluca*, with external gamete fusion, and two genera of Heliozoa with internal autogamous fusion, in one case approaching gamontogamy). We can further reason that, among gamontogamous organisms, the site of meiosis will differ according to whether competition is more severe between gametes or between gamonts. No prediction is possible for taxa which are either isogametic and isogamontic or anisogametic and anisogamontic, but taxa which are anisogametic and iso-gamontic should have a gametic meiosis and those which are isogametic and anisogamontic a zygotic meiosis. Unfortunately the data are not sufficiently extensive to test this second-order prediction, but what data there are do not support it: the anisogamontic ciliates have a gametic meiosis, whilst the peculiar *Stylocephalus* (Gregarinida) is anisogametic and isogamontic with a zygotic meiosis.

The gametic hypothesis correctly predicts the state of higher organisms, all of which are fundamentally oogametic diplonts, but it scarcely leads one to expect the absence of virtually any variation in life cycle to correspond with what must be enormous differences in gamete competition generated by differ-ences in gamete morphology, the method of fertilization or the occurrence of autogamy. One cannot be sure that this actually constitutes a contradiction, but it is sufficiently puzzling to suggest that an historical dimension may be an essential feature of any complete theory of the alternation of generations. It may be easy to pass from vegetative haploidy to vegetative diploidy, but very difficult to move in the reverse direction – for all the conventional reasons involving the unmasking of deleterious recessive genes. I still find it vaguely disquieting that, whilst vegetative haploidy has evolved several times among multicellular animals, it has never evolved in both sexes.

There is one intriguing piece of evidence from higher animals which, on the face of it, appears to give powerful support to the gametic hypothesis. If the hypothesis holds for higher animals, one expects that gametes should be more highly diversified when they are produced in greater excess. On the face of it, this risky prediction is strongly supported by the remarkable correlation between gamete redundancy and chiasma frequency during spermatogenesis discovered by Cohen (1967, 1973, 1975) and illustrated in Figure 5.8. Unfortunately, it is notorious that there is no evidence for an effect of the haploid genome in the sperm of higher animals; the properties of the sperm seem all to be determined by the diploid tissue of the male. I am therefore reluctant to invoke Figure 5.8 as support for the gametic hypothesis.

Clearly, neither the vegetative nor the gametic hypotheses have been decisively falsified, nor is the balance of evidence sufficiently skew for us unequivocally to prefer one hypothesis to the other. The reader can make up his own mind; for myself, I am at present inclined to believe that the gametic hypothesis offers much the better interpretation of the alternation of generations. The predictions made by the vegetative hypothesis often seem vague and logically dubious; the only major success of the hypothesis is the almost total restriction of a zygotic meiosis to freshwater rather than marine algae, and I would be surprised if this

Figure 5.8: Chiasma Frequency and Gamete Redundancy. The lower-case letters represent estimates of sperms per fertilization for a variety of animals; the upper-case letters give analogous data for oocyte redundancy.

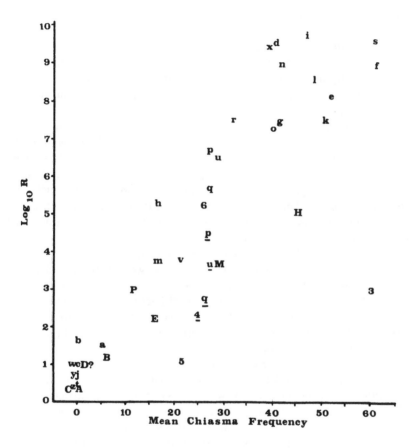

Source: Cohen (1975), p. 103, Figure 1, which should be consulted for further details.

does not turn out to be the indirect consequence of correlations, perhaps concerning vegetative structure, that I have not noticed. The gametic hypothesis, it is true, sometimes disappoints; but its predictions are refreshingly precise and vulnerable, and have proven more or less successful across a very wide range of algae and protozoans. It is the only hypothesis which seems capable of offering a detailed functional interpretation of the site of meiosis, group by group throughout the Algae. At the same time I need hardly point out that the account of the hypothesis given here amounts as much to an advocacy as to an objective and ruthless assessment and, had the hypothesis first been put forward by someone else, I might have rejected it on the failure of its first prediction, regarding the crude correlation between gamete dimorphism and the site of meiosis.

5.4 Gamete Dimorphism

Amphimictic animals, other than protozoans, invariably have two sexes, which are defined by their production of self-incompatible gametes of different size, the male typically producing minute motile microgametes. Essentially the same situation is found among higher plants, where the behaviour and function of a much-reduced gametophyte is almost wholly gametic. These facts are so thoroughly familiar that any exception to the rule would appear monstrous: and yet amongst algae, fungi and protozoans gender is not invariably bipolar; nor do individuals of different gender always produce gametes of different size; nor, if gametes of different size do occur, is there an invariable prohibition on fusion between microgamete and microgamete or between macrogamete and macrogamete. Gamete dimorphism and the bipolarity of gender are therefore among the most pervasive epiphenomena of sexuality, and require functional explanation rather than mere acquiescence.

Aside from antique theories of gender (briefly reviewed by Ghiselin, 1974a), the first attempts to provide such an explanation were mathematical proofs that, when gametes were released into the external medium, the rate of fusion was maximized when the gametes were very different in size (Kalmus, 1932; Scudo, 1967; see also Parker, 1971, for an incorrect argument of the same general sort). The evolution of gamete dimorphism was therefore held to follow from the greater production of zygotes by dimorphic populations – a process of group selection. Until quite recently, the only authors to invoke individual natural selection as an explanation for gamete dimorphism were Kalmus and Smith (1960) and Ghiselin (1974a), who offered verbal arguments couched in terms of a physiological division of labour between microgamete and macrogamete, and who were concerned more with oogamety than with any lesser degree of dimorphism. The only theory of individual selection sufficiently simple to put into mathematical form is that advanced by Parker *et al.* (1972) and subsequently elaborated by Bell (1978b), Charlesworth (1978) and Maynard Smith (1978), as the result of a correspondence between the three of us. Since the details of the mathematical theory can be obtained from these publications, I shall only sketch in the outline here.

Evolution of Pseudoanisogamety. In the first place, we are concerned only to explain the evolution of any degree of dimorphism among gametes, regardless of whether or not this is accompanied by a differentiation of gender. The state in which gametes of two sizes occur but in which there are no restrictions on fusion is called *pseudoanisogamety*.

If we take individual natural selection to be axiomatic, then selection must

Figure 5.9: Logic of Selection Scheme for Gamete Dimorphism. (a) It is necessary to generate disruptive selection for gamete size. (b) An obvious effect on fitness is through the number of gametes produced: doubling the size of gametes means halving their number. (c) Gamete size may also effect zygote fitness; three possible curves are shown. (d) It is obvious that a disruptive fitness scheme is generated only if some region of curve (c) is positively accelerated (concave upwards).

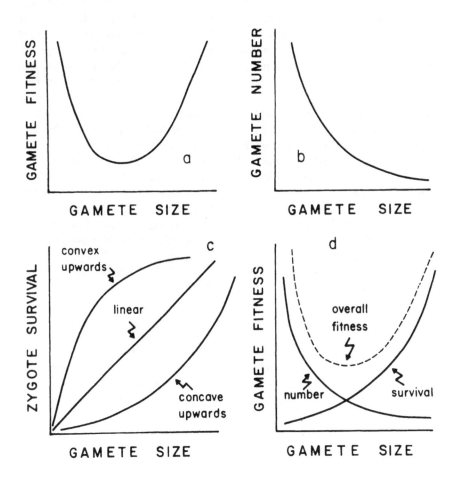

be disruptive; that is, gametes of intermediate size must be the least fit (Figure 5.9a). Then the formal mathematical problem reduces to generating a plausible scheme of disruptive selection, and the general theoretical problem will be to relate the terms of the mathematical argument to patterns in nature. The most obvious effect of gamete size on fitness is through gamete number: as gamete size increases, the maximum number of gametes that can be elaborated from some fixed mass of material declines. This relationship is inevitable and its form is precisely defined: each doubling of gamete mass causes a halving of gamete number (Figure 5.9b). If the effect on gamete number were the only way in which gamete size affected fitness, then the optimal strategy is easily found: it is to produce an indefinitely large number of indefinitely small gametes. At some point, however, this argument must fail, if only because an indefinitely small gamete could not carry out the essential gametic function of transmitting a finite quantity of genetic material. More generally, we can state that fitness will vary with gamete size because the size of the gamete will help to determine the size of the zygote, and in turn the size of the zygote will influence successful germination and growth. The survival of the zygote to some arbitrary stage will therefore be a function of gamete size; it might be no more than a step function, with zygotes of less than a certain critical size being lethal and those exceeding this size, by whatever margin, all being equally viable, but a smooth function is both more general and much more tractable analytically. This function might be of any shape (Figure 5.9c), but it is self-evident that, if gamete number and zygote survival are the only ways in which gamete size influences fitness, the curve of zygote survival on gamete size must be positively accelerated in order for selection to be disruptive (Figure 5.9d). We conclude that gamete dimorphism will evolve only if a unit increase in gamete size causes a disproportionate increase in zygote survival, at least over part of the range of gamete size.

It will be helpful in later analysis to provide an algebraic proof of this proposition. To do so, we must first define the population model more precisely. An arbitrary gametic type i occurs with frequency p_i and each vegetative individual producing type-i gametes produces a number n_i, each of mass m_i, from some fixed mass M of gametic material. Zygotes which are formed by fusion between a type-i and a type-j gamete survive to some given census stage at rate s_{ij}. The expected survival of an arbitrary zygote formed by fusion between a type-i gamete and another gamete chosen at random is $s_i = \sum_k p_k s_{ik}$ and we can appropriately define the fitness of type-i individuals as $w_i = n_i s_i$. Gamete number and gamete mass are related to one another as $n_i = M/m_i$. With these definitions, there will be disruptive selection for gamete size only if there exists some value of m_i, say m_i', for which both the following relationships are true:

$$\left. \frac{dw_i}{dm_i} \right|_{m_i = m_i'} = 0$$

and

$$\left. \frac{d^2 w_i}{dm_i^2} \right|_{m_i = m_i'} > 0.$$

Since we know the relationship between gamete size and gamete number, the solution to the problem is the relationship beween gamete size and zygote survival — given the assumption of random gamete fusion — that satisfies these two conditions. This relationship turns out to be that

$$\sum_k p_k \frac{\mathrm{d} s_{ik}}{\mathrm{d} m_i} = (1/m_i') \left[p_i s_{ii} + (1 - p_i) \sum_k p_k s_{ik} \right]$$

and

$$\sum_k p_k \frac{\mathrm{d}^2 s_{ik}}{\mathrm{d} m_i^2} > (2 p_i / m_i') \left[\frac{\mathrm{d} s_{ii}}{\mathrm{d} m_i} - \frac{s_{ii}}{m_i'} \right],$$

all derivatives being evaluated at $m_i = m_i'$. Whether or not disruptive selection will occur depends in part on the relationship between gamete size and zygote survival, and in part on the frequencies of the different gametic types. The simplest situation, however, can be analyzed without reference to gametic frequencies. Suppose that the initial population consists exclusively of type-j individuals, but a very few type-i individuals arise by mutation. Since the frequency of the type-i individuals can be neglected, the conditions for disruptive selection become approximately that

$$\mathrm{d} s_{ij} / \mathrm{d} m_i = s_{ij} / m_i'$$

and

$$\mathrm{d}^2 s_{ij} / \mathrm{d} m_i^2 > 0.$$

Provided that the conditions under which this approximation is valid are satisfied (they might not be, for example, if a single microgamete-producer arose in a finite population of macrogamete-producers, since p_i would then be initially nonnegligible), the crucial condition for the evolution of pseudoanisogamety is clearly that zygote survival should increase more steeply than linearly with gamete mass, confirming our original, less rigorous argument. The geometry of this situation has been discussed by Maynard Smith (1978), and is described at greater length below.

   This is a model of a selection pressure; we can guess that, if gamete size were determined genetically, then a dimorphic population would evolve if the conditions derived above were satisfied. To prove this proposition, we must harness the selection model to a specific model of inheritance. The simplest realistic model, which is known to be a good description of a number of chlorophyte algae (see review by Wiese, 1976), is that gamete size is determined by alternative alleles at a single locus acting in the gametophytic generation of a haplont with zygotic meiosis. Consider a zygote which on germination gives rise to four gametophytes by meiosis, and whose fitness is defined through the number of gametes produced by these gametophytes and the survival of the zygotes which they in turn form; fitness is thus the rate of increase of a zygotic type over a single zygote-to-zygote cycle, and we can write

$$w_{ii} = 2n_is_i,$$
$$w_{ij} = n_is_i + n_js_j$$

and

$$w_{jj} = 2n_js_j.$$

There will be a genetic equilibrium if the fitnesses of these three genotypes are equal, and this will be true only if $n_is_i = n_js_j$. Now, values of s_k are means weighted by gametic frequencies, and thus determined by the frequency of each gametophytic type and the number of gametes produced by that type; more precisely,

$$s_i = \sum_k p_k s_{ik} = \sum_k n_k x_k s_{ik},$$

where x_k is the frequency of the kth allele among gametophytes. The equality $n_is_i = n_js_j$ is therefore equivalent to:

$$2x_in_in_js_{ij} + (1 - x_i)n_j^2 s_{jj} = x_in_i^2 s_{ii} + n_in_js_{ij}.$$

The frequency of the i allele in the gameophyte population at which all zygotes have equal fitness is therefore

$$\hat{x}_i = (n_in_js_{ij} - n_j^2 s_{jj})/(2n_in_js_{ij} - n_i^2 s_{ii} - n_j^2 s_{jj}).$$

This value represents a stable equilibrium only if a mutant ij zygote can invade a population fixed for the j allele (i.e., $w_{ij} > w_{jj} | x_i \to 0$) or a population fixed for the i allele (i.e., $w_{ij} > w_{ii} | x_i \to 1$). These stability criteria are related to the relative size of the two gamete morphs and to the way in which the survival of the zygote varies with its size. To express them more concretely, first define $\theta = m_i/m_j = n_j/n_i$, where, by arbitrarily defining the i gametes to be the smaller, we constrain θ to lie between zero and unity. Next, suppose that the relationships between the survival and the size of the zygote is of the general form

$$s_{ij} = \alpha m_{ij}^\beta,$$

where $m_{ij} \equiv (m_i + m_j)$. The conditions for stability are then that

$$2^{-\beta}\theta^{-1}(1 + \theta)^\beta > 1$$

and

$$2^{-\beta}\theta^{1-\beta}(1 + \theta)^\beta > 1.$$

The upper condition specifies the maximum value of β for given θ at which microgametes can invade a population of macrogametes; if β exceeds this value only macrogamete-producers should persist. The lower condition specifies the minimum value of β for given θ at which macrogametes can invade a population of microgametes; at smaller values of β only microgametes should persist. The conditions therefore define a range of values for β within which the degree of gamete dimorphism measured by θ will be maintained by disruptive natural selection. The conditions are shown graphically in Figure 5.10. Precisely analogous conditions apply if meiosis is sporic or gametic.

Figure 5.10: Conditions for the Evolution of Pseudoanisogamety.

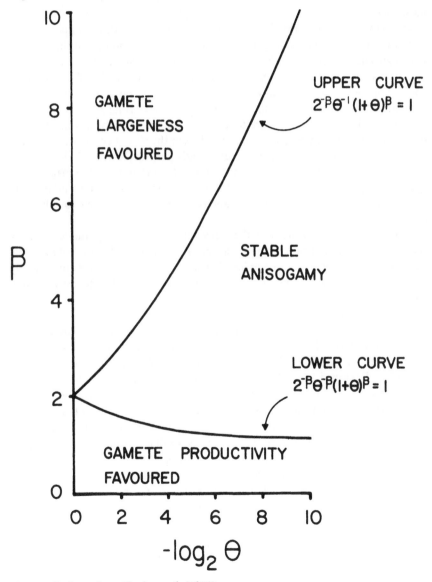

Source: Redrawn from Charlesworth (1978).

Basis for Comparative Analysis. What I have described is an exceedingly simple mechanistic theory of gamete dimorphism, and if we apply it without comment or modification we can expect to be able to interpret only the broadest patterns in nature. The theory hinges on the relationship between gamete size and gamete number and on that between gamete size and zygote survival; and since the former is an invariable mathematical law — setting aside cases in which some residual mass of material remains after gametogenesis — any predictions we make must follow from the latter relationship, as expressed by the parameter β.

It seems obvious enough that β has something to do with the importance of stored food reserves in fuelling the early growth of the zygote; plausible to infer that high values of β would reflect a necessity for substantial growth and differentiation after germination; legitimate then to argue that there will be a positive correlation between the value of β and the size and complexity of the adult organism. To illustrate the principles involved, I have gone so far as to map a number of algal taxa onto Figure 5.10 by guessing a value for β from the level of vegetative organization and a value for θ from the size of antheridial cells and the number of products of microgametogenesis per cell (Figure 5.11). This is an exercise in science fiction, of course; Figure 5.11 differs from the maps of the old cartographers only in the size and ferocity of the dragons. I would not care to defend it in detail, nor can an exercise of this sort be used to falsify an hypothesis.

The problem with using β as the basis for testing the hypothesis is that a simple power law is a poor representation of the way in which zygote survival varies with size; for $\beta > 1$ it implies that zygote survival increases indefinitely with size, which cannot be true. It seems more reasonable to assume that, although survival initially increases with size, and may do so in a positively accelerated manner over a certain range of sizes, it will eventually reach a limit beyond which further increase in size has no effect on survival. The graph of survival on size would then have an inflection point at which the second derivative changes sign, and an algebraic treatment would be tedious; fortunately, Maynard Smith has developed a simple geometrical argument that I shall outline here. It treats the case of a population fixed for a single gamete type, into which some alternative type is introduced as a rare mutant, and the argument is therefore based on the approximation to the original geometrical result given above: gamete dimorphism will evolve if there exists a value of gamete mass $m_i{}^*$ such that at this point $\mathrm{d}s_{ij}/\mathrm{d}m_i = s_{ij}/m_i{}^*$ and $\mathrm{d}^2 s_{ij}/\mathrm{d}m_i{}^2 > 0$.

For simplicity, suppose that each individual undergoes n cytokineses per cell during gametogenesis, and that, if $n = k$, gametes of a minimal size δ are produced, with gametes smaller than δ being nonfunctional. The size of any gamete is then a multiple of δ. The graph of zygote survival on gamete size is an S-shaped curve, which will be defined by three parameters: the asymptotic value of zygote survival at large size, the position of the point of inflection and the slope of the curve at this point. In Figure 5.12, I have tried to show within the compass of a single diagram the effect of altering the point of inflection and the slope of

Figure 5.11: Highly Speculative Mapping of Gamete Dimorphism in Various Taxa of Free-living Gametogamous Algae and Protozoans. Reference numbers of families (F) and order (O) are those of Table 5.7. The position of a taxon on the plane is guessed from entries in the GAM, ISO, MIC, SIZ, DIF and LAB columns of Table 5.7. See text.

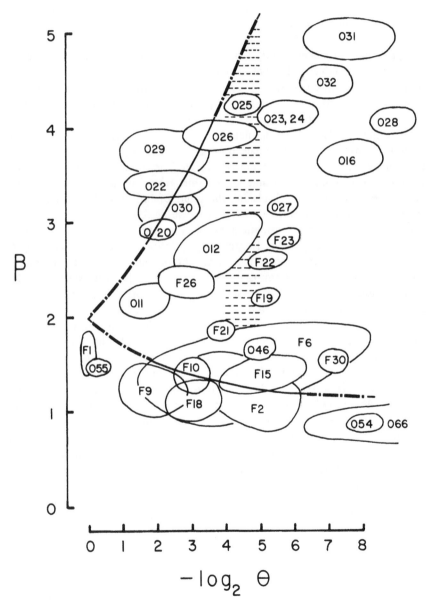

the curve while holding the asymptote constant. The three broken lines represent three curves with different slopes but the same point of inflection; curve A is the shallowest and C the steepest, with B intermediate. The size axis is scaled in multiples of δ, and the point of inflection of the curve is altered by changing the value of the constant multiplier x. For example, take the intermediate curve B and set $x = 1$. We first find the locally stable size for isogametes, $\hat{m}$, by constructing the line of greatest slope $s_{\hat{m}}/\hat{m}$ on curve B; the diagram is drawn so that in this case $\hat{m} = 2\delta$. A population of 2δ-producers is locally stable, in that no type which produces slightly larger or slightly smaller gametes can invade. It is also globally stable: individuals producing gametes of size 4δ or larger cannot invade since, for instance, $s_{6\delta}/4\delta < s_{4\delta}/2\delta$; and δ-producers cannot invade since $s_{3\delta}/\delta < s_{4\delta}/2\delta$. At the evolutionary equilibrium, therefore, the population is isogametic and consists exclusively of 2δ-producers. But suppose we set $x = 8$, so that the locally stable isogamete size is 16δ. Although such a population cannot be invaded by individuals producing much larger gametes, it can be invaded by δ-producers since $s_{9\delta}/\delta > s_{16\delta}/8\delta$. The equilibrium population will therefore be markedly anisogametic, the shift from isogamety to anisogamety having been procured by displacing the point of inflection to the right. If a similar exercise is performed on curve A it will be found that δ-producers can invade the locally stable isogametic population whether $x = 1$ or 8; contrariwise, for curve C the locally stable isogametic population cannot be invaded by δ-producers in either case. The effect of making the curve more steep is thus to reduce the likelihood that anisogamy will evolve. Finally, if we shift the point of inflection far to the right by setting $x = 64$, δ-producers can invade isogametic populations represented by any of the three curves.

(As an aside, I should mention that analysis does not always yield such clear-cut results. Suppose we set $x = 1.3$ on curve B, and follow the fate of an initial population of δ-producers. This will be invaded by 2δ-producers, since $s_{3\delta}/2\delta > s_{2\delta}/\delta$; and since $s_{3\delta}/\delta < s_{4\delta}/2\delta$ the δ-producers cannot reinvade, and are excluded. However, the 2δ-population can be invaded by 4δ-producers, since $s_{6\delta}/4\delta > s_{4\delta}/2\delta$. No larger gamete can invade; but the appearance of 4δ-producers gives the δ-producers another opportunity, since $s_{5\delta}/\delta > s_{8\delta}/4\delta$. Moreover, the 4δ-producers are unable completely to exclude 2δ-producers, since $s_{6\delta}/2\delta > s_{8\delta}/4\delta$. The dynamics of the process cannot be followed further by this sort of analysis, but it seems certain that the equilibrium population will include all three gametic types; and, more generally, some curves of zygote survival on size will generate pseudoanisogametic populations in which gametes of all sizes within certain limits are maintained by selection.)

I have tried to summarize these results in Table 5.14. They lead to three propositions concerning the relationship between gamete evolution and the graph of zygote survival on size:

(1) amongst isogametic forms, the size of isogametes increases as the point of inflection increases (moves to the right);

Figure 5.12: Interpretation of More Realistic Description of Zygote Survival. See text for explanation.

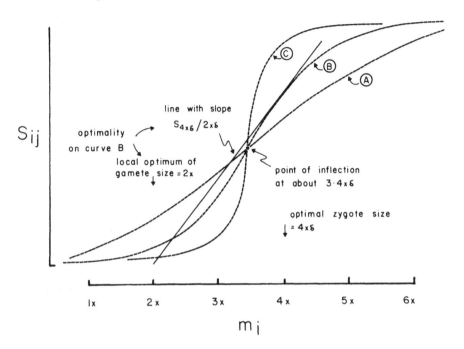

(2) the likelihood that gamete dimorphism will evolve increases as the slope decreases;

(3) the likelihood that gamete dimorphism will evolve increases as the point of inflection increases.

To use any of these propositions for the purpose of prediction requires that the slope and the point of inflection of the graph be given biological meaning.

Curves with very steep slopes begin to look like step functions, and describe the fate of zygotes if almost all survived, provided that they exceeded some critical minimum size. Shallower curves describe a more gradual, probabilistic increase in the rate of survival with size. For instance, internally fertilized zygotes brooded by a parent until capable of full vegetative function might have a very steep curve, and zygotes fertilized and developing autonomously in the external medium a much shallower curve. However, I shall not press this argument here, since the interpretation of the point of inflection seems much more straightforward. The point of inflection will increase as optimal zygote size increases; optimal zygote size will increase with the necessity for postzygotic enlargement, growth or differentiation before the embryo is vegetatively self-sufficient. The point of inflection will thus occur at small gamete size among

Table 5.14: Evolution of Gamete Dimorphism in Relation to the Slope and the Point of Inflection of the Curve of Zygote Survival on Size. Two entries in each cell: the first is the answer to the question, is the locally stable isogametic population liable to invasion by microgamete-producers? The second is a description of the population at equilibrium. See Figure 5.12 and text.

	SLOPE		
	A	B	C
	shallow	intermediate	steep
high c. 200δ	yes; oogamety	yes; oogamety	yes; oogamety
intermediate c. 27δ	yes; oogamety	yes; anisogamety	no; macroisogamety
low c. 3·4δ	yes; anisogamety	no; microisogamety or range of small gametes	no; microisogamety

(left axis label: POINT OF INFLECTION)

protists in which the zygote or the product of zygote germination is a fully functional vegetative cell; it will shift to the right if the zygote must grow before full vegetative function is established; and it will shift much further to the right among large multicellular organisms dependent for their early growth on reserves supplied by the zygote.

We can now test the hypothesis by making two empirically vulnerable predictions. The first is that amongst isogametic forms the size of isogametes will increase with vegetative size and complexity. I shall not follow up this prediction, partly because of lack of data and partly because it is not directly relevant to the topic of gamete dimorphism. The second prediction is that anisogamous organisms should tend to be large and complex whilst isogamous organisms should be smaller and simpler. The same prediction could be made from the simpler algebraic theory in which the graph of zygote survival on size is an uninflected power curve — the two theories, being based on the same principle, are of course complementary rather than antagonistic — but the interpretation of the point of inflection seems less equivocal than that of the exponent β, and therefore provides a sounder basis for comparative analysis. It should be noted that the prediction concerns a general tendency; it will be falsified by observing the opposite tendency, or by failing to observe any tendency, but it is not falsified even by a substantial number of counterexamples, since gamete dimorphism is held to be controlled by the slope as well as by the point of inflection of the curve of zygote survival on size.

Comparative Analysis. The general basis for the prediction developed above is that size is often crucial to the future survival of the zygote. The two main lines of evidence in favour of this opinion are the low variability of macrogamete size relative to macrogamete number in most animals and plants, and the fact that macrogametes but not microgametes often have the capacity to develop parthenogenetically. Unfortunately, there is a fundamental obstacle in the way of taking the obvious step of measuring the effect of size on zygote survival directly; if zygote size has been rather strictly optimized by natural selection, most of its variance will have been eliminated, and the range of material we need to make the measurement will not exist. We are thus driven back to the indirect technique of predicting a correlation between gamete dimorphism and vegetative organization.

The specific basis for this prediction is that optimal zygote size should increase with vegetative size and complexity. This link in the chain of argument itself constitutes an empirically vulnerable hypothesis which should be tested. If it were true, we would expect the number of products of germination of the zygote to be greater in unicellular than in multicellular haplontic algae. Because the size of the zygote as well as the extent of its subdivision may vary, this tendency might be a weak one, but certainly we would be disturbed to find the opposite trend. There is only one product of germination in *Peridinium*, but more than one in all the other unicellular forms for which data is presented in Table 5.7 (*Chlamydomonas*, desmids, *Ochrosphaera*, *Ceratium* and *Hypnodinium*). In multicellular forms the occurrence of a single product of germination is more usual (*Eudorina*, *Volvox*, Zygnemataceae, Charales, *Vaucheria*), but several products occur in such vegetatively advanced chlorophytes as *Ulothrix*, *Coleochaete* and *Sphaeroplea*. There is a hint in Volvocales that the more advanced genera (*Eudorina*, *Platydorina*, *Volvox*) may have a single product of germination and the less advanced genera (*Stephanosphaera*, *Gonium*) four products.

The proposition is thus plausible but not beyond dispute; accepting it for the time being, it leads us to believe that the hypothesis predicts a correlation between the degree of gamete dimorphism and the level of vegetative size and complexity. On a very broad phylogenetic scale, this is of course the case: one of the patterns we are primarily concerned to interpret is the invariable oogamety of higher plants and animals, with isogamety being restricted to smaller and simpler forms. We are likely to be convinced by the hypothesis, however, only if it correctly predicts the correlates of gamete dimorphism among those groups of organisms some of whose members produce dimorphic gametes whilst others do not. Knowlton (1974) pointed out that the expected correlation is indeed displayed by the volvocine algae, where there is quite a pretty transition between the simpler isogametic and the more advanced anisogametic or oogametic genera. However, Bell (1978b) observed that the correlation was much less distinct in other series of chlorophyte algae. In Table 5.15, I have used the vegetative index as a rough quantitative measure of vegetative development, and shown that there is a clear correlation between the level of vegetative organization and the degree of gamete dimorphism at all levels amongst algae and protozoans. This correlation is not sensitive to the particular method used to calculate the index, nor is it jeopardized by adding other groups of lower plants to those included in Table 5.7. The rhodophytes are vegetatively advanced algae which are invariably oogametic; among fungi, isogamy is found only in the vegetatively simple Chytridiales and Blastocladiales. The major prediction made by the hypothesis is therefore accepted.

Table 5.15: Relationship Between the Level of Vegetative Organization and the Degree of Gamete Dimorphism. Data and abbreviations from Table 5.9. The 'vegetative index' (VEG) given here is VEG = SIZ × DIF × LAB. The analysis is at three levels: generic (G), familial (F) and ordinal (O).

| VEG | GAM | | | | | | | | |
| | I | | | A, PA | | | O | | |
	G	F	O	G	F	O	G	F	O
1	73	4	19	11	0	1	17	1	2
2–10	17	4	2	5	0	0	7	0	0
11–25	11	6	2	2	1	0	1	0	0
26–50	9	4	1	4	3	1	4	1	0
51–100	17	4	5	13	6	1	6	4	2
101–50	4	0	0	6	1	2	12	0	4
>150	0	0	0	0	0	0	13	0	2

It would be very satisfying if the hypothesis also enabled us to interpret gamete structure at a much lower phyletic level; a particularly sensitive test would concern the fate of the zygote in comparable isogametic and anisogametic unicells. This is a task which should properly be left to a theoretical phycologist; but since this breed of men seems excessively rare, I shall chance my arm by adding a few comments on diatoms.

The diatom frustule is shaped like a box with a tight-fitting lid, the larger 'lid' valve (epivalve) overlapping the smaller 'box' valve (hypovalve). Following vegetative cell division, both valves of the parental cell function as epivalves in the new cells: the result is that one division product is slightly smaller than the other and, as time goes on, a population reproducing exclusively by vegetative fission, especially if maintained in unfavourable conditions, experiences a reduction in average cell size. This reduction is made good only by sexual reproduction, sexuality being confined to small cells and the zygote ('auxospore') increasing in size after fertilization. The hypothesis therefore predicts that diatoms, despite being unicellular, will tend to be anisogametic. This prediction is accepted for

Centrales, which are oogametic. A hint of the same phenomenon can be found amongst unicellular chlorophyte algae, where either all haploid cells are potential gametes, or else the potential for sexual fusion is restricted to smaller cells, being lost as the cell enlarges prior to vegetative fission: the isogametic species of *Chlamydomonas* and members of the isogametic genera *Polytoma* and *Dunaliella* belong to the former group, the non-isogametic species of *Chlamydomonas* and *Chlorogonium* to the latter. But the othe major taxon of diatoms, the Pennales, are isogametic, and therefore seem to nullify the support given to the hypothesis by Centrales. However, while the Centrales are gametogamous, the Pennales typically have gamentangial fusion with amoeboid isogametes – the few gameto-gamous members of the order, such as *Rhabdonema*, are anisogametic. The hypothesis is therefore straightforwardly confirmed by gametogamous diatoms but not by those with gametangial fusion. But the hypothesis is not designed to deal with gametangial fusion, so it is not certain that the Pennales represent a falsifying instance.

Suppose that two gamonts or gametangia fuse, one of them producing iso-gametes of a certain size representing the locally stable optimum. The other, let us say, produces much smaller microgametes. Then most of these gametes will be wasted, since they are not permitted to fuse, except within the common space of the fused gamont shells. Those which do fuse will form smaller zygotes than would have been the case had they been larger. Therefore δ-producers will be unable to invade a population fixed for the locally stable isogamete. Macro-gamete-producers will also fail to invade, since the increase in the size of the zygotes to which they give rise will be more than balanced by the decrease in their number. It follows that the gamontogamous relatives of gametogamous anisogametic organisms should be isogametic; and the Pennales then fall into place as confirming a specific case of the hypothesis, rather than contradicting the general case. This argument hinges on the proposition that the production of microgametes is wasteful when the assumption of unrestricted gamete fusion is not met. It can therefore be tested by the specific prediction that gamonto-gamous forms should produce fewer, larger gametes than their gametogamous relatives. This appears to be true. In the oogametic Coccidia, gametogamous forms produce many (*Eimeria*), twelve to 32 (*Eococcidium*) or four to eight (*Haemoproteus, Plasmodium*) microgametes, but gamontogamous forms only two to four gametes (*Klossia, Karyolysis, Adelina*). In the isogametic Foramini-fera, *Iridia* (gametic fusion) produces about 10^7 gametes, *Discorbis* (gametic and gamontic fusion) about 250, and *Patellina* (gamontic fusion) only ten.

A secondary prediction can also be made. The argument above applies only to isogamontic forms, in which the fusing gamonts are of the same size. In aniso-gamontic forms where fusion occurs only between microgamonts and macro-gamonts, the optimal strategy for the microgamont is to produce the same number of gametes as its larger partner, so that all of its partner's gametes will be fertilized, and this implies that the gametes produced by microgamonts will be smaller than those produced by macrogamonts. It follows that, among gamontogamous organisms, anisogamety should be associated with anisogamonty. Table 5.16 shows that this is the case.

Thus, by tailoring the original hypothesis to particular contexts it is possible to fashion derivative hypotheses which have predictive power beyond the situations they were designed to explain.

Table 5.16: Support for the Predicted Correlation Between Anisogamonty and Anisogamety. Data and abbreviations from Table 5.9. Statistical analysis is not valid, since there is no guarantee that genera are chosen randomly from higher categories; for what it is worth, for genera the uncorrected $\chi^2 = 12.1$, with $P < 0.001$.

| | GAM | | | |
| | I | | A, PA, O | |
FUS	G	O	G	O
Gia, Col, Gon, Gon I	36	4	3	0
Gon PA, Gon A, Dm	7	3	7	1

Oogamety. So far, I have treated gamete dimorphism as though it were a category of phenomena within which the various cases differed only in degree. It is usually imagined, however, that the occurrence of a massive immotile macrogamete and a minute motile microgamete represents a qualitatively distinct category, to be dignified by the term 'oogamety'. Several authors have explained the evolution of oogamety from anisogamety as a process of increasing specialization of the two gamete types to different functions (Kalmus and Smith, 1960; Parker *et al.*, 1972; Ghiselin, 1974a). If macrogametes could direct the pattern of fusion, they would choose to fuse only with other macrogametes, in order to maximize the size and thus the survival of the zygote. However, any barriers erected by macrogametes to prevent fusion by microgametes are likely to be broken down faster than they can be mended, both because the selection acting on microgametes is more severe (since microgamete-to-microgamete fusion is very likely to produce wholly inviable zygotes) and because microgametes are far more numerous and thus have in the aggregate far more genetic variation. Oogamety then evolves from anisogamety as the macrogametes evolve immotility in order to conserve the food reserves needed by the zygote and the microgametes evolve motility to the utmost in order to maximize the probability of fusion.

These are simple and eminently plausible ideas, but they are rather difficult to test. I can only suggest that the most effective way of conserving the reserves of a macrogamete will be to retain it within the gametangium, where it can be nourished and protected until fertilized by a sperm. If this is a correct inference from the theory, one expects internal fertilization to occur more frequently and external fertilization less frequently amongst oogametic gametogamous genera than amongst isogametic or anisogametic gametogamous genera. Table 5.17 shows that this prediction is accepted.

Table 5.17: Support for the Prediction that among Gametogamous Taxa Oogamety and Internal Fertilization will be Associated. Data and abbreviations from
Table 5.7. With the same caveat as Table 5.16, χ^2 (for genera) = 39.4, $P \ll 0.001$;
the result is also superficially significant for families and orders.

	GAM					
	I, PA, A			O		
FUS	G	F	O	G	F	O
Get, Get E	89	25	19	20	1	5
Get I, E; Get (I); Get I	7	1	0	24	4	5

Sexual Differentiation. The mathematical theory developed above assumes that
gamete fusion is random; that is, it assumes that the difference in size between
two gamete morphs does not coincide with any consistent difference in gender.
This state of pseudoanisogamety is known from a number of algae. And yet in
oogametic organisms, and probably in all organisms where there is a marked
difference in size between the gamete morphs, it is an invariable rule that the
two gamete morphs have different gender, with only disassortative fusion being
permitted. Despite the importance of the problem, very little theoretical work
on the relationship between gamete and gender differentiation has been published,
and the few comments I can make do not justify a comparative analysis.

   In isogametic organisms there seems to be no definite limit to the number of
sexes. For instance, imagine an isogametic population in which there are two
equally frequent sexes, say, *A* and *B*. A few individuals of a third sex *C* are then
introduced, having the property of being able to fertilize either *A* or *B*. There
is a tendency for the third sex to spread, since the gametes produced by *A* or *B*
individuals will be able to fuse with only about half the gametes they encounter
at random, whilst any *C* gamete can fuse with virtually any other gamete, so long
as *C* is rare. It is elementary to show that at equilibrium the three sexes are
equally frequent; but a fourth sex can now invade, and so forth. Suppose that in
a three-sex system an allele arises at some locus other than that determining
gender, and has the effect of causing the gametes of individuals which bear it
not to fuse with *C* gametes. If this allele tends to spread when rare then the
three-sex population is after all structurally unstable. However, I have found by
simulation, using several different sets of rules to govern the fusion of gametes,
that such an allele will not tend to spread when rare; it will spread, and produce
a permanent change in the population, only if it is initially extremely common
or if some extrinsic advantage is invoked, such as an ability of gametes carrying
the mutant allele to fuse before normal gametes. There is, then, no reason to
expect the familiar sexual bipolarity of higher plants and algae always to apply
amongst isogametic protists, and it is comforting to find that it does not. Multipolar systems are usual among ciliates, and reach a limit in the foraminiferan
Metarotaliella parva, where each gamont is sexually distinct and the only limi-
tation on gamete fusion is that autogamy is not permitted.

Some isogamontic ciliates (e.g., *Paramecium aurelia*) and, probably, several isogametic algae (see Wiese, 1976) are sexually bipolar, and it is not easy to see why this should be. Anisogametic algae seem invariably to be bipolar, whether or not gamete fusion is disassortative. Charlesworth (1978) studied the evolution of disassortative fusion in a bipolar anisogametic haplont, using a two-locus model in which alternative alleles determined gender at one locus and gamete size at the other. He concluded that, if anisogamy is stable, linkage disequilibrium will develop, with gametes of the same size tending to have the same gender. When the two loci are initially unlinked, however, selection is very slow.

If the disparity in gamete size is rather large, it is easy enough to see what should happen. No third sex of intermediate gamete size can invade, so that if the two gamete morphs are not sexually distinct there must be sexual differentiation within each morph. If this necessitated assortative fusion it would be counterselected, since microgamete-to-microgamete fusion would be lethal or nearly so, whilst macrogamete-to-macrogamete fusion would be very infrequent; the argument is the same as that used in the discussion of oogamety. On the other hand, there might be two microgamete sexes, each of which was capable of fusing with only one of the two macrogamete sexes. But if we begin with a bipolar population with sexes M and F, and then introduce two rare sexes M^* and F^*, then the fitness of the M^* sex will be proportional to the frequency of the F^* sex among macrogametes, and the fitness of the F^* sex proportional to the frequency of the M^* sex among microgametes, so that neither will tend to spread when rare. It is thus difficult to see how more than two sexually distinct types can be maintained within anisogametic species, and in fact such a condition does not seem to have evolved in algae or protozoans. However the brevity of this discussion will serve to underline the need for more rigorous theoretical work directed towards predicting the state of sexual systems in bipolar and multipolar algae, ciliates and fungi.

Conclusion. It has been known for some time that the familiar state of higher animals and plants, with their bipolar oogametic sexuality organized around a gametic meiosis in a basically diplontic life cycle, is foreshadowed among protists and thallophytes by broad general correlations between the level of vegetative structure, the site of meiosis and the degree of gamete dimorphism. It has been too common in the past to view these correlations in a clandestinely orthogenetic fashion, as reflecting evolution from a 'lower' (isogamety, zygotic meiosis, simple habit) to a 'higher' (oogamety, gametic meiosis, complex habit) phylogenetic state, and to use them solely as taxonomic criteria since, of all characters, they seem the furthest removed from immediate functional significance. There may indeed be an historical component in their evolution which cannot be ignored: it may be much easier to evolve from vegetative haploidy to vegetative diploidy than to revert to haploidy. What I have tried to do in the last two sections is to take the contrary view, and show that the site of meiosis and the nature of the gametes are functional characters evolving under individual natural selection, about which it is possible to erect and to test explicit and falsifiable hypotheses. The hypotheses I have advanced are no doubt inadequate; but they are not narrow or trivial, because their object is to gain one of the glittering prizes of biology, an understanding of the fundamental attributes of sexual organisms. Accepted, amended or falsified by others, they may bring us closer to one of the crowning achievements of evolutionary thought.

6 METAGENETICS

As this essay progressed, my thoughts seemed to be cohering at some distance from any recognized academic discipline – at some distance, that is, from anything that would normally be taught as an undergraduate course. It is no doubt pompous to imagine that one's own petty speciality constitutes a major and hitherto unconsolidated area of thought, but I shall risk the charge. I began this essay with the axiom that all characters are wholly functional, in the sense that their present states can be explained wholly in terms of their immediate consequences to the individual. I do not now believe this to be true, in part because, despite my own inclination, I have been forced to invoke an historical component in the evolution of automixis, of diploidy and of oogamety, and even of sex itself, albeit only as a drag upon a fundamental functionalism. But more importantly, my concept of function itself has changed, in a way which seems to set the study of genetic systems a little apart from the rest of evolutionary biology. This insight is by no means original; it corresponds roughly to the theme of Dawkins' recent popularization, *The Selfish Gene*, itself a recrudescence of ideas long current in the literature (Dawkins, 1976).

Broadly speaking, what I propose to call *metagenetics* is the study of how the machinery of inheritance, which mediates all evolution, itself evolves. It is, so to speak, the 'why' science of genetics, which attempts to explain what geneticists, having discovered, have long taken for granted. The province of metagenetics includes any attempt to interpret or explain, rather than merely to describe, the organization of genetic material into chromosomes; the number and structure of these chromosomes; the rules regulating their behaviour during mitosis and meiosis, notably their reassortment and recombination; the nature of sexual systems; the alternation of ploidy in the life cycle; and the many kindred epiphenomena of sex discussed or referred to in the body of this essay, besides others I have ignored or forgotten. I have tried to sketch this province in Figure 6.1.

Of course, it is possible to denominate a 'meta' science corresponding to any descriptive or nonexplanatory area of research: thus we might recognize metacolouration, metarespiration and any number of others. My reason for proposing metagenetics as a distinct field of enquiry is that, defined more tersely, metagenetics is the study of autoselection. A gene which affects, say, colouration will be favourably selected if its effects contribute to the harmonious functioning of the organism as a whole, perhaps by rendering it more cryptic or by advertizing its unpalatability. Any such gene, affecting colouration or size or cold resistance or locomotion, or indeed any aspect of anatomy or physiology, will affect the ability of the organism to survive, the number of germ cells it produces, and the competence of these germ cells to fertilize, to be fertilized or otherwise to develop. Only if the sum of its effects tends on the whole to increase the fitness of the organism will it be favourably selected. For some characters one might

498

Figure 6.1: Metagenetics

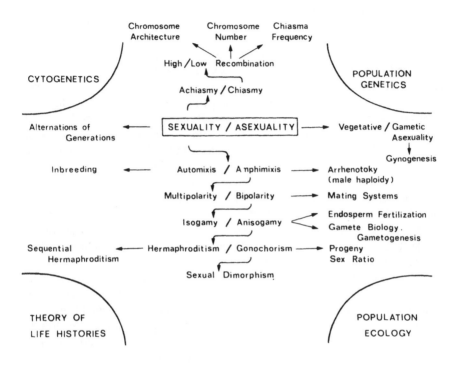

want to invoke the fitness of the sibship or the local population or some other group, but the principle remains the same, and the process is one of natural or sexual selection. But if the gene in question changes the machinery of inheritance so as to favour its own replication, it will tend to spread even though it contributes nothing to the economy of the individual or the population or any other entity; it will tend to spread even though it degrades normal function and reduces fitness. The purest instances of autoselected systems, such as the segregation-distorter chromosomes of *Drosphila* and the mutator genes of prokaryotes, evolve despite the fact that they are antagonistic to the normal function of the organism.

Autoselected elements usually spread so far and no further. The reason is that, by adjusting the machinery of inheritance, they alter the context in which alleles at other loci are expressed, and therefore necessarily alter the context in which they themselves are expressed, relative to some alternative element. In this way they are sooner or later exposed to natural selection acting through the genetic constellation they have themselves helped to create, and their fate is eventually decided by a tension between autoselection and natural selection.

Another way of thinking about function is expressed in the old query, what is the unit of selection? Well, the unit of selection is the differential element: that element between variants of which selection discriminates. It may be a gene, a group of genes, an individual organism, a colony or a part of a colony, a clone or a group of clones, a population of sexual individuals or any other entity showing variation and reproduction. In general it is several or all of these simultaneously, although at different levels selection may act with very different intensities and even in different directions. It was a hard lesson to learn that organic evolution is caused by the blind selection of characters arising by chance mutation and subsequently shuffled at random, but at least we were comforted by the reflection that character states are selected by virtue of the contribution they make towards the harmonious functioning of a greater whole, since this assuages our notions of order and propriety. With autoselection this comfort is denied. Selection is wholly at the level of the gene or some comparable element of the genome; and it is not concerned in any way with the well-being of the whole, but only with the ruthless efficiency of its own blind replication. It is the starkest of Darwinian concepts.

If autoselection is recognized as a distinct category of selection, then metagenetics will be recognized as a distinct field within genetics; the field, moreover, at the heart of genetics, where the reasons for the most fundamental features of living organisms are debated. I need hardly say that metagenetics and autoselection are new names for old concepts. Weismann and Fisher knew of autoselection; forty years ago Darlington (1939) wrote a book whose title, *The Evolution of Genetic Systems*, epitomizes metagenetics; and the last decade has seen a stirring of interest in topics such as the evolution of sex, recombination and reassortment, which had long lain dormant, or been regarded as mere curiosities. My reason for creating two neologisms is to point out the relationship between apparently disparate problems, to argue that they are quite different from apparently related problems and so to focus attention on an area of profound importance to biology as a whole.

GLOSSARY OF TERMS

The Glossary is intended to be merely a reminder of the meaning of certain technical terms. Most familiar terms are excluded; so are most unfamiliar terms which are defined in the text on their first and only appearance. All taxon names and most anatomical terms are excluded. Words which are printed in bold type in the definitions are themselves defined in the Glossary, with a few exceptions where the definition is implicit.

Acentric. Acentric fragment: a fragment of a **chromosome** which lacks a **centromere** and therefore does not **segregate** regularly during nuclear division.

Achene. A single-seeded fruit.

Achiasmate. Achiasmate (achiasmatic) **meiosis**: a **meiosis** in which **homologous chromosomes** pair without **crossing over.**

Achiasmy. The state of possessing an **achiasmate meiosis.**

Acyclic. Reproducing by obligate thelytoky; without sexual periods (Cladocera).

Agamete. Broadly, any reproductive propagule which develops into a new **individual** without **syngamy.** Usually intended to mean a **haploid** propagule which, while having the capacity for **syngamy,** develops without it (**azygospore**).

Agametic. (Reproduction) without **gametes.**

Agamogony. Asexual (vegetative) reproduction of unicells, proceeding by a mitosis followed by **cytokinesis** to yield two products or by several **mitoses** followed by **fragmentation** into uninucleate propagules (**agametes**).

Agamont. An individual which gives rise to **agametes.**

Akinete. A dormant structure formed by the encystment of a vegetative cell (Algae).

Allele (Allelomorph). One of the two or more **genes** which may occupy the same site (locus) on a **chromosome.**

Allopatry. Having separate geographical distributions.

Alloploidy. Having one or more sets of **chromosomes** from a parent of one species and another set or sets from a parent of another species. Hence allodiploidy, allopolyploidy.

Alternation of generations. The alternation of ploidy enforced in all sexual organisms by the alternation of **meiosis** and **syngamy.**

Amixis. The absence of mixis; asexuality.

Amphigony. Amphimixis.

Amphimixis. The occurrence of **syngamy** between **gametes** produced by different **individuals. Amphimict:** an organism reproducing by **amphimixis.**

Amphogeny. The production of both **male** and **female** progeny, usually in more or less equal numbers. In Cecidomyidae, the production of both **thelytokous female** larvae and sexual **male** larvae.

501

Anabiotic. Having an extremely low metabolic rate, as in the dormant adults of bdelloids, tardigrades and nematodes.

Ancestrula. The overwintering structure of gymnolaemate bryozoans.

Androgenic. A larva which gives rise to **sexual male** imagos (Cecidomyidae).

Aneuploid. Having one or more whole **chromosomes** more or less than the regular complement; having a **chromosome** complement not a simple multiple of a basic **haploid** number.

Anholocyclic. Reproducing by obligate **thelytoky**; without a **sexual** generation (Aphids).

Anisogamety. The occurrence of **gametes** of different size, those of like size having the same **gender**.

Anisogamonty. The occurrence of unequal **gamonts**.

Anisogamy. The occurrence of **gametes** of different size; more broadly, the occurrence of any differences in size, shape, structure or behaviour between gametes. Cf. **anisogamety, isogamety, oogamety, pseudoanisogamety**.

Antheridium. The **gametangium** in which **microgametes** are produced. Cf. **oogonium**.

Aplanospore. An asexual propagule which does not realize a potential for motility (Algae).

Apomixis. The absence of both **meiosis** and **syngamy** amongst organisms which reproduce by **eggs**.

Architomy. **Vegetative fission** giving rise to **individuals** of which one or more must undergo considerable growth and differentiation before in turn reproducing.

Arrhenogeny. The production of **male** progeny exclusively (especially in isopods).

Arrhenotoky. The production of **haploid males** from unfertilized **eggs** and **diploid females** from **fertilized eggs**. **Haplodiploidy**. In 'diploid arrhenotoky' the **males** are made **diploid** but **homozygous** by an **endomitosis** during early **cleavage**.

Asexual. Lacking **sexuality**; **amictic**.

Asymptote. A value which is approached in the limit.

Asynaptic. Asynaptic meiosis: a **meiosis** in which **homologous chromosomes** do not pair.

Atoke. A vegetative **individual** which gives rise to a **sexual** individual (**epitoke**) (Polychaeta).

Autocolony. A miniature of the parental **colony** produced by **mitosis**.

Autogamy. **Syngamy** between **gametes** (usually) or gametic pronuclei (in unicells; cf. **automixis**) from the same **individual**. **Self-fertilization**.

Autogeny. In the sense used in this book, the occurrence of an adult which does not feed (Diptera).

Automixis. **Syngamy** between **meiotically** reduced nuclei descending immediately from the same **zygote**, whether or not distinct **gametic** cells are formed.

Autoploidy. The occurrence of two or more sets of **chromosomes**, all of which derive from members of the same species. Hence, **autodiploidy, autopolyploidy**.

Autosegregation. Segregation caused by crossing-over during **mitosis**.

Autoselection. The selection of genes according to the efficiency with which they are transmitted without reference to their phenotypic effects.

Autosomal. An autosomal chromosome (autosome) is any chromosome other than a sex chromosome.

Autospore. An immotile asexual propagule smaller than, but otherwise identical to the parental cell (Algae).

Auxospore. The zygote of a diatom.

Azygospore. A haploid propagule which having the capacity for syngamy develops without it (Cf. agamete). Parthenospore.

Azygote. A haploid propagule not having the capacity for syngamy which develops directly into a new organism.

Bipolar. Of mating systems; having two genders, either of gametes alone or of both individuals and gametes. Cf. Multipolar.

Bivalent. The entity formed by the pairing of two homologous chromosomes during meiosis. Hence trivalent, quadrivalent, etc., when more than two homologous chromosomes exist and come into association.

Bud. See budding.

Budding. A loose term used to denominate a process of vegetative fission, typically but not always resulting in the formation of a larger and one or more smaller individuals. It has been used to describe certain kinds of protozoan agamogony, the paratomy of *Hydra*, the architomy of naiad and aelosomatid oligochaetes and (most often) the proliferation of zooids in hydrozoans, endoprocts, bryozoans, tunicates and other colonial animals. These are all instances of external budding; structures such as the gemmules of sponges and the statoblasts of bryozoans are sometimes referred to as internal buds.

Capricious. A term used here to describe an environment in which the correlation between adaptively important features frequently changes sign. Equivalently, a situation in which the coefficient of linkage disequilibrium between two given loci often changes sign. Note that there is a prior usage of the term by Lewontin (1966) to mean a random process with limited memory.

Carrying capacity. The density of a density-regulated population at equilibrium.

Centric fusion. (1) Fusion by symmetrical translocation of two acrocentric chromosomes to yield a large metacentric and a small fragment. (2) Automictic fusion of the two central non-sister polar bodies. The second sense is the one used here.

Centromere (Kinetochore). The region of a chromosome with which the spindle fibres become associated during nuclear division and which is responsible for orderly segregation. Diffuse centromere: the spindle fibres are attached along the whole length of the chromosome, so that a fragment of any size will segregate regularly.

Chemostat. An apparatus designed to maintain a constant chemical environment for laboratory cultures of algae or bacteria.

Chiasma. A region of contact between homologous chromosomes during meiosis in which there is an exchange of homologous parts between non-sister chromatids. Cf. crossing-over.

Chromatid. One of the two identical replicates of a **chromosome** formed during nuclear division, at first closely associated with its sister chromatid but separating during anaphase.

Chromosome. In **eukaryotes**, a visible microscopic entity within the nucleus bearing genetic information arranged in a linear sequence. Thus, a sequence of **linked genes**, a **linkage group**. The chromosome of prokaryotes is a circular structure not localized within a nuclear membrane.

Cleavage. (1) The division of the cytoplasm following nuclear division; see **cytokinesis**. (2) The early cell divisions of the **zygote**. The second sense is the one used here.

Clone. Broadly, any assemblage of genetically identical organisms. More specifically, an assemblage of **individual** organisms which are identical by descent for every **allele** at every **locus**.

Coccoid. Unicellular, without flagella or motility.

Coenocyte. A multinucleate structure without cross-walls.

Colony. Broadly, an assemblage of organisms of the same species living in the same place. Never used in that sense here. More specifically, an assemblage of genetically identical organisms descending **amictically** from the same parent which remain in physical continuity or in any other way lack **individuality**.

Cross-fertilization. **Syngamy** of **gametes** produced by different **individuals**. Often used to denote syngamy of **gametes** produced by genetically differing **individuals** (in opposition to **self-fertilization**) or by unrelated individuals (**outcrossing**, in opposition to **inbreeding**).

Crossing-over. Usually, the exchange of genetic material between non-sister **chromatids** of **homologous chromosomes** by symmetrical breakage and reunion during **meiosis**. See **chiasma**.

Cyclical parthenogenesis. A life history in which a sequence of one or more thelytokous generations is followed by a single arrhenotokous or amphimictic generation; usually applied to cases in which the alternation is more or less regular and predictable. **Heterogony**.

Cyst. A dormant structure formed by thickening the wall of a cell, a zooid or a **colony**.

Cytokinesis. The division of the cytoplasm and the consequent segregation of daughter nuclei into different cells following nuclear division.

Cytotype. Any assemblage of individuals of the same species defined on cytological grounds.

Density-regulated (Density-dependent). The rate of increase of a density-regulated population is inversely proportional to population density, and is negative for densities which exceed a certain value.

Desynapsis. The failure of chromosome pairing during **meiosis**; Cf. **asynaptic**.

Deuterotoky. The **parthenogenetic** development of both **males** and **females** from unfertilized **eggs**.

Diacmic. Having two population maxima in the same growing season (Cladocera).

Diallelic. Having two possible **allelic** states.

Diapause. Spontaneous dormancy.

Dicentric. Of **chromosomes**, having two **centromeres**.

Dicyclic. Having two sexual periods in the same growing season (Rotifera, Cladocera).

Dikaryon. A cell with two nuclei; or a mycelium comprising such cells.

Dioecious. With **males** and **females** as separate and distinct **individuals**; **gonochoric**. Usually applied to plants.

Diploid. Having two sets of homologous chromosomes, one of maternal and the other of paternal descent.

Diploid arrhenotoky. See **arrhenotoky**.

Diplont. An organism with a **gametic meiosis**, in which the diploid generation is vegetatively **dominant**. Cf. **Haplont**.

Disruptive selection. **Selection** favouring extreme variants at the expense of modal individuals.

Dominant. (1) An **allele** is said to be **dominant** if (or to the extent that) its **phenotypic** manifestation is apparent in **heterozygotes**. (2) Loosely, to mean larger or longer-lived, as 'vegetatively dominant' (see definition of **Diplont**).

Double fertilization. The **fertilization** by the two pollen-tube cell nuclei of both an **ovule** nucleus and an endosperm nucleus in flowering plants.

Drift. See **genetic drift**.

Eclosion. Hatching from an **egg**.

Egg. A **fertilized ovum** or **ovule**; a **zygote**; the **unfertilized** but unreduced propagule of **apomicts**.

Electrophoresis. The separation of molecules of differing net charge by virtue of their differing mobilities under an applied electric potential.

Endomitosis. A mitosis all of whose products are retained within the same nuclear membrane; thus a means of increasing **ploidy**.

Ephippium. The dormant encysted embryo of cladocerans, usually the result of **sexual** reproduction.

Epigamy. The modification of an **asexual** individual (**atoke**) into a **sexual** individual (**epitoke**) without any intervening process of **reproduction** (Polychaeta).

Epistasis. Any interaction with **phenotypic** effect between **nonallelic genes**.

Epitoke. The **sexual individual** of polychaete annelids.

Equilibrium. Any state of a system which once attained tends to persist indefinitely. **Stable equilibrium**: if a system in equilibrium is perturbed then the equilibrium is stable if it tends to be restored when the perturbing force has been removed (example: a ball in a trough). **Unstable equilibrium**: if the system continues to move away from the equilibrium when the perturbing force is removed, the equilibrium is unstable (example: a ball on a ridge). **Neutral equilibrium**: if the system conserves its new state after the removal of the perturbing force, the equilibrium is neutral (neutrally stable) (example: a ball on a plane).

Eukaryote. An organism with linear **chromosomes** localized within a nuclear membrane, typically showing both **mitosis** and **meiosis**.

Female. The **gender** or an **individual** of the **gender** producing macrogametes.

Exephippial female. A thelytokous female cladoceran which has hatched from an **ephippium**.

Fertilization. The union of **gametes**; syngamy.

Fission. The **vegetative amictic** division of the body to form two or more distinct individuals. **Binary fission**: division into two individuals; cf. **budding**. **Multiple fission**: division into several or many individuals; cf. **budding, fragmentation**. **Transverse fission**: division at right-angles to the main axis of the body. **Longitudinal fission**: division along the main axis of the body. **Equal fission**: division into two or more parts of the same size. **Unequal fission**: division into two or more parts not all of the same size. Cf. **architomy, paratomy**.

Fitness. A rate of increase; usually the rate of increase of some type relative to that of an alternative type in the same population over a specified period of time. Often used loosely.

Floatoblast. A bryozoan **statoblast** provided with an annulus of gas-filled cells and thus floating in water.

Fragmentation. Vegetative fission into a large number of small parts. **Colony fragmentation**: the casual or regular detachment of small portions of the colonies of sponges, corals, bryozoans and other organisms.

Frequency-dependent selection. Any scheme of **selection** in which the fitness of a type varies with its frequency.

Frustule. (1) An **asexual** propagule of hydrozoans, developing into a polyp. (2) One of the two valves of a diatom.

Gametangium. A structure in which **gametes** are produced.

Gamete. A cell capable of undergoing **syngamy** with another cell to form a **zygote**.

Gametic. Descriptive of behaviour promoting **syngamy**: thus pollen grains are **gametophytes** with largely gametic properties.

Gametic meiosis. A **meiosis** whose products function immediately as **gametes**.

Gametogenesis. The process of differentiation of **gametes**.

Gametophyte. The **haploid gamete**-producing generation of plants.

Gemmiparous stolonization. A chain of zooids, later separated by fission as distinct **individuals**, formed by repeated intercalary **budding** in certain polychaetes.

Gemmule. An **asexual** propagule of sponges.

Gender. The set of **individuals** or **gametes** of the same species incapable of fertilizing one another, e.g., **males** or **females** in **gonochores**, or **microgametes** and **macrogametes** in **anisogametic amphimicts**.

Gene. The unit of inheritance; a small part of a **chromosome** producing a certain protein and separated from adjacent parts producing different proteins.

Genetic drift. The stochastic change in **gene** frequency caused by random sampling of a finite population.

Genetic recombination. The change in the relationship between **loci** on the same chromosome caused by **crossing-over**.

Genome. The totality of genetic material transmitted to progeny.

Genotype. (1) The allelic state of any specified number of loci in a given individual. (2) The genetic potential realized in the phenotype.

Germ line. The lineage of gamete-producing cells.

Gonochoric. With males and females as separate and distinct individuals; dioecious. Usually applied to animals.

Gynogenesis. Thelytokous reproduction requiring pseudogamy.

Haplodiploidy. See arrhenotoky.

Haploid. Possessing a single set of chromosomes, none of which are homologous to one another. Cf. diploid.

Haplont. An organism with a zygotic meiosis in which the haploid generation is vegetatively dominant. Cf. diplont.

Hardy-Weinberg equilibrium. The frequencies of genotypes at a given diploid locus determined by random gamete fusion, reached in a single generation and thereafter conserved.

Hemiclone. A lineage of hybridogenetic animals with a common maternal haploid genome.

Hemizygous. Haploid at a specified locus or loci; applied when the same locus or other loci are diploid in the same or in other individuals of the same species.

Hermaphroditic. (1) In animals, the formation of macrogametes and microgametes by the same individual; the simultaneous possession of both male and female function (simultaneous hermaphroditism). Sequential hermaphroditism: alternation from one gender to the other during the life of an individual. (2) In plants, the occurrence of male and female organs in the same ('perfect') flower of monoecious species.

Heterochromatin. Any segment of a chromosome which becomes heteropycnotic (densely staining) at any stage during the cell cycle.

Heteroecy. The habit of feeding on two or more different plant hosts (Aphids).

Heterogenetic. Applied to parasites with a complex life cycle.

Heterogony. Cyclical parthenogenesis.

Heteromorphic. Heteromorphic alternation of generations: a life cycle in which the haploid and diploid generations are morphologically dissimilar.

Heterosis. The superior performance of heterozygotes relative to homozygotes.

Heterotrichous. Having both erect and prostrate filaments.

Heterozygous. A given locus in a diploid (or polyploid) individual is heterozygous if it bears two (two or more) different alleles. A multilocus genotype is heterozygous if (or to the extent that) any of the loci bear different alleles. Cf. homozygous.

Hibernaculum. An asexual resting structure, e.g., of freshwater gymnolaemate bryozoans.

Holocyclic. Having sexual periods (Aphids).

Hologamety. The direct transformation of a haploid vegetative cell into a gamete.

Homogenetic. Applied to parasites with a single host. Cf. heterogenetic.

Homologous chromosomes. The paternally and maternally derived chromosomes which bearing the same sequence of loci pair during meiosis and may cross over.

Homozygous. A given **locus** in a **diploid** (or **polyploid**) organism is **homozygous** if the two (or more) **genes** present are identical in state. A multilocus **genotype** is **homozygous** if (or to the extent that) all **loci** bear identical **alleles**. Cf. **heterozygous.**

Hygrophilous. Inhabiting moist places.

Hybridogenesis. A mode of **thelytoky** in which a **sperm genome** donated by a male of a related **amphimictic** species is received and expressed by an **hybridogenetic zygote** but is not transmitted to the progeny of that **zygote.**

Hypnospore. A dormant **aplanospore.**

Inbreeding. Mating between related **amphimicts.**

Inbreeding depression. The inferior performance of inbred relative to outbred progeny, attributed to some consequence of their greater **homozygosity.**

Individual. An elementary and therefore difficult concept expressing the extent to which one living thing is separate from or independent of other living things. Widely discussed in the text (e.g., Section 1.2 and 3.4).

Inflection. The point on a curve at which the acceleration of the slope changes sign; the point at which the second derivative is equal to zero.

Infusorigen. The dispersive infectious stage of the life cycle of dicyemid mesozoans.

Intrameiotic restitution. The **automictic** restoration of ploidy by the suppression of one of the two **meiotic** divisions.

Inversion. A chromosome segment in which the ordering of **loci** is reversed relative to that of some standard sequence.

Isogamety. The state in which **gametes** of different gender have the same size.

Isogamy. **Isogamety.**

Isomorphic. Isomorphic **alternation of generations**: a life cycle in which the **haploid** and **diploid** generations are morphologically indistinguishable or nearly so.

Karyotype. The **chromosome** complement: the number and appearance of the **chromosomes.**

Laceration. See **pedal laceration.**

Lethal. A **gene** which is fatal (at some arbitrary but early stage) to its carrier if expressed.

Limit cycle. A stable oscillation: an oscillation which endlessly passes through the same sequence of points and which if perturbed returns to that sequence after the removal of the perturbing force.

Linkage. The propensity of **nonallelic genes** to be transmitted together, for instance because they occur near to one another on the same **chromosome.**

Linkage disequilibrium. The nonrandom association of **genes** at different **loci**; the extent to which gametic frequencies differ from those computed from allelic frequencies assuming complete independence of gene effect. **Linkage equilibrium** is the limiting case in which the frequencies of multilocus **haploid genotypes** are their random expectations.

Linkage group. A block of **genes** which tend to be transmitted together, e.g., a **chromosome.**

Locus. A site on a chromosome at which different **genes** are allelic to one another.

Macrogamete. The larger **gametic** type of **anisogamous** organisms, typically an **ovum** or **ovule**. Cf. **microgamete**.

Male. The **gender** or an **individual** of the **gender** which produces **microgametes**.

Mating type. A **gender**, especially of **isogamous** protists with **multipolar** mating systems. Cf. **bipolar, gender**.

Meiosis. A single or (much more often) a sequence of two nuclear divisions during which **ploidy** is halved, typically involving **genetic recombination** through **crossing-over** between **homologous chromosomes**.

Meiotic drive. The propensity for one allele or chromosome rather than its alternative to be incorporated into **gametes**.

Mendelism. A theory of inheritance based on the independent **assortment** and random **recombination** of particulate, conserved hereditary determinants, now widely accepted in a modified form for **eukaryotes**.

Merogony. (1) **Monogeny**. (2) Development of cytoplasm containing a **sperm** nucleus but not an **ovum** nucleus; not used here in this sense.

Mesic. Moist; strictly, intermediate between dry (**xeric**) and wet (hydric).

Microgamete. The smaller gametic type of **anisogamous** organisms, typically a **sperm** or a pollen grain (strictly, pollen-tube cell nucleus).

Mitosis. A single nuclear division typically resulting in the exact replication of the **genome**.

Mixis. Sexuality. The rearrangement of genetic material through **meiosis** or **syngamy** or (usually) both, almost always resulting in the production of one or more new organisms differing genetically from one another and from their parents.

Modifier (Modifying gene). A **gene** which has no **phenotypic** effect other than to modify the phenotypic expression of genes at another **locus**.

Monacmic. Having a single population maximum during the growing season (Cladocera).

Monocyclic. Having a single **sexual** period during the growing season. (Monogononta, Cladocera, and other **heterogonic** organisms).

Monoecious. (1) With the **sexes** united; with the same **individual** possessing both **male** and **female** function and producing both **microgametes** and **macrogametes**. Usually applied to plants. (2) In aphids, restricted to a single plant host.

Monogenetic. Of parasites, having a single host species. **Digenetic** parasites have two hosts.

Monogeny (Merogony). The production of progeny all of which are of the same sex. Cf. **arrhenogeny, thelygeny, amphogeny**.

Monomorphic. Uniform; the possession of the same **phenotype** by all members of a population. Cf. **polymorphic**.

Monospore. An asexual propagule formed singly in a **sporangium**.

Morphotype. A variety defined according to morphological criteria.

Mosaic. An individual some of whose somatic cells differ genetically from one another.

Multipolar. Of mating systems, having three or more **genders** of isogametes. Cf. **bipolar.**

Mutation. An abrupt heritable change in the genetic material, especially in a single **gene (gene mutation).**

Nematogen. An infestive stage in the life cycle of dicyemid mesozoans.

Neutral allele. A **gene** which differs chemically from another **gene** at the same **locus** but which has indistinguishable effects on the **phenotype.**

Niche. A term perhaps best left undefined. I have used it in the sense of the intimate local circumstances of an individual or genotype, as distinct from the general habitat or environment of the population.

Nurse cells. Cells intimately investing and contributing to the support of **oocytes, ova** or **eggs.**

Oligopyrene. Oligopyrene **sperm** lack a large part of the normal **haploid genome** and are incapable of **fertilizing ova. Apyrene** sperm lack a **genome** altogether.

Oocyte. A cell immediately ancestral to the **ovum.** The primary oocyte undergoes the first **meiotic** division to give rise to the **secondary oocyte** (and the first **polar body**) the **secondary oocyte** undergoing the **second** meiotic division to yield the **ovum** (and the second **polar body**).

Oogamety. An extreme gamete dimorphism, characterized by the presence of a massive immotile **ovum** and a minute **sperm.**

Oogenesis. The process of formation of **ova** from primary **oogonia.**

Oogonium. (1) The germ cell which gives rise to the **oocyte.** Primary oogonia proliferate **mitotically** to give rise to secondary **oogonia,** which become transformed into primary **oocytes.** (2) The female **gametangium.**

Optimum. The value taken by an independent variable when a given dependent variable is maximized or minimized. In evolutionary biology, independent variables such as fecundity, gamete size or parental care are usually optimized under the constraint that **fitness** is maximized.

Orthogenesis. The notion that once evolution has begun to proceed in a certain direction it tends to persist in that direction by virtue of its own momentum.

Overdominance. A locus is overdominant if the performance (for a given metric on a given scale) of the **heterozygote** exceeds that of either **homozygote.** It is **underdominant** if the performance of the **heterozygote** falls below that of both **homozygotes.**

Oviparous. Reproducing by eggs which hatch outside the body of the parent. Cf. **viviparous.**

Ovule. The megasporangium of angiosperms, analogous to the **ovum** of animals.

Ovum. The **macrogamete** of animals.

Paedogenesis. Reproduction by young and especially by larval individuals.

Palmelloid. Consisting of a number of non-motile cells in a gelatinous matrix.

Paracentric. A paracentric **inversion** does not include the **centromere.** Cf. **pericentric.**

Paratomy. Vegetative **fission**, the products of which are morphologically complete individuals. For **slow, rapid, stylarian** and **naiadian paratomy**, see Section 3.24. Cf. **architomy, fission, budding.**

Parthenogenesis. The production of **eggs** which develop without **fertilization** by another **individual.** See Section 1.2.2. Cf. **amphimixis, automixis, apomixis, arrhenotoky, thelytoky, deuterotoky, gynogenesis, hybridogenesis.**

Pedal laceration. The casting-off of small pieces of tissue which are able to develop into new individuals from the pedal discs of certain anthozoans and ctenophores. Cf. **architomy.**

Penetrance. The degree to which an allele is expressed in the phenotype of **heterozygotes.**

Pericentric. A pericentric **inversion** includes the **centromere.** Cf. **paracentric.**

Perigynium. The envelope surrounding the fruit in *Carex.*

Phenotype. Any measurable characteristic of an organism. Used conventionally in opposition to **genotype**, which then either means the chemistry of the genetic material or becomes an abstraction.

Pistillate. Bearing pistils, the **female** reproductive structures.

Ploidy. The number of **haploid chromosome** complements in the **genome**. Cf. **haploid, diploid, polyploid.**

Podocyst. An **asexual** dormant propagule of scyphozoans.

Polar body. Small vesicles containing chromatin expelled from the **ovum** or **egg** and representing those nuclei other than the **ovum** nucleus descending by **meiosis** from primary oocytes.

Polycyclic. Having several sexual periods during the same growing season (Monogononta, Cladocera).

Polyembryony. The **amictic** proliferation of embryos before birth.

Polymorphic. Diverse: the occurrence of different **phenotypes** among members of the same population.

Polyploid. Possessing several haploid chromosome complements; hence **triploid**, with three complements, **tetraploid** with four, **pentaploid** with five, etc.

Postmeiotic restitution. The automictic restoration of ploidy by **endomitosis** during early **cleavage.**

Premeiotic restitution. The **automictic** doubling of ploidy prior to **meiosis.**

Prokaryote. An organism with a circular **chromosome** not localized in a nuclear envelope and lacking **mitosis** or **meiosis**; a bacterium or blue-green alga.

Pseudoanisogamety. The occurrence of **gametes** of different size, the difference in size not correlated with a difference in **gender.**

Pseudofertilization. Pseudogamy.

Pseudogamy. The initiation of **parthenogenetic** development by the penetration of a **sperm** into an **ovum**, without the **sperm genome** contributing genetic information to the **zygote.** Cf. **gynogenesis.**

Recessive. A recessive **allele** is not expressed in **heterozygotes.** Cf. **dominant, penetrance.**

Recombination. See **genetic recombination.**

Recombination index. A measure of the rate of **genetic recombination**; an attempt to estimate the number of blocks of **genes** which are transmitted independently. It is equal to the sum of the **haploid** complement and the number of **chiasmata** per nucleus.

Reduction. The halving of **chromosome** number during **meiosis**.

Regenerand. Fragment produced by the **architomical fission** of the posterior part of enteropneusts.

Rhombogen. Phase in the life cycle of dicyemid mesozoans, formed by accumulation of food reserves in a **nematogen**.

Schizogamy. Creation of **sexual** individuals by **fission** in polychaetes.

Schizogony. Multiple fission of protozoans.

Schizont. A stage in the life history of sporozoans, reproducing within the host by **schizogony**.

Segregation. The separation of pairs of **alleles** at **meiosis** and their passage into different **haploid cells**. The **segregation ratio** is the proportion of **genotypes** in **zygotes** following **segregation** and **syngamy**.

Segregation distortion. Any deviation from equiprobability in the **segregation ratio**.

Selection. The change in the proportions of self-replicating entities caused by differential reproduction; the logical consequence of differences in **fitness** in a finite world. If differential reproduction has a genetic basis **selection** causes a permanent genetic change, or **evolution**.

Self-compatible. Able to achieve **self-fertilization**.

Self-fertilization. See **autogamy**.

Self-incompatible. Not able to achieve **self-fertilization**.

Sessoblast. A bryozoan **statoblast** which adheres to the substrate.

Sex. (1) **Mixis.** A process of genetic reorganization through **meiosis** and **syngamy**, usually associated with the formation of several or many reproductive propagules (**sexual reproduction**). (2) **Gender**; one of the two **genders** of **bipolar amphimicts**.

Sperm (Spermatozoon). The **microgamete** of **oogametic** animals. **Spermatogenesis** consists of the mitotic proliferation of secondary from primary **spermatogonia**, the secondary **spermatogonia** being transformed into primary **spermatocytes** which give rise to two secondary **spermatocytes** at the completion of the first **meiotic** division; the second meiotic division results in the formation of four **spermatids** from the two secondary **spermatocytes**, and these then differentiate into **spermatozoa**.

Spike. An inflorescence bearing sessile flowers along its axis.

Spore. An asexual reproductive propagule, especially of plants, formed in a **sporangium**.

Sporic meiosis. Meiosis occurring in a **diploid spore**, whose products are haploid gametophytes.

Sporocyte. A **diploid** cell which gives rise to **spores** by **meiosis**.

Sporophyte. The **diploid spore**-producing generation of plants. Cf. **gametophyte**.

Staminate. Bearing **stamens**, the male reproductive structures.

Statoblast. The **asexual** dormant propagule of bryozoans.

Statospore. The bipartite dormant **spore** of certain algae.

Strobilation. Reproduction by **architomical fragmentation**, especially in scyphozoans, cestodes and tunicates.

Subitaneous. Hatching without **diapause**.

Swarmer. A motile **asexual** propagule (Algae).

Sympatry. Having overlapping or coincident geographical distributions.

Syngamy. Fertilization. The union of **gametes**, followed by nuclear fusion.

Tetrad. The quartet of **haploid** cells resulting from **meiosis**.

Tetraspore. One of the **tetrad** of **spores** formed by some algae.

Thallus. The body of a nonvascular plant.

Thelygeny. The production of **female** progeny exclusively (especially in isopods). Cf. **arrhenogeny, monogeny, amphogeny.**

Tun. The dormant phase of bdelloid rotifers and tardigrades, a lemon-shaped or barrel-shaped encysted **anabiotic** adult.

Tychoparthenogenesis. The occasional production of **parthenogenetic** offspring. **Tychothelytoky:** the occasional production of **parthenogenetic** female offspring.

Vegetative reproduction. Fission, fragmentation. Any mode of reproduction not involving the elaboration of **eggs.**

Viviparous. In animals, the practice of giving birth to young able to survive without stored reserves of food.

Xeric. Of habitats, dry.

Zoospore. A motile asexual propagule (Algae).

Zygote. The **diploid** cell resulting from **syngamy.**

Zygotic meiosis. Meiosis occurring in a **zygote**, whose products are gametophytes.

BIBLIOGRAPHY

Abbot, R. T. *Handbook of Medically Important Molluscs of the Orient and West Pacific*, Bull. Mus. Comp. Zool. Harvard Univ. 100 (1948).

Abbott, D. P. 'Asexual Reproduction in the Colonial Ascidian *Metandrocarpa taylori* Huntsman', Univ. Calif. Publ. Zool., vol. 61 (1953), pp. 1-78.

Abdel-Hameed, F., Rootham, D. L. and Flinn, R. R. 'Structural and Numerical Aberrations in Natural Populations of the Grasshopper, *Melanoplus differentialis differentialis*', Genetics, vol. 64, suppl. 1 (1970).

Abdullah, N. F. and Charlesworth, B. 'Selection for Reduced Crossing-over in *Drosophila melanogaster*', Genetics, vol. 76 (1974), pp. 447-51.

Abramoff, P., Darnell, R. M. and Balsond, J. S. 'Electrophoretic Demonstration of the Hybrid Origin of the Gynogenetic Teleost *Poecilia formosa*', Amer. Natur., vol. 102 (1968), pp. 555-8.

Acker, T. S. '*Craspedacusta sowerbii*: An Analysis of an Introduced Species', in G. O. Mackie (ed.), *Coelenterate Ecology and Behaviour* (Plenum, New York, 1976), pp. 219-25.

Acker, T. S. and Muscat, A. M. 'Ecology of *Craspedacusta sowerbii* Lankester, a Freshwater Hydrozoan', Amer. Midl. Nat., vol. 95 (1975), pp. 323-36.

Acton, A. B. 'An Unsuccessful Attempt to Reduce Recombination by Selection' Amer. Natur., vol. 95 (1961), pp. 119-20.

Agassiz, A. 'Embryology of the Ctenophora', Mem. Amer. Acad. Arts Sci., vol. 10 (1874), pp. 356-98.

Aldrich, J. M. 'Notes on Diptera', Psyche (1918), p. 25.

Allan, J. D. 'An Analysis of Seasonal Dynamics of a Mixed Population of *Daphnia* and the Associated Cladoceran Community', Freshwater Biology, vol. 7 (1977), pp. 505-12.

Allard, R. W. 'Evidence for Genetic Restriction of Recombination in the Lima Bean', Genetics, vol. 48 (1963), pp. 1389-95.

Allen, E. J. 'Regeneration and Reproduction of the Syllid *Procerastea*', Phil. Trans. Roy. Soc. B, vol. 211 (1921), pp. 131-77.

Alm, G. 'Beitrage zur Kenntnis der Nordlichen und Arktischen Ostracodenfauna', Ark f. Zool., vol. 9 (1914), pp. 1-20.

—, 'Monographie der Swedischen Süsswasserostracoden', Zool. Bidrag Uppsala, vol. 4 (1916), pp. 1-249.

Alvarino, A. 'Chaetognaths', Ann. Rev. Oceanogr. Mar. Biol., vol. 3 (1965), pp. 115-94.

Ammermann, D. 'Parthenogenese bei dem Tardigraden *Hypsibius dujardini* (Doy.)', Naturwiss., vol. 49 (1962), p. 115.

—, 'Die Cytologie der Parthenogenese bei dem Tardigraden *Hypsibius dujardini*', Chromosoma, vol. 23 (1967), pp. 203-13.

Ananthakrishnan, T. N. 'Biosystematics of Thysanoptera', Ann. Rev. Entomol., vol. 24 (1979), pp. 159-83.

Anderson, J. M. 'Sexual Reproduction without Cross-copulation in the Freshwater Triclad Turbellarian, *Curtisia foremanii*', Biol. Bull., vol. 102 (1952a), pp. 1-8.

—, 'A Further Report on Sexual Reproduction without Cross-copulation in the Freshwater Triclad Turbellarian, *Curtisia foremanii*', Anat. Rec., vol. 113 (1952b), p. 601.

Anderson J. M. and Johann J. C. 'Some Aspects of Reproductive Biology in the Freshwater Triclad Turbellarian *Curtisia foremanii*', Biol. Bull., vol. 115 (1958), pp. 375-83.

Anderson, L. E. and Lemmon, B. E. 'Gene-flow Distances in the Moss, *Weissia controversa* Hedw', J. Hattori Bot. Lab., vol. 38 (1974), pp. 67-90.

Anderson, R. H. 'The Laying Worker in the Cape Honeybee *Apis mellifera capensis*', J. Apic. Res., vol. 2 (1963), pp. 85-92.

André, F. 'Mise en Évidence et Modalité de l'Autofécondation chez le Ver du Fumier *Eisenia foetida* Sav', CR Acad. Sci. Paris, vol. D254 (1962a), pp. 3442-3.

—, 'Une Fécondation Interne Peut-elle Être la Cause de l'Autofécondation chez le Lombricien *Eisenia foetida* Sav.?', CR Soc. Biol., vol. 156 (1962b), pp. 621-2.

—, 'Contribution à l'Étude Expérimentale de la Reproduction des Lombriciens', Bull. Biol. France Belgique, vol. 97 (1963), pp. 4-101.

André, F. and Davant, N. 'L'Authofécondation chez les Lombriciens. Observation d'un Cas d'Autoinsémination chez *Dendrobaena rubida* f. *subrubicunda* Eisen', Bull. Soc. Zool. France, vol. 97 (1972), pp. 725-8.

Andrewartha, H. G. 'The Bionomics of *Otiorrhynchus cribricollis*', Bull. Entomol. Res., vol. 24 (1933), pp. 373-84.

Angus, R. B. '*Helophorus orientalis* (Coleoptera: Hydrophilidae), a Parthenogenetic Water Beetle from Siberia and North America, and a British Pleistocene Fossil', Can. Entomol., vol. 102 (1970), pp. 129-43.

Ankel, W. 'Neuere Arbeiten zur Zytologie der Naturlichen Parthenogenese der Tiere', Z. Abstamm. Lehre, vol. 45 (1927), pp. 232-78.

—, 'Neuere Arbeiten zur Zytologie der Naturlichen Parthenogenese der Tiere', Z. Abstamm. Lehre, vol. 52 (1929), pp. 318-70.

Annandale, N. *Fauna of British India, Part III. Freshwater Sponges, Hydroids and Polyzoa* (1911).

—, 'Fauna of the Chilka Lake Sponges', Mem. Indian Mus., vol. 5 (1915), pp. 23-54.

Anon. 'Two Letters from a Gentleman in the Country', Phil. Trans. Roy. Soc., vol. 23 (1704), pp. 283-8.

Arcangeli, A. 'Ermafroditismo e Partenogenesi negli Isopodi Terrestri (Prima Nota)', Monit. Zool. Ital., vol. 36 (1925), pp. 105-22.

Armitage, K. B. and Smith, B. B. 'Population Studies of Pond Zooplankton', Hydrobiologia, vol. 32 (1968), pp. 384-416.

Ar-Rushdi, A. H. 'The Cytology of Achiasmatic Meiosis in the Female *Tigriopus* (Copepoda)', Chromosoma, vol. 13 (1963), pp. 526-39.

Artom, C. 'L'Origine e l'Evoluzione della Partenogenesi Attraverso i Differenti Biotopi di una Specie Collectiva (*Artemia salina* L), con Speciale Riferimento al Biotipo Diploide Partenogenetico di Sète', Mem. R. Acc. Ital., Classe Scienze, vol. 2 (1931), pp. 1-57.

Asher, J. H. 'Parthenogenesis and Genetic Variability. II. One-locus Models for Various Diploid Populations', Genetics, vol. 66 (1970), pp. 369-91.

Astaurov, B. L. 'Artificial Parthenogenesis in the Silkworm — (*Bombyx mori* L.)', (Akad. Nauk SSSR, Moscow, 1940).

—, 'The Present State of Problems of Artificial Parthenogenesis in the Silkworm. Diploid and Polyploid Parthenogenesis in *Bombyx mori* L., *B. mandarini* Moore and in the Hybrids', Sympos. Genet. et Biol.Italica,vol.9(1962)pp.1-20.

Atkins, D. 'The Loxosomatidae of the Plymouth Area, Including *L. obesum* sp. nov.', Quart. J. Microscop. Sci., vol. 75 (1932), pp. 321-91.

Ax, P. and Schulz, E. 'Ungeschlechtliche Fortpflanzung durch Paratomie bei Acoelen Turbellarien', Biol. Zentralblatt, vol. 78 (1959), pp. 613-22.

Axtell, R. W. 'Geographic Distribution of the Unisexual Whiptail *Cnemidophorus neomexicanus* (Sauris: Telidae), Present and Past', Herpetologica, vol. 22 (1966) pp. 241-53.

Ayonoadu, U. and Rees, H. 'The Influence of B-chromosomes on Chiasma Frequencies in Black Mexican Sweet Corn', Genetica, vol. 39 (1968), pp. 75-81.

Baccetti, B. 'Notulae Orthopterologicae. IX. Osservazioni Cariologiche sulle *Dolichoptera* Italiane', Redia, vol. 43 (1958), pp. 315-27.

Bacci, G., Cognetti, G. and Vaccari, A. M. 'Endomeiosis and Sex Determination in *Daphnia pulex*', Experientia, vol. 17 (1961), pp. 501-6.

Badonnel, A. 'Psocoptéres', Faune de France, vol. 42 (1943), pp. 1-164.

—, 'Ordre des Psocoptéres', in P. P. Grassé (ed.), *Traité de Zoologie*, vol. 10 (Masson et Cie., Paris, 1951), pp. 1301-40.

von Baehr, W. B. 'Die Oogenese bei Einigen Viviparen Aphididen und die Spermatogenese von *Aphis saliceti* mit Besonderer Berucksichtigung der Chromosomenverhaltnisse', Arch. Zellforsch., vol. 3 (1909), pp. 269-333.

—, 'Recherches sur la Maturation des Oeufs Parthenogenetiques dans l'*Aphis palmae*', Cellule, vol. 30 (1920), pp. 315-54.

Baker, B. S. and Hall, J. C. 'Meiotic Mutants: Genic Control of Meiotic Recombination and Chromosome Segregation', in M. Ashburner and E. Novitski (eds.), *The Biology and Genetics of Drosophila*, vol. 1a (Academic, London, 1977), pp. 352-435.

Baker, F. C. 'Self-fertilization and nidification in *Physa halei*', Nautilus, vol. 47 (1933), p. 35.

Baker, R. R. and Parker, G. A. 'The Origin and Evolution of Sexual Reproduction up to the Evolution of the Male-female Phenomenon', Acta Biotheoretica, vol. 22, no. 2 (1973), pp. 1-77.

Balduf, W. V. 'The Bionomics of *Dinocampus coccinellae* Schrank', Ann. Entomol. Soc. Amer., vol. 19 (1926), pp. 465-98.

Ball, I. R. 'A Contribution to the Phylogeny and Biogeography of the Freshwater Triclads (Platyhelminthes: Turbellaria)', in N. W. Riser and M. P. Morse (eds.), *Biology of the Turbellaria* (McGraw-Hill, New York, 1974), pp.339-99.

Banta, A. M. 'A Thelytokous Race of Cladocera in Which Pseudosexual Reproduction Occurs', Z. Indukt. Abstamm. Vererb., vol. 40 (1926), pp. 28-41.

Banta, A. M. and Brown, L. A. 'Control of Sex in Cladocera. I. Crowding the Mothers as a Means of Controlling Male Production', Physiol. Zool., vol. 2 (1929a), pp. 80-92.

—, 'Control of Sex in Cladocera. III. Localization of the Critical Period for Control of Sex', Proc. Nat. Acad. Sci. US, vol. 15 (1929b), pp. 77-81.

—, 'Control of Male and Sexual-egg Production', in A. M. Banta (ed.) *Studies on the Physiology, Genetics and Evolution of Some Cladocera*, Carnegie Inst. Washington Publ. 513, Paper No. 39 (Department of Genetics, 1939), pp. 106-30.

Banta, A. M. and Wood, T. R. 'General Studies in Sexual Reproduction', in A. M. Banta (ed.) *Studies on the Physiology, Genetics and Evolution of Some Cladocera*. Carnegie Inst. Washington Publ. 513, Paper No. 39 (Department of Genetics, 1939), pp. 131-81.

Barash, D. P. 'What Does Sex Really Cost?', Amer. Natur., vol. 119 (1976), pp. 894-7.

Barber, H. S. 'The Remarkable Life-history of a New Family (Micromalthidae) of Beetles', Proc. Biol. Soc. Washington, vol. 26 (1913), pp. 185-90.

Barghoorn, E. S. 'The Oldest Fossils', Scient. Amer., vol. 224 (1971), pp. 30-43.

Barigozzi, C. 'I fenomeni cromosomici delle cellule germinale in *Artemia salina* Leach', Chromosoma, vol. 2 (1944), pp. 549-75.

—, 'Differenciation des Génotypes et Distribution Geographique d'*Artemis salina*: Données et Problèmes', Année Biol., vol. 33 (1951), pp. 241-50.

Barker, J. F. 'Variation of Chiasma Frequency in and between Natural Populations of Acrididae', Heredity, vol. 14 (1960), pp. 211-14.

—, 'Climatological Distribution of a Grasshopper Supernumary Chromosome', Evolution, vol. 20 (1966), pp. 655-67.

Barlow, P. W. and Vosa, C. G. 'The Effect of Supernumary Chromosomes on Meiosis in *Pushkinia libanotica* (Liliaceae)', Chromosoma, vol. 30 (1970), pp. 344-55.

Barnes, H. F. 'On the Gall-midges Injurious to the Cultivation of Willows. II. The "Shot-hole" Gall Midges (*Rhabdophaga* spp.)', Ann. Appl. Biol., vol. 22 (1935), pp. 86-105.

—, 'Investigations on the Raspberry Cane Midge, 1943-1944', J. Roy. Hort. Soc., vol. 69 (1944), pp. 370-5.

Barnes, H. and Crisp, D. J. 'Evidence of Self-fertilization in Certain Species of Barnacles', J. Mar. Biol. Ass., vol. 35 (1956), pp. 631-9.

Basch, P. F. 'Studies on the Development and Reproduction of the Freshwater Limpet *Ferrissia shimekii* (Pilsbry)', Trans. Amer. Microscop. Soc., vol. 78 (1959), pp. 269-76.

Baskin, D. G. and Golding, D. W. 'Experimental Studies on the Endocrinology and Reproductive Biology of the Viviparous Polychaete Annelid, *Nereis limnicola* Johnson', Bio. Bull., vol. 139 (1970), pp. 416-76.

Basrur, V. R. and Rothfels, K. H. 'Triploidy in Natural Populations of the Black Fly *Cnephia mutata* (Malloch)', Can. J. Zool., vol. 37 (1959), pp. 571-89.

Baud, F. 'Biologie et Cytologie de Cinq Espèces du Genre *Lonchoptera* Meig. (Dipt.) Dont l'Une Est Parthénogénètique et les Autres Bisexuées, avec Quelques Remarques d'Ordre Taxonomique', Rev. Suisse Zool., vol. 80 (1973), pp. 473-515.

Bauer, H. 'Uber die Chromosomen der Bisexuellen und der Parthenogenetischen Rasse des Ostracoden *Heterocypris incongruens* Ramd.', Chromosoma, vol. 1 (1940), pp. 620-37.

—, 'Gekoppelte Vererbung bei *Phryne fenestralis* und die Beziehung zwischen Faktorenaustausch und Chiasmabildung', Biol. Zentralblatt., vol. 65 (1946), pp. 108-15.

Bauer, W. and Beerman, W. 'Der Chromosomenzyklus der Orthocladiinen (Nematocera: Diptera)', Z. Naturf., vol. 76 (1952), pp. 557-63.

Baumann, H. 'Uber den Lebenslauf und die Lebensweise von *Milnesium tardigradum* Doyère (Tardigrada)', Veroff. Uberseums Bremen, vol. 3 (1964), pp. 161-71.

Bayly, I. A. E. and Williams, W. D. *Inland Waters and Their Ecology* (Longman, Camberwell, Victoria, Australia, 1973).

Bayreuther, K. 'Holokinetische Chromosomen bei *Haematopinus suis* (Anoplura, Haematopinidae)', Chromosoma, vol. 7 (1955a), pp. 260-70.

—, 'Die Oogenese der Tipuliden', Chromosoma, vol. 7 (1955b), pp. 508-57.

Bazin, M. J. 'Sexuality in a Blue-green Alga: Genetic Recombination in *Anacystis nidulans*', Nature, vol. 28 (1968), pp. 282-3.

Beadle, L. C. *The Inland Waters of Tropical Africa* (Longman, London, 1974).

Beatty, R. A. 'Parthenogenesis in Vertebrates', in C. B. Mertz and A. Monroy (eds.), *Fertilization: Comparative Morphology, Biochemistry and Immunology*, vol. 1 (Academic, New York, 1967), pp. 413-40.

Bedford, G. O. 'Biology and Ecology of the Phasmotodea', Ann. Rev. Entomol., vol. 23 (1978), pp. 125-49.

Beermann, S. 'Cromatin-Diminution bei Copepoden', Chromosoma, vol. 10 (1959), pp. 504-14.

Beermann, W. 'Weibliche Heterogametie bei Copepoden', Chromosoma, vol. 6 (1954), pp. 381-96.

Bei-Bienko, G. Y. 'Orthoptera, Tettigoniidae; Subfamily Phaneropterinae' in *Fauna of the USSR 2*, part 2 (1954), pp. 301-4.

Belar, K. 'Ueber den Chromosomenzyklus von Parthenogenetischen Erdnematoden', Biol. Zentralblatt, vol. 43 (1923), pp. 513-18.

Bell, A. W. '*Enchytraeus fragmentosus*, a New Species of Naturally Fragmenting Oligochaete Worm', Science, vol. 129 (1959), p. 1278.

Bell, G. 'Group Selection in Structured Populations', Amer. Natur., vol. 112 (1978a), pp. 389-99.

—, 'The Evolution of Anisogamy', J. Theoret. Biol., vol. 73 (1978b), pp. 247-70.

Benazzi, M. 'Ginogenesi in Tracladi d'Acqua Dolce', Chromosoma, vol. 3 (1950), pp. 474-82.

—, 'Fissioning in Planarians from a Genetic Standpoint', in N. W. Riser and M. P. Morse (eds.), *Biology of the Turbellaria* (McGraw-Hill, New York, 1974).

Benazzi, M. and Pulcinella, I. 'Analisi comparativa del cariogramma dei biotipi di *Dugesia lugubris*', Atti Ass Genet Italia, vol. 6 (1961), pp. 419-26.

Benazzi-Lentati, G. 'Considerazioni sul Determinismo dei Cicli Cromosomici in Ibridi di Planarie', Caryologia, vol. 14 (1961), pp. 271-7.

—, 'Due Modalita di Sviluppo dello Stesso Tipo di Uova in Ibridi Interraziali di Planarie', Acta Embryol. Morph. Exp., vol. 5 (1962), pp. 145-60.

—, 'La Polisomia nelle Planarie', Atti Soc. Toscana Sci. Nat., vol. 71 (1964), pp. 44-51.

—, 'Gametogenesis and Egg Fertilization in Planarians', Int. Rev. Cytol., vol. 27, (1970), pp. 101-79.

Benazzi-Lentati, G. and Bertini, V. 'Sul Determinismo della "Asynapsi Femminile" in Ibridi di *Dugesia benazzii* (Tricladida, Paludicola)', Atti Soc. Toscana Sci. Nat., vol. B68 (1961), pp. 83-112.

Bennett, E. 'New Zealand Sea Stars', Rec. Canterbury Mus., vol. 3 (1927). Not seen: cited by L. H. Hyman (1955).

Berg, K. 'A Faunistic and Biological Study of Danish Cladocera', Vidensk Medd. Dansk. naturh. Foren, vol. 88 (1929), pp. 31-111.

—, 'Studies on the Genus *Daphnia* O. F. Muller with Special Reference to the Mode of Reproduction' Vidensk. Meddr. Dansk. Naturh. Foren, vol. 92 (1931), pp. 1-222.

—, 'Cyclical Reproduction, Sex Determination and Depression in the Cladocera' Biol. Rev., vol. 9 (1934), pp. 139-74.

Berg. L. S. 'On Unisexual Reproduction among Goldfish', in L. S. Berg (ed.), *Ichthyology*, vol. 4 (Academy of Science Press, Moscow, 1961).

Berger, E. 'Heterosis and the Maintenance of Enzyme Polymorphism', Amer. Natur., vol. 110 (1976), pp. 823-39.

Bergerard, J. 'Parthénogénèse Facultative de *Clitumnus extradentatus* Br. (Phasmidae)', Bull. Soc. Zool. France, vol. 79 (1954), pp. 169-75.

—, 'Etude de la Parthénogénèse Facultative de *Clitumnus extradentatus* Br. (Phasmidae)', Bull. Biol. France Belg., vol. 92 (1958), pp. 87-182.

—, 'Parthenogenesis in the Phasmidae'. Endeavour, vol. 21 (1962), pp. 137-43.

Bergerard, J. and Seugé, J. 'La Parthénogénèse Accidentale chez *Locusta migratoria* L.', Bull. Biol. France Belgique, vol. 93 (1959), pp. 16-37.

Bergquist, P. R., Sinclair, M. E. and Hogg J. J. 'Adaptation to Intertidal Existence: Reproductive Cycles and Larval Behaviour in Demospongiae' in W. G. Fry (ed.) *Biology of the Porifera*, Symp. Zool. Soc. London no. 25 (Academic, New York, 1970), pp. 247-71.

Berkeley, E. and Berkeley, C. 'Notes on the Life History of the Polychaete *Dodecaceria fewkesi* (nom. nov.)', J. Fish Res. Bd. Canada, vol. 11 (1953), pp. 326-34.

Berland, L. 'Un Cas Probable de Parthénogénèse Géographique chez *Leucopsis gigas* (Hymenoptére)', Bull. Soc. Zool. France, vol. 59 (1934), pp. 172-5.

Bernard, H. M. 'Hermaphroditism in the Apodidae', Nature, vol. 43 (1889), p. 343.

Berrill, N. J. 'Studies in Tunicate Development. IV. Asexual Reproduction', Phil. Trans. Roy. Soc. Ser. B, vol. 225 (1935), pp. 327-79.

—, 'Developmental Analysis of Scyphomedusae', Biol. Rev., vol. 24 (1949), pp. 393-410.

—, 'Development and Medusa-bud Formation in the Hydromedusae', Quart. Rev. Biol., vol. 25 (1950), pp. 292-316.

—, 'Regeneration and Budding in Tunicates', Biol. Rev., vol. 26 (1951), pp. 456-75.

—, 'Growth and Form in Gymnoblastic Hydroids. II. Sexual and Asexual Reproduction in *Rathkea*', J. Morphol., vol. 90 (1952a), p. 1.

—, 'Regeneration and Budding in Worms', Biol. Rev., vol. 27 (1952b), pp.401-38.

—, 'Growth and Form in Gymnoblastic Hydroids. VII. Growth and Reproduction in *Syncoryne* and *Coryne*', J. Morphol., vol. 92 (1953), pp. 273-302.

—, 'Chordata: Tunicata' in A. C. Giese and J. S. Pearse (eds.), *Reproduction of Marine Invertebrates*, vol. 2 (Academic, New York, 1975).

Berry, A. J. and Kadri, A. bin H. 'Reproduction in the Malayan Freshwater Cerithiacean Gastropod *Melanoides tuberculata*', J. Zool. London, vol. 172 (1974), pp. 369-81.

Bertolani, R. 'Cytology and Systematics in Tardigrada'. Mem. Ist. Ital. Idrobiol., vol. 32 (suppl.) (1975), pp. 17-35.

Bezy, R. L. (1969). Paper presented at AAIH meeting, cited by Maslin (1971).

Bier, K. 'Zur Scheinbaren Thelytokie der Ameisengattung *Lasius*', Naturwissenschaften, vol. 18 (1952), p. 433.

Bigelow, H. B. 'The Medusae', Mem. Mus. Comp. Zool., vol. 37 (1909), pp. 9-245.

Birge, E. A. 'Plankton Studies on Lake Mendota. II. The Crustacea of the Plankton from July 1894 to December 1896', Trans. Wisconsin Acad. Sci. Arts Lett., vol. 11 (1898), pp. 274-451.

Birky, C. W. 'Studies on the Physiology and Genetics of the Rotifer, *Asplanchna*. I. Methods and Physiology', J. Exp. Zool., vol. 155 (1964), pp. 273-92.

—, 'Studies on the Physiology and Genetics of the Rotifer, *Asplanchna*. III. Results of Outcrossing, Selfing and Selection', J. Exp. Zool., vol. 164 (1967), pp. 105-16.

Birky, C. W. and Gilbert, J. J. 'Parthenogenesis in Rotifers: The Control of Sexual and Asexual Reproduction', Amer. Zool., vol. 11 (1971), pp. 245-66.

Black, L. M. and Oman, P. W. 'Parthenogenesis in a Leafhopper, *Agallia quadripunctata*. (Provancher) (Homoptera: Cicadellidae)', Proc. Entomol. Soc. Washington, vol. 49 (1947), pp. 19-20.

Blackwelder, R. E. *Classification of the Animal Kingdom* (Southern Illinois Univ. Press, Carbondale, Illinois, 1975).

Block, K. 'Chromosome Variation in the Agromyzidae. II. *Phytomeza crassiseta* Zetterstedt – A Parthenogenetic Species', Hereditas, vol. 62 (1969), pp. 357-81.

Bloomer, H. H. 'On Experiments on Self-fertilization in *Anodonta cygnea* (L)', Proc. Malacol. Soc. London, vol. 24 (1940), pp. 113-21.

Boardman, R. S. and Cheetham, A. H. 'Degrees of Colony Dominance in Steno-laemate and Gymnolaemate Bryozoa', in R. S. Boardman, A. H. Cheetham and W. A. Oliver (eds.), *Animal Colonies* (Dowden, Hutchinson and Ross, Stroudsberg, Pa., 1974), pp. 121-220.

Bobin, G. 'Interzooecial Communications and the Funicular System', in R. M. Woollacott and R. L. Zimmer (eds.), *Biology of Bryozoans* (Academic, New York, 1974), pp. 307-33.

Bocher, J. 'Preliminary Studies on the Biology and Ecology of *Chlamydatus pullus* (Reuter) (Heteroptera:Miridae)', Meddr. Grönland, vol. 191 (1971), pp. 1-29.

Bodmer, W. F. and Edwards, A. W. F. 'Natural Selection and the Sex-ratio', Ann. Hum. Genet., vol. 24 (1960), pp. 239-44.

Bodmer, W. F. and Parsons, P. A. 'Linkage and Recombination in Evolution', Advance Genet., vol. 11 (1962), pp. 1-100.

Boecker, E. 'Die Geschlechtliche Fortpflanzung der Deutschen Susswasser-polypen', Biol. Zentralblatt, vol. 38 (1918), pp. 479-99.

de Boer, P. and van den Hoeven, F. A. 'Son-sire Regression-based Heritability Estimates of Chiasma Frequency, Using T70H Mouse Translocation Hetero-zygotes, and the Relation between Univalence, Chiasma Frequency and Sperm Production', Heredity, vol. 39 (1977), pp. 335-43.

Bogoslovsky, A. C. 'Observation on the Reproduction of *Conochiloides coeno-basis*'. Zool. Zh., vol. 39 (1960), pp. 670-7 (in Russian).

Bold, H. C. and Wynne, M. J. *Introduction to the Algae* (Prentice-Hall, Engle-wood Cliffs, NJ, 1978).

Bonnemaison, L. 'Contribution à l'Étude des Facteurs Provoquant l'Apparition des Formes Ailées et Sexuées chez les Aphidinae', Ann. Epiphyties, vol. 2 (1951), pp. 1-380.

Bonnemaison, L. 'Facteurs d'Apparition des Formes Sexuparés ou Sexuées chez le Puceron Cendré du Pommier (*Sappaphis plantaginea*)', Ann. Epiphyties, vol. 3 (1958), pp. 331-55.

Bonner, J. T. *Size and Cycle. An Essay on the Structure of Biology* (Princeton University Press, Princeton, NJ, 1965).

Bonnet C. *Traité d'Insectologie* (1745). See 'Observations sur les Pucerons' in *Oeuvres d'Histoire Naturelle et de Philosophie*, vol. 1; or *Oeuvres Com-plètes*, vol. 1 (Neuchatel, 1799).

Boorman, S. A. and Levitt, P. R. 'Group Selection on the Boundary of a Stable Population', Theoret. Pop. Biol., vol. 4 (1973), pp. 85-128.

Borg, F. 'Studies on Recent Cyclostomatous Bryozoa', Zool. Bidrag, vol. 10 (1926), pp. 181-507.

Borgia, G. 'Evolution of Haplodiploidy: Models for Inbred and Outbred Systems'. Theoret. Pop. Biol. vol. 17 (1980), pp. 103-28.

Borragan, P. J. and Callan, H. G. 'Chiasma Formation in Spermatocytes and Oocytes of the Turbellarian *Dendrocoelum lacteum*', J. Genet., vol. 50 (1952), pp. 449-54.

Borror, D. J. and DeLong, D. M. *An Introduction to the Study of Insects*. 3rd edn. (Holt, Rinehart and Winston, New York, 1971).

Bounoure, L. *Continuité Germinale et Reproduction Agame* (Gauthier-Villais, Paris, 1940).

Bournier, A. 'Contribution à l'Étude de la parthénogenèse des Thysanoptères et de sa Cytologie', Arch. Zool. Exp. Gén., vol. 93 (1956), pp. 219-317.

Bowen, S. T. 'The Genetics of *Artemia salina*. I. The Reproductive Cycle', Biol. Bull., vol. 122 (1962), pp. 25-32.

Bowen, W. R. and Stern, V. M. 'Effect of Temperature on the Production of Males and Sexual Mosaics in a Uniparental Race of *Trichogamma semifumatum* (Hymenoptera:Trichogrammatidae)', Ann. Entomol. Soc. Amer., vol. 59 (1966), pp. 823-34.

Boycott, A. E. 'Where Is the Male of *Paludestrina jenkisoni?*', J. Conchol., vol. 15 (1917), p. 216.

—, 'Parthenogenesis in *Paludestrina jenkinsoni*', J. Conchol., vol. 16 (1919), p. 54.

Boycott, A. E., Diver, C., Garstang, S. and Turner, F. M. 'The Inheritance of sinistrality in *Limnaea pereger* (Mollusca, Pulmonata)', Phil. Trans. Roy. Soc., vol. B219 (1931), pp. 51-131.

Boyden, A. A. 'Comparative Evolution with Special Reference to Primitive Mechanisms'. Evolution, vol. 7 (1953), pp. 21-30.

Boyden, A. A. 'The Significance of Asexual Reproduction', Syst. Zool., vol. 3 (1954), pp. 26-37.

Brandham, P. E. and Bhattarai, S. 'The Effect of B-chromosome Number on Chiasma Frequency within and between Individuals of *Gibasis linearis* (Commelinaceae)', Chromosoma, vol. 64 (1977), pp. 343-8.

Brattstrom, H. '*Phoronis ovalis*' Lunds Univ. Aarskrift Ard2, vol. 39, no. 2 (1943).

Braverman, M. H. 'Studies in Hydroid Differentiation. I. *Podocoyne carnea* Culture Methods and Carbon Dioxide Induced Sexuality', Exp. Cell Res., vol. 27 (1962), pp. 301-6.

—, 'Studies on Hydroid Differentiation. III. Colony growth and the Initiation of Sexuality', J. Embryol. Exp. Morphol., vol. 11 (1963), pp. 239-53.

Bresslau, E. 'Die Sommer- und Wintereier der Rhabdocolen des süssen Wassers', Verhandl. Deutsch. Zool. Gesell., vol. 13 (1903).

Brien, P. 'Reproduction Asexuée des Phylactolémales', Mém. Mus. Hist. Natur. Belgique, Ser. 2, fasc. 3 (1936).

—, 'La Reproduction Asexuée (Tunicates)', Ann. Biol., vol. 34 (1958), pp. 241-62.

—, 'Formation des Statoblastes dans le Genre *Potamolepis: P. symoensi* (Brien),

P. pechuelli (Marshall), *P. schoutedeni* (Burton)', Bull. Acad. Roy. Belgique, vol. 53 (1967), pp. 573-91.

Brien, P. and Reniers-Decoen, M. 'La Croissance, la Blastogénèse, l'Ovogénènese chez *H. lusca* (Pallas)', Bull. Biol. France Belgique, vol. 83 (1949), pp. 293-386.

Briggs, T. S. 'Relict Harvestmen from the Pacific Northwest', Pan-Pacific Entomol., vol. 47 (1971), pp. 165-78.

Brinkhurst, R. O. and Jamieson, B. G. M. *Aquatic Oligochaeta of the World* (University of Toronto Press, Toronto, 1971).

Britt, N. W. 'Biology of Two Species of Lake Erie Mayflies, *Ephoron album* (Say) and *Ephemera simulans* Walker', Bull. Ohio Biol. Survey, vol. 1 (1962), pp. 1-70.

Broadhead, E. 'New Species of *Liposcelis* Motschulsky (Corrodentia, Liposcelidae) in England', Trans. Roy. Entomol. Soc. London, vol. 98 (1947), pp. 41-58.

—, 'A New Parthenogenetic Psocid from Stored Products with Observations on Parthenogenesis in Other Psocids', Entomol. Monthly Mag., vol. 90 (1954), pp. 10-16.

Brock, T. D. *Thermophilic Microorganisms and Life at High Temperatures* (Springer-Verlag, New York, 1978).

Brookes, C. H. 'The Life Cycle of *Proteroiulus fuscus* (Am Stein) and *Isobates varicornis* (Koch) with Notes on the Anamorphosis of Blaniulidae', Symp. Zool. Soc. London, vol. 32 (1974), pp. 485-501.

Brooks, J. L. 'Cladocera' in Ward and Whipple's *Freshwater Biology* (ed. W. T. Edmondson), 2nd edn (Wiley, New York, 1978), Chap. 27.

Brooks, W. K. 'The Life History of the Hydromedusae: A Discussion of the Origin of Medusae and of the Significance of Metagenesis', Mem. Boston Nat. Hist. Soc., vol. 3 (1886), pp. 359-430.

Brown, C. J. D. 'A Limnological Study of Certain Freshwater Polyzoa, with Special Reference to Their Statoblasts', Trans. Amer. Microscop. Soc., vol. 52 (1933), pp. 271-316.

Brown, L. M. and Jones, R. N. 'B-chromosome Effects at Meiosis in *Crepis capillaris*', Cytologia, vol. 41 (1976), pp. 493-506.

Brown, S. W. 'Chromosomal Survey of the Armoured and Palm Scale Insects (Coccoidea: Diaspididae and Phoenicococcidae)', Hilgardia, vol. 36 (1965), pp. 189-214.

Brown, T. F. 'The Biology of *Physa anatina*, a Snail Living in a Sewage Treatment Plant', Amer. Midl. Nat., vol. 18 (1937), pp. 251-9.

Brown, W. M. and Wright, J. W. 'Mitochondrial DNA Analyses and the Origin and Relative Age of Parthenogenetic Lizards (genus *Cnemidophorus*)', Science, vol. 203 (1979), pp. 1247-9.

Brues, C. T. 'Animal Life in Hot Springs', Quart. Rev. Biol., vol. 2 (1927), pp. 181-203.

Brues, C. T. 'A Note on the Genus *Pelecinus*', Psyche, vol. 35 (1928), pp. 205-9.

Brunson, R. B. 'The Life History and Ecology of Two North American Gastrotrichs', Trans. Amer. Microscop. Soc., vol. 68 (1949), pp. 1-20.

Bryant, E. H. 'On the Adaptive Significance of Enzyme Polymorphisms in Relation to Environmental Variability', Amer. Natur., vol. 108 (1974), pp. 1-19.

Bryden, R. R. 'Ecology of *Pelmatohydra oligactis* in Kirkpatricks Lake, Tennessee', Ecol. Monogr., vol. 22 (1952), pp. 45-68.

Buchner, H. 'Experimentelle Untersuchungen über den Generationswechsel der Radertiere. II', Zool. Jahrb. Aügem. Zool. Physiol. Tiere, vol. 60 (1941), pp. 279-344.

Buchner, H. and Kiechle, H. 'Die Determination der Heterogonen Fortpflanzungsurten bei den Radertieren', Naturwissenschaften, vol. 52 (1965), p. 647.

Buchner, H., Mutsculer, C. and Kiechle, H. 'Die Determination der Mannchen- und Daueriproduktion', Biol. Zentralblatt, vol. 86 (1967), pp. 599-621.

Buffa, P. 'Studi Intorno al Ciclo Parthenogenetico dell'*Heliothrips haemorrhoidalis*', Redia 7 (1911).

Bull, J. J. 'An Advantage for the Evolution of Male Haploidy and Systems with Similar Genetic Transmission', Heredity, vol. 43 (1979), pp. 361-81.

Bullini, L. 'Richerche sulle Caratteristiche Biologiche dell'Anfigonia e della Partenogenesi in una Populazione Bisessuata di *Bacillus rossius* (Rossi) (Cheleutoptera=Phasmoidea)', Riv. Biol., vol. 58 (1965), pp. 189-216.

—, 'Spanandria e Partenogenesi Geografica in *Bacillus rossius* (Rossi)', Acc. Naz. Lincei Rend. Sci., vol. 40 (1966), pp. 926-32.

Bulmer, M. G. 'The Sib Competition Model for the Maintenance of Sex and Recombination', J. Theoret. Biol., vol. 82 (1980), pp. 335-45.

Burnett, A. L. 'A Model of Growth and Cell Differentiation in *Hydra*', Amer. Natur., vol. 100 (1966), pp. 165-90.

—, *Biology of Hydra* (Academic, New York, 1973).

Burnett, A. L. and Diehl, N. A. 'Initiation of Sexuality in *Hydra*', J. Exp. Zool., vol. 157 (1964), pp. 237-49.

Bushnell, J. H. 'Environmental Relationships of Michigan Ectoprocta and Dynamics of Natural Populations of *Plumatella repens*', Ecol. Monogr., vol. 36 (1966), pp. 95-123.

Bushnell, J. H. and Rao, K. S. 'Dormant or Quiescent Stages and Structures among the Ectoprocta: Physical and Chemical Factors Affecting Viability and Germination of Statoblasts', Trans. Amer. Microscop. Soc., vol. 93 (1974), pp. 524-43.

Butcher, R. W. *A New Illustrated British Flora*, vol. 2 (Leonard Hill, London, 1961).

Cable, R. M. 'Parthenogenesis in Parasitic Helminths', Amer. Zool., vol. 11 (1971), pp. 267-72.

Cain, A. J. 'The Perfection of Animals', Viewpoints in Biology, vol. 3 (1965), pp. 36-63.

—, 'Breeding System of a Sessile Animal', Nature, vol. 247 (1974), pp. 289-90.

Calow, P., Beveridge, M. and Sibly, R. 'Heads and Tails: Adaptional Aspects of Asexual Reproduction in Freshwater Triclads', Amer. Zool., vol. 20 (1980), pp. 715-27.

Calow, P. and Woollhead, A. S. 'The Relationship between Ration, Repro-ductive Effort and Age-specific Mortality in the Evolution of Reproductive Strategies – Some Observations on Freshwater Triclads', J. Anim. Ecol., vol. 46 (1977), pp. 765-81.

Le Calvez, J. 'Morphologie et Comportement des Chromosomes dans la Spermato-génèse de Quelques Mycetophilides'. Chromosoma, vol. 3 (1947), pp. 137-65.

Camenzind, R. 'Untersuchungen über Bisexuelle Fortpflanzung einer Paedo-genetischen Gallmücke', Rev. Suisse Zool., vol. 69 (1962), pp. 377-84.

——, 'Die Zytologie der Bisexuellen und Parthenogenetischen Fortpflanzung von *Heteropeza pygmaea* Winnertz, einer Gallmücke mit pädogenetischer Verme-hrung', Chromosoma, vol. 18 (1966), pp. 123-52.

Cameron, F. M. and Rees, H. 'The Influence of B-chromosomes on Meiosis in *Lolium*', Heredity, vol. 22 (1967), pp. 446-50.

Campbell, R. D. 'Cnidaria', in A. C. Giese and J. S. Pearse (eds.), *Reproduction of Marine Invertebrates*, vol. 1 (Academic, New York, 1974), pp. 133-99.

Cannon, H. G. 'A Further Account of the Spermatogenesis of Lice', Quart. J. Microscop. Sci., vol. 66 (1922), pp. 657-67.

Cannon, L. R. G. PhD Thesis (University of Toronto, 1970); cited by Cable (1971).

Cappe de Baillon, P. CR Acad. Sci. Paris, vol. 199 (1939), pp. 1069-70.

Cappe de Baillon, P. and de Vichet, G. 'La Parthénogénèse des Espèces du genre *Leptynia* Pant. (Orthoptera Phasmidae)', Bull. Biol. France Belgique, vol. 74 (1940), pp. 43-87.

Cappe de Baillon, P., Favrelle, M. and de Vichet, G. 'Parthénogénèse et Variation chez les Phasmes. I. *Baculum artemis* Westw., *Carausius thieseni* n.sp.', Bull. Biol. France Belgique, vol. 68 (1934), pp. 109-66.

——, 'Parthénogénèse et Variation chez les Phasmes. II. *Carausius furcillatus* Pant., *Menexenus semiarmatus*, Westw.', Bull. Biol. France Belgique, vol. 69 (1935), pp. 1-46.

——, 'Parthénogénèse et Variation chez les Phasmes. III. *Bacillus rossi* Rossi, *Epibacillus lobipes* Luc, *Phobaeticus sinetyi* Bt., *Parasosibia parva* Redt, *Carausius rotundato-lobatus* Br', Bull. Biol. France Belgique, vol. 71 (1937a), pp. 129-89.

——, 'Parthénogénèse et Variation chez les Phasmes. IV. Discussion de Faits', Bull. Biol. France Belgique, vol. 72 (1937b), pp. 1-47.

Carlin, B. 'Die Planktonrotatorien des Motalastrom: Zur Taxonomie und Oko-logie der Planktonrotatorien', Medd. Lunds Univ. Limnol. Inst., vol. 5 (1943), pp. 1-255.

Carlsson, G. 'Studies on Scandanavian Blackflies', Opuscula Entomol., suppl. 21 (1962), pp. 1-280.

Carneil, L. 'Untersuche in der Chiasma-frequenz bei Makro- und Mikro-spore Mutterzellen von *Rhoeo discolor*', Osten Bot. Zeit., vol. 107 (1960), pp. 241-4.

Carrada, C. C. and Sacchi, C. F. 'Ecology of *Victorella*', Vie et Milieu, vol. 15 (1964), pp. 389-426.

Carson, H. L. 'Rare Parthenogenesis in *Drosophila robusta*', Amer. Natur., vol. 95 (1961), pp. 81-6.

——, 'Fixed Heterozygosity in a Parthenogenetic Species of *Drosophila*', in *Studies in Genetics II*, University of Texas Publ. 6205 (1962), pp. 55-62.

——, 'Selection for Parthenogenesis in *Drosophila mercatorum*', Genetics, vol. 55 (1967), pp. 157-71.

Carson, H. L., Wheeler, W. R. and Heed, W. B. *A Parthenogenetic Strain of Drosophila mangabeiri Malagowkin*, University of Texas Publ. 5721 (1957), pp. 115-22.

Carter, C. R. and Smith-White, S. 'The Cytology of *Brachycome lineariloba*. 3. Accessory Chromosomes', Chromosoma, vol. 39 (1972), pp. 361-79.

Cassagnau, P. 'Parthénogénèse Géographique et Polyploidie chez *Neanura muscorum* (Templeton), Collembole Nearnuride', CR Hebd. Séanc. Acad. Sci. Paris, Ser D, vol. 274 (1972), pp. 1846-8.

Castle, W. A. 'The Life History of *Planaria velata*', Biol. Bull., vol. 53 (1927), pp. 139-44.

——, 'An experimental and histological study of the life-cycle of *Planaria velata*', J. Exp. Zool. vol. 51 (1928), pp. 417-83.

Catcheside, D. G. 'The Control of Recombination in *Neurospora crassa*', in W. J. Peacock and R. D. Brock (eds), *Replication and Recombination of Genetic Material* (Australian Acad. Sci. Canberra, 1968), pp. 216-26.

——, 'Occurrence in Wild Strains of *N. crassa* of Genes Controlling Recombination', Aust. J. Biol. Sci., vol. 28 (1975), pp. 213-25.

——, *The Genetics of Recombination* (University Park Press, Baltimore, Md., 1977).

Caullery, M. and Lavallée, A. 'Recherches sur le Cycle Évolutif des Orthonectides', Bull. Sci. France Belgique, vol. 46 (1912), pp. 139-71.

Caullery, M. and Mesnil, F. 'Sur l'Existence de la Multiplication Asexuée (Scissiparité Normale) chez Certains Sabelliens (*Potomilla torelli* Malmg. et *Myxicola dinardensis* St. Jos.), CR Acad. Sci. Paris, vol. 171 (1920), pp. 683-5.

Cavalier-Smith, T. 'The Origin of Nuclei and of Eukaryotic Cells', Nature, vol. 256 (1975), pp. 463-8.

Cernosvitov, L. 'Die Selbstbefruchtung bein den Oligochaeten', Biol. Zentralblatt, vol. 47 (1927), pp. 587-95.

Chadwick, H. C. 'Notes on *Cucumaria planci*', Trans. Liverpool Biol. Soc. vol. 5 (1981); cited by L. H. Hyman (1955).

Chamisso, A. von. 'De Animalibus Quibusdam e Classe Vermium Linneana in Circumnavigatione Terre' (Berolini, 1819).

Champ, P. and Pourriot, R. 'Reproductive Cycle in *Sinantherina socialis*', Arch. Hydrobiol. Beih. Ergenbn. Limnol., vol. 8 (1977), pp. 184-6.

Chanter, D. O. and Owen, D. F. 'The Inheritance and Population Genetics of Sex Ratio in the Butterfly *Acraea encedon*', J. Zool. (London), vol. 166 (1972), p. 363.

Chapman, A. 'Terrestrial Ostracoda of New Zealand', Nature, vol. 185 (1960), p. 121.

——, 'Terrestrial Ostracod of New Zealand, *Mesocypris audax*', Crustaceana, vol. 2 (1961), pp. 255-61.

Chapman, D. M. 'Evolution of the Scyphistoma', Symp. Zool. Soc. London, vol. 16 (1966), pp. 51-75.

——, 'Structure, Histochemistry and Formation of the Podocyst and Cuticle of *Aurelia aurita* (L.)', J. Mar. Biol. Ass. UK, vol. 48 (1968), pp. 187-208.

Charlesworth, B. 'Recombination Modification in a Fluctuating Environment', Genetics, vol. 83 (1976), pp. 181-95.

——, 'The Population Genetics of Anisogamy', J. Theoret. Biol., vol. 73 (1978), pp. 347-57.

——, 'The Cost of Sex in Relation to Mating System', J. Theoret. Biol., vol. 84 (1980a), pp. 655-71.

——, 'The Cost of Meiosis with Alternations of Sexual and Asexual Generations', J. Theoret. Biol. vol. 87 (1980b), pp. 517-28.

Charlesworth, B. and Charlesworth, D. 'Selection of New Inversions in Multilocus Genetic Systems', Genet. Res., Cambridge, vol. 23 (1973), pp. 167-83.

Charlesworth, B. and Hartl, D. 'Population Dynamics of the Segregation Distorter Polymorphism of *Drosophila melanogaster*', Genetics, vol. 87 (1978), pp. 171-92.

Charlesworth, D., Charlesworth, B. and Strobeck, C. 'Effects of Selfing on Selection for Recombination', Genetics, vol. 86 (1977), pp. 213-26.

Charniaux-Cotton, H. 'Déterminisme des Phénomènes d'Intersexualité chez *Orchestia gammarella*. Premiers Resultats', CR Acad. Sci. Paris, vol. 245 (1957), pp. 1665-9.

——, 'Sex Determination', in T. H. Waterman (ed.), *The Physiology of Crustacea*, vol. 1 (Academic, New York, 1960), chap. 13, pp. 411-48.

Charnov, E. L. 'Sex Ratio Selection in an Age-structured Population', Evolution, vol. 29 (1975), pp. 366-8.

Charnov, E. L. and Krebs, J. R. 'The Evolution of Alarm Calls: Altruism or Manipulation?', Amer. Natur., vol. 109 (1975), pp. 107-12.

Cherfas, N. B. 'Natural Triploidy in Females of the Unisexual Form of Silver Carp (*Carussius auralus gibelio* Bloch)', Genetika, vol. 5 (1966), pp. 16-24.

Chia, F.-S. 'Sea Anemone Reproduction: Patterns and Adaptive Radiation', in G. O. Mackie (ed.), *Coelenterate Ecology and Behaviour* (Plenum, New York, 1976), pp. 261-70.

Chia, F.-S. and Rostron, M. A. 'Some Aspects of the Reproductive Biology of *Actinia equina* (Cnidaria: Anthozoa)', J. Mar. Biol. Ass. UK, vol. 50 (1970), pp. 253-64.

Child, C. M. 'The Asexual Cycle of *Planaria velata* in Relation to Senescence and Rejuvenescence', Biol. Bull., vol. 25 (1913), pp. 181-203.

——, 'Asexual Breeding and Prevention of Senescence in *Planaria velata*', Biol. Bull., vol. 26 (1914), pp. 286-93.

Chinnici, J. P. 'Modification of Recombination Frequency in *Drosophila*. I. Selection for Increased and Decreased Crossing-over', Genetics, vol. 69 (1971), pp. 71-83.

Chitwood, B. G. and Chitwood, M. B. *Nematology* (University Park Press, Baltimore, Md., 1974).

Chopard, L. 'Description d'une Espèce Nouvelle de Genre *Myrmecophila* (Orth. Gryllidae) et Remarques sur la Sexualité chez les Espèces de ce Genre', Bull. Soc. Biol. France, vol. 43 (1919), pp. 339-46.

—, 'Les Orthoptères Cavernicoles de la Faune Paléarctique', Arch. Zool. Expér. Gén., vol. 74 (1932), pp. 263-86.

— 'La Parthénogénèse chez les Orthopteroides', Année Biol. Sér. 3, vol. 24 (1948), pp. 15-22.

Christiansen, B. 'Asexual Reproduction in the Enchytraeidae (Olig.)', Nature, vol. 184 (1959), pp. 1159-60.

—, 'A Comparative Cytological Investigation of the Reproductive Cycle of an Amphimictic Diploid and a Parthenogenetic Triploid from *Lumbricillus lineatus* (O.F.M.) (Oligochaeta: Enchytraeidae)', Chromosoma, vol. 11 (1960), pp. 365-79.

—, 'Studies on Cytotaxonomy and Reproduction in the Enchytraeidae. With Notes on Parthenogenesis and Polyploidy in the Animal Kingdom', Hereditas, vol. 47 (1961), pp. 387-450.

Christiansen, B. and Jensen, J. 'Sub-amphimictic Reproduction in a Polyploid Cytotype of *Enchytraeus lacteus* Neilsen & Christiansen (Oligochaeta Enchytraeidae)', Hereditas, vol. 52 (1964), pp. 106-18.

Christiansen, B. and O'Connor, B. F. 'Pseudo-fertilization in the Genus *Lumbricillus* (Enchybaeidae)', Nature, vol. 181 (1958), pp. 1085-6.

Christiansen, B., Berg, U. and Jelnes, J. 'A Comparative Study on Enzyme Polymorphism in Sympatric Diploid and Triploid Forms of *Lumbricillus lineatus* (Enchytraeidae, Oligochaeta)', Hereditas, vol. 84 (1976), pp. 41-8.

Christiansen, J. L. and Ladman, A. J. 'The Reproductive Morphology of *Cnemidophorus neomexicanus* X *C. inornatus* hybrid males', J. Morphol., vol. 125 (1968), pp. 367-78.

Christiansen, K. 'Bionomics of Collembola', Ann. Rev. Entomol., vol. 9 (1964), pp. 147-78.

Christie, J. R. 'Some Observations on Sex in the Mermithidae', J. Exp. Zool., vol. 54 (1929), pp. 59-76.

Christie, J. R. 'Life History (Zooparasitica)', in B. G. Chitwood and M. B. Chitwood (eds.), *Nematology*. (University Park Press, Baltimore, Md., 1974), pp. 246-66.

Church, K. 'Meiosis in the Grasshopper. III. Chiasma Frequencies in Females after Elevated Temperature', Heredity, vol. 32 (1974), pp. 159-64.

Church, K. and Wimber, D. E. 'Meiosis in the Grasshopper. Chiasma Frequency after Elevated Temperature and X-rays', Can. J. Genet. Cytol., vol. 11 (1969), pp. 209-16.

Chvala, M. 'The Tachydromiinae (Dipt. Empididae) of Fennoscandia and Denmark', Fauna Entomol. Scand., vol. 3 (1975), pp. 1-336.

Cimino, M. C. 'Meiosis in a Triploid All-female Fish (*Poeciliopsis*, Poeciliidae)', Science, vol. 175 (1971), pp. 1484-6.

Cimino, M. C. and Schultz, R. J. 'Production of a Diploid Male Offspring by a Gynogenetic Triploid Fish of the Genus *Poeciliopsis*', Copeia, vol. 4 (1970), pp. 760-3.

Clanton, W. 'An Unusual Situation in the Salamander *Ambystoma jeffersonianum* (Green)', Occ. Pap. Mus. Zool. Univ. Mich., vol. 290 (1934), pp. 1-14.

Clark, E. 'Functional Hermaphroditism and Self-fertilization in a Serranid Fish', Science, vol. 129 (1959), pp. 215-16.

Clark, H. L. 'Autotomy in *Linckia*', Zool. Anz., vol. 43 (1913), pp. 156-9.

Clark, W. C. 'The Ecological Implications of Parthenogenesis', in A. D. Lowe (ed.), *Perspectives in Aphid Biology*, Bulletin No. 2 of the Entomological Society of New Zealand (1973), pp. 103-13.

Clarke, B. C. 'The Evolution of Genetic Diversity' Proc. Roy. Soc. London B, vol. 205 (1979), pp. 453-74.

van Cleave, H. J. and Altringer, D. A. 'Studies on the Life Cycle of *Campeloma rufum*, a Parthenogenetic Snail', Amer. Natur., vol. 71 (1937), pp. 167-84.

Clemens, W. A. 'A Parthenogenetic Mayfly (*Ameletus ludens* Needham)', Can. Entomol., vol. 54 (1922), pp. 77-8.

Clément, P. and Pourriot, R. 'Influence de la Densité de Population sur la Production de Femelles Mictiques Induites par Photopériode chez *Notammata copeus* (Rotifère)', CR Acad. Sci. Paris, vol. 276 (1973a), pp. 3151-4.

—, 'Mise en Evidence d'un Effet de Masse et d'un Effet de Groupe dans l'Apparition de Phases de Reproduction Sexuée chez le Rotifère *Notommata copeus*', CR Acad. Sci. Paris, vol. 277 (1973b), pp. 2533-6.

—, 'Influence de Groupement et de la Densité de Population sur le Cycle de Reproduction de *Notommata copeus* (Rotifère)', Arch. Zool. Exp. Gén., vol. 117 (1975), pp. 5-13.

Cleveland, L. R. 'Brief Accounts of the Sexual Cycles of the Flagellates of *Cryptocercus*', J. Protozool., vol. 3 (1956), pp. 161-80.

Cloud, P. E., Licari, G. R., Wright, L. A. and Troxel, B. W. 'Proterozoic Eukaryotes from Eastern California', Proc. Nat. Acad. Sci. US, vol. 62 (1969), pp. 623-30.

Coe, W. R. 'Asexual Reproduction in Nemerteans', Physiol. Zool., vol. 3 (1930), pp. 297-308.

—, 'Sexual Differentiation in Molluscs. I. Pelecypods', Quart. Rev. Biol., vol. 18 (1943), pp. 154-64.

—, 'Sexual Differentiation in Molluscs. II. Gastropods, Amphineurans, Scaphopods, and Cephalopods', Quart. Rev. Biol., vol. 19 (1944), pp. 85-97.

Cognetti, G. 'Selezione per la Forme Attere in Linee Parthenogenetiche di *Myzodes persicae* Sulzer (Homoptera Aphididae)', Bull. Zool., vol. 27 (1960), pp. 107-11.

—, 'Citogenetica della Partenogenesi negli Afidi', Arch. Zool. Italiana, vol. 46 (1961a), pp. 89-122.

—, 'Endomeiosis in Parthenogenetic Lines of Aphids', Experientia, vol. 17 (1961b), pp. 168-9.

Cognetti, G. and Delavault, R. 'La Sexualité des Astérides', Cahiers Biol. Marine, vol. 3 (1962), pp. 157-82.

Cohen, J. 'Correlation between Chiasma Frequency and Sperm Redundancy', Nature, vol. 215 (1967), pp. 862-3.

—, 'Cross-overs, Sperm Redundancy and Their Close Association', Heredity, vol. 31 (1973), pp. 408-13.

—, 'Gamete Redundancy – Wastage or Selection?', in D. L. Mulcahy (ed.), *Gamete Competition in Plants and Animals* (North-Holland, Amsterdam, 1975), pp. 99-112.

Coil, W. H. 'Studies on Dioecious Tapeworms', J. Parasitol., vol. 54 (1970), p. 59.

Cole, C. J. 'Evolution of Parthenogenetic Species of Reptiles', in R. Reinboth (ed.), *Intersexuality in the Animal Kingdom* (Springer-Verlag, New York, 1975), pp. 340-55.

——, 'Parthenogenetic Lizards', Science, vol. 201 (1978), pp. 1154-5.

Cole, G. A. 'Desert Limnology', in G. W. Brown (ed.) *Desert Biology* (Academic, New York, 1968), pp. 424-87.

Colton, H. S. 'Self-fertilization in *Lymanaea*', Proc. Nat. Sci. Philadelphia, vol. 1912 (1912), pp. 173-183.

——, 'Self-fertilization in the Air-breathing Pond Snails', Biol. Bull., vol. 35 (1918), pp. 48-9.

—, 'Five Years of a Self-fertilized Line of *Lymnaea columella*', Proc. Amer. Soc. Zool., in Anat. Rec., vol. 23 (1922), p. 97 (abstr.).

Colton, H. S. and Pennypacker, M. 'The Results of Twenty Years of Self-fertilization in the Snail *Lymnaea columella* Say', Amer. Natur., vol. 68 (1934), pp. 129-36.

Comfort, A. 'The Longevity and Mortality of a Fish (*Lebistes reticulatus* Peters) in Captivity', Gerontologia, vol. 5 (1961), pp. 209-22.

Comrie, L. C. 'Biological and Cytological Observations on Tenthredinid Parthenogenesis', Nature, vol. 142 (1938), pp. 877-8.

Connell, J. H. 'Population Ecology of Reef-building Corals', in O. Jones and R. Endean (eds.), *Biology and Geology of Coral Reefs*, vol. II, *Biology 1* (Academic, New York, 1973), pp. 205-45.

—, 'Competitive Interactions and the Species Diversity of Corals', in G. O. Mackie (ed.), *Coelenterate Ecology and Behaviour* Plenum, New York, 1976), pp. 51-8.

Cooke, W. J. in G. O. Mackie (ed.), *Coelenterate Ecology and Behaviour* (Plenum, New York, 1976), pp. 281-8.

Corbet, P. J. 'Parthenogenesis in Caddisflies (Trichoptera)', Can. J. Zool., vol. 44 (1966), pp. 981-2.

Couzin, D. A. and Fox, D. P. 'Variation in Chiasma Frequency during Tulip Anther Development', Chromosoma, vol. 46 (1974), pp. 173-9.

Cox, E. C. and Gibson, T. C. 'Selection for High Mutation Rates in Chemostats', Genetics, vol. 77 (1974), pp. 169-84.

Crabb, E. D. 'Genetic Experiments with Pond Snails, *Lymnaea* and *Physa*', Amer. Natur., vol. 61 (1927), pp. 54-67.

Creighton, M. 'Parthenogenesis in *Chorthippus curtipennis*', Genetics, vol. 23 (1938), p. 145 (abstr.).

Creighton, M. and Robertson, W. R. B. 'Genetic Studies on *Chorthippus longicornis*' J. Heredity, vol. 32 (1941), pp. 339-41.

Cresp, J. Études Expérimentales et Histologiques sur la Régénération et le Bourgeonnement chez les Serpulides *Hydroides norwegica* (Gunns) et *Salmacina incrustans* (Clap.)', Bull. Biol. France Belgique, vol. 98 (1964), pp. 3-152.

Crew, F. A. E. and Koller, P. C. 'The Sex Incidence of Chiasma Frequency and Genetical Crossing-over in the Mouse', J. Genetics, vol. 26 (1932), pp. 359-82.

Crow, J. F. and Kimura, M. 'Evolution in Sexual and Asexual Populations', Amer. Natur., vol. 99 (1965), pp. 439-50.

—, 'Evolution in Sexual and Asexual Populations', Amer. Natur., vol. 103 (1969), pp. 89-91.

Crowe, J. H. 'The Physiology of Cryptobiosis in Tardigrades', Mem. Ist. Ital. Idrobiol., vol. 32 (suppl.) (1975), pp. 37-59.

Crowe, J. H. and Madin, K. A. 'Anhydrobiosis in Tardigrades and Nematodes', Trans. Amer. Microscop. Soc., vol. 43 (1974), pp. 573-24.

Crozier, W. J. 'Multiplication by Fission in Holothurians', Amer. Natur., vol. 51 (1917), pp. 560-6.

—, 'Notes on Some Problems of Adaptation. 2. On the Temporal Relations of Asexual Propagation and Gametic Reproduction in *Coscinasterias tenuispina*: With a Note on the Direction of Progression and on the Significance of the Madrepores', Biol. Bull. Mar. Lab. Woods Hole, vol. 34 (1920a), pp. 116-29.

—, 'Multiplication by Fission in a Balanoglossid', Anat. Rec., vol. 20 (1920b), p. 186.

Cuellar, O. 'Additional Evidence for True Parthenogenesis in Lizards of the Genus *Cnemidophorus*', Herpetologica, vol. 24 (1968), pp. 146-50.

—, 'Reproduction and the Mechanism of Meiotic Restitution in the Parthenogenetic Lizard *Cnemidophorus uniparens*', J. Morphol., vol. 133 (1971), pp. 139-65.

—, 'On the Origin of Parthenogenesis in Vertebrates: The Cytogenetic Factors', Amer. Natur., vol. 108 (1974), pp. 625-48.

—, 'Intraclonal Histocompatibility in Parthenogenetic Lizard: Evidence of Genetic Homogeneity', Science, vol. 193 (1976), pp. 150-3.

—, 'Animal parthenogenesis', Science, vol. 197 (1977), pp. 837-43.

—, 'Parthenogenetic lizards', Science, vol. 201 (1978), p. 1155.

—, 'On the Ecology of Coexistence in Parthenogenetic and Bisexual Lizards of the Genus *Cnemidophorus*', Amer. Zool., vol. 19 (1980), pp. 773-86.

Cuellar, O. and Kluge, A. G. 'Natural Parthenogenesis in the Gekkonid Lizard *Lepidodactylus lugubris*', J. Genet., vol. 61 (1972), pp. 14-26.

Curtis, W. C. 'The Life History, the Normal Fission and the Reproductive Organs of *Planaria maculata*', Proc. Boston Nat. Hist. Soc. (1902), p. 30.

Custance, D. R. N. 'Light as an Inhibitor of Strobilation in *Aurelia aurita*', Nature, vol. 204 (1964), pp. 1219-20.

—, 'The Effect of a Sudden Rise in Temperature on Strobilae of *Aurelia aurita*', Experientia, vol. 22 (1966), pp. 588-9.

—, 'Studies on Strobilation in the Scyphozoa', J. Biol. Educ., vol. 1 (1967), pp. 79-81.

Cuvier, G. L. *The Animal Kingdom*, vol. 13 (Whittaker, London, 1833).

Daggett, R. F. and Davis, C. C. 'A Seasonal Quantitative Study of the Littoral Cladocera and Copepoda in a Bog Pond and an Acid Marsh in Newfoundland', Int. Rev. Ges. Hydrobiol., vol. 59 (1974), pp. 667-83.

Dahm, A. G. *Taxonomy and Ecology of Five Species Groups in the Family Planariidae* (Nyo Litograten, Malmo, 1958).

Dallot, S. 'Observations Préliminaires sur la Réproduction en Élevage du Chaeto-gnathe Planctonique *Sagitta setosa* Muller', Rapp. Commun. Int. Mer. Medit., vol. 19 (1968), pp. 521-3.

Daly, H. V. 'Biological Studies on *Ceratina dallotorranea*, an Alien Bee in California Which Reproduces by Parthenogenesis. (Hymenoptera:Apoidea)', Ann. Entomol. Soc. Amer., vol. 59 (1966), pp. 1138-54.

Darevsky, I. S. 'Natural Parthenogenesis in Certain Subspecies of Rock Lizard, *Lacerta saxicola* Eversmann', Dokl. Biol. Sci. Sect., vol. 122 (1958), pp. 877-9.

—, 'On the Origin and Biological Role of Natural Parthenogenesis in a Poly-morphic Group of Caucasian Rock Lizards, *Lacerta saxicola* Eversmann', Zool. Zh. USSR, vol. 41 (1962), pp. 397-408.

—, 'Natural Parthenogenesis in a Polymorphic Group of Caucasian Rock Lizards Related to *Lacerta saxicola* Eversmann', J. Ohio Herpet. Soc., vol. 5 (1966), pp. 115-52.

Darevsky, I. S. and Kulikova, V. N. 'Natural Triploidy in a Polymorphic Group of Rock Lizards, *Lacerta saxicola* Eversmann, Resulting from Hybridization between Bisexual and Parthenogenetic Forms of the Species', Dokl. Akad. Nauk USSR, vol. 158 (1964), pp. 202-5.

Darlington, C. D. 'The Origin and Behaviour of Chiasmata. VIII. *Secale cereale* ($n = 8$)', Cytologia, vol. 4 (1933), pp. 444-52.

—, *The Evolution of Genetic Systems* (Oliver and Boyd, Edinburgh, 1939).

Darnell, R. M., Lamb, E. and Abramoff, P. 'Matroclinous Inheritance and Clonal Structure of a Mexican Population of the Gynogenetic fish, *Poecilia formosa*', Evolution, vol. 21 (1967), pp. 168-73.

Darwin, C. R. 'On the Agency of Bees in the Fertilization of Papilionaceous Flowers, and on the Crossing of Kidney Beans', Ann. Mag. Nat. Hist., vol. 2 (1858), pp. 459-65.

——, *The Effects of Cross and Self Fertilization in the Vegetable Kingdom* (Murray, London, 1888).

Davenport, C. B. 'On *Urnatella gracilis*', Bull. Mus. Comp. Zool. Harvard Univ., vol. 24 (1893), pp. 1-44.

Davidson, J. 'On the Occurrence of the Parthenogenetic and Sexual Forms in *Aphis rumieis* L., with Special Reference to the Influence of Environmental Factors', Ann. Appl. Biol., vol. 16 (1929), pp. 104-34.

Davies, D. M. and Petersen, B. V. 'Observations on the Mating, Feeding, Ovarian Development and Oviposition of Adult Black Flies (Simuliidae, Diptera)', Can. J. Zool., vol. 34 (1956), pp. 614-55.

Davies, E. D. G. and Jones, G. H. 'Chiasma Variation and Control in Pollen Mother Cells and Embryo-sac Mother Cells of Rye', Genet. Res., vol. 23 (1974), pp. 185-90.

Davies, E. W. 'Cytology, Evolution and Origin of the Aneuploid Series in the Genus *Carex*', Hereditas, vol. 42 (1956), pp. 349-65.

Davies, L. 'Observations on *Prosimulium ursinum* Edw. at Holandsfjord, Norway' Oikos, vol. 5 (1954), pp. 94-8.

Dawkins, R. *The Selfish Gene* (Oxford University Press, 1976).

Dawydoff, C. 'Multiplication Asexuée, par Lacération, chez les *Ctenoplana*', CR Acad. Sci. Paris, vol. 206 (1938), pp. 127-8.

——, 'Phenomène d'Épitoquie dans le Groupe des Archiannélides', CR Acad. Sci. Paris, vol. 229 (1949), pp. 96-7.

DeBach, P. 'Uniparental, Sibling and Semispecies in Relation to Taxonomy and Biological Control', Israel J. Entomol., vol. 4 (1969), pp. 11-28.

Degrange, C. 'Recherches sur la Reproduction des Ephéméroptères', Trav. Lab. Hydrobiol. Grenoble, vol. 51 (1960), pp. 7-183.

Dehorne, A. 'La Schizometamérie et les Segments Tetragemmes du *Dodecaceria caulleryi* sp. n.', Bull. Biol. France Belgique, vol. 67 (1933), pp. 298-326.

Dehorne, L. 'Les Naidomorphes et leur Reproduction Asexuée', Arch. Zool. Exp. Gén., vol. 56 (1916), pp. 25-157.

Deichmann, E. 'On Some Cases of Multiplication by Fission and of Coalescence in Holothurians: With Notes on the Synonymy of *Actinopyga parvula*', Vidensk. Medd. Dansk. Naturh. Foren Kobenhaven, vol. 73 (1921), pp. 199-215.

Delage, Y. and Goldsmith, M. *La Parthénogénèse Naturelle et Expérimentale* (Flammarion, Paris, 1913).

Delavault, R. 'L'Autofécondation chez les Métazoaires', Ann. Biol., vol. 34 (1958), pp. 5-16.

——, 'Determinism of Sex', in R. A. Boolootian (ed.), *Physiology of Echinodermata* (Wiley, New York, 1966), pp. 615-38.

Delorme, L. D. 'Freshwater Ostracodes of Canada. Part I. Subfamily Cypridinae', Can. J. Zool., vol. 48 (1970a), pp. 153-68.

Delorme, L. D. 'Freshwater Ostracodes of Canada. Part II. Subfamily Cypridopsinae and Herpetocypridinae, and Family Cyclocyprididae', Can. J. Zool., vol. 48 (1970b), pp. 252-66.

Delorme, L. D. 'Freshwater Ostracodes of Canada. Part III. Family Candonidae', Can. J. Zool., vol. 48 (1970c), pp. 1099-127.

Delorme, L. D. 'Freshwater Ostracodes of Canada. Part IV. Families Ilyocyprididae, Notodromadidae, Darwinulidae, Cytherideidae and Entocytheridae' Can. J. Zool., vol. 48 (1970d), pp. 1251-9.

Delorme, L. D. 'Freshwater Ostracods of Canada. Part V. Families Limnocytheridae, Oxochonchidae', Can. J. Zool., vol. 49 (1971), pp. 49-64.

Despax, R. 'Trichoptères Observés dans les Pyrénées Françaises', Bull. Soc. Nat. Toulouse (1928), p. 57.

Detlefson, J. A. and Roberts, E. 'Studies on Crossing-over. I. The Effect of Selection on Crossover Values', J. Exp. Zool., vol. 32 (1921), pp. 333-54.

DeWitt, R. M. 'Reproductive Capacity in a Pulmonate Snail (*Physa gyrina* Say)', Amer. Natur., vol. 88 (1954), pp. 159-64.

DeWitt, R. M. and Sloan, W. C. 'The Innate Capacity for Increase in Numbers in the Pulmonate Snail, *Lymnaea columella*', Trans. Amer. Microscop. Soc., vol. 77 (1958), pp. 290-4.

Dexter, R. W. 'Studies on North American Fairy Shrimps with the Description of Two New Species', Amer. Midl. Nat., vol. 49 (1953), pp. 751-71.

Dhandapani, P. 'Some Observations on Self-Fertilization among Doliolida (Pelagic Tunicata)', Pubbl. Staz. Zool. Napoli, vol. 39 (Suppl.) (1975), pp. 108-13.

Diver, W. L. 'Precambrian Microfossils of Carpenterian Age from Bungle Bungle Dolomite of Western Australia', Nature, vol. 247 (1974), pp. 361-2.

Dixon, A. F. G. 'Aphid Ecology: Life Cycles, Polymorphism and Population Regulation', Ann. Rev. Ecol. Syst., vol. 8 (1977), pp. 329-53.

Dobbin, C. N. 'Freshwater Ostracoda from Washington and Other Western Localities', Univ. Washington Publ. Biol., vol. 4 (1941), pp. 175-246.

Dobers, E. 'Biologie der Bdelloidea', Int. Rev. Ges. Hyrobiol., vol. 6, supp. 7 (1915).

Dodds, G. S. 'Mayflies from Colorado', Trans. Amer. Entomol. Soc., vol. 49 (1923), pp. 93-114.

Dodds, K. S. 'Chromosome Numbers and Spermatogenesis in Some Species of the Hymenopterous Family Cynipidae', Genetica, vol. 20 (1937), pp. 67-84.

—, 'Oogenesis in *Neuroterus baccarum* L', Genetica, vol. 21 (1939), pp. 177-90.

Dogiel, V. A. *General Protozoology* (Clarendon Press, Oxford, 1965).

Domontay, J. S. 'Autotomy in Holothurians', Nat. Appl. Sci. Bull., vol. 1 (1931), pp. 389-404.

Doncaster, L. 'On the Maturation of the Unfertilized Eggs and the Fate of the Polar Bodies in the Tenthredinidae (Saw-flies)', Quart. J. Microscop. Sci., vol. 49 (1906), pp. 561-89.

—, 'Gametogenesis and Fertilization in *Nematus ribesii*', Quart. J. Microscop. Sci., vol. 51 (1907), pp. 101-14.

—, 'Gametogenesis of the Gall-fly, *Neuroterus lenticularis*. Part I', Proc. Roy. Soc. London, Ser. B, vol. 82 (1910), pp. 88-113.

—, 'Gametogenesis of the Gall-fly, *Neuroterus lenticularis*. Part II', Proc. Roy. Soc. London, Ser. B83 (1911), pp. 476-89.

—, 'Gametogenesis of the Gall-fly, *Neuroterus lenticularis*. Part III', Proc. Roy. Soc. London, Ser. B89 (1916), pp. 183-200.

Doncaster, L. and Cannon, H. G. 'On the Spermatogenesis of the Louse (*Pediculus humanus* and *P. capitis*) with Observations on the Maturation of the Egg', Quart. J. Microscop. Sci., vol. 64 (1919), pp. 303-28.

Dorjes, J. '*Paratomella unichaeta* nov. gen. nov. spec., Vertreter einer Neuen Familie der Turbellaria Acoela mit Asexueller Fortpflanzung durch Paratomie', Veroff. Inst. Meeresforsch. Bremerhaven, vol. 2 (1966), pp. 187-200.

Dougherty, E. C. 'Comparative Evolution and the Origin of Sexuality', Syst. Zool., vol. 4 (1955), pp. 145-69, 190.

Dougherty, E. C. and Harris, L. G. 'Antarctic Micrometazoa: Freshwater Species in the McMurdo Sound Area', Science, vol. 140 (1963), pp. 497-8.

Doutt, R. L. and Smith, R. A. 'Males and Intersexes in a Normally Thelytokous Insect, *Tropidophryne melvillei* Compere (Hymenoptera: Encyrtidae)', Can. Entomol., vol. 82 (1950), pp. 165-70.

Downes, J. A. 'The Food-habits and Description of *Atrichopogon pollinivorus*, sp. nov. (Diptera: Ceratopogonidae)', Trans. Roy. Ent. Soc. London, vol. 106 (1955), pp. 439-53.

—, 'The Feeding Habits of Biting Flies and Their Significance in Classification', Ann. Rev. Entomol., vol. 3 (1958), pp. 249-66.

—, 'Arctic Insects and Their Environment', Can. Entomol., vol. 96 (1964), pp. 279-307.

—, 'Adaptations of Insects in the Arctic', Ann. Rev. Entomol., vol. 10 (1965), pp. 257-74.

Dowrick, G. J. 'The Effect of Temperature on Meiosis', Heredity, vol. 11 (1957), pp. 37-49.

Drosopoulos, S. 'Triploid Pseudogamous Biotype of the Leafhopper, *Muellerianella fairmairei*', Nature, vol. 263 (1976), pp. 499-500.

Duerden, J. E. 'Aggregated Colonies in Madreporian Corals', Amer. Natur., vol. 36 (1902), pp. 461-7.

Duncan, C. J. 'The Anatomy and Physiology of the Reproductive System of the Freshwater Snail *Physa fontinalis* (L.)', Proc. Zool. Soc. London, vol. 131 (1958), pp. 55-84.

—, 'Reproduction', in V. Fretter and J. Peake (eds.) *Pulmonates*, vol. 1 (Academic, London, 1975), pp. 309-66.

Durchon, M. 'Sur l'Existence de la Parthénogénèse chez *Nereis diversicolor* O. F. Muller (Annélide polychète)', Ann. Sci. Nat. Zool., Sér. 11, vol. 19 (1957), pp. 49-57.

Dybas, H. S. 'Evidence for Parthenogenesis in the Featherwing Beetles, with a Taxonomic Review of a New Genus and Eight New Species (Coleoptera: Ptiliidae)', Fieldiana: Zoology, vol. 51 (1966), pp. 11-52.

Eason, E. H. *Centipedes of the British Isles* (Warne, London, 1964).

Eaton, A. E. 'Parthenogenesis in *Orgyia antiqua*', Entomol. Monthly Mag. vol. 2 (1866).

Eckardt, M. J. and Whimster, I. W. 'Skin Homografts in the All-female Gekkonid Lizard *Hemidactylus garnotii*', Copeia (1971), pp. 152-4.

Edmondson, C. H. 'Autotomy and Regeneration in Hawaiian Starfishes', Occ. Papers Bernice P. Bishop Mus. Hawaii, vol. 11, no. 8 (1935).

Edmondson, W. T. 'The Seasonal Life History of *Daphnia* in an Arctic Lake', Ecology, vol. 30 (1955), pp. 439-55.

Edward, D. H. D. 'The Biology of a Parthenogenetic Species of *Lundstroemia* (Diptera: Chironomidae), with Descriptions of the Immature Stages', Proc. Roy. Entomol. Soc. London, Ser. A, vol. 38 (1963), pp. 165-70.

Edwards, C. A. and Lofty, J. R. *Biology of Earthworms* (Chapmen and Hall, London, 1972).

Edwards, F. W. 'Some Parthenogenetic Chironomidae', Ann. Mag. Nat. Hist., vol. 9 (1919), pp. 222-8.

Eertmoed, G. 'The Life History of *Peripsocus quadrifasciatus* (Psocoptera: Peripsocidae), J. Kansas Entomol. Soc., vol. 39 (1966), pp. 54-65.

El-Bayoumi, A. S. 'A Study of Meiosis in Some Wild Plant Species from Kuwait', Cytologia, vol. 38 (1973), pp. 357-61.

Elgmork, K. 'Dynamics of Zooplankton Communities in Some Small Inundated Ponds', Folia Limnol. Scand., vol. 12 (1964), p. 1-83.

Elliott, C. G. 'The Effect of Temperature on Chiasma Frequency', Heredity, vol. 9 (1955), pp. 385-98.

—, 'Environmental Effects on the Distribution of Chiasmata among Nuclei and Bivalents and Correlation between Bivalents', Heredity, vol. 12 (1958), pp. 429-39.

Elofson, O. 'Neue und Wenig Bekannte Cytheriden von der Schwedischen Westkuste', Arkiv. f. Zool., vol. 30, no. 21 (1938), pp. 1-22.

—, 'Zur Kenntnis der Marinen Ostracoden Schwedens mit Besonderer Berucksichtigung des Skageraks', Zool. Bidr. Uppsala, vol. 19 (1941), pp. 215-534.

Emlen, J. M. *Ecology: An Evolutionary Approach* (Addison-Wesley, Reading, Mass., 1973).

Engelmann, F. *The Physiology of Insect Reproduction*. (Pergamon, Oxford, 1970).

Enghoff, H. 'Parthenogenesis in Insects, Myriapods, Arachnids and Terrestrial Isopods', Ent. Meddr., vol. 44 (1976a), pp. 31-64 (in Danish; English summary).

—, 'Parthenogenesis and Bisexuality in the Millipede *Nemasoma varicorne* C. L. Koch, 1847 (Diplopoda: Blaniulidae). Morphological, Ecological and Biogeographical Aspects', Vidensk. Meddr. Dansk. Naturh. Foren, vol. 139 (1976b), pp. 21-59.

—, 'Morphological Comparison of Bisexual and Parthenogenetic *Polyxenus lagurus* (Linne 1758) (Diplopoda, Polyxenidae) in Denmark and Southern Sweden, with Notes on Taxonomy, Distribution and Ecology', Entomol. Medd., vol. 44 (1976c), pp. 161-82.

—, '*Cylindrodesmus laniger* Schubart, a Widespread, Probably Parthenogenetic Millipede (Diplopoda, Polydesmida: Haplodesmidae)', Entomol. Scand., vol. 9 (1978), p. 80.

Entwhistle, P. F. 'Inbreeding and Arrhenotoky in the Ambrosia Beetle *Xyleborus compactus* (Eichh.) (Coleoptera: Scolytidae)', Proc. Roy. Entomol. Soc. London, vol. 39 (1964), pp. 83-88.

Eshel, I. 'On the Neighbour Effect and the Evolution of Altruistic Traits', Theoret. Pop. Biol., vol. 3 (1972) pp. 258-77.

—, 'Selection on Sex-ratio and the Evolution of Sex-determination', Heredity, vol. 34 (1975), pp. 351-61.

Eshel, I. and Feldman, M. W. 'On the Evolutionary Effect of Recombination', Theoret. Pop. Biol., vol. 1 (1970), pp. 88-100.

Evans, A. A. F. and Fisher, J. M. 'Some Factors Affecting the Number and Size of Nematodes in Populations of *Aphelenchus avenae*', Nematologica, vol. 16 (1970), pp. 295-304.

Evans, A. C. and Guild, W. J. McL. 'Studies on the Relationships between Earthworms and Soil Fertility. IV. On the Life Cycles of Some British Lumbricidae', Ann. Appl. Biol., vol. 35 (1948), pp. 471-84.

Evans, D. 'The Bisexual and Agamic Generations of *Besbicus mirabilis* (Hymenoptera: Cynipidae), and Their Associated Insects', Can. Ent., vol. 99 (1967), pp. 187-96.

Fage, L. 'Les Amphipodes Pélagique du Genre *Rhabdosoma*' CR Acad. Sci. Paris, vol. 239 (1954), pp. 661-3.

Farstad, C. W. 'Thelytokous Parthenogenesis in *Cephus cinctus* (Hymenoptera Cephidae)', Can. Entomol., vol. 70 (1938), pp. 206-7.

Faulkner, G. H. 'The Anatomy and the Histology of Bud-formation in the Serpulid *Filograna implexa*, together with Some Cytological Observations on the Nuclei of the Neoblasts', Biol. J. Linn. Soc. London, vol. 37 (1930), pp. 109-90.

Fedorov, A. A. (ed.) 'Chromosome Numbers of Flowering Plants'. Original in Russian: translation published by Otto Koeltz Science Publishers, Koenigstein, West Germany, 1974.

Feldman, M. W. 'Selection for Linkage Modification. 1. Random Mating Populations', Theoret. Pop. Biol., vol. 3 (1972), pp. 324-46.

Fell, P. E. 'Diapause in the Gemmules of the Marine Sponge *Haliclona loosanoffi*, with a Note on the Gemmules of *Haliclona oculata*', Biol. Bull., vol. 147 (1974a), pp. 333-51.

—, 'Porifera', in A. C. Giese and J. S. Pearse (eds.), *Reproduction of Marine Invertebrates*, vol. 1 (Academic, New York, 1974b), pp. 51-132.

Felsenstein, J. 'The Evolutionary Advantage of Recombination', Genetics, vol. 78 (1974), pp. 737-56.

Felsenstein, J. and Yokoyama, S. 'The Evolutionary Advantage of Recombination. II. Individual Selection for Recombination', Genetics, vol. 83 (1976), pp. 845-59.

Ferguson, E. 'The Ostracod Genus *Potamocypris* with the Description of a New Species', Proc. Biol. Soc. Washington, vol. 72 (1959), pp. 133-8.

—, 'The Ostracod (Crustacea) Genus *Cypridopsis* in North America, and a Description of *Cypridopsis howei* sp nov.', Trans. Amer. Microscop. Soc., vol. 83 (1964), pp. 380-4.

Filipponi, A. 'Experimental Taxonomy Applied to the Macrochelidae (Acari: Mesotigmata)', in G. O. Evans (ed.), *Proceedings of Second International Congress of Acarology* (Akedemiai Kiado, Budapest, 1964), pp. 92-100.

Fincham, J. R. S. 'Heterozygous Advantage as a Likely General Basis for Enzyme Polymorphisms', Heredity, vol. 28 (1972), pp. 387-91.

Fisher, R. A. 'The Correlation between Relatives on the Supposition of Mendelian Inheritance', Trans. Roy. Soc. Edinburgh, vol. 52 (1918), pp. 399-433.

—, *The Genetical Theory of Natural Selection* (Oxford University Press, 1930).

Fisher, W. K. 'Asexual Reproduction in the Starfish, *Sclerasterias*', Biol. Bull. Mar. Lab. Woods Hole, vol. 48 (1925), pp. 171-5.

Flanders, S. E. 'The Role of Mating in the Reproduction of Parasitic Hymenoptera', J. Econ. Entomol., vol. 36 (1943), pp. 802-3.

—, 'Control of Sex and Sex-limited Polymorphism in the Hymenoptera', Quart. Rev. Biol., vol. 21 (1946), pp. 135-43.

—, 'Aphelinid Biologies with Implications for Taxonomy', Ann. Entomol. Soc. Amer., vol. 46 (1953), pp. 86-94.

Fogel, S. and Roth, R. 'Mutations Affecting Meiotic Gene Conversion in Yeast', Mol. Gen. Genetics, vol. 130 (1974), pp. 189-201.

Fogwill, M. 'Differences in Crossing-over and Chromosome Size in the Sex Cells of *Lilium* and *Fritillaria*', Chromosoma, vol. 9 (1958), pp. 493-504.

Foster, S. W. and Jones, P. R. *The Life-history and Habits of the Pear Thrips in California*, US Dept Agric. Bull., 173 (1915). Not seen; cited by A. Vandel (1931).

Fox, H. 'The Urinogenital System of Reptiles', in C. Gans and T. S. Parsons (eds.), *Biology of the Reptilia*, vol. 6 (Academic, London and New York, 1977).

Francis, L. 'Intraspecific Aggression and Its Effect on the Distribution of *Anthopleura elegantissima* and Some Related Sea Anemones', Biol. Bull., vol. 144 (1973), pp. 73-92.

—, 'Social Organization within Clones of the Sea Anemone *Anthopleura elegantissima*', Biol. Bull., vol. 150 (1976), pp. 361-76.

—, 'Contrast Between Solitary and Clonal Lifestyles in the Sea Anemone *Anthopleura elegantissima*', Amer. Zool., vol. 19 (1979), pp. 669-81.

Freeman, G. 'Studies on Regeneration in the Creeping Ctenophore, *Vallicula multiformis*', J. Morphol., vol. 123 (1967), pp. 71-84.

Freidenfelt, T. 'Zur Biologie von *Daphnia longiremis* G. O. Sars und *Daphnia cristata* G. O. Sars', Int. Rev. Ges. Hydrobiol. Hydrogr., vol. 6 (1913), pp. 229-42.

Fretter, V. 'The Structure and Life History of Some Minute Prosobranchs of

Rock Pools: *Skeneopsis planorbis* (Fabricius), *Omalogyra atomus* (Philippi), *Rissoella diaphana* (Alder) and *Rissoella opalina* (Jeffreys)', J. Mar. Biol. Ass. UK, vol. 27 (1948), pp. 597-632.

Fretter, V. and Graham, A. 'Reproduction', in K. M. Wilbur and C. M. Yonge (eds.), *Physiology of Mollusca*, vol. 1 (Academic, New York, 1964), pp. 127-64.

Freudenthal, H. '*Symbiodinium* gen. nov. and *Symbiodinium microadriaticum* sp. nov., a Zooxanthella: Taxonomy, Life Cycle and Morphology', J. Protozool., vol. 9 (1962), pp. 45-52.

Frick, K. E. 'Parthenogenetic Reproduction in *Phytomyza plantaginis* R.D., the Second Reported Case in the Family Agromyzidae', Science, vol. 114 (1951), p. 576.

Fritsch, F. E. *The Structure and Reproduction of the Algae*, vols. I and II (Cambridge University Press, 1935, 1945).

Fritts, T. H. 'The Systematics of the Parthenogenetic Lizards of the *Cnemidophorus cozumela* Complex', Copeia (1969), pp. 519-35.

Furtos, N. C. 'The Ostracoda of Ohio', Ohio Biol. Surv. Bull., vol. 29 (1933), pp. 413-524.

—, 'Freshwater Ostracoda from Massachusetts', J. Wash. Acad. Sci., vol. 25 (1935), pp. 530-44.

—, 'Freshwater Ostracoda from Florida and North Carolina', Amer. Midl. Natur., vol. 17 (1936), pp. 491-522.

Gabritschevsky, E. 'Sénescence Embryonnaire, Rejeunissement et Déterminisme des Formes Larvaires de *Miastor metraloas* (Cecidomyidae, Diptera). Étude Expérimentale', Bull. Biol. France Belgique, vol. 62 (1928), pp. 478-524.

—, 'Der Umkehrbare Entwicklungscyclus bei *Miastor metraloas*', Arch. Entwicklungsmech. Organ., vol. 121 (1930), pp. 450-65.

Gadgil, M. 'Evolution of Social Behaviour through Interpopulation Selection', Proc. Nat. Acad. Sci. US, vol. 72 (1975), pp. 1199-1201.

Gagné, W. C. and Howarth, F. G. 'The Cavernicolous Fauna of Hawaiian Larva Tubes. 6. Mesoveliidae or Water Treaders (Heteroptera)', Pacific Insects, vol. 16 (1975), pp. 399-413.

Galbraeth, G. J. 'The Evolution of Monozygotic Polyembryony in *Dasypus*', In G. G. Montgomery (ed.), *Evolution and Ecology of the Xenarthra* (Smithsonian Institute, 1980).

Gale, M. D. and Rees, H. 'Genes Controlling Chiasma Frequency in *Hordeum*', Heredity, vol. 25 (1970), pp. 393-410.

Ganning, B. 'On the Ecology of *Heterocypris salinus, H. incongruens* and *Cypridopsis aculeata* (Crustacea: Ostracoda) from Baltic Brackish-water Rock Pools', Mar. Biol., vol. 8 (1971), pp. 271-9.

Garber, E. D. 'The Genus *Collinsia*. I. Chromosome Number and Chiasma Frequency of Species of Two Sections', Bot. Gaz., vol. 118 (1956), pp. 71-2.

Gates, G. E. 'On a Taxonomic Puzzle and the Classification of the Earthworms', Bull. Mus. Comp. Zool. Harvard, vol. 121 (1959), pp. 229-61.

—, 'On Oligochaeta Gonads', Megadrilogica, vol. 1, no. 9 (1974), pp. 1-4.

Gavrilov, K. 'Contributions à l'Étude de l'Autofécondation chez les Oligochètes', Acta Zool., Stockholm, vol. 16 (1935), pp. 21-64.

Geddes, P. E. and Thomson, J. A. *The Evolution of Sex* (Walter Scott, London, n.d., ?1901).

von Gelei, J. 'Existiert eine Selbst-befruchtung bei den Planairien?', Biol. Zentralblatt, vol. 44 (1924), pp. 295-9.

Gerritson, J. 'Sex and Parthenogenesis in Sparse Populations', Amer. Natur., vol. 115 (1980), pp. 718-42.

Gersch, M. 'Der Entwicklungszyklus der Dicyemiden', Z. Wiss. Zool., vol. 151 (1938), pp. 515-605.

Gershenson, S. 'A New Sex-ratio Abnormality in *Drosophila obscura*', Genetics, vol. 13 (1928), pp. 488-507.

Ghirardelli, E. 'Some Aspects of the Biology of Chaetognaths', Adv. Mar. Biol., vol. 6 (1968), pp. 271-375.

Ghiselin, M. T. *The Economy of Nature and the Evolution of Sex* (University of California Press, Berkeley, Ca., 1974a).

—, 'A Radical Solution to the Species Problem', Syst. Zool., vol. 23 (1974b), pp. 536-44.

Ghose, K. C. 'Observations on the Mating and Oviposition of Two Land Pulmonates, *Achatina fulica* Bowdich and *Macrochlamys indica* Goodwin-Austin', J. Bombay Nat. Hist. Soc., vol. 56 (1959), pp. 183-7.

Gibson, P. H. and Clark, R. B. 'Reproduction of *Dodecaceria caulleryi* (Polychaeta: Cirratulidae)', J. Mar. Biol. Ass. UK, vol. 56 (1976), pp. 649-74.

Gibson, R. *Nemerteans* (Hutchinson, London, 1972).

Giese, A. C. and Pearse, J. S. *Reproduction of Marine Invertebrates*, vols 1-5 (Academic Press, New York, 1974-79).

Gilbert, J. J. 'Mictic Female Production in the Rotifer *Brachionus calyciflorus*', J. Expt. Zool., vol. 153 (1963), pp. 113-23.

—, 'Control of Sexuality in the Rotifer *Asplanchna brightwelli* by Dietary Lipids of Plant Origin'. Proc. Nat. Acad. Sci. US, vol. 57 (1967), pp. 1218-25.

—, 'Dietary Control of Sexuality in the Rotifer *Asplanchna brightwelli* Gosse', Physiol. Zool., vol. 41 (1968), pp. 14-43.

—, 'Dormancy in Rotifers', Trans. Amer. Microscop. Soc., vol. 93 (1974), pp. 490-513.

—, 'Field Experiments on Gemmulation in the Freshwater Sponge *Spongilla lacustris*', Trans. Amer. Microscop. Soc., vol. 94 (1975), pp. 347-56.

—, 'Mictic-female Production in Monogonont Rotifers', Arch. Hydrobiol. Beih. Ergebn. Limnol., vol. 8 (1977), pp. 142-55.

Gilbert, J. J. and Thompson, G. A. 'Alpha Tocopherol Control of Sexuality and Polymorphism in the Rotifer *Asplanchna*', Science, vol. 159 (1968), pp. 734-8.

Gilchrist, F. G. 'Budding and Locomotion in the Scyphistomas of *Aurelia*', Biol. Bull. Mar. Lab., Woods Hole, vol. 72 (1937), pp. 99-124.

Gilchrist, J. 'Reproduction by Transverse Fission in *Phoronopsis*', Quart. J. Microscop. Sci., vol. 63 (1919), pp. 493-507.

——, 'A Form of Dimorphism and Asexual Reproduction in *Ptychodera capensis*', J. Linn. Soc. London, vol. 35 (1923), pp. 393-8.

Gilpin, M. E. *Group Selection in Predator-Prey Communities* (Princeton University Press, Princeton, NJ, 1975).

Ginsburger-Vogel, T. 'Détermination Génétique du Sexe, Monogénie et Intersexualité chez *Orchestia gammarella* Pallas (Crustacés Amphipodes Talitridae). I. Phénomènes de Monogénie dans la Population de Penzé', Arch. Zool. Exp. Gen., vol. 114 (1973a), pp. 397-438.

——, 'Déterminisme Génétique du Sexe, Monogénie et Intersexualité chez *Orchestia gammarella* Pallas (Crustacés Amphipodes Talitridae). II. Étude des Rélations entre la Monogénie et l'Intersexualité; Influence de la Température', Arch. Zool. Exp. Gen., vol. 115 (1973b), pp. 96-128.

Glesener, R. R. 'Recombination in a Simulated Predator-Prey Interaction', Amer. Zool., vol. 19 (1979), pp. 763-71.

Glesener, R. R. and Tilman, D. 'Sexuality and the Components of Environmental Uncertainty: Clues from Geographic Parthenogenesis in Terrestrial Animals', Amer. Natur., vol. 112 (1978), pp. 659-73.

Godward, M. B. E. *The Chromosomes of the Algae* (Edward Arnold, London, 1966).

Goetghebuer, M. 'Un Cas de Parthénogénèse Observe chez un Diptère Tentipèdide (*Coryneura celeripes* Winnertz)', Bull. Classe Sci. Acad. Roy. Belgique, vol. 40 (1913), pp. 231-3.

Goetsch, W. 'Versuche über Selbsbefruchtung bei Planarien', Biol. Zentralblatt, vol. 44 (1925), pp. 667-71.

——, 'Die Gestechtsverhältnisse der Süsswasserhydroiden und Ihre Experiment. Beeinflussung', Wilhelm Roux' Arch. Entw.-Mech., vol. 111 (1927), pp. 173-249.

Goldschmidt, E. 'Polyploidy and Parthenogenesis in the Genus *Saga*', Nature, vol. 158 (1946), p. 587.

——, 'Fluctuation in Chromosome Number in *Artemia salina*', J. Morph., vol. 91 (1952), pp. 111-13.

Goldschmidt, R. 'On a Case of Facultative Parthenogenesis in the Gypsy Moth *Lymantria dispar* L. with a Discussion of the Relation of Parthenogenesis to Sex', Biol. Bull., vol. 32 (1917), pp. 35-43.

Gontcharoff, M. 'Sur la Reproduction Sexuée chez *Lineus sanguineus* (*Lineus ruber* B)', CR Acad. Sci. Paris, vol. 230 (1950), pp. 233-4.

——, 'Biologie de la Régénération et de la Reproduction chez Quelques Lineidae de France', Ann. Sci. Natur. (Zool.), vol. 13 (1951), pp. 149-235.

Gooch, J. L. and Schopf, T. J. M. 'Population Genetics of Marine Species of the Phylum Ectoprocta', Biol. Bull, vol. 138 (1970), pp. 138-56.

——, 'Genetic Variability in the Deep Sea: Relation to Environmental Variability', Evolution, vol. 26 (1973), pp. 545-52.

Goodrich, E. S. 'Notes on *Protodrilus*', Quart. J. Microscop. Sci., vol. 74 (1931), pp. 303-19.

Goss, R. J. 'Ovarian Development and Oogenesis in the Booklouse, *Liposcelis divergens* Badonnel (Psocoptera: Peripsocidae)', Ann. Entomol. Soc. Amer., vol. 47 (1954), pp. 190-207.

Goto, H. E. 'Facultative Parthenogenesis in Collembola', Nature, vol. 188 (1960), pp. 958-9.

Gould-Somero, M. 'Echiura', in A. C. Giese and J. S. Pearse (eds.), *Reproduction of Marine Invertebrates*, vol. 3 (Academic, New York, 1975), pp. 227-311.

Gowen, J. W. 'A Biometrical Study of Crossing-over', Genetics, vol. 4 (1919), pp. 205-50.

Grant, V. 'The Regulation of Recombination in Plants', Cold Spring Harbour Symp. Quant. Biol., vol. 23 (1958), pp. 377-63.

Grassé, P.-P. 'Note sur la Biologie d'un Collembole: *Hypogastrura armata* (Nicolet)', Ann. Soc. Entomol. France, vol. 91 (1922), pp. 190-2.

——, 'La Parthénogénèse', in *Faune de France*, vol. 8 fasc. (V.-A. Masson, Paris, 1966), pp. 291-408.

Green, C. D. 'The Life History and Fecundity of *Folsomia candida* (Willem) var. *distincta* (Bagnall) (Collembola: Isotomidae)', Proc. Roy. Entomol. Soc. London A, vol. 39 (1964), pp. 125-8.

Green, J. 'Studies on a Population of *Daphnia magna*', J. Anim. Ecol., vol. 24 (1955), pp. 84-97.

Grell, K. G. 'Sexual Reproduction in Protozoa', Res. Protozool., vol. 2 (1967), pp. 147-213.

——, *Protozoology* (Springer, New York, 1973).

Greve, W. 'Cultivation Experiments on North Sea Ctenophores', Helgoländer Wiss. Meeresunters., vol. 20 (1970), pp. 304-17.

Griffiths, J. T. and Tauber, O. E. 'Fecundity, Longevity, and Parthenogenesis of the American Roach, *Periplaneta americana* L', Physiol. Zool., vol. 15 (1942), pp. 196-209.

Gross, J. 'Uber die Gonadenbildung bei Süsswasserpolypen Auslösenden Bedingungen', Naturwissenschaften, vol. 26 (1925), pp. 580-1.

Grosvenor, G. H. and Smith, G. 'The Life Cycle of *Moina rectiostris*', Quart. J. Microscop. Sci., vol. 58 (1913), pp. 511-22.

Guenther, C. *Darwinism and the Problems of Life. A Study of Familiar Animal Life*, translated by J. McCabe (A. Owen, London, 1906).

Gueutal, J. 'D'un Cas de Parthénogénèse chez un Opillion *Phalangium opilis*', CR Soc. Biol. (1943), p. 137.

Gunther, K. and Herter, K. 'Dermaptera (Ohrwürmer)', in *Handbuch der Zoologie*, vol. 4 (1974), pp. 1-158.

Gupta, M. L. 'A Preliminary Account of the Meiotic Mechanism in Nineteen Species of the Indian Mantids', Res. Bull. Punjab Univ., vol. 17 (1966), pp. 421-2.

Gustaffson, A. 'The Origin and Properties of the European Blackberry Flora', Hereditas, vol. 28 (1942), pp. 249-77.

——, 'Apomixis in the Higher Plants. I. The Mechanism of Apomixis', Lunds Univ. Arsskr. NF Avd. 2, vol. 42 (1946), pp. 1-66.

Hadfield, M. G. 'Hemichordata', in A. C. Giese and J. S. Pearse (eds.), *Reproduction of Marine Invertebrates*, vol. 2 (Academic, New York, 1975), pp. 185-240.

Halbach, V. and Halbach-Keup, G. 'Einfluss von Aussenfaktoren auf den Fortpflanzungsmodus Heterogonor Rotatorien', Oecologia, vol. 9 (1972), pp. 203-14.

Haldane, J. B. S. *The Causes of Evolution* (Harper and Row, New York, 1932).

Hale, W. G. 'Experimental Studies on the Taxonomic Status of Some Members of the *Onychiurus armatus* Species Group', Rev. Ecol. Biol. Sol., vol. 1 (1964), pp. 501-10.

Halkka, O. 'Recombination in Six Homopterous Families', Evolution, vol. 18 (1964), pp. 81-8.

Hall, D. J. 'An Experimental Approach to the Dynamics of a Natural Population of *Daphnia galatea mendotae*', Ecology, vol. 45 (1962), pp. 94-112.

Hall, W. P. 'Three Probable Cases of Parthenogenesis in Lizards (Agamidae, Chameleontidae, Gekkonidae)', Experientia, vol. 26 (1970), pp. 1271-3.

Hamilton, A. G. 'Thelytokous Parthenogenesis for Four Generations in the Desert Locust (*Schistocerca gregaria* Forsk) (Acrididae)', Nature, vol. 172 (1953), pp. 1153-4.

—, 'Parthenogenesis in the Desert Locust (*Schistocerca gregaria* Forsk.) and Its Possible Effects on the Maintenance of the Species', Proc. Roy. Entomol. Soc. London, vol. 30 (1955), pp. 103-14.

Hamilton, W. D. 'Extraordinary Sex Ratios', Science, vol. 156 (1967), pp. 477-88.

—, 'Gamblers since Life Began: Barnacles, Aphids, Elms', Quart. Rev. Biol., vol. 50 (1975), pp. 175-80.

Hamilton, W. D., Henderson, P. A. and Moran, N. A. 'Fluctuation of Environment and Coevolved Antagonist Polymorphism as Factors in the Maintenance of Sex', in R. D. Alexander and D. W. Tinkle (eds.), *Natural Selection and Social Behaviour* (1981, in press).

Hand, C. 'The Sea Anemones of Central California. III. The Acontiarian Anemones', Wasman J. Biol., vol. 13 (1955), pp. 189-251.

Hansen, E., Yarwood, E. A. and Buecher, E. J. 'Temperature Effects on Sex Differentiation in *Aphelenchus avenae*', J. Nematol., vol. 3 (1971), p. 311 (abstr).

Hanson, E. D. 'Asexual Reproduction in Acoelous Turbellaria', Yale J. Biol. Med., vol. 33 (1960), pp. 107-11.

Harding, J. P. 'The First Known Example of a Terrestrial Ostracod, *Mesocypris terrestris*, sp. nov.', Ann. Natal. Mus., vol. 12 (1953), pp. 359-65.

Harmer, S. F. 'Origin of the Embryos of Cyclostomatous Polyzoa', Proc. Cambridge Phil. Soc., vol. 7 (1890).

—, 'On the Occurrence of Embryonic Fission in Cyclostomatous Polyzoa', Quart. J. Microscop. Sci., vol. 34 (1893), pp. 199-241.

—, 'On *Phoronis ovalis*', Quart. J. Microscop. Sci., vol. 62 (1917), pp. 115-48.

Harrington, R. W. 'Oviparous Hermaphroditic Fish with Internal Self-fertilization', Science, vol. 134 (1961), pp. 1749-50.

——, 'How Ecological and Genetic Factors Interact to Determine When Self-fertilizing Hermaphrodites of *Rivulus marmoratus* Change into Functional Secondary Males, with a Reappraisal of the Modes of Intersexuality among Fishes', Copeia (1971), pp. 389-432.

Harrington, R. W. and Kallman, K. D. 'The Homozygosity of Clones of the Self-fertilizing Hermaphroditic Fish *Rivulus marmoratus* Poey (Cyprinodontidae, Atheriniformes)', Amer. Natur., vol. 102 (1968), pp. 337-43.

Harris, H. 'Enzyme Polymorphisms in Man', Proc. Roy. Soc. London B, vol. 164 (1966), pp. 298-310.

Harrison, F. W. 'Sponges (Porifera: Spongillidae)', in C. W. Hart and S. L. H. Fuller (eds.), *Pollution Ecology of Freshwater Invertebrates* (Academic, New York, 1974).

Harrison, F. W. and Crowden, R. R. *Aspects of Sponge Biology* (Academic, New York, 1976).

Hart, D. G. and Hart, C. W. 'The Ostracod Family Entocytheridae', Monogr. Acad. Nat. Sci., vol. 18 (1974), pp. 1-239.

Hartl, D. L. and Brown, S. W. 'The Origin of Male Haploid Genetic Systems and Their Expected Sex Ratio', Theoret. Pop. Biol., vol. 1 (1970), pp. 165-90.

Hartman, W. D. 'Natural History of the Marine Sponges of Southern New England', Bull. Peabody Mus. Nat. Hist., vol. 12 (1958), pp. 1-155.

Hartnoll, R. G. 'Reproductive Strategy in Two British Species of *Alcyonium*', in B. F. Keegan, P. O. Céidigh and P. J. S. Boaden (eds.), *Biology of Benthic Organisms* (Pergamon, New York, 1977).

Haskell, G. 'Polyploidy, Ecology and the British Flora', J. Ecol., vol. 40 (1952), pp. 265-82.

Haskins, C. P. and Enzmann, E. V. 'On the Occurrence of Impaternate Females in the Formicidae', J. New York Entomol. Soc., vol. 53 (1945), pp. 263-77.

Hauschteck, E. 'Uber die Zytologie der Parthenogenese und der Geschlechts-bestimmung bei der Gallmücke *Oligarces paradoxus* Mein.', Experientia, vol. 15 (1959), pp. 260-5.

——, 'Die Zytologie der Paedogenese und der Geschlechtsbestimmung einer Heterogonen Gallmücke', Chromosoma, vol. 13 (1962), pp. 163-82.

Haven, N. D. 'Temporal Patterns of Sexual and Asexual Reproduction in the Colonial Ascidian *Metandrocarpa taylori* Huntsman', Biol. Bull., vol. 140 (1971), pp. 400-15.

Haydak, M. H. 'Influence of the Protein Level of the Diet on the Longevity of Cockroaches', Ann. Ent. Soc. Amer., vol. 46 (1953), pp. 547-60.

Hayman, D. L. and Parsons, P. A. 'The Effect of Temperature, Age and an Inversion on Recombination Values and Interference in the X-chromosome of *Drosophila melanogaster*', Genetica, vol. 31 (1961), pp. 1-15.

Hebert, P. D. N. 'Ecological Differences between Genotypes in a Natural Population of *Daphnia magna*', Heredity, vol. 33 (1974a), pp. 327-37.

——, 'Enzyme Variability in Natural Populations of *Daphnia magna*. II. Genotypic Frequencies in Permanent Populations', Genetics, vol. 77 (1974b), pp. 323-34.

—, 'Enzyme Variability in Natural Populations of *Daphnia magna*. III. Genotypic Frequencies in Intermittent Populations', Genetics, vol. 77 (1974c), pp. 335-41.

—, 'Enzyme Variability in Natural Populations of *Daphnia magna*. I. Population Structure in East Anglia', Evolution, vol. 28 (1975), pp. 546-56.

—, 'Enzyme Variability in Natural Populations of *Daphnia magna*. IV. Ecological Differentiation and Frequency Changes of Genotypes at Audley End', Heredity, vol. 36 (1976), pp. 331-41.

—, 'A Revision of the Taxonomy of the Genus *Daphnia* (Crustacea, Daphnidae) in Southeastern Australia', Aust. J. Zool., vol. 25 (1977), pp. 371-98.

—, 'The Population Biology of *Daphnia* (Crustacea, Daphnidae)', Biol. Rev., vol. 53 (1978), pp. 387-426.

Hebert, P. D. N. and Ward, R. D. 'Inheritance during Parthenogenesis in *Daphnia magna*', Genetics, vol. 71 (1972), pp. 639-42.

Hedrick, P. W., Ginevan, M. E. and Ewing, E. P. 'Genetic Polymorphism in Heterogeneous Environments', Ann. Rev. Ecol. Syst., vol. 7 (1976), pp. 1-32.

Hegner, R. W. 'The History of the Germ Cells in the Paedogenetic Larvae of *Miastor*', Science, vol. 36 (1912).

—, 'Studies on Germ Cells. Parts I and II', J. Morph., vol. 25 (1914), pp. 375-509.

Heilborn, O. 'Chromosome Studies in Cyperaceae' Hereditas, vol. 11 (1928), pp. 182-92.

—, 'Aneuploidy and Polyploidy in *Carex*', Svensk. Bot. Tidsskr., vol. 26 (1932), pp. 137-46.

Heinemann, R. L. and Hughes, R. D. 'The Cytological Basis for Reproductive Variability in the Anoetidae (Sarcoptiformes: Acari)', Chromosoma, vol. 28 (1969), pp. 346-56.

Helle, W. and Overmeer, W. P. J. 'Variability in Tetranychid Mites', Ann. Rev. Entomol., vol. 18 (1973), pp. 97-120.

Henderson, S. A. 'Chiasma Distribution at Diplotene in a Locust', Heredity, vol. 18 (1963), pp. 173-90.

Henderson, S. A. and John, B. 'Asynapsis and Polyploidy in *Schistocerca paranensis*', Chromosoma, vol. 13 (1962), pp. 111-47.

Henley, C. 'Platyhelminthes (Turbellaria)', in A. C. Giese and J. S. Pearse (eds.), *Reproduction of Marine Invertebrates*, vol. 1 (Academic, New York, 1974).

Henry, J.-P. 'Remarques sur l'Aselle Psammique *Proasellus walteri* (Chappuis 1948) (Crustacea, Isopoda, Asellota)', Int. J. Spéléol., vol. 8 (1976), pp. 75-80.

Henslee, E. D. 'Sexual Isolation in a Parthenogenetic Strain of *Drosophila mercatorum*', Amer. Natur., vol. 100 (1966), pp. 191-7.

Herfs, A. 'Studien an dem Steinnussborkenkafer, *Coccotrypes tanganus* Eggers', Hofchen-Briefe (1950), pp. 1-57.

Hering, M. 'Nachgewiesene Parthenogenetische Fortpflanzung bei einer Blattminierenden Acalyptraten Muscide (Dipt.)', Zool. Anz., vol. 68 (1926), pp. 283-7.

Herlant-Meewis, H. 'La Gemmulation chez *Suberites domuncula* (Olivi)', Nardo.

Arch. Anat. Microscop., vol. 37 (1948), pp. 289-322.

—, 'La Réproduction Asexuée chez les Annelides', Année Biol., vol. 34 (1958), pp. 133-66.

Heron, R. J. 'Studies on the Starvation of Last-instar Larvae of the Larch Sawfly, *Pristiophora erichsonii* (Htg) (Hymenoptera: Tenthredinidae)', Can. Entomol., vol. 87 (1955), pp. 417-27.

Hérouard, E. 'Relations entre la Depression et la Formation de Pseudoplanula Tentaculaire chez le Scyphistome', CR Acad. Sci. Paris, vol. 156 (1913), pp. 1043-5.

Hertwig, P. 'Abweichende Form der Parthenogenese bei einer Mutation von *Rhabditis pellio*', Arch. Mikroskop. Anat., vol. 94 (1922), pp. 303-37.

Heuch, I. 'Maintenance of Butterfly Populations with All-female Broods under Recurrent Extinction and Recolonization', J. Theoret. Biol., vol. 75 (1978), pp. 115-22.

Hewitt, G. M. 'Population Cytology of British Grasshoppers. I. Chiasma Variation in *Chorthippus bruneus*, *Chorthippus parallelus* and *Omocestus viridulus*', Chromosoma, vol. 15 (1964), pp. 212-30.

——, 'Population Cytology of British Grasshoppers. II. Annual Variation in Chiasma Frequency', Chromosoma, vol. 16 (1965), pp. 679-90.

—, 'A New Hypothesis for the Origin of the Parthenogenetic Grasshopper *Morabo virgo*', Heredity, vol. 34 (1975), pp. 117-36.

Hewitt, G. M. and John, B. 'The B-chromosome System of *Myrmeleotettix maculatus* (Thunb.). III. The Statistics', Chromosoma, vol. 21 (1967), pp. 140-62.

—, 'Parallel Polymorphism for Supernumerary Segments in *Chorthippus parallelus* (Zetterstedt). I. British Populations', Chromosoma, vol. 25 (1968), pp. 319-42.

Hewitt, G. M. and Ruscoe, C. 'Changes in Microclimate Correlated with a Cline for B-chromosomes in a Grasshopper *Myrmeleotettix maculatus* (Thunb.) (Orthoptera: Acrididae)', J. Anim. Ecol., vol. 40 (1971), pp. 735-65.

Hidu, H. 'Gregarious Settling in the American Oyster *Crassostrea virginica* Gmelin', Chesapeake Sci., vol. 10 (1969), pp. 35-92.

Higgins, R. P. 'Kinorhyncha', in A. C. Giese and J. S. Pearse (eds.), *Reproduction in Marine Invertebrates*, vol. 1 (Academic, New York, 1974), pp. 507-18.

Hill, R. 'Parthenogenesis bei *Nagara modesta* Dollf. (Isopoda)', Chromosoma, vol. 3 (1948), pp. 232-56.

Hill, W. G. and Robertson, A. 'The Effect of Linkage on Limits to Artificial Selection', Genet. Res., Cambridge, vol. 8 (1966), pp. 269-94.

Hille Ris Lambers, D. 'Aphididae', in *Zoology of Iceland*, vol. III, part 52a (Munksgraad, Copenhagen and Reykjavik, 1955).

—, 'Additions to the Aphid Fauna of Greenland', Medd. Groenland, vol. 159 (1960), pp. 1-18.

—, 'Polymorphism in Aphididae', Ann. Rev. Entomol., vol. 11 (1966), pp. 47-78.

Hindle, E. 'Sex Inheritance in *Pediculus humanus* var *corporis*', J. Genet., vol. 8 (1919), pp. 267-77.

Hindle, E. and Pontecorvo, G. 'Mitotic Divisions Following Meiosis in *Pediculus humanus* Males' Nature, vol. 149 (1942), p. 668.

Hinds, W. E. 'Contribution to a Monograph of the Insects of the Order Thysanoptera Inhabiting North America', Proc. US Nat. Mus., vol. 26 (1903), pp. 79-242.

Hirschmann, H. 'Reproduction of Nematodes' in J. N. Sasser and W. R. Jenkins (eds.), *Nematology* (University of North Carolina Press, Chapel Hill, North Carolina, 1960).

Hoff, C. C. 'The Ostracods of Illinois', Illinois Biol. Monogr., vol. 19 (1942), pp. 1-196.

Hoffmann, R. J. 'Genetics and Asexual Reproduction of the Sea Anemone *Metridium senile*', Biol. Bull., vol. 151 (1976), pp. 478-88.

Hofmeister, W. *Vergleichende Untersuchungen der Keimung, Entfaltung und Fruchtbildung Höherer Kryptogamen und der Samenbildung der Coniferen* (Leipzig, 1851).

Hogben, L. T. 'Studies on Synapsis. Parts I and II', Proc. Roy. Soc. London, Ser B, vol. 91 (1920), pp. 268-93, 305-29.

Honda, H. 'Experimental and Cytological Studies on Bisexual and Hermaphrodite Free-living Nematodes, with Special Reference to Problems of Sex', J. Morphol., vol. 40 (1925), pp. 191-233.

Hoogstral, H., Roberts, F. S. H., Kohls, G. M. and Tipton, V. J. 'Review of *Haemaphysalis* (*Kaiseriana*) *longicornis* Neumann (Resurrected) of Australia, New Zealand, New Caledonia, Fiji, Japan, Korea and Northeastern China and USSR, and Its Parthenogenetic and Bisexual Populations', J. Parasitol., vol. 54 (1968), pp. 1197-213.

Hope, W. D. 'Nematoda', in A. C. Giese and J. S. Pearse (eds.), *Reproduction in Marine Invertebrates*, vol. 1 (Academic, New York, 1974), pp. 391-469.

Hopping, G. R. 'The Sex Ratio in *Ips tridens* (Mannerheim) (Coleoptera: Scolytidae)', Can. Entomol., vol. 94 (1962), p. 506.

Horstmann, H.-J. 'Untersuchungen zur Physiologie der Begattung und Befruchtung der Schlammschnecke *Lymnaea stagnalis* L', Z. Morph. Ökol. Tiere, vol. 44 (1955), pp. 222-68.

Hovasse, R. 'Quelques Données Nouvelles sur la Cochenille *Marchalina hellenica* (Genn.)', CR Acad. Sci. Paris, vol. 190 (1930), pp. 1025-6.

Howard, H. W. 'The Genetics of *Armadillidium vulgare* Latr. I. A General Survey of the Problems', J. Genet., vol. 40 (1940), pp. 83-108.

—, 'The Genetics of *Armadillidium vulgare* Latr. II. Studies on the Inheritance of Monogeny and Amphogeny', J. Genet., vol. 44 (1942), pp. 143-59.

Howell, A. B. 'The Involved Genetics of Fish', Science, vol. 77 (1933), pp. 389-90.

Hsu, W. S. 'Oogenesis in the Bdelloid Rotifer *Philodina roseola* Ehrenberg', Cell. Rec. Cytol. Histol., vol. 57 (1956a), pp. 283-96.

—, 'Oogenesis in *Habrotrocha tridens* (Milne)', Biol. Bull., vol. 111 (1956b), pp. 364-74.

Hubbell, T. H. and Norton, R. M. 'The Systematics and Biology of the Cave Crickets of the North American Tribe Hadenoecini (Orthoptera: Saltatoria: Ensifera: Rhaphidophoridae: Dolichopodinae)', Misc. Publ. Mus. Zool. Univ. Michigan, vol. 155 (1978), pp. 1-124.

Hubbs, C. L. 'Hybridization Between Fish Species in Nature', Syst. Zool., vol. 4 (1955), pp. 1-20.

—, 'Interactions between a Bisexual Fish Species and Its Gynogenetic Sexual Parasite', Bull. Texas Mem. Mus., vol. 8 (1964), pp. 1-72.

Hubbs, C. L. and Hubbs, L. C. 'Apparent Parthenogenesis in Nature in a Form of Fish of Hybrid Origin', Science, vol. 76 (1932), pp. 628-30.

—, 'Breeding Experiments with the Invariable Female, Strictly Matroclinous Fish, *Mollienisia formosa*', Rec. Genetics Soc. Amer., vol. 14 (1946), p. 48.

Hubby, J. L. and Lewontin, R. C. 'A Molecular Approach to the Study of Genic Heterozygosity in Natural Populations. I. The Number of Alleles at Different Loci in *Drosophila pseudoobscura*', Genetics, vol. 54 (1966), pp. 577-94.

Hubendick, B. 'Recent Lymnaeidae, Their Variation, Morphology, Taxonomy, Nomenclature and Distribution', Kgl. Svenska Vetenkapsatrad. Handl., vol. 3 (1951), pp. 1-223.

Hughes-Schrader, S. 'Cytology of Hermaphroditism in *Icerya purchasi* (Coccidae)' Z. Zellforsch., vol. 2 (1925), pp. 264-92.

—, 'Meiosis Without Chiasmata in Diploid and Tetraploid Spermatocytes of the Mantid *Callimantis antillarum* Saussure', J. Morphol., vol. 73 (1943), pp. 111-41.

—, 'Cytology of Coccids (Coccoidea-Homoptera)', Advances Genet., vol. 2 (1948), pp. 127-203.

—, 'The Chromosomes of Mantids (Orthoptera: Manteidae) in Relation to Taxonomy', Chromosoma, vol. 4 (1950), pp. 1-55.

—, 'Supplementary Notes on the Cytotaxonomy of Mantids (Orthopteroidea: Mantoidea)', Chromosoma, vol. 6 (1953), pp. 79-90.

Hughes-Schrader, S. and Monahan, D. F. 'Hermaphroditism in *Icerya zeteki* Cockerell, and the Mechanism of Gonial Reduction in Iceryine Coccids (Coccoidea: Margarodidae)', Chromosoma, vol. 20 (1966), pp. 15-31.

Hughes-Schrader, S. and Tremblay, E. '*Gueriniella* and the Cytotaxonomy of Iceryine Coccids (Coccoidea: Margarodidae)', Chromosoma, vol. 19 (1966), pp. 1-13.

Hummon, W. D. 'Morphology, Life History, and Significance of the Marine Gastrotrich, *Chaetonotus testiculophorus* n. sp.', Trans. Amer. Microscop. Soc., vol. 85 (1966), pp. 450-7.

—, 'Gastrotricha' in A. C. Giese and J. S. Pearse (eds.) *Reproduction in Marine Invertebrates*, vol. 1 (Academic, New York, 1974), pp. 485-505.

Hunter, R. W. 'Life Cycles of Four Freshwater Snails in Limited Populations in Loch Lomond, with a Discussion of Infraspecific Variation', Proc. Zool. Soc. London, vol. 137 (1961), pp. 135-71.

—, 'Physiological Aspects of Ecology in Nonmarine Molluscs', in K. M. Wilbur and C. M. Yonge (eds.), *Physiology of Mollusca*, vol. 1 (Academic, New York, 1964).

Husain, M. A. and Mathur, C. B. 'Studies on *Schistocerca gregaria* Forsk. XIII. Sexual Life', Indian J. Entomol., vol. 7 (1945), pp. 89–101.

Husson, R. and Palévody, C. 'La Parthénogénèse chez les Collemboles', Ann. Soc. Entomol. France NS, vol. 3 (1967), pp. 631–3.

Hutchinson, G. E. *A Treatise on Limnology*, vol. II, *Introduction to Lake Biology and the Limnoplankton* (Wiley, New York, 1967).

Huxley, T. H. 'Note on the Reproductive Organs of the Cheilostome Polyzoa', Quart. J. Microscop. Sci., vol. 4 (1856), pp. 191–2.

Hyman, L. H. 'Some Cave Planarians of the United States', Trans. Amer. Microscop. Soc., vol. 56 (1937), pp. 457–77.

—, *The Invertebrates: Protozoa through Ctenophora* (McGraw-Hill, New York, 1940).

—, *The Invertebrates: Platyhelminthes and Rhynchocoela, The Acoelomate Bilateria* (McGraw-Hill, New York, 1951a).

—, *The Invertebrates: Acanthocephala, Aschelminthes, and Entoprocta – The Pseudocoelomate Bilateria* (McGraw-Hill, New York, 1951b).

—, *The Invertebrates: Echinodermata, the Coelomate Bilateria* (McGraw-Hill, New York, 1955).

—, *The Invertebrates: Smaller Coelomate Groups – Chaetognatha, Hemichordata, Pogonophora, Phoronida, Ectoprocta, Brachiopoda, Sipunculida – The Coelomate Bilateria* (McGraw-Hill, New York, 1959).

—, *The Invertebrates: Mollusca I* (McGraw-Hill, New York, 1967).

Ikeda, H. and Carson, H. L. 'Selection for Mating Reluctance in Females of a Diploid Parthenogenetic Strain of *Drosophila mercatorum*', Genetics, vol. 75 (1973), pp. 541–55.

Ikeda, K. 'Studies on Hermaphroditism in Pulmonata. II. Cytogenetic Studies on the Self-fertilization of *Philomycus bilineatus*', J. Sci. Hiroshima Univ. B, vol. 5 (1937), pp. 66–123.

Jack, R. W. 'Parthenogenesis amongst the Workers of the Cape Honeybee', Trans. Roy. Entomol. Soc. London, vol. 64 (1916), pp. 396–403.

Jacob, J. 'Cytological Studies of Melaniidae (Mollusca) with Special Reference to Parthenogenesis and Polyploidy. I. Oogenesis of the Parthenogenetic Species of *Melanoides* (Prosobranchia-Gastropoda)', Trans. Roy. Soc. Edinburgh, vol. 63 (1957), pp. 341–56.

—, 'Cytological Studies of Melaniidae (Mollusca) with Special Reference to Parthenogenesis and Polyploidy. II. A Study of Meiosis in the Rare Males of the Polyploid Race of *Melanoides tuberculatus* and *Melanoides lineatus*', Trans. Roy. Soc. Edinburgh, vol. 63 (1958), pp. 433–44.

Jaenike, J. 'An Hypothesis to Account for the Maintenance of Sex within Populations', Evol. Theory, vol. 3 (1978), pp. 191–4.

Jaenike, J., Parker, E. D., and Selander, R. K. 'Clonal Niche Structure in the

Parthenogenetic Earthworm *Octolasion tyrtaeum*', Amer. Natur., vol. 116 (1980), pp. 196-205.

Jägersten, G. 'Für Kenntnis der Physiologie der Zeugung bei *Sagitta*', Zool. Bidr. Uppsala, vol. 18 (1940), pp. 397-413.

—, 'Morphology and Reproduction of Endoproct Larvae', Zool. Bidr. Uppsala, vol. 36 (1964), pp. 295-314.

Jagiello, G., Ducayen, M., Fang, J-S. and Graffeo, J. 'Cytological Observations in Mammalian Oocytes', Chromosomes Today, vol. 5 (1976), pp. 43-64.

Jahn, E. 'Über Parthenogenese bei Forstschädlichen *Otiorrhynchus* Arten in den während der Eiszeit Vergletscherten Gebieten der Ostalpen', Z. Angew. Entomol., vol. 28 (1941), pp. 366-72.

Jankowski, A. W. 'Morphology and Evolution of Ciliophora. IV. Sapropelebionts of the Family Loxocephalidae fam. nova, Their Taxonomy and Evolutionary History', Acta Protozoologica, vol. 2 (1964), pp. 33-58.

Jenkin, P. M. 'Cladocera from the Rift Valley Lakes in Kenya', Ann. Mag. Nat. Hist. Ser. 10, vol. 13 (1934), pp. 137-60, 281-308.

John, B. and Freeman, M. 'The Cytogenetic Systems of Grasshoppers and Locusts. III. The Genus *Tolgadia* (Oxyinae: Acrididae)', Chromosoma, vol. 55 (1976), pp. 105-19.

John, B. and Hewitt, G. M. 'The B-chromosome System of *Myrmeleotettix maculatus* (Thunb.). I. The Mechanics', Chromosoma, vol. 16 (1965a), pp. 548-78.

—, 'The B-chromosome System of *Myrmeleotettix maculatus* (Thunb.). II. The Statics', Chromosoma, vol. 17 (1965b), pp. 121-38.

Johnson, G. 'Contribution à l'Étude de la Différenciation Sexuelle Mâle chez les Oniscoides: Phénomènes d'Hermaphroditisme et de Monogénie', Bull. Biol. France Belgique, vol. 95 (1961), pp. 177-272.

Jolicoeur, H. and Topsent, E. 'Études sur l'Écrivain ou Gribouri (*Adoxus vitis* Kirby)', Mem. Soc. Zool. France, vol. 5 (1892), pp. 723-30.

Jones, A. W. and Mackiewicz, J. S. 'Naturally Occurring Triploidy and Partheno-genesis in *Atractolytocestus huronensis* (Cestoidea: Caryophyllidea) from *Cyprinus carpio* L. in North America', J. Parasitol., vol. 55 (1969), pp. 1105-18.

Jones, P. A. and Gilbert, J. J. 'Male Haploidy in Rotifers: Relative DNA Content of Nuclei from Male and Female *Asplanchna*', J. Exp. Zool., vol. 198 (1976), pp. 281-5.

Jordan, K. 'Anatomie und Biologie der Physapoda', Z. Wiss. Zool., vol. 47 (1888), pp. 544-620.

Juchault, P. and Legrand, J. J. 'Intersexualité et Monogénie chez les Crustaces Isopodes Terrestres: Induction de la Thelygénie chez *Armadillidium vulgare*; Facteurs Contrôlant le Psuedo-hermaphroditisme Masculin Externe chez *Porcellio dilatatus*', Ann. Endocrinol. Paris, vol. 31 (1970), pp. 525-30.

Judge, F. D. 'Polymorphism in a Subterranean Aphid, *Pemphigus bursarius*. I. Factors Affecting the Development of Sexuparae', Ann. Entomol. Soc. America, vol. 61 (1968), pp. 819-27.

Kaestner, A. *Invertebrate Zoology*, vol. 3, translated by H. W. Levi and L. R. Levi (Interscience, New York, 1970).

Kahan, D. 'The Fauna of Hot Springs', Verh. Int. Verein. Theor. Angew. Limnol., vol. 17 (1969), pp. 811-16.

——, '*Cyclidium citrullus*, a Ciliate from the Hot Springs of Tiberias (Israel)', J. Protozool., vol. 19 (1972), pp. 593-7.

Kahle, W. 'Die Pädogenesis der Cecidomyiden', Zoologica, vol. 21 (1908), pp. 1-80.

Kallman, K. D. 'Population Genetics of the Gynogenetic Teleost, *Molliensia formosa* (Girard)', Evolution, vol. 16 (1962), pp. 497-504.

——, 'Homozygosity in a Gynogenetic Fish – *Poecilia formosa*', Genetics, vol. 50 (1964), pp. 260-1.

Kallman, K. D. and Harrington, R. W. 'Evidence for the Existence of Homozygous Clones in the Self-fertilizing Hermaphroditic Teleost *Rivulus marmoratus*', Biol. Bull., vol. 126 (1964), pp. 101-14.

Kalmus, H. 'Über den Erhaltungswert der Phänotypischen (Morphologischen) Anisogamie und die Entstehung der Ersten Geschlechtsunterschiede', Biol. Zentralblatt, vol. 52 (1932), pp. 716-26.

Kalmus, H. and Smith, C. A. B. 'Evolutionary Origin of Sexual Differentiation and the Sex-ratio', Nature, vol. 186 (1960), pp. 1004-6.

Kanaev, I. I. *Hydra: Essays on the Biology of Freshwater Polyps* (Soviet Academy of Sciences, Moscow, 1952) (in Russian); translation from Russian by E. T. Burrows and H. M. Lenhoff, edited by H. M. Lenhoff, published privately.

Karlin, S. 'Sex and Infinity; A Mathematical Analysis of the Advantages and Disadvantages of Recombination', in M. S. Bartlett and R. W. Hiorns (eds.), *The Mathematical Theory of the Dynamics of Natural Populations* (Academic, London, 1973), pp. 155-94.

Karlin, S. and McGregor, J. L. 'Towards a Theory of the Evolution of Modifier Genes', Theoret. Pop. Biol., vol. 5 (1974), pp. 59-103.

Karsten, H. *Parthenogenesis und Generations-Weschel im Thier und Pflanzenreiche* (Berlin, 1888).

Katheriner, L. 'Ueber die Entwicklung von *Gyrodactylus elegans* v.Nrdm', Zool. Jahrb., vol. 70 (1904), pp. 519-50.

Kawakatsu, M. 'On the Ecology and Distribution of Freshwater Planarians in the Japanese Islands, with Special Reference to Their Vertical Distribution', Hydrobiologia, vol. 26 (1965), pp. 349-408.

——, 'Further Studies on the Vertical Distribution of Freshwater Planarians in the Japanese Islands', in N. W. Riser and M. P. Morse (eds.), *Biology of the Turbellaria* (McGraw-Hill, New York, 1974), pp. 291-338.

Kawakatsu, M., Teshirogi, A. W. and Yagihashi, M. 'Report on the Ecological Survey of Freshwater Planarians in the Western Part of Aomori Prefecture (Hirosaki City, the Mt. Iwaki and the Juniko districts), Honshu', Japan. J. Ecol., vol. 17 (1967a), pp. 34-40.

Kawakatsu, M., Yamada, T. and Iwaki, S. 'Environment and Reproduction in Japanese Freshwater Planarians', Japan. J. Ecol., vol. 17 (1967b), pp. 263-6.

Keeler, C. E. 'Thelytoky in *Scleroderma immigrans*', Psyche, vol. 36 (1929a), pp. 41-4.

—, 'Critical Data upon Thelytoky in *Scleroderma immigrans*', Psyche, vol. 36 (1929b), pp. 121-2.

Kenk, R. 'Studies on Virginian Triclads', J. Elisha Mitchell Sci. Soc., vol. 51 (1935), pp. 79-126.

—, 'Sexual and Asexual Reproduction in *Euplanaria tigrini* (Girard)', Biol. Bull., vol. 73 (1937), pp. 280-94.

Kennedy, J. S. and Stroyan, H. L. G. 'Biology of Aphids', Ann. Rev. Entomol., vol. 4 (1959), pp. 139-60.

Kennett, C. E. 'Some Predaceous Mites of the Subfamilies Phytoseiinae and Aceosejinae (Acarina: Phytoseiidae, Aceosejidae) from Central California with Descriptions of New Species', Ann. Entomol. Soc. America, vol. 51 (1958), pp. 471-9.

Kenten, J. 'The Effect of Photoperiod and Temperature on Reproduction in *Acyrthosiphon pisum* (Harris) and on the Forms Produced', Bull. Entomol. Res., vol. 46 (1955), pp. 599-624.

Keyl, H-G. 'Beobachtungen uber die ♂ – Meiose der Muschel *Sphaerium corneum*', Chromosoma, vol. 8 (1956), pp. 12-17.

—, 'Zur Karyologie die Hydrachnellen (Acarina)', Chromosoma, vol. 8 (1957), pp. 719-29.

Kidwell, M. G. 'Genetic Changes of Recombination Value in *Drosophila melanogaster*. I. Artificial Selection for High and Low Recombination and Some Properties of Recombination-modifying Genes', Genetics, vol. 70 (1972a), pp. 419-32.

—, 'Genetic Changes of Recombination Value in *Drosophila melanogaster*. II. Simulated Natural Selection', Genetics, vol. 70 (1972b), pp. 433-43.

Kimura, M. 'A Model of a Genetic System Which Tends to Closer Linkage by Natural Selection', Evolution, vol. 10 (1956), pp. 278-87.

—, 'On the Evolutionary Adjustment of Spontaneous Mutation Rates', Genet. Res., Cambridge, vol. 9 (1967), pp. 23-34.

King, C. E. 'Comparative Survivorship and Fecundity of Mictic and Amictic Female Rotifers', Physiol. Zool., vol. 43 (1970), pp. 206-12.

—, 'Adaptation of Rotifers to Seasonal Variation', Ecology, vol. 53 (1972), pp. 408-18.

—, 'Genetics of Reproduction, Variation and Adaptation in Rotifers', Arch. Hydrobiol. Beih. Ergebn. Limnol., vol. 8 (1977), pp. 187-201.

King, C. E. and Snell, T. W. 'Sexual Recombination in Rotifers', Heredity, vol. 39 (1977), pp. 357-60.

King, M. and Hayman, D. 'Seasonal Variation of Chiasma Frequency in *Phyllodactylus marmoratus* (Gray) (Gekkonidae-Reptilia)', Chromosoma, vol. 69 (1978), pp. 131-54.

King, R. L. and Slifer, E. H. 'Genetic Analysis of Parthenogenesis in Nabours' Grouse Locusts', Amer. Natur., vol. 67 (1933), pp. 80-1.

——, 'Insect Development. VIII. Maturation and Early Development of Un-fertilized Grasshopper Eggs', J. Morph., vol. 56 (1934), pp. 603-20.

Kinzelbach, R. 'Strepsiptera (Facherflugler)', Handbuch Zool., vol. 4, no. 2 (1971), pp. 1-68.

Kitzmiller, J. B. 'Parthenogenesis in *Culex fatigans*', Science, vol. 129 (1959), pp. 837-8.

Klie, W. 'Ostracoda', Biologie der Tiere Deutschlands, vol. 16, no. 22 (1926), pp. 1-56.

Kluge, A. G. and Eckardt, M. J. '*Hemidactylus garnotti* Duméril and Bibron, a Triploid All-female Species of Gekkonid Lizard', Copeia (1969), pp. 651-64.

Knight, A. Phil. Trans. Roy. Soc. London (1799). Not seen.

Knight Jones, E. W. 'Gregariousness and Some Other Aspects of the Settling Behaviour of Spirorbis', J. Mar. Biol. Ass. UK, vol. 30 (1951), pp. 201-22.

——, 'Laboratory Experiments in Gregariousness during Settling in *Balanus balanoides* and Other Barnacles', J. Exp. Biol., vol. 30 (1953), pp. 584-98.

Knoll, A. H. and Barghoorn, E. S. 'Archaean Microfossils Showing Cell Division from the Swaziland System of South Africa', Science, vol. 198 (1977), pp. 396-8.

Knowlton, N. 'A Note on the Evolution of Gamete Dimorphism', J. Theoret. Biol., vol. 46 (1974), pp. 283-5.

Korotnev, A. 'Zoologische Paradoxen (*Cunoctantha* und *Gastrodes*)', Z. Wiss. Zool., vol. 51 (1891), pp. 613-28.

Korschelt, E. 'Weiteres uber die Dauer der Ungeschlechtlichen Fortpflanzung des *Ctenodrilus monostylos*', Zool. Anz., vol. 137 (1942), pp. 162-6.

Koscielski, B. 'Polyembryony in *Dendrocoelum lacteum* O. F. Muller', in Shapira *et al.*, *Biology of Turbellaria: Experimental Advances* (MSS Information Corp., New York, 1973), pp. 103-7.

Koul, A. K. and Wakhlu, A. K. 'B-chromosomes in *Aster novae-angliae* L. Caryo-logia', vol. 29 (1976), pp. 369-76.

Kramp, P. L. 'Freshwater Medusae in China', Proc. Zool. Soc. London, vol. 120 (1950), pp. 165-84.

Krempf, A. '*Coeloplana gonoctena*: Biologie, Organisation, Développement', Bull. Biol. France Belgique, vol. 54 (1921), pp. 252-312.

Kristensen, I. 'Competition in Three Cyprinodont Fish Species in the Netherlands Antilles', Stud. Fauna Curaçao Carib. Isl., vol. 32, no. 119 (1970), pp. 82-101.

Krüger, E. 'Fortpflanzung und Keimzellenbildung von *Rhabditis aberrans* nov. sp.', Z. Wiss. Zool., vol. 105 (1913), pp. 87-124.

Kudo, R. R. *Protozoology*, 5th edn. (Charles C. Thomas, Springfield, Illinois, 1966).

Lack, D. 'Trichoptera, Lepidoptera and Coleoptera from Bear Island', Ann. Mag. Nat. Hist., vol. 12 (1933), pp. 205-10.

——, 'Some Insects from the Scoresby Sound Region, East Greenland, with an account of the Fauna of a Nunatak', Ann. Mag. Nat. Hist., vol. 14 (1934), pp. 599-606.

Lamb, R. Y. and Willey, R. B. 'Are Parthenogenetic and Related Bisexual Insects Equal in Fertility?', Evolution, vol. 33 (1979), pp. 774-5.

Lambert, F. J. 'Jellyfish. The Difficulties of Study of Their Life History and Other Problems', Essex Natur., vol. 25 (1936), pp. 70-86.

Lameere, A. 'L'Histoire Naturelle des Dicyémides', Bull. Acad. Belgique Ol. Sci., vol. 8 (1922), pp. 779-92.

Lams, H. 'Note sur la Biologie Sexuelle d'un Gasteropode Pulmone (*Arion empiricorum*)', CR Hebd. Séanc. Mem. Soc. Biol. Paris, vol. 62 (1907), pp. 255-7.

van der Land, J. 'Priapulida', in A. C. Giese and J. S. Pearse (eds.), *Reproduction of Marine Invertebrates*, vol. 2 (Academic, New York, 1975). pp. 55-65.

Lanier, G. N. 'Interspecific Mating and Cytological Studies of Closely Related Species of *Ips* DeGeer and *Orthotomicus* Ferrari (Coleoptera: Scolytidae)', Can. Entomol., vol. 98 (1966), pp. 175-88.

Lanier, G. N. and Oliver, J. H. ' "Sex-ratio" condition: unusual mechanism in bark beetles', Science, vol. 153 (1966), pp. 208-9.

Lanier, G. N. and Wood, D. L. 'Controlled Mating, Morphology and Sex-ratio in the *Dendroctonus ponderosae* Complex', Ann. Entomol. Soc. America, vol. 61 (1968), pp. 517-26.

Lantz, L. A. and Cyren, D. O. 'Contribution à la Connaissance de *Lacerta saxicola* Eversmann', Bull. Soc. Zool. France, vol. 61 (1936), pp. 159-81.

de Larambergue, M. 'Étude de l'Autofécondation chez les Gastéropodes Pulmonés. Recherches sur l'Aphallie et la Fécondation chez *Bulinus* (*Isidora*) *contortus* Michaud', Bull. Biol. France Belgique, vol. 73 (1939), pp. 21-231.

Lasserre, P. 'Clitellata', in A. C. Giese and J. S. Pearse (eds.), *Reproduction of Marine Invertebrates*, vol. 3 (Academic, New York, 1975), pp. 215-75.

Law, C. N. 'An Effect of Potassium on Chiasma Frequency and Recombination', Genetica, vol. 33 (1963), pp. 313-29.

Lawrence, C. W. 'The effect of Irradiation of Different Stages of Microsporogenesis on Chiasma Frequency', Heredity, vol. 16 (1961), pp. 83-9.

Lawrence, J. F. 'Biology of the Parthenogenetic Fungus Beetle *Cis fuscipes* Mellié (Coleoptera: Ciidae)', Breviora, vol. 258 (1967), pp. 1-14.

Lawrence, R. F. 'Whipscorpions (Uropygi) from Angola, the Belgian Congo and Mossambique', Publiçoes Cult Co. Diam Angola, vol. 40 (1958), pp. 71-9.

Ledoux, A. 'La Cycle Évolutif de la Fourmi Fileuse (*Oecophylla longinoda* Latr.)', CR, vol. 729 (1949), pp. 246-8.

Lee, V. H. 'Parthenogenesis and Autogeny in *Culicoides bambusicola* Lutz (Ceratopogonidae, Diptera)', J. Med. Entomol., vol. 5 (1968), pp. 91-3.

Lees, A. D. 'The Role of Photoperiod and Temperature in the Determination of Parthenogenetic and Sexual Forms in the Aphid *Megoura viciae* Buckton. I. The Influence of These Factors on Apterous Virginoparae and Their Progeny', J. Insect Physiol., vol. 3 (1959), pp. 92-117.

——, 'The Role of Photoperiod and Temperature in the Determination of the Parthenogenetic and Sexual Forms in the Aphid *Megoura viciae* Buckton.

II. The Operation of the "Interval Timer" in Young Clones', J. Insect Physiol., vol. 4 (1960a), pp. 154-75.

—, 'Some Aspects of Animal Photoperiodism', Cold Spring Harbour Symp. Quant. Biol., vol. 25 (1960b), pp. 261-8.

—, 'Clonal Polymorphism in Aphids', Symp. Roy. Entomol. Soc. London, vol. 1 (1961), pp. 68-79.

—, 'The Role of Photoperiod and Temperature in the Determination of Parthenogenetic and Sexual Forms of the Aphid *Megoura viciae* Buckton. III. Further Properties of the Maternal Switching Mechanism in Apterous Aphids', J. Insect Physiol., vol. 9 (1963), pp. 153-64.

—, 'The Location of the Photoperiodic Receptor in the Aphid *Megoura viciae* Buckton', J. Exp. Biol., vol. 41 (1964), pp. 119-33.

—, 'The Control of Polymorphism in Aphids', Adv. Insect Physiol., vol. 3 (1966), pp. 207-77.

von Leeuwenhoek, A. Phil. Trans. Roy. Soc. London (1702). Not seen.

Legrand, J. J. and Juchault, P. 'La Déterminisme de la Monogénie chez les Oniscoides', CR Acad. Sci. Paris, vol. 268 (1969), pp. 1774-7.

Legrand, J. J. and Legrand-Hamelin, E. 'Déterminisme de l'Intersexualité et de la Monogénie chez les Crustacés Isopodes', Pubbl. Staz. Zool. Nappoli, vol. 29 (suppl.) (1975), pp. 443-61.

Lei, C. and Clifford, H. F. *Field and Laboratory Studies of Daphnia schodleri Sars from a Winterkill Lake of Alberta*, Nat. Mus. Canada Publ. Zool., no. 9 (1974).

Leigh, E. G. 'Natural Selection and Mutability', Amer. Natur., vol. 104 (1970), pp. 301-5.

—, 'The Evolution of Mutation Rates', Genetics (suppl.), vol. 73 (1973), pp. 1-18.

Lenhoff, H. M. and Loomis, W. F. *The Biology of Hydra* (University of Miami Press, Coral Gables, Fla., 1961).

Leuckart, A. 'Zeugung', in Wagner (ed.), *Handworterbuch der Physiologie*, (1853).

Levene, H. 'Genetic Equilibrium When More Than One Ecological Niche is Available', Amer. Natur., vol. 87 (1953), pp. 131-3.

Levin, B. R. 'Coexistence of Two Asexual Strains on a Single Resource', Science, vol. 175 (1972), pp. 1272-4.

Levin, B. R. and Kilmer, W. L. 'Interdemic Selection and the Evolution of Altruism: A Computer Simulation Study', Evolution, vol. 28 (1975), pp. 527-45.

Levin, D. A. 'Pest Pressure and Recombination Systems in Plants', Amer. Natur., vol. 109 (1975), pp. 437-51.

Levine, R. P. 'Chromosome Structure and the Mechanism of Crossing-over', Proc. Nat. Acad. Sci. US, vol. 41 (1955), pp. 727-30.

Levins, R. 'Theory of Fitness in a Heterogeneous Environment. VI. The Adaptive Significance of Mutation', Genetics, vol. 56 (1967), pp. 163-78.

—, *Evolution in Changing Environments* (Princeton University Press, Princeton, NJ, 1968).

—, 'Extinction', in M. Gerstenhaber (ed.), *Some Mathematical Problems in Biology* (American Mathematical Society, Providence, RI, 1970), pp. 77-107.

Lewis, K. R. and John, B. *The Meiotic System. Protoplasmatologia VI FI* (Springer-Verlag, Vienna, 1965).

Lewis, K. R. and John, B. *The Matter of Mendelian Heredity*, 2nd edn. (Longman, London, 1972).

Lewontin, R. C. Bioscience, vol. 16 (1966), p. 25.

—, 'The Effect of Genetic Linkage on the Mean Fitness of a Population', Proc. Nat. Acad. Sci. US, vol. 68 (1971), pp. 984-6.

—, *The Genetic Basis of Evolutionary Change* (Colombia University Press, New York, 1974).

Lewontin, R. C. and Hubby, J. L. 'A Molecular Approach to the Study of Genic Heterozygosity in Natural Populations. II. Amount of Variation and Degree of Heterozygosity in Natural Populations of *Drosophila pseudoobscura*', Genetics, vol. 54 (1966), pp. 595-609.

Liberman, U. 'Theory of Meiotic Drive: Is Mendelian Segregation Stable?', Theoret. Pop. Biol., vol. 10 (1976), pp. 127-32.

Liberman, U. and Feldman, M. W. 'On the Evolutionary Significance of Mendel's Ratios', Theoret. Pop. Biol., vol. 17 (1980), pp. 1-15.

Lieder, U. 'Männchenmangel und Natürliche Parthenogenese bei der Silberkarausche *Carassius auratus gibelio* (Vertebrata, Pisces)', Naturwissenschaften, vol. 42 (1955), p. 590.

—, 'Über die Entwicklung bei Mannchenlosen Stammen der Silberkarausche *Carassius auratus gibelio* (Bloch)', Biol. Zentralblatt, vol. 78 (1959), pp. 284-91.

Light, S. F. 'Parthenogenesis in Termites of the Genus *Zootermopsis*', Univ. Calif. Publ. Zool., vol. 43 (1944), pp. 405-12.

Lindeberg, B. 'A Parthenogenetic Race of *Monotanytarsus boreoalpinus* Th. (Dipt., Chironomidae) from Finland', Ann. Entomol. Fenn., vol. 24 (1958), pp. 35-8.

Linderoth, C. H. 'Experimentelle Beobachtungen an Parthenogenestischem und Bisexuellem *Otiorrhynchus dubius* Stroem (Col., Curculionidae)', Entomol. Tidskrift, vol. 75 (1954), pp. 111-16.

Loan, C. '*Pygostolus falcatus* (Nees) (Hymenoptera, Braconidae), A Parasite of *Sitona* Species (Coleoptera, Curculionidae)', Bull. Entomol. Res., vol. 52 (1961), pp. 473-88.

Loeb, J. 'On the Nature of the Process of Fertilization and the Artificial Production of Normal Larvae (Plutei) from the Unfertilized Eggs of the Sea Urchin', Amer. J. Physiol., vol. 3 (1899), pp. 135-8.

Lokki, J. 'Genetic Polymorphism and Evolution in Parthenogenetic Animals. VII. The Amount of Heterozygosity in Diploid Populations', Hereditas, vol. 83 (1976a), pp. 57-64.

—, 'Genetic Polymorphism and Evolution in Parthenogenetic Animals. VIII. Heterozygosity in Relation to Polyploidy', Hereditas, vol. 83 (1976b), pp. 65-72.

Lokki, J., Saura, A., Lankinen, P. and Suomaleinen, E. 'Genetic Polymorphism and Evolution in Parthenogenetic Animals. V. Triploid *Adoxus obscurus* (Coleoptera: Chrysomelidae)', Genet. Res., Cambridge, vol. 28 (1976a), pp. 27-36.

—, 'Genetic Polymorphism and Evolution in Parthenogenetic Animals. VI. Diploid and Triploid *Polydrosus mollis* (Coleoptera: Curculionidae)', Hereditas, vol. 82 (1976b), pp. 209-16.

Lokki, J., Suomalainen, E., Saura, A. and Lankinen, P. 'Genetic Polymorphism and Evolution in Parthenogenetic Animals. II. Diploid and Polyploid *Solenobia triquetrella* (Lepidoptera: Psychidae)', Genetics, vol. 79 (1975), pp. 513-25.

Lomnicki, A. and Slobodkin, L. B. 'Floating in *Hydra littoralis*', Ecology, vol. 47 (1966), pp. 881-9.

Longhurst, A. R. 'Reproduction in Notostraca', Nature, vol. 173 (1954), pp. 781-2.

—, 'The Reproduction and Cytology of the Notostraca (Crustacea, Phyllopoda)', Proc. Zool. Soc. London, vol. 125 (1955), pp. 671-80.

Loomis, W. F. 'Sexual Differentiation in *Hydra*: Control by Carbon Dioxide Tension', Science, vol. 126 (1957), pp. 735-9.

—, 'The Sex Gas of Hydra', Sci. Amer., vol. 200 (1959), pp. 145-56.

—, 'Feedback Factors Affecting Sexual Differentiation in *Hydra littoralis*', in H. M. Lenhoff and W. F. Loomis (eds.), *The Biology of Hydra* (University of Miami Press, Coral Gables, Florida, 1961), pp. 337-60.

Loomis, W. F. and Lenhoff, H. M. 'Growth and Sexual Differentiation of Hydra in Mass Culture', J. Exp. Zool., vol. 132 (1956), pp. 555-68.

Lowe, C. H. and Wright, J. W. 'Evolution of Parthenogenetic Species of *Cnemidophorus* (Whiptail Lizards) in Western North America', J. Ariz. Acad. Sci., vol. 4 (1966), pp. 81-7.

Lowe, C. H., Wright, J. W., Cole, C. J. and Bezy, R. L. 'Chromosomes and Evolution of the Species Groups of *Cnemidophorus* (Reptilia: Teiidae)', Syst. Zool., vol. 19 (1970), pp. 128-41.

Lowndes, A. G. 'The Sperms of Freshwater Ostracods', Proc. Zool. Soc. London, (1935), pp. 35-48.

Lyon, E. P. 'Experiments in Artificial Parthenogenesis', Amer. J. Physiol., vol. 9 (1903).

Lytle, C. F. 'Patterns of Budding in the Freshwater Hydroid *Craspedacusta*', in H. M. Lenhoff and W. F. Loomis (eds.), *The Biology of Hydra* (University of Miami Press, Coral Gables, Florida, 1961), pp. 317-36.

Maas, O. 'Erlidigte und Strittige Fragen der Schwammerentwicklung', Biol. Zentralblatt, vol. 16 (1896), pp. 231-9.

MacArthur, R. H. 'Ecological Consequences of Natural Selection', in T. H. Waterman and H. J. Morowitz (eds.), *Theoretical and Mathematical Biology* (Blaisdell, New York, 1965), pp. 388-97.

MacArthur, R. H. *Geographical Ecology. Patterns in the Distribution of Species* (Harper and Row, New York, 1972).

MacBride, E. W. 'The Development of *Asterina gibbosa*', Quart. J. Microscop. Soc., vol. 38 (1896), pp. 339-412.

McCafferty, W. P. and Huff, B. L. 'Parthenogenesis in the Mayfly *Stenonema femoratum* (Say) (Ephemeroptera: Heptageniidae)', Entomol. News, vol. 85 (1974), pp. 76-80.

McClary, A. 'The Effect of Temperature on Growth and Reproduction in *Craspedacusta sowerbyi*', Ecology, vol. 40 (1959), pp. 158-62.

McConnaughey, B. H. 'The Life Cycle of the Dicyemid Mesozoa', Univ. Calif. Publ. Zool., vol. 55 (1951), pp. 295-336.

McDowell, S. B. 'A Catalogue of the Snakes of New Guinea and the Solomons, with Special Reference to Those in the Bernice P. Bishop Museum. Part 1. Scolecophidia', J. Herpetol., vol. 8 (1974), pp. 1-57.

MacGillivray, M. E. and Anderson, G. B. 'The Effect of Photoperiod and Temperature on the Production of Gamic and Agamic Forms in *Macrosiphum euphorbiae* (Thomas)', Can. J. Zool., vol. 24 (1964), pp. 491-510.

MacGregor, H. C. and Uzzell, T. M. 'Gynogenesis in Salamanders Related to *Ambystoma jeffersonianum*', Science, vol. 143 (1964), pp. 1043-5.

MacGregor, I. M., Truswell, J. F. and Eriksson, K. A. 'Filamentous Algae from the 2,300 m.y. Old Transvaal Dolomite', Nature, vol. 247 (1974), pp. 538-9.

Machado, A. de B. 'Ochryoceratidae Nouveaux d'Afrique (Araneae)', Ann. Natal Mus., vol. 16 (1964), pp. 215-30.

McIntosh, W. C. 'On a Remarkable *Syllis* Dredged by HMS *Challenger*', J. Linn. Soc. London, vol. 14 (1879), pp. 720-4.

McKay, F. E. 'Behavioural Aspects of Population Dynamics in Unisexual-bisexual *Poeciliopsis* (Pisces: Poeciliidae)', Ecology, vol. 52 (1971), pp. 778-90.

Mackensen, O. 'The Occurrence of Parthenogenetic Females in Some Strains of Honey Bees', J. Econ. Entomol., vol. 36 (1943), pp. 465-7.

Mackenzie, K. K. 'Cyperaceae' in *North American Flora*, vol. 18, (New York Botanical Garden, Bronx Park, New York, 1931-5), parts 1-7.

McKinney, C. O., Fenton, R. K. and Anderson, R. A. 'A New All-female Species of the Genus *Cnemidophorus*', Herpetologica, vol. 29 (1973), pp. 361-6.

Macklin, M. 'Analysis of Growth Factor Gradients in *Hydra*', J. Cell. Comp. Physiol., vol. 72 (1968), pp. 1-8.

MacLachlan, R. *A Monographic Revision and Synopsis of the Trichoptera of the European Fauna* (London, 1874-80).

McNaughton, S. J. 'r- and k-selection in *Typha*', Amer. Natur., vol. 109 (1975), pp. 251-61.

Maeda, T. 'On the Configurations of Gemini in the Pollen Mother Cells of *Vicia faba* L', Mem. Coll. Sci. Kyoto Imp. Univ., vol. 5 (1930), pp. 125-37.

Magnusson, W. E. 'Production of an Embryo by an *Acrochordus javanicus* Isolated for Seven Years', Copeia (1979), pp. 744-5.

Mahendra, B. C. and Sharma, S. 'Classification of the Modes of Animal Reproduction', Ann. Zool. (Agra), vol. 1 (1955), pp. 85-6.

Makino, S. *An Atlas of the Chromosome Numbers in Animals*, 2nd edn (Iowa State College Press, Amer Iowa, 1965).

Makino, S. and Momma, E. 'An Idiogram Study of the Chromosomes in Some Species of Reptiles', Cytologia, vol. 15 (1949), pp. 96-108.

Malaquin, A. 'Recherches sur les Syllidiens: Morphologies, Anatomies, Réproduction, Développement', Mem. Soc. Sci. Arts Lille (1893), pp. 1-477.

—, 'La Formation du Schizozoite dans la Scissiparité chez les Filigranes et les Salmacines', CR Acad. Sci. Paris, vol. 121 (1895), pp. 953-5.

Malik, C. P. and Tripathi, R. C. 'B-chromosomes and Meiosis in *Festuca mariei*', Zeits. Biol., vol. 116 (1970), pp. 321-6.

Manga, V. 'Chiasma Frequencies in Primary Trisomics of Pearl Millet', Can. J. Genet. Cytol., vol. 18 (1976), pp. 11-15.

Manning, J. T. 'Is Sex Maintained to Facilitate or Minimize Mutational Advance?', Heredity, vol. 36 (1976), pp. 351-7.

Marcovitch, S. 'Plant Lice and Light Exposure', Science, vol. 58 (1923), pp. 537-8.

—, 'The Migration of the Aphididae and the Appearance of the Sexual Forms as Affected by the Relative Length of Daily Light Exposure', J. Agric. Res., vol. 27 (1924), pp. 513-22.

Marcus, E. and Macrae, W. 'Architomy in a Species of *Convoluta*', Nature, vol. 173 (1954), p. 130.

Mariscal, R. N. 'Entoprocta', in A. C. Giese and J. S. Pearse (eds.), *Reproduction of Marine Invertebrates*, vol. 2 (Academic, New York, 1975), pp. 1-41.

Marshall, A. C. 'A Qualitative and Quantitative Study of the Trichoptera of Western Lake Erie (as Indicated by Light Trap Material)', Ann. Entomol. Soc. Amer., vol. 32 (1939), pp. 665-88.

Marshall, V. G. and Kevan, D. K. McE. 'Preliminary Observations on the Biology of *Folsomia candida* Willem 1902 (Collembola: Isotomidae)', Can. Entomol., vol. 94 (1962), pp. 575-86.

Martin, E. A. 'Polymorphism and Methods of Asexual Reproduction in the Annelid *Dodecaceria* of Vineyard Sound', Biol. Bull., vol. 65 (1933), pp. 99-105.

Martin, G., Juchault, P. and Legrand, J. J. 'Mise en Évidence d'un Micro-organisme Intracytoplasmique Symbiote de l'Oniscoide *Armadillidium vulgare* Latr. dont la Présence Accompagne l'Intersexualité ou la Féminization Totale des Mâles Génétiques de la Lignée Thélygène', CR Acad. Sci. France, vol. 276 (1973), pp. 2313-16.

Maslin, T. P. 'All Female Species of the Lizard Genus *Cnemidophorus*', Science, vol. 135 (1962), pp. 212-13.

—, 'The Sex of Hatchlings of Five Apparently Unisexual Species of Whiptail Lizards (*Cnemidophorus*, Teiidae)', Amer. Midl. Natur., vol. 76 (1966), pp. 369-78.

——, 'Skin-grafting in the Bisexual Teiid Lizard *Cnemidophorus sexlineatus* and in the Unisexual *C. tesselatus*', J. Exp. Zool., vol. 166 (1967), pp. 137-49.

——, 'Taxonomic Problems in Parthenogenetic Vertebrates', Syst. Zool., vol. 17 (1968), pp. 219-31.

——, 'Parthenogenesis in Reptiles', Amer. Zool., vol. 11 (1971), pp. 361-80.

Mather, K. 'Crossing Over and Heterochromatin in the X Chromosome of *Drosophila melanogaster*', Genetics, vol. 24 (1939), pp. 413-35.

——, *Genetical Structure of Populations* (Chapman and Hall, London, 1973).

Mathias, P. 'Biologie des Crustacés Phyllopodes', Actual. Scient. Ind., vol. 447 (1937), pp. 1-106.

Matthey, R. 'Étude Biologique et Cytologique de *Saga pedo* (Orthoptéres-Tettigoniidae)', Rev. Suisse Zool., vol. 48 (1941), pp. 92-142.

——, 'Contribution à l'Étude Cytologique de la Parthénogénèse chez les Orthoptères: *Pycnoscelus surinamensis* L. (Blatt. Panchloridae)', Actes Soc. Helvetique Sci. Nat., vol. 122 (1942), p. 135.

——, 'L'Ovogénese et la Cytologie de la Parthénogénèse chez la Blatte *Pycnoscelus surinamensis* L. (Blatt. Panchlorinae)', Arch. Julius Klaus-Stift, vol. 18 (1943), pp. 683-7.

——, 'Les Processus de la Maturation chez *Pycnoscelus surinamensis* L. (Blattidae Panchlorinae)', Arch. Julius Klaus-Stift, vol. 19 (1944), pp. 529-32.

——, 'Cytologie de la Parthénogénèse chez *Pycnoscelus surinamensis*', Rev. Suisse Zool., vol. 52 (suppl. 1) (1945), pp. 1-109.

——, 'Demonstration du Charactère Géographique de la Parthenogénèse de *Saga pedo* Pallas et de Sa Polyploidie, par Comparison avec les Espèces Bisexuées *S. ephippigera* Fisch. et *S. gracilipes* Uvar.', Experientia, vol. 2 (1946), pp. 260-1.

Matthieson, F. A. 'Parthenogenesis in Scorpions', Evolution, vol. 16 (1962), p. 255.

——, 'The Breeding of *Tityus serrulatus* Lutz and Mello 1927 in Captivity (Scorpiones: Buthidae)', Rev. Bras. Pesqui Med. Biol., vol. 4 (1971), pp. 299-300.

Mattox, N. T. 'Oogenesis of *Campeloma rufum*, a Parthenogenetic Snail', Z. Zellforsch., vol. 27 (1938), pp. 455-64.

Mattox, N. T. and Velardo, J. T. 'Effect of Temperature on the Development of the Eggs of a Conchostracan Phyllopod *Caenestheriella gynecia*', Ecology, vol. 31 (1950), pp. 497-506.

Maupas, E. 'Sur la Déterminisme de la Sexualité chez l'*Hydatina senta*', Comt. Rend., vol. 63 (1891), pp. 388-90.

——, 'Modes et Formes de Reproduction des Nématodes', Ann. Zool. Exp. Gen., Ser. 3, vol. 8 (1900), pp. 463-624.

May, R. M. *Stability and Complexity in Model Ecosystems* (Princeton University Press, Princeton, NJ, 1973).

Maynard Smith, J. 'Evolution in Sexual and Asexual Populations', Amer. Natur., vol. 102 (1968), pp. 469-73.

—, 'What Use Is Sex?', J. Theoret. Biol., vol. 30 (1971a), pp. 319–35.

—, 'The Origin and Maintenance of Sex', in G. C. Williams (ed.), *Group Selection* (Aldine-Atherton, Chicago, 1971b).

—, 'Recombination and the Rate of Evolution', Genetics, vol. 78 (1974), pp. 299–305.

—, 'Group Selection', Quart. Rev. Biol., vol. 51 (1976a), pp. 277–83.

—, 'A Short Term Advantage for Sex and Recombination through Sib-competition', J. Theoret. Biol., vol. 63 (1976b), pp. 245–58.

—, *The Evolution of Sex* (Cambridge University Press, 1978).

Maynard Smith, J. and Williams, G. C. 'Reply to Barash', Amer. Natur., vol. 110 (1976), p. 897.

deMeijere, J. C. H. 'Die Lonchopteren des paläarktischen Gebietes', Tijdschr. Entomol., vol. 49 (1906), pp. 44–98.

Metcalf, R. A., Marlin, J. C. and Whitt, G. S. 'Low Levels of Genetic Heterozygosity in Hymenoptera', Nature, vol. 257 (1975), pp. 792–4.

Metz, C. W. and Nonidez, J. F. 'Spermatogenesis in the Fly, *Asilus sericeus* Say', J. Exp. Zool., vol. 32 (1921), pp. 165–85.

—, 'Spermatogenesis in *Asilus notatus* Wied. (Diptera)', Arch. Zellforsch., vol. 17 (1923), pp. 438–9.

Meyer, H. 'Investigations Concerning the Reproductive Behaviour of *Molliensia formosa*', J. Genet., vol. 36 (1938), pp. 329–66.

Micoletzky, H. 'Freilebende Nematoden von der Sargassosee', Mitt. Zool. Staatsinst., Zool. Mus. Hamburg., vol. 39 (1922).

Mikulska, I. 'Cytological Studies upon Genus *Otiorrhynchus* (Curculionidae, Coleoptera) in Poland', Experientia, vol. 5 (1940), pp. 473–5.

—, 'Cytological Studies upon Genus *Otiorrhynchus* (Curculionidae, Coleoptera) in Poland', Experientia, vol. 5 (1949), p. 473.

—, 'The Chromosomes of the Parthenogenetic and Thelytokian Weevil *Eusomus ovulum* Germ. (Curculionidae, Coleptera)', Bull. Acad. Pol. Sci. Lett., vol. B2 (1953), pp. 269–76.

—, 'New Data to the Cytology of the Parthenogenetic Weevils of the Genus *Otiorrhynchus* Germ. (Curculionidae, Coleoptera) from Poland', Cytologia, vol. 25 (1960), pp. 322–33.

Milkman, R. 'Electrophoretic Variation in *Escherichia coli* from Natural Sources', Science, vol. 182 (1973), pp. 1024–6.

Miller, D. E. 'A Limnological Study of *Pelmatohydra* with Special Reference to Their Quantitative Seasonal Distribution', Trans. Amer. Microscop. Soc., vol. 55 (1936), pp. 123–93.

Miller, R. M. and Schultz, R. J. 'All Female Strains of the Teleost Fishes of the Genus *Poeciliopsis*', Science, vol. 130 (1959), pp. 1656–7.

Milliron, H. E. 'A Parthenogenetic New Species of the Genus *Perimegatoma* Horn (Coleoptera: Dermestidae)', Ann. Entomol. Soc. Amer., vol. 32 (1939), pp. 570–4.

Minasian, L. L. 'Characteristics of Asexual Reproduction in the Sea-anemone,

Haliplanella luciae (Verrill), Reared in the Laboratory', in G. O. Mackie (ed.), *Coelenterate Ecology and Behaviour* (Plenum, New York, 1976), pp. 289-98.

Minchin, E. A. 'The Position of Sponges in the Animal Kingdom', Sci. Progr., vol. 6 (1897), pp. 426-60.

Minton, S. A. 'Observations on Amphibians and Reptiles of the Big Bend Region of Texas', Southwest Natur., vol. 3 (1958), pp. 28-54.

Mitchell, C. W. 'Sex-determination in *Asplachna amphora*', J. Exp. Zool., vol. 15 (1913), pp. 225-55.

Mitter, C., Futuyma, D. J., Schneider, J. C. and Hare, J. D. 'Genetic Variation and Host Plant Relations in a Parthenogenetic Moth', Evolution, vol. 33 (1979), pp. 777-90.

Mockford, E. L. 'Studies on the Reuterelline Psocids (Psocoptera)', Proc. Entomol. Soc. Washington, vol. 57 (1955), pp. 97-108.

——, 'Parthenogenesis in Psocids (Insecta: Psocoptera)', Amer. Zool., vol. 11 (1971), pp. 327-39.

Moffet, A. A. 'The Origin and Behaviour of Chiasmata. 13. Diploid and Tetraploid *Culex pipiens*', Cytologia, vol. 7 (1936), pp. 184-97.

El Mofty, M. M. and Smyth, J. D. 'Endocrine Control of Sexual Reproduction in *Opalina rananum* Parasitic in *Rana temporaria*', Nature, vol. 186 (1959), p. 559.

Monks, S. 'Variability and Autonomy of *Linckia*', Proc. Acad. Sci. Philadelphia, vol. 56 (1904), pp. 596-600.

Monterosso, B. '*Liposcelis divinatorius* (Muller). II. Partenogenesi Constante e Completa', Boll. Accad. Gioenia Sci. Nat. Catania, Ser. IV, vol. 2 (1952), pp. 151-67.

Moore, B. P., Woodruffe, G. E. and Sanderson, A. R. 'Polymorphism and Parthenogenesis in Ptinid Beetle', Nature, vol. 177 (1956), pp. 847-8.

Moore, W. S. 'Stability of Small Unisexual-Bisexual Populations of *Poeciliopsis* (Pisces: Poeciliidae)', Ecology, vol. 56 (1975), pp. 791-808.

——, 'Components of Fitness in the Unisexual Fish *Poecilipsis monacha-occidentalis*', Evolution, vol. 30 (1976), pp. 564-78.

——, 'A Histocompatibility Analysis of Inheritance in the Unisexual Fish *Poeciliopsis 2 monacha-lucida*', Copeia (1977), pp. 213-23.

——, 'Coexistence of Clones in a Heterogeneous Environment', Science, vol. 199 (1978), pp. 549-52.

Morgan, T. H. 'Some Further Experiments on Self-fertilization in *Ciona*', Biol. Bull., vol. 8 (1905), pp. 313-30.

——, 'Sex Determination and Parthenogenesis in Phylloxerans and Aphids', Science, vol. 29 (1909), pp. 234-7.

——, 'The Predetermination of Sex in Phylloxerans and Aphids', J. Exp. Zool., vol. 19 (1915), pp. 285-321.

——, 'Removal of the Block to Self-fertilization in the Ascidian *Ciona*', Proc. Nat. Acad. Sci. USA, vol. 9 (1923), pp. 170-1.

——, 'The Genetic and Physiological Problems of Self-sterility in *Ciona*. I. Data on Self and Cross-fertilization', J. Exp. Zool., vol. 78 (1938), pp. 271-318.

——, 'The Genetic and Physiological Problems of Self-sterility in *Ciona*. V. The Genetic Problem', J. Exp. Zool., vol. 90 (1942), pp. 199-228.

——, 'The Genetic and Physiological Problems of Self-sterility in *Ciona*. VI. Theoretical Discussion of Genetic Data', J. Exp. Zool., vol. 96 (1944), pp. 37-59.

——, 'The Conditions That Lead to Normal or Abnormal Development of *Ciona*', Biol. Bull., vol. 88 (1945), pp. 50-62.

Morris, M. 'A Cytological Study of Artificial Parthenogenesis in *Cumingia*', J. Exp. Zool., vol. 22 (1917), pp. 1-51.

——, (M. Morris Hoskins) 'Further Experiments on the Effect of Heat on the Eggs of *Cumingia*', Biol. Bull., vol. 35 (1918), pp. 260-72.

Mortensen, T. 'Biological Observations on Ophiuroids', Vidensk. Medd. Dansk. Naturh. Foren, vol. 93 (1933), pp. 171-94.

Mortimer, C. H. 'Experimentelle und Cytologische Untersuchungen über den Generationswechsel der Cladoceran', Zool. Jahrbuch, vol. 56 (1936), pp. 323-88.

Moyer, S. E. 'Selection for Modification of Recombination Frequency of Linked Genes', PhD thesis (University of Minnesota); cited by Kidwell (1972).

Mrazek. A. 'Einige Bemerkungen über die Knospungen und Geschlechtliches Fortpflanzung bei *Hydra*', Biol. Zentralblatt, vol. 27 (1907), pp. 392-6.

Mukherjee, A. S. 'Effect of Selection on Crossing-over in the Males of *Drosophila ananassae*', Amer. Natur., vol. 95 (1961), pp. 57-9.

Muldal, S. 'The Chromosomes of the Earthworms. I. The Evolution of Polyploidy', Heredity, vol. 6 (1952), pp. 55-76.

Muller, H. J. 'Some Genetic Aspects of Sex', Amer. Natur., vol. 66 (1932), pp. 118-38.

——, 'Evolution by Mutation', Bull. Amer. Math. Soc., vol. 64 (1958), pp. 137-60.

——, 'The Relation of Recombination to Mutational Advance', Mutat. Res., vol. 1 (1964), pp. 2-9.

Muller, T. H. 'Why Polyploidy Is Rarer in Animals than in Plants', Amer. Natur., vol. 59 (1925), pp. 346-53.

Mulvey, R. H. 'Parthenogenesis and the Role of the Male in Reproduction', in J. N. Sasser and W. R. Jenkins (eds.), *Nematology* (University of North Carolina Press, Chapel Hill, NC, 1960), pp. 331-5.

Murdy, W. H. and Carson, H. L. 'Parthenogenesis in *Drosphila mangabeiri*', Amer. Natur., vol. 93 (1959), pp. 355-63.

Murray, B. G. 'The Cytology of the Genus *Briza* L. (Gramineae). II. Chiasma Frequency, Polyploidy and Interchange Heterozygosity', Chromosoma, vol. 57 (1976), pp. 81-93.

Murray, J. *Genetic Diversity and Natural Selection* (Oliver and Boyd, Edinburgh, 1972).

Muus, K. 'Notes on the Biology of *Protohydra leuckarti* Greef (Hydroidea, Protohydridae)', Ophelia, vol. 3 (1966), pp. 141-50.

Nabours, R. K. 'Parthenogenesis and Crossing-over in the Grouse Locust *Apotettix*', Amer. Natur., vol. 53 (1919), pp. 131-42.

Nabours, R. K. and Forster, M. E. 'Parthenogenesis and the Inheritance of Colour Patterns in the Grouse Locust *Paratettix texanus* Handcock', Biol. Bull., vol. 54 (1929), pp. 129-55.

Nair, K. K. N. 'Observation on the Biology of *Cyclestheria hislopi* (Baird) (Conchostraca: Crustacea)', Arch. Hydrobiol., vol. 65 (1968), pp. 96-9.

Nankivell, R. N. 'Karyotype Differences in the *crenaticeps* Group of *Atractomorpha* (Orthoptera, Acridoidea, Pyrgomorphidae)', Chromosoma, vol. 56 (1976), pp. 127-42.

Narbel, M. 'La Cytologie de la Parthénogénèse chez *Apterona helix* Sieb. (Lep. Psychidae)', Rev. Suisse Zool., vol. 53 (1946), pp. 625-81.

Narbel-Hofstetter, M. 'La Cytologie de la Parthénogénèse chez *Solenobia* sp. *lichinella* (L.) (Lep. Psych.)', Chromosoma, vol. 4 (1950), pp. 56-90.

——, 'La Cytologie de la Parthénogénèse chez *Luffia ferchaultella* Steph. (Lepid. Psych.). Communication Préliminaire', Rev. Suisse Zool., vol. 61 (1954), pp. 416-19.

——, 'La Pseudogamie chez *Luffia lapidella* Goeze (Lepid. Psychidae)', Rev. Suisse Zool., vol. 62 (1955), pp. 224-9.

——, 'La Cytologie des *Luffia*: Le Croisement de l'Éspèce Parthénogénètique avec l'Éspèce Bisexuée', Rev. Suisse Zool., vol. 63 (1956), pp. 203-8.

——, 'La Surmaturation des Oeufs de *Luffia* (Lepidoptera, Psychidae)', Rev. Suisse Zool., vol. 67 (1960), pp. 238-44.

——, 'Cytologie Comparée de l'Éspèce Parthénogénètique *Luffia ferchaultella* Steph. et de l'Éspèce Bisexuée *L. lapidella* Goeze. (Lépid. Psych.)', Chromosoma, vol. 12 (1961), pp. 505-52.

——, 'Cytologie de la Pseudogamie chez *Luffia lapidella* Goeze (Lépid. Psych.)', Chromosoma, vol. 13 (1963), pp. 623-45.

——, 'Les Altérations de la Méiose ches les Animaux Parthénogénetiques', *Protoplasmatologia*, VI, F2 (Springer-Verlag, 1964).

Narzikulov, M. N. 'New Data on the Aphids (Homoptera, Aphidinea) of Soviet Central Asia', Entomol. Rev., vol. 49 (1970), pp. 216-21.

Nauwerck, A. 'Die Beziehungen zwischen Zooplankton und Phytoplankton im See Erken', Symb. Bot. Uppsala, vol. 17 (1963), pp. 1-163.

Neaves, W. B. 'Adenosine Deaminase Phenotypes among Sexual and Parthenogenetic Lizards in the Genus *Cnemidophorus* (Teiidae)', J. Exp. Zool., vol. 171 (1969), pp. 175-84.

Neaves, W. B. and Gerald, P. S. 'Lactate Dehydrogenase Isozymes in Parthenogenetic Teiid Lizards (*Cnemidophorus*)', Science, vol. 160 (1968), pp. 1004-5.

——, 'Gene Dosage at the Lactate Dehydrogenase B Locus in Triploid and Diploid Teiid Lizards', Science, vol. 164 (1969), pp. 557-8.

Needham, J. G., Traver, J. R. and Hsu, Y. C. *The Biology of Mayflies* (Comstock, Ithaca, NY, 1935).

Nei, M. 'Modification of Linkage Intensity by Natural Selection', Genetics, vol. 57 (1967), pp. 625-41.

——, 'Linkage Modification and Sex Differences in Recombination', Genetics, vol. 63 (1969), pp. 681-99.

Nevo, E. 'Genetic Variation in Natural Populations: Pattern and Theory', Theoret. Pop. Biol., vol. 13 (1978), pp. 121-77.

New, T. R. 'The Flight Activity of Some British Hemerobiidae and Chrysopidae, as Indicated by Suction-trap Catches (Neuroptera)', Proc. Roy. Entomol. Soc., London, vol. 42 (1967), pp. 93-100.

Newman, H. H. 'On the Development of the Spontaneously Parthenogenetic Eggs of *Asterina (Patinia) miniata*', Biol. Bull., vol. 40 (1921), pp. 105-17.

Nicholas, G. 'Nocturnal Migration of *Hadenoecus subterraneus*', Nat. Speleol. Soc. News, vol. 20 (1962).

Nicholas, W. L. *The Biology of Free Living Nematodes* (Clarendon Press, Oxford, 1976).

Nielson, C. 'Some Loxosomatidae (Entoprocta) from the Atlantic Coast of the United States', Ophelia, vol. 3 (1966a), pp. 249-75.

——, 'On the Life Cycle of Some Loxosomatidae (Entoprocta)', Ophelia, vol. 3 (1966b), pp. 221-47.

Nigon, V. 'Les Modalités de la Reproduction et Déterminisme du Sexe chez Quelques Nématodes Libres', Ann. Sci. Nat. Zool., vol. 11 (1949), pp. 1-32.

Nikolei, E. 'Vergleichende Untersuchungen zur Fortpflanzung Heterogoner Gallmücken unter Experimentellen Bedingungen', Z. Morph. Ökol. Tiere, vol. 50 (1961), pp. 281-329.

Nishihira, M. 'Observations of the Algal Selection by Larvae of *Sertularella miurensis* in Nature', Bull. Mar. Biol. Stat. Asamustri, vol. 13 (1967), pp. 35-48.

——, 'Experiments on the Algal Selection by the Larvae of *Coryne uchidai* Stechow (Hydrozoa)', Bull. Mar. Biol. Stat. Asamustri, vol. 13 (1968), pp. 83-9.

Noda, S. 'Achiasmate Bivalent Formation by Parallel Pairing in PMCs of *Fritillaria amabilis*', Bot. Mag., vol. 81 (1968), pp. 344-5.

Nolte, D. J. 'The Nuclear Phenotype of Locusts', Chromosoma, vol. 15 (1964), pp. 367-88.

Nouvel, H. 'Les Dicyémides. Première Partie: Systématique, Génerations Vermiformes, Infusorigène et Sexualité', Arch. Biol., vol. 58 (1947), pp. 59-219.

Novikov, N. P. 'On the Possibility of Gynogenesis in the Fish (*Atheresthes stomias* Jord. et Gilb.) from the Bering Sea', Dokl. Akad. Nauk SSSR, vol. 147 (1962), pp. 215-16. English translation: Dokl. Biol. Sci. Sect., vol. 147 (1963), pp. 1285-6.

Novitski, E., Peacock, W. J. and Engel, J. 'Cytological Basis of "sex-ratio" in *Drosophila pseudoobscura*', Science, vol. 148 (1965), pp. 516-17.

Nur, U. 'Meiotic Parthenogenesis and Heterochromatization in a Soft Scale, *Pulvinaria hydrangae* (Coccoidea: Homoptera)', Chromosoma, vol. 14 (1963), pp. 123-39.

—, 'Parthenogenesis in Coccids (Homoptera)', Amer. Zool., vol. 11 (1971), pp. 301-8.

—, 'Diploid Arrhenotoky and Automictic Thelytoky in Soft Scale Insects (Lecaniinae: Coccoidea: Homoptera)', Chromosoma, vol. 39 (1972), pp. 381-401.

Nussbaum, M. 'Die Entstehung der Geschlechts bei *Hydatina*', Arch. Microscop. Anat., vol. 49 (1897), pp. 227-308.

Nyholm, K. 'Development and Larval Form of *Labidoplax buski*', Zool. Bidrag., vol. 29 (1921), pp. 239-54.

Oehler, D. Z., Schopf, J. W. and Kvenvolden, K. A. 'Carbon Isotopic Studies of Organic Matter in Precambrian Rocks', Science, vol. 175 (1972), pp. 1246-8.

Ohmachi, F. 'Facultative Parthenogenesis in *Loxoblemmus*', Proc. Imp. Acad. Japan, vol. 5 (1929), pp. 357-9.

Okada, Y. K. 'Regeneration and Fragmentation in the Syllidean Polychaetes', Wilhelm Roux Arch. Entw. Org., vol. 115 (1929), pp. 542-600.

—, 'Stolonization in *Myrianida*', J. Mar. Biol. Ass. UK, vol. 20 (1935), pp. 93-8.

Oldham, C. 'Auto-fecundation and Duration of Life in *Limax cinereo-niger*', Proc. Malacol. Soc. London, vol. 25 (1942), pp. 9-10.

Oliver, J. H. 'Parthenogenesis in Mites and Ticks (Arachnida: Acari)', Amer. Zool., vol. 11 (1971), pp. 283-99.

—, 'Cytogenetics of Mites and Ticks', Ann. Rev. Entomol., vol. 22 (1977), pp. 407-29.

Oliver, J. H. and Tanaka, K. 'Reproduction and Chromosomes of a Haemaphysalid Tick', Bull. Entomol. Soc. America, vol. 15 (1969), pp. 221.

Olofsson, P. O. 'Studien über die Süsswasser Fauna Spitzbergens', Zool. Bdg. Uppsala., vol. 6 (1913), pp. 183-646.

Olsen, M. W. 'Nine-year Summary of Parthenogenesis in Turkeys', Proc. Soc. Expt. Biol. Med., vol. 105 (1960), pp. 279-81.

—, 'Twelve-year Summary of Selection for Parthenogenesis in Beltsville Small White Turkeys', Brit. Poultry Sci., vol. 6 (1965), pp. 1-6.

Olsen, M. W. and Marsden, S. J. 'Natural Parthenogenesis in Turkey Eggs', Science (1954), pp. 545-6.

Omodeo, P. 'Cariologia dei Lumbricidae', Inst. Biol. Zool. Gen. Univ. Siena, vol. 4 (1952), pp. 173-275.

Ottaway, J. R. and Kirby, G. C. 'Genetic Relationships between Brooding and Brooded *Actinia tenebrosa*', Nature, vol. 255 (1975), pp. 221-3.

Otto, D. 'Zur Erscheinung der Arbeiterinnenfertilität und Parthenogenese bei der Kahlückigen Roten Waldameise (*Formica polyctena* Först) (Hym.)', Deutsch. Entomol. ZNF, vol. 7 (1960), pp. 1-9.

van den Ouden, H. 'A Note on Parthenogenesis and Sex Determination in *Heterodera rostochiensis*', Nematologica, vol. 5 (1960), pp. 215-16.

Owen, R. *Parthenogenesis; Or, the Successive Production of Procreating Individuals from a Single Ovum* (London, 1849).

Packard, A. 'Asexual Reproduction in *Balanoglossus*', Proc. Roy. Soc. London, Ser. B, vol. 171 (1968), pp. 261-72.

Packard, C. E. 'Observations on the Gastrotricha Indigenous to New Hampshire', Trans. Amer. Microscop. Soc., vol. 55 (1936), pp. 422-7.

Pagliai, A. 'L'Endomeiosi in *Toxoptera aurantiae* (Boyer de Foscolombe) (Homoptera, Aphididae)', Rend. Accad. Naz. Lincei, Ser. 8, vol. 31 (1961), pp. 455-7.

——, 'La maturazione dell'uovo partenogenetico e dell'uovo anfigonico in *Brevicoryne brassicae*', Caryologia, vol. 15 (1962), pp. 537-44.

Painter, R. H. 'Observations on the Biology of the Hessian Fly', J. Econ. Entomol., vol. 23 (1930), pp. 326-8.

Palmén, E. 'The Diplopoda of Eastern Fennoscandia', Ann. Zool. Fenn., vol. 13 (1961), pp. 1-54.

Panelius, S. 'Germ Line and Oogenesis during Paedogenetic Reproduction in *Heteropeza pygmaea* Winnertz (Diptera: Cecidomyidae)', Chromosoma, vol. 23 (1968), pp. 333-45.

——, 'Male Germ-line, Spermatogenesis and Karyotypes of *Heteropyza pygmaea* Winnertz (Diptera: Cecidomyidae)', Chromosoma, vol. 32 (1971), pp. 295-331.

Pantulu, J. V. and Manga, U. 'Cytogenetics of 16-chromosome Plants in Pearl Millet', Cytologia, vol. 37 (1972), pp. 389-94.

Paraense, W. L. 'One-sided Reproductive Isolation between Geographically Remote Populations of a Planorbid Snail', Amer. Nat., vol. 93 (1959), pp. 93-101.

Parias, P. and Basak, S. L. 'Genotypic Control of Chromosome Behaviour in *Corchorus capsularis* L.', Nucleus, vol. 16 (1973), pp. 210-15.

Park, H. D. 'Apparent Rhythmicity in Sexual Differentiation of *Hydra littoralis*' in H. M. Lenhoff and W. F. Loomis (eds.), *The Biology of Hydra* (University of Miami Press, Coral Gables, Florida, 1961), pp. 363-70.

Park, H. D., Mecca, C. and Ortmeyer, A. 'Sexual Differentiation in *Hydra* in Relation to Population Density', Nature, vol. 191 (1961), pp. 92-3.

Parker, B. C. 'On the Evolution of Isogamy to Oogamy', in B. C. Parker and R. M. Brown (eds.), *Contributions to Phycology* (privately published by the students of Harold C. Bold, 1971), pp. 47-52.

Parker, E. D. 'Phenotypic Consequences of Parthenogenesis in *Cnemidophorus* Lizards. I. Variability in Parthenogenetic and Sexual Populations', Evolution, vol. 33 (1979a), pp. 1150-66.

——, 'Ecological Implications of Clonal Diversity in Parthenogenetic Morphospecies', Amer. Zool., vol. 19 (1979b), pp. 753-62.

Parker, E. D. and Selander, R. K. 'The Organization of Genetic Diversity in the Parthenogenetic Lizard *Cnemidophorus tesselatus*', Genetics, vol. 84 (1976), pp. 791-805.

Parker, E. D., Selander, R. K. Hudson, R. O. and Lester, I. J. 'Genetic Diversity in Colonizing Parthenogenetic Cockroaches', Evolution, vol. 31 (1978), pp. 836-42.

Parker, G. A., Baker, R. R. and Smith, V. G. F. 'The Origin and Evolution of Gamete Dimorphism and the Male-female Phenomenon', J. Theoret. Biol., vol. 36 (1972), pp. 529-53.

Parker, G. H. 'The Effects of the Winter of 1917-1918 on the Occurrence of *Sagartia luciae* Verrill', Amer. Natur., vol. 53 (1919), pp. 280-1.

Parker, J. S. 'Chromosome-specific Control of Chiasma Formation', Chromosoma, vol. 49 (1975), pp. 391-406.

Parsons, P. A. 'Selection for Increased Recombination in *Drosophila melanogaster*', Amer. Natur., vol. 92 (1958), pp. 255-6.

Patil, A. M. 'The Occurrence of a Male of the Prosobranch *Potamopyrgus jenkinsoni* (Smith) var. *carinata* Marshall, in the Thames at Sonning, Berks.', Ann. Mag. Nat. Hist., vol. 13 (1958), pp. 232-40.

Patterson, J. T. 'Polyembryonic Development in *Tatusia novemcincta*', J. Morphol., vol. 24 (1913), pp. 559-684.

—, 'Functionless Males in Two Species of *Neuroterus*', Biol. Bull., vol. 54 (1928), pp. 196-200.

Peacock, A. D. and Sanderson, A. R. 'The Cytology of the Thelytokous Parthenogenetic Sawfly *Thrinax macula* Kl', Proc. Roy. Soc. Edinburgh, vol. 59 (1939), pp. 647-60.

Peacock, A. D. and Weidman, V. Przeglad Zool., vol. 5 (1961), pp. 5-27 (in Polish, English summary).

Pearman, J. V. 'Biological Observations on British Psocoptera', Entomol. Monthly Mag., vol. 64 (1928), pp. 263-8.

Pearse, A. S. 'Autotomy in Holothurians', Biol. Bull. Mar. Lab., Woods Hole, vol. 15 (1909), pp. 259-88.

Pearse, V. B. and Muscatine, L. 'Role of Symbiotic Algae (Zooxanthellae) in Coral Calcification', Biol. Bull., vol. 141 (1971), pp. 350-63.

Pehani, H. 'Die Geschlechtszellen der Phasmiden. Zugleich ein Beitrag zur Fortpflanzungsbiologie der Phasmiden', Z. Wiss. Zool., vol. 125 (1925), pp. 167-238.

Pelseneer, P. 'Mollusca' in E. Ray Lankester (ed.), *A Treatise on Zoology* (1906), part 5.

Pennak, R. W. *Fresh-water Invertebrates of the United States*, 2nd edn (Wiley-Interscience, New York, 1978).

Penney, J. T. 'Reduction and Regeneration in Fresh-water Sponges (*Spongilla discoides*)', J. Exp. Zool., vol. 65 (1933), pp. 475-97.

Penney, J. T. and Racek, A. A. 'Comprehensive Revision of a Worldwide Collection of Freshwater Sponges (Porifera: Spongillidae)', Bull. US Nat. Mus., vol. 272 (1968), pp. 1-184.

Pennock, L. A. 'Triploidy in Parthenogenetic Species of the Teiid Lizard, Genus *Cnemidophorus*', Science, vol. 149 (1965), pp. 539-40.

Perrot, J. L. 'La Spermatogénèse et l'Ovogénèse du Mallophage *Goniodes stylifer*', Quart. J. Microscop. Sci., vol. 76 (1934), pp. 353-77.

Perry, P. E. and Jones, G. H. 'Male and Female Meiosis in Grasshoppers. I. *Stethophyma grossum*', Chromosoma, vol. 47 (1974), pp. 227-36.

Petersen, H. 'Parthenogenesis in Two Common Species of Collembola: *Tullbergia krausbaueri* (Börner) and *Isotoma notabilis* Schäffer', Rev. Ecol. Biol. Sol., vol. 8 (1971), pp. 133-8.

Petersen, J. A. and Ditardi, A. S. F. 'Asexual Reproduction in *Glossobalanus crozieri* (Ptychoderidae, Enteropneusta, Hemichordata)', Mar. Biol., vol. 9 (1971), pp. 78-85.

Phillips, J. H. H. 'Biological and Behavioural Differences between *Lecanium cerasifex* Fitch and *Lecanium putmani* Phillips (Homoptera: Coccoidea)', Can. Entomol., vol. 97 (1965), pp. 303-9.

Pianka, H. D. 'Ctenophora' in A. C. Giese and J. S. Pearse (eds.), *Reproduction of Marine Invertebrates*, vol. 1 (Academic, New York, 1974), pp. 201-65.

Picchi, V. D. 'Parthenogenetic Reproduction in the Silverfish *Nicoletia meinerti* (Thysanura)', J. New York Entomol. Soc., vol. 80 (1972), pp. 2-4.

Pictet, A. 'Contribution à l'Étude de la Parthénogénèse chez les Lépidoptères', Bull. Soc. Lépidoptérologie Genève, vol. 5 (1924). Not seen; cited by P.-P. Grassé, *Traité de Zoologie*.

Pieronek, B. 'The Biology and Morphology of Larval *Fenusa dohrnii* Tischbein (Tenthredinidae, Hymenoptera)', Acta Zool. Cracov, vol. 18 (1973), pp. 41-72 (in Polish; English summary).

Pijnacker, L. P. 'The Maturation Divisions of the Parthenogenetic Stick Insect *Carausius morosus* Br. (Orthoptera, Phasmidae)', Chromosoma, vol. 19 (1966), pp. 99-112.

—, 'Oogenesis in the Parthenogenetic Stick Insect *Sipyloidea sipylus* Westwood (Orthoptera, Phasmidae)', Genetica, vol. 38 (1968), pp. 504-15.

—, 'Automictic Parthenogenesis in the Stick Insect *Bacillus rossius* Rossi (Cheleutoptera, Phasmidae)', Genetica, vol. 40 (1969), pp. 343-9.

Pijnacker, L. P. and Ferwerda, M. A. 'Maturation Divisions with Double the Somatic Chromosome Number in the Privet Mite *Brevipalpus obovatus*', Experientia, vol. 31 (1975), pp. 421-2.

Piza, S de T. 'Meiosis in the Male of the Brazilian Scorpion *Tityus bahiensis*', Rev. Agric. Sao Paulo, vol. 18 (1943), pp. 249-76. Not seen; cited by John and Lewis (1965).

—, 'Interessante Comportamento dos Cromossômios na Espermatogenese do Escorpiao *Isometrus maculatus* de Geer', An. Esc. Sup. Agric. L. de Queiroz, vol. 4 (1947), pp. 117-82.

Plough, H. H. 'The Effect of Temperature on Crossing-over in *Drosophila*', J. Exp. Zool., vol. 24 (1917), pp. 147-209.

Pollack, L. W. 'Reproductive Anatomy of Some Marine Heterotardigrada', Trans. Amer. Microscop. Soc., vol. 89 (1970), pp. 308-16.

Pomeyrol, R. 'La Parthénogénèse des Thysanoptères. La Maturation des Oeufs Parthénogénètiques chez l'*Heliothrips haemorrhoidalis*', Bull. Biol. France Belgique, vol. 62 (1928), pp. 1-20.

Poole, H. K., Healey, W. V., Russell, P. A. and Olsen, M. W. 'Evidence of Heterozygosity in Parthenogenetic Turkeys from Homograft Responses', Proc. Soc. Exp. Biol. Med., vol. 113 (1963), pp. 503.

Porter, D. L. 'Oogenesis and Chromosomal Heterozygosity in the Thelytokous Midge *Lundstroemia parthenogenetica* (Diptera, Chironomidae)', Chromosoma, vol. 32 (1971), pp. 333-42.

Possompès, B. Dévellopement Ovarien Après Ablation du Corpus Allatum Juvenile chez *Calliphora erythrocephala* Meig. (Diptére), et chez *Sipyloidea sipylus* W. (Phasmoptére)', Ann. Sci. Nat., Zool. Sér. 11, vol. 18 (1956), pp. 313-14.

Potts, E. A. 'Notes on the Free-living Nematodes. I. The Hermaphrodite Species', Quart. J. Microscop. Sci., vol. 55 (1910), pp. 433-84.

Potts, F. A. 'Methods of Reproduction in the Syllids', Ergeb. Fortschr. Zool., vol. 3 (1911), pp. 1-72.

—, 'Polychaeta from the Northeast Pacific. The Chaetopteridae. With an Account of the Phenomenon of Asexual Reproduction in *Phyllochaetopterus* and the Description of Two New Species of Chaetopteridae from the Atlantic', Proc. Zool. Soc. London (1914), pp. 955-94.

Poulsen, E. M. 'Ostracoda-Myodocopa. Part I. Cypridiniformes-Cypridinidae', Dana-report, vol. 57 (1962), pp. 1-414.

—, 'Ostracoda-Myodocopa. Part II. Cypridiniformes-Rutidermatidae, Sarsiellidae and Asteropidae', Dana-report, vol. 65 (1965), pp. 1-484.

Pourriot, R. 'Influence du Rythme Nycthéméral sur le Cycle Sexuel de Quelques Rotifères', CR Acad. Sci. Paris, vol. 256 (1963), pp. 5216-17.

Pourriot, R. and Clément, P. 'Influence de la Durée de l'Éclairement Quotidien sur le Taux de Femelles Mictiques chez *Notommata copeus* Ehr (rotifère)', Oecologia, vol. 22 (1975), pp. 67-77.

Prasad, S. K. and Webster, J. M. 'Effect of Temperature on the Rate of Development of *Naccobus serendipiticus* in Excised Tomato Roots', Nematologica, vol. 13 (1967), pp. 85-90.

Prehn, I. M. and Rasch, E. M. 'Cytogenetic Studies of *Poecilia* (Pisces). I. Chromosome Numbers of Naturally Occurring Poeciliid Species and Their Hybrids from Eastern Mexico', Can. J. Genet. Cytol., vol. 11 (1969), pp. 880-95.

Price, D. J. 'Variation in Chiasma Frequency in *Cepaea nemoralis*', Heredity, vol. 32 (1974), pp. 211-16.

Pringle, J. A. 'A Contribution to the Knowledge of *Micromalthus debilis* LeC. (Coleoptera)', Trans. Roy. Entomol. Soc. London, vol. 87 (1938), pp. 271-86.

Prior, R. N. B. and Stroyan, H. L. G. 'On a New Collection of Aphids from Iceland', Entomol. Medd., vol. 29 (1960), pp. 266-93.

Proctor, V. W. 'Genetics of Charophyta', in R. A. Lewin (ed.), *The Genetics of Algae* (Blackwell, Oxford, 1976), pp. 210-18.

Proszynska, M. 'The Annual Cycle in Occurrence of Cladocera and Copepoda in Small Water Bodies', Polski Arch. Hydrobiol., vol. 10, no. 23 (1962), pp. 379-422.

Prout, T., Bundgaard, J. and Bryant, S. 'Population Genetics of Modifiers of Meiotic Drive. I. The Solution of a Special Case and Some General Implications', Theoret. Pop. Biol., vol. 4 (1973), pp. 446-65.

Purchon, R. D. *The Biology of the Mollusca* (Pergamon, Oxford, 1968).

Putman, W. L. 'Life History and Behaviour of the Predaceous Mite *Typhlodromus (T.) caudiglans* Schuster (Acarina: Phytoseiidae) in Ontario, with Notes on the Prey of Related Species', Canad. Entomol., vol. 94 (1962), pp. 163-77.

Quick, H. E. 'Parthenogenesis in *Paludestrina jenkinsoni* from Brackish Water', J. Conchol., vol. 16 (1920), pp. 97-100.

Rajulu, G. S. 'Asexual Reproduction by Budding in the Sipuncula', in *Proceedings of an International Symposium on the Biology of Sipuncula and Echiuria* (Naueno Pelo Press, Belgrade, 1975).

Rajulu, G. S. and Krishnan, N. 'Occurrence of Asexual Reproduction by Budding in Sipunculida', Nature, vol. 223 (1969), pp. 187-8.

Rangnow, H. 'Parthenogenesis bei *Orgyia dubia*', Int. Entomol. Z., vol. 5 (1912).

Rantala, M. 'Sex Ratio and Periodomorphosis of *Proteroiulus fuscus* (Am Stein)', Symp. Zool. Soc. London, vol. 32 (1974), pp. 463-70.

Rasch, E. M., Darnell, R. M., Kallman, K. D. and Abramoff, P. 'Cytophotometric Evidence for Triploidy in Hybrids of the Gynogenetic fish, *Poecilia formosa*', J. Exp. Zool., vol. 160 (1965), pp. 155-70.

Rasch, E. M., Prehn, L. M. and Rasch, R. W. 'Cytogenetic Studies of *Poecilia* (Pisces). II. Triploidy and DNA Levels in Naturally Occurring Populations Associated with the Gynogenetic Teleost, *Poecilia formosa* (Girard)', Chromosoma, vol. 31 (1970), pp. 18-40.

Rasmont, R. 'Gemmulation in Freshwater Sponges', in D. Rudnick (ed.), *Regeneration, 20th Growth Symposium* (Ronald Press, New York, 1962), pp. 3-25.

——, 'Some New Aspects of the Physiology of Freshwater Sponges', Symp. Zool. Soc. London, vol. 25 (1970), pp. 415-22.

Rasmussen, E. 'Asexual Reproduction in *Pygospio elegans* Clap. (Polychaeta Sedentaria)', Nature, vol. 171 (1953), pp. 1161-2.

Rattenbury, J. 'Reproduction in *Phoronopsis viridis*. The Annual Cycle in the Gonads, Maturation and Fertilization of the Ovum', Biol. Bull., vol. 104 (1953), pp. 182-96.

Reed, E. B. 'Records of Freshwater Crustacea from Arctic and Subarctic Canada', Nat. Mus. Canada Bull., vol. 199 (1964), pp. 29-59.

Rees, H. and Thompson, J. B. 'Genotypic Control of Chromosome Behaviour in Rye. III. Chiasma Frequency in Homozygotes and Heterozygotes', Heredity, vol. 10 (1956), pp. 409-24.

Reeve, M. R. and Cosper, T. C. 'Chaetognatha' in A. C. Giese and J. S. Pearse (eds.), *Reproduction of Marine Invertebrates*, vol. 2 (Academic, New York, 1975), pp. 157-83.

Reeve, M. R. and Walter, M. A. 'Observations and Experiments on Methods of Fertilization in the Chaetognath *Sagitta cuspida*', Biol. Bull., vol. 143 (1972), pp. 207-14.

Reinboth, R. 'Morphologische und funktionelle zweigeschlechtlichkeit bei marinen Teleostiern (Serranidae, Sparidae, Centracanthidae, Labridae)', Zool. Jahrb. Physiol., vol. 69 (1962), pp. 405-80.

Reisa, J. J. 'Ecology', in A. L. Burnett (ed.), *Biology of Hydra* (Academic, New York, 1973), pp. 59-105.

Reisinger, E. 'Die Süsswassermeduse *Craspedacusta sowerbii* Lankester und Ihr Vorkommen in Flussgebiet von Rhein und Maas', Natur am Niederrhein, vol. 10 (1934), pp. 33-43.

——, 'Die Cytologische Grundlage der Parthenogenetischen Dioogonie', Chromosoma, vol. 1 (1940), pp. 531-53.

——, 'Zur Entwicklungsgeschichte und Entwicklungsmechanik von *Craspedacusta* (Hydrozoa, Limnotrachylina)', Z. Morph. Ökol. Tiere, vol. 45 (1957), pp. 656-98.

Reitberger, A. 'Die Cytologie des Pädogenetischen Entwicklungszyklus der Gallmücke *Oligarces paradoxus* Mein', Chromosoma, vol. 1 (1939), pp. 391-473.

Remane, A. 'Beiträge zur Systematik von Süsswasser-gastrotrichen', Zool. Jahrb. (Abt. Systematik), vol. 53 (1927), pp. 269-320.

Reynolds, J. W. 'The Earthworms of Maryland, Oligochaeta: Acanthodrilidae, Lumbricidae, Megascolecidae and Sparganophilidae', Megadrilogica, vol. 1, no. 11 (1974), pp. 1-12.

Reynoldson, T. B. 'Environment and Reproduction in Freshwater Triclads', Nature, vol. 189 (1961), pp. 329-30.

Rhein, A. 'Diploide Parthenogenese bei *Hydrobia jenkinsoni* Smith', Naturwissenschaften, vol. 23 (1935), p. 100.

Ribaga, C. 'La Partenogenesi nei Copeognathi', Redia, vol. 11 (1905), pp. 33-6.

Rice, M. E. 'Asexual Reproduction in a Sipunuculan Worm', Science, vol. 167 (1970), pp. 1618-20.

——, 'Sipuncula', in A. C. Giese and J. S. Pearse (eds.), *Reproduction of Marine Invertebrates*, vol. 2 (Academic, New York, 1975), pp. 67-127.

Richards, C. S. and Ferguson, F. F. '*Plesiophysa hubendicki*, a New Puerto Rican Planorbid Snail', Trans. Amer. Microscop. Soc., vol. 81 (1962), pp. 251-6.

Richards, W. R. 'The Aphididae of the Canadian Arctic', Can. Entomol., vol. 95 (1963), pp. 449-64.

——, 'The Scale Insects of the Canadian Arctic', Can. Entomol., vol. 97 (1965), pp. 143-76.

Rifaat, O. M. 'Effect of Temperature on Crossing-over in *Neurospora crassa*', Genetica, vol. 30 (1959), pp. 312-23.

Riley, R. 'Genetics and Regulation of Meiotic Chromosome Behaviour', Sci. Progress, vol. 54 (1966), pp. 193-207.

Riley, R. and Law, C. N. 'Genetic Variation in Chromosome Pairing', Adv. Genet., vol. 13 (1965), pp. 57-114.

Riser, N. W. 'Nemertinea', in A. C. Giese and J. S. Pearse (eds.), *Reproduction of Marine Invertebrates*, vol. 1 (Academic, New York, 1974), pp. 359-89.

Risler, H. and Kempter, E. 'Die Haploidie der Mannchen und die Endopolyploidie in Einigen Geweben von *Haplothrips*', Chromosoma, vol. 12 (1962), pp. 351-61.

Robertson, A. 'Embryology and Embryonic Fission in *Crisia*', Univ. California Publ. Zool., no. 1 (1903).

Robertson, J. G. 'The Chromosomes of Bisexual and Parthenogenetic Species of *Calligrapha* (Coleoptera: Chrysomelidae) with Notes on Sex Ratio, Abundance and Egg Number', Can. J. Genet. Cytol., vol. 8 (1966), pp. 695-732.

Robertson, R. B. 'Chromosome Studies. V. Diploidy and Persistent Relations in Partheno-produced Tettigidae (*Apotettix eurycephalus* and *Paratettix texanus*)', J. Morphol. a. Physiol. (Am.), vol. 50 (1930), pp. 209-57.

—, 'On the origin of partheno-produced males in Tettigidae (*Apotettix* and *Paratettix*)', Genetics, vol. 16 (1931), pp. 353-6.

Robotti, C. 'Chromosome Complement and Male Haploidy of *Asplanchna priodonta* Gosse 1850 (Rotatoria)', Experientia, vol. 31 (1975), pp. 1270-2.

Rogers, W. A. and Ulmer, M. J. 'Effects of Continued Selfing on *Hymenolepis nana* (Cestoda)', Iowa Acad. Sci., vol. 69 (1962), pp. 557-71.

Ross, E. S. 'A Revision of the Embioptera, or Web-spinners, of the New World', US Nat. Mus., vol. 94 (1944), pp. 401-504.

—, 'Parthenogenetic African Embioptera', Wasmann J. Biol., vol. 18 (1961), pp. 297-304.

Rossi, L. 'Sexual Races in *Cereus pedunculatus* (Boad.)', Pubbl. Staz. Zool. Napoli, vol. 39 (1975), pp. 462-70.

Roth, L. M. 'Sexual Isolation in Parthenogenetic *Pycnoscelus surinamensis* and Application of the Name *Pycnoscelus indicus* to Its Bisexual Relative (Dictyoptera: Blattaria: Blaberidae: Pycnoscelinae)', Ann. Entomol. Soc. Amer., vol. 60 (1967), pp. 774-9.

—, 'Evolution and Taxonomic Significance of Reproduction in Blattaria', Ann. Rev. Entomol., vol. 15 (1970), pp. 75-96.

—, 'Reproductive Potential of Bisexual *Pycnoscelus indicus* and Clones of Its Parthenogenetic Relative *Pycnoscelus surinamensis*', Ann. Ent. Soc. Amer., vol. 67 (1974), pp. 215-23.

Roth, L. M. and Cohen, S. H. 'Chromosomes of the *Pycnoscelus indicus* and *P. surinamensis* Complex', Psyche, vol. 75 (1968), pp. 54-76.

Roth, L. M. and Willis, E. R. 'Parthenogenesis in Cockroaches', Ann. Entomol. Soc. Amer., vol. 49 (1956), pp. 195-204.

—, 'A Study of Bisexual and Parthenogenetic Strains of *Pycnoscelus surinamensis* (Blattaria: Epilamprinae)', Ann. Ent. Soc. Amer., vol. 54 (1961), pp. 12-25.

Roth, R. and Fogel, S. 'A System Selective for Yeast Mutants Deficient in Meiotic Recombination', Molec. Gen. Genetic, vol. 112 (1971), pp. 295-305.

Rowe, H. J. and Westerman, M. 'Population Cytology of the Genus *Phaulacridium*. I. *Phaulacridium vittatum* (Sjost): Australian Mainland Populations', Chromosoma, vol. 46 (1974), pp. 197-205.

Rowlands, D. G. 'The Control of Chiasma Frequency in *Vicia faba* L.', Chromosoma, vol. 9 (1958), pp. 176-84.

Russell, F. S. *The Medusae of the British Isles*, vol. 1, *Anthomedusae, Leptomedusae, Limnomedusae, Trachymedusae, and Narcomedusae* (Cambridge University Press, 1953).

——, *The Medusae of the British Isles. II. Pelagic Scyphozoa, with a Supplement to the First Volume on Hydromedusae* (Cambridge University Press, 1970).

Russell, F. S. and Rees, W. J. 'A Viviparous Scyphomedusa, *Stygiomedua fabulosa* Russell', J. Mar. Biol. Ass. UK, vol. 39 (1960), pp. 303-17.

Ruttner-Kolisko, A. 'Das Zooplankton der Binnengewasser, I. Rotatoria', Die Binnengewasser, vol. 26 (1972), pp. 99-234.

——, 'Amphoteric Reproduction in a Population of *Asplanchna priodonta*', Ann. Hydrobiol. Beih. Ergebn. Limnol., vol. 8 (1977), pp. 178-81.

Ryland, J. S. *Bryozoans* (Hutchinson, London, 1970).

Ryland, J. S. and Austin, A. P. 'Three Species of Kamptozoa New to Britain', Proc. Zool. Soc. London, vol. 133 (1960), pp. 423-33.

Sabbadin, A. 'Self- and Cross-fertilization in the Compound Ascidian, *Botryllus schlosseri*', Develop. Biol., vol. 24 (1971), pp. 379-91.

Sacks, M. 'Observations on the Embryology of an Aquatic Gastrotrich *Lepidodermella squammata* (Dujardin, 1841)', J. Morphol., vol. 96 (1955), pp. 473-95.

——, 'Life History of an Aquatic Gastrotrich', Trans. Amer. Microscop. Soc., vol. 83 (1964), pp. 358-62.

Salmon, J. T. 'The Genus *Acanthoxyla* (Phasmidae)', Trans. Roy. Soc. New Zealand (Zoology), vol. 82 (1955), pp. 1149-56.

Sandberg, P. A. 'Degree of Individuality in Cheilostome Bryozoa: Skeletal Criteria', in R. S. Boardman, A. H. Cheetham and W. A. Oliver (eds.), *Animal Colonies* (Dowden Hutchingson and Ross, Stroudsberg, Pa., 1974), pp. 305-15.

Sanderson, A. R. 'The Cytology of Parthenogenesis in the Snail *Potamopyrgus jenkinsi* Smith', Adv. Sci., vol. 1 (1939), p. 46.

——, 'Maturation in the Parthenogenetic Snail *Potamopyrgus jenkinsi* (Smith) and in the Snail *Peringia ulvae* (Pennant)', Proc. Zool. Soc. London, vol. 110 (1940), pp. 11-15.

——, 'Maturation in the Parthenogenetic Weevil, *Listroderes costirostris* Schonh. (*obliquus* Gyll)', Proc. 14th Int. Congr. Zool. (1953), pp. 185-6.

——, 'Maturation in the Parthenogenetic Weevil, *Listroderes costirostris* Schonh. (*obliquus* Gyll.)', in *Proceedings of Fourteenth International Congress on Zoology* (Copenhagen, 1956).

——, 'The Cytology of a Diploid Bisexual Spider Beetle, *Ptinius clavipes* Panzer, and Its Triploid Gynogenetic Form *mobilis*', Proc. Roy. Soc. Edinburgh, Ser. B, vol. 67 (1960), pp. 333-50.

——, 'The Cytology of the Parthenogenetic Australian Weevil *Listroderes costirostris* Schönh.', Trans, Roy. Soc. Edinburgh, vol. 69 (1973), pp. 71-89.

Sars, G. O. 'An Account of the Crustacea of Norway', Bergen Mus. (1928), pp. 1-277.

Saura, A., Lokki, J., Lankinen, P. and Suomalainen, E. 'Genetic Polymorphism and Evolution in Parthenogenetic Animals. III. Tetraploid *Otiorhynchus scaber* (Coleoptera: Curculionidae)', Hereditas, vol. 82 (1976a), pp. 79-100.

—, 'Genetic Polymorphism and Evolution in Parthenogenetic Animals. IV. Triploid *Otiorrhynchus salicis* Ström (Coleoptera: Curculionidae)', Ent. Scand., vol. 7 (1976b), pp. 1-6.

Sax, K. 'Variation in Chiasma Frequencies in *Secale*, *Vicia* and *Tradescantia*', Cytologia, vol. 6 (1935), pp. 289-93.

Scali, V. 'Obligatory Parthenogenesis in the Stick Insect *Bacillus rossius* (Rossi)', Acc. Naz. Lincei Rend. Sci., vol. 49 (1970), pp. 307-14.

—, 'La Citologia della Partenogenesi de *Bacillus rossius*', Boll. Zool., vol. 39 (1972), pp. 567-73.

Van der Schalie, H. 'Observations on the Sex of *Campeloma* (Gastropoda: Viviparidae)', Occ. Pap. Mus. Zool. Univ. Michigan, vol. 641 (1965), pp. 1-9.

Schall, J. J. 'Reproductive Strategies in Sympatric Whiptail Lizards (*Cnemidophorus*)', Copeia (1978), pp. 108-16.

Scharff, R. 'On *Ctenodrilus parvulus*, n.sp.', Quart. J. Microscop. Sci., vol. 27 (1887), pp. 591-604.

Scheller, U. 'The Pauropoda of Ceylon', Ent. Scand., Supp. 1 (1970), pp. 5-97.

Schimmer, F. 'Beitrag zu einer Monographie der Gryllodengattung *Myrmecophila* Latr.', Z. Wiss. Zool., vol. 93 (1909), pp. 409-534.

Schleip, W. 'Die Reifung des Eies von *Rhodites rosae* L. und einige allgemeine Bemerkungen uber die Chromosomen bei parthenogenetischer Fortpflanzung', Zool. Anz., vol. 35 (1909), pp. 203-13.

Schmid, F. 'Contribution à l'Étude de la Sous-famille des Apataniinae (Trichoptera, Limnophilidae). II', Tijdschr. Entomol., vol. 97 (1954), pp. 1-74.

—, 'Ergebnisse der zoologischen Forschungen von Dr. Z. Kaszab in der Mongolei. 63. Trichoptera', Reichenbachia, Dresden, vol. 7 (1965), pp. 201-3.

Schmidt, H. 'A Note on the Sea Anemone *Bunodactis verrucosa* Pennant', Pubbl. Staz. Zool. Napoli, vol. 35 (1967), pp. 252-3.

—, '*Anthopleura stellula* (Actinaria, Actiniidae) and Its Reproduction by Transverse Fission', Mar. Biol., vol. 5 (1970), pp. 245-55.

Schneider, H. 'Vergleichende Untersuchungen über Parthenogenese und Entwicklungsrhythmen bei Einheimischen Psocopteren', Biol. Zentralblatt, vol. 74 (1955), pp. 273-310.

Scholl, H. 'Ein Beitrag zur Kenntnis der Spermatogenese der Mallophagen', Chromosoma, vol. 7 (1955), pp. 271-4.

—, 'Die Chromosomen Parthenogenetischer Mücken', Naturwiss., vol. 43 (1956), pp. 91-2.

—, 'Die Oogenese Einiger Parthenogenetischer Orthocladiinen (Diptera)', Chromosoma, vol. 11 (1960), pp. 380-401.

Schopf, J. W. 'PreCambrian Microorganisms and Evolutionary Events Prior to the Evolution of Vascular Plants', Biol. Rev. Cambridge Phil. Soc., vol. 45 (1970), pp. 319-52.

—, 'Are the Oldest, "Fossils", Fossils?' Origins of Life, vol. 7 (1976), pp. 19-36.

Schopf, J. W. and Barghoorn, E. S. 'Alga-like Fossils from the Early Precambrian of South Africa', Science, vol. 156 (1967), pp. 508-12.

Schopf, J. W. and Fairchild, T. R. 'Late Precambrian Microfossils: A New Stromatolitic Biota from Boorthanna, South Australia', Nature, vol. 242 (1973), pp. 537-8.

Schopf, J. W. and Oehler, D. Z. 'How Old are the Eukaryotes?', Science, vol. 193 (1976), pp. 47-9.

Schopf, T. J. M. 'Ergonomics of Polymorphism: Its Relation to the Colony as the Unit of Natural Selection in Species of the Phylum Ectoprocta', in R. S. Boardman, A. H. Sheetham and W. A. Oliver (eds.), *Animal Colonies* (Dowden, Hutchinson and Ross, Stroudsberg, Pa., 1974a), pp. 247-94.

—, 'Survey of Genetic Differentiation in a Coastal Zone Invertebrate: The Ectoproct *Schizoporella errata*', Biol. Bull., vol. 146 (1974b), pp. 78-87.

Schrader, F. 'Sex Determination in the White Fly (*Trialeurodes vaporariorum*)', J. Morph., vol. 34 (1920), pp. 267-305.

—, 'The Cytology of Pseudosexual Eggs in a Species of *Daphnia*', Z. Indukt. Abstamm. Vererb., vol. 40 (1925), pp. 1-36.

—, 'Notes on the English and American Races of the Greenhouse White Fly (*Trialeurodes vaporariorum*)', Ann. Appl. Biol., vol. 13 (1926), pp. 189-96.

—, 'Notes on Reproduction in *Aspidiotus hederae* (Coccidae)', Psyche, vol. 36 (1929), pp. 232-6.

Schrader, F. and Hughes-Schrader, S. 'Haploidy in *Icerya purchasi*', Z. Wiss. Zool., vol. 128 (1926), pp. 182-200.

—, 'Haploidy in Metazoa', Quart. Rev. Biol., vol. 6 (1931), pp. 411-38.

Schroeder, L. 'Population Growth Efficiencies of Laboratory *Hydra pseudo-oligactis* Hyman Populations', Ecology, vol. 50 (1969), pp. 81-6.

Schroeder, P. C. and Hermans, C. O. 'Annelida: Polychaeta' in A. C. Giese and J. S. Pearse (eds.), *Reproduction of Marine Invertebrates*, vol. 3 (Academic, New York, 1975), pp. 1-213.

Schroeter, D. L. 'Pericentric Inversion Polymorphism in *Trimerotropis helferi* (Orthoptera: Acrididae) and Its Effect on Chiasma Frequency', PhD Thesis (University of California at Davis, 1968). Cited by Weissman (1976).

Schultz, E. '*Dinophilus rostratus* (nov. spec.)', Wiss. Meeresunters. Komm. Helgoland, vol. 5 (1902), pp. 1-10.

Schultz, R. J. 'Reproductive Mechanisms in Unisexual and Bisexual Strains of the Viviparous Fish *Poeciliopsis*', Evolution, vol. 15 (1961), pp. 302-25.

—, 'Hybridization Experiments with an All-female Fish of the Genus *Poeciliopsis*', Biol. Bull., vol. 130 (1966), pp. 415-29.

—, 'Gynogenesis and Triploidy in the Viviparous Fish *Poeciliopsis*', Science, vol. 157 (1967), pp. 1564-7.

—, 'Hybridization, Unisexuality and Polyploidy in the Teleost *Poeciliopsis* (Poeciliidae) and Other Vertebrates', Amer. Natur., vol. 103 (1969), pp. 605-19.

—, 'Special Adaptive Problems Associated with Unisexual Fishes', Amer. Zool., vol. 11 (1971), pp. 351-60.

——, 'Unisexual Fish: Laboratory Synthesis of a "Species", Science, vol. 179 (1973), pp. 180-1.

Schultz, J. R. and Kallman, K. D. 'Triploid Hybrids between the All Female Teleost *Poecilia formosa* and *P. sphenops'*, Nature, vol. 219 (1968), pp. 280-2.

Scott, A. C. 'Haploidy and Aberrant Spermatogenesis in a Coleopteran, *Micromalthus debilis* Le Conte', J. Morph., vol. 59 (1936), pp. 485-515.

——, 'Paedogenesis in the Coleoptera', Z. Morphol. Ökol. Tiere, vol. 33 (1938), pp. 633-53.

Scott, J. W. 'Morphology of the Parthenogenetic Development of *Amphitrite'*, J. Exp. Zool., vol. 3 (1906), pp. 49-98.

Scudo, F. M. 'The Adaptive Value of Sexual Dimorphism. I. Anisogamy', Evolution, vol. 21 (1967), pp. 285-91.

Sebastyén, O. 'Cladocera Studies in Lake Balaton. II. Littoral Cladocera from the North Western Shores of the Tihany Peninsula', Arch. Biol. Hung., vol. 18 (1948), pp. 101-16.

Seetharam, A. and Srinivasachar, D. 'Cytomorphological Studies in the Genus *Linum'*, Cytologia, vol. 37 (1972), pp. 661-71.

Séguy, E. *La Biologie des Diptères* (Lechevalier, Paris, 1950).

Seiler, J. 'Geschlechtschromosomenuntersuchungen an Psychiden. IV. Die Parthenogenese der Psychiden', Z. Indukt. Abstam. Vererb., vol. 31 (1923), pp. 1-99.

——, 'Ergebnisse aus der Kreuzung parthenogenetische und zweigeschlechtlicher Schmetterlinge', Biol. Zentralblatt, vol. 47 (1927), pp. 426-46.

——, 'Neue Ergebnisse aus der Kreuzung parthenogenetischer Schmetterlinge mit Mannchen zweigeschlechtlicher Rassen', Verh. Deutsch Zool. Ges., vol. 38 (1936), pp. 147-50.

——, 'Uber den Ursprung der Parthenogenese und Polyploidie bei Schmetterlingen', Arch. Julius Klaus-Stift, vol. 18 (1943), pp. 691-9.

——, 'Die Verbreitungsgebiete der verschiedenen Rassen von *Solenobia triquetrella* (Psychidae) in der Schweiz', Rev. Suisse Zool., vol. 53 (1946), pp. 529-33.

——, 'Die Zytologie eines Parthenogenetischen Rüsselkäfers, *Otiorrhynchus sulcatus* F.', Chromosoma, vol. 3 (1947), pp. 88-109.

——, 'Untersuchungen über die Entstehung der Parthenogenese bei *Solenobia triquetrella* F.R. (Lepidoptera, Psychidae). I. Die Zytologie der Bisexuellen *S. triquetrella*, Ihr Verhalten und Ihr Sexualverhältnis', Chromosoma, vol. 10 (1959), pp. 73-114.

——, 'Untersuchungen über die Entstehung der Parthenogenese bei *Solenobia triquetrella* F.R. (Lepidoptera, Psychidae). III. Die Geographische Verbreitung der Drei Rassen und in Angrenzenden Ländern und die Beziehungen zur Eiszeit. Bemerkungen über die Entstehung der Parthenogenese', Z. Vererbungslehre, vol. 92 (1961), pp. 261-316.

——, 'Untersuchungen über die Entstehung der Parthenogenese bei *Solenobia triquetrella* F.R. (Lepidoptera, Psychidae). IV. Wie Besamen Begattete

Diploid und Tetraploid Parthenogenetische Weibchen von *Solenobia trique-trella* Ihre Eier? Schicksal der Richtungskorper im Besamten und Unbesamten Ei', Z. Vererbungslehre, vol. 94 (1963), pp. 29-66.

——, 'Untersuchungen über die Entstehung der Parthenogenese bei *Solenobia triquetrella* F.R. (Lepidoptera, Psychidae). V. Biologische und Zytologische Beobachtungen zum Übergang von der Diploiden zur Tetraploiden Parthenogenese', Chromosoma, vol. 15 (1964), pp. 503-39.

Seiler, J. and Schaffer, K. 'Untersuchungen über die Enstehung der Parthenogenese bei *Solenobia triquetrella* F R (Lepidoptera, Psychidae). II. Analyse der Diploid Parthenogenetischen *S. triquetrella*. Verhalten, Aufzuchtresulte und Zytologie', Chromosoma, vol. 11 (1960), pp. 29-102.

Sekera, E. 'Ueber die Verbreitung der Selbstbefruchtung bei den Rhabdocoeliden', Zoo. Anz., vol. 30 (1906), pp. 142-53.

Selander, R. K. and Hudson, R. O. 'Animal Population Structure under Close Inbreeding: The Land Snail *Rumina* in Southern France', Amer. Natur., vol. 110 (1976), pp. 695-718.

Selander, R. K. and Kaufman, D. W. 'Self-fertilization and Genetic Population Structure in a Colonizing Land Snail', Proc. Nat. Acad. Sci. US, vol. 70 (1973), pp. 1186-90.

Selander, R. K., Parker, E. D. and Browne, R. A. 'Clonal Variation in the Parthenogenetic Snail *Campeloma decisa*', Veliger, vol. 20 (1977), pp. 349-51.

Sethi, H. L. and Swenson, K. G. 'Formation of Sexuparae in the Aphid *Eriosoma pyricola* on Pear Roots', Entomol. Exp. Appl., vol. 10 (1967), pp. 97-102.

Sharpe, R. W. 'Further Report on the Ostracoda of the United States National Museum', Proc. US Nat. Mus., vol. 35 (1908), pp. 399-430.

——, 'The Ostracoda', in H. B. Ward and G. C. Whipple (eds.), *Freshwater Biology* (Wiley, New York, 1918), pp. 790-827.

Shaw, D. D. 'Genetic and Environmental Components of Chiasma Control. II. The Response to Selection in *Schistocerca*', Chromosoma, vol. 37 (1972), pp. 297-308.

——, 'Genetic and Environmental Components of Chiasma Control. III. Genetic Analysis of Chiasma Frequency Variation in Two Selected Lines of *Schistocerca gregaria* Forsk', Chromosoma, vol. 46 (1974), pp. 365-74.

Shaw, R. F. 'The Theoretical Genetics of the Sex Ratio', Genetics, vol. 43 (1958), pp. 149-63.

Shick, J. M. 'Ecological Physiology and Genetics of the Colonizing Actinian *Haliplanella luciae*', in G. O. Mackie (ed.), *Coelenterate Ecology and Behaviour* (Plenum, New York, 1976), pp. 137-46.

Shick, J. M. and Lamb, A. N. 'Asexual Reproduction and Genetic Population Structure in the Colonizing Sea Anemone *Haliplanella luciae*', Biol. Bull., vol. 153 (1977), pp. 604-17.

Shick, J. M., Hoffmann, R. J. and Lamb, A. N. 'Asexual Reproduction, Population Structure and Genotype-environment Interactions in Sea-anemones', Amer. Zool., vol. 19 (1979), pp. 699-713.

Shull, A. F. 'Biology of the Thysanoptera. II. Sex and the Life Cycle', Amer. Natur., vol. 48 (1914), pp. 236-47.

—, 'Duration of Light and the Wings of the Aphid *Macrosiphum solanifolii*', Arch. Entwicklungsmech. Organ., vol. 113 (1928), pp. 210-39.

—, 'The Effect of Intensity and Duration of Light and of Duration of Darkness, Partly Modified by Temperature, upon Wing Production in Aphids', Arch. Entwicklungsmech. Organ., vol. 115 (1929), pp. 825-51.

von Siebold, C. T. *Beitrage zur Parthenogenesis der Arthropoden* (Leipzig, 1871).

Silén, L. 'Automized Tentacle Crowns as Reproductive Bodies in *Phoronis*', Acta Zoologica, vol. 36 (1955), pp. 159-65.

—, 'On the Fertilization Problem in the Gymnolaematous Bryozoa', Ophelia, vol. 3 (1966), pp. 113-40.

Simchen, G. 'Genetic Control of Recombination and the Incompatibility System in *Schizophyllum commune*', Genet. Res., vol. 9 (1967), pp. 195-210.

Simchen, G. and Stamberg, J. 'Fine and Coarse Controls of Genetic Recombination', Nature, vol. 222 (1969), pp. 329-32.

Simpson, T. L. and Fell, P. E. 'Dormancy in the Porifera: Gemmule Formation and Germination in Fresh-water and Marine Sponges', Trans. Amer. Microscop. Soc., vol. 93 (1974), pp. 544-77.

Simpson, T. L. and Gilbert, J. J. 'Gemmulation, Gemmule Hatching and Sexual Reproduction in Fresh-water Sponges. I. The Life Cycle of *Spongilla lacustris* and *Tubella pennsylvanica*', Trans. Amer. Microscop. Soc., vol. 92 (1973), pp. 422-33.

—, 'Gemmulation, Gemmule Hatching, and Sexual Reproduction in Freshwater Sponges. II. Life Cycle Events in Young, Larva-produced Sponges of *Spongilla lacustris* and an Unidentified Species', Trans. Amer. Microscop. Soc., vol. 93 (1974), pp. 39-45.

Simroth, H. 'Anatomie and Schizogonie der *Ophiactis virens* Sars.', Z. Wiss. Zool., vol. 28 (1877), pp. 419-526.

Sinha, S. S. N. and Roy, H. 'Cytological Studies in the Genus *Phaseolus*. II. Meiotic Analysis of Sixteen Species', Cytologia, vol. 44 (1979), pp. 201-9.

Sivaramakrishnan, V. R. 'Early Development and Regeneration in Indian Marine Sponges', Proc. Indian Acad. Sci., vol. 34 (1951), pp. 273-310.

Sjödin, J. 'Induced Asynaptic Mutants in *Vicia faba* L.', Hereditas, vol. 66 (1970), pp. 215-32.

Slatkin, M. 'Gene Flow and Selection in a Two-Locus System', Genetics, vol. 81 (1975), pp. 787-802.

Slobodchikoff, C. N. and Daly, H. V. 'Systematic and Evolutionary Implications of Parthenogenesis in the Hymenoptera', Amer. Zool., vol. 2 (1971), pp. 273-82.

Slobodkin, L. B. 'Population Dynamics in *Daphnia obtusa* Kurtz.', Ecol. Monogr., vol. 24 (1954), pp. 69-88.

—, 'Experimental Populations of Hydrida', J. Anim. Ecol., vol. 33 (suppl.) (1964), pp. 69-101.

Smith, D. A. 'A Mutant Affecting Meiosis in *Neurospora*', Genetics, vol. 80 (1975), pp. 125-33.

Smith, M. A. *The Fauna of British India including Ceylon and Burma*, vol. 2, *Sauria* (Taylor and Francis, London, 1935).

Smith, M. Y. and Fraser, A. 'Polymorphism in a Cyclic Parthenogenetic Species: *Simocephalus serrulatus*', Genetics vol. 84 (1976), pp. 631-7.

Smith, N. and Lenhoff, H. M. 'Regulation of Frequency of Pedal Laceration in a Sea Anemone', in G. O. Mackie (ed.), *Coelenterate Ecology and Behaviour* (Plenum, New York, 1976), pp. 117-25.

Smith, R. I. 'Embryonic Development in the Viviparous Nereid Polychaete, *Neanthes lighti* Hartman', J. Morphol., vol. 87 (1950), pp. 417-66.

—, 'On Reproductive Pattern as a Specific Characteristic among Nereid Polychaetes', Syst. Zool., vol. 7 (1958), pp. 60-73.

Smith, S. G. 'Thelytokous Parthenogenesis in *Cephus cinctus*: A Criticism', Can. Entomol., vol. 70 (1938), pp. 259-60.

—, 'A New Form of Spruce Sawfly Identified by Means of Its Cytology and Parthenogenesis', Sci. Agric., vol. 21 (1941), pp. 243-305.

—, 'Cytogenetics of Obligatory Parthenogenesis', Can. Entomol., vol. 87 (1955), pp. 131-5.

—, 'Cytogenetic Pathways in Beetle Speciation', Can. Entomol., vol. 94 (1962), pp. 941-55.

—, 'Parthenogenesis and Polyploidy in Beetles', Amer. Zool., vol. 11 (1971), pp. 341-9.

Smith, T. L. 'Genetical Studies in the Wax Moth *Galleria mellonella* Linn.', Genetics, vol. 23 (1938), pp. 115-37.

Smithers, C. N. 'On a Small Collection of Psocoptera from Britain', Entomol. Monthly Mag., vol. 105 (1969), p. 54.

Snell, T. W. 'Lifespan of Male Rotifers', Arch. Hydrobiol. Beih. Ergebn. Limnol., vol. 8 (1977), pp. 65-6.

Snell, T. W. and King, C. E. 'Amphoteric Reproduction in *Asplanchna girodi*', Arch. Hydrobiol. Beih. Ergebn. Limnol., vol. 8 (1977), pp. 182-3.

Snyder, T. P. 'Lack of Allozymic Variability in Three Bee Species', Evolution, vol. 28 (1974), pp. 687-9.

Solbrig, O. 'Chromosomal Cytology and Evolution in the Family Compositae', in V. H. Heywood and B. L. Turner (eds.), *The Biology and Chemistry of the Compositae* (Academic, New York, 1976), pp. 267-81.

Sommermann, K. M. 'Description and Bionomics of *Caecilius manteri* n. sp. (Corrodentia)', Proc. Entomol. Soc. Washington, vol. 45 (1943), pp. 29-39.

Soulié, J. 'Des Considérations Écologiques Peuvent-elles Apporter une Contribution à la Connaissance du Cycle Biologique des Colonies de *Crematogaster* (Hymenoptera-Formicoidea)', Insectes Sociaux, vol. 7 (1960), pp. 283-95.

Southward, E. C. 'Pogonophora', in A. C. Giese and J. S. Pearse (eds.), *Reproduction of Marine Invertebrates*, vol. 2 (Academic, New York, 1975), pp. 129-55.

Spangenberg, D. B. 'Thyroxine-induced Metamorphosis in *Aurelia*', J. Exp. Zool., vol. 178 (1971), pp. 183-94.

Speicher, B. R. 'Oogenesis, Fertilization and Early Cleavage in *Habrobracon*', J. Morph., vol. 59 (1936), pp. 401-21.

——, 'Oogenesis in a Thelytokous Wasp, *Nemeritis canescens* (Grav.)', J. Morphol., vol. 61 (1937), pp. 453-72.

Spurway, H. 'Spontaneous Parthenogenesis in a Fish', Nature, vol. 171 (1953), pp. 437.

——, 'Hermaphroditism with Self-fertilization, and the Monthly Extrusion of Unfertilized Eggs, in the Viviparous Fish *Lebistes reticulatus*', Nature, vol. 180 (1957), pp. 1248-51.

Stalker, H. D. 'Parthenogenesis in *Drosophila*', Genetics, vol. 39 (1954), pp. 4-34.

——, 'A Case of Polyploidy in Diptera', Proc. Nat. Acad. Sci. US, vol. 42 (1956a), pp. 194-9.

——, 'On the Evolution of Parthenogenesis in the Lonchoptera (Diptera)', Evolution, vol. 10 (1956b), pp. 345-59.

Stanley, S. M. 'Clades Versus Clones in Evolution: Why we have Sex', Science, vol. 190 (1976), pp. 382-3.

Stebbing, A. R. D. 'Aspects of the Reproduction and Life Cycle of *Rhabdopleura compacta*', Mar. Biol., vol. 5 (1970), pp. 205-12.

Stebbins, G. L. *Variation and Evolution in Plants* (Columbia University Press, New York, 1950).

——, 'Longevity, Habitat and the Release of Genetic Variability in the Higher Plants', Cold Spring Harbour Symp. Quant. Biol., vol. 23 (1958), pp. 365-78.

Steenstrup, J. J. S. *Über den Generations-wechsel* (Copenhagen, 1842). English translation published by Ray Society (1845).

Stefani, R. 'Il Problema dell Partenogenesi in *Haploembia solieri* Ramb. (Embioptera-Oligotomidae)', Atti Accad. Naz. Lincei, Ser. 8, vol. 5 (1956), pp. 127-201.

——, 'La Maturazione dell'Uovo nell'*Artemia salina* di Sète', Riv. Biol., vol. 60 (1967), pp. 599-615.

Steffan, A. W. 'Die Stammes- une Siedlungsgeschichte des Artenkreises *Sacchiphantes viridis* (Ratzeburg 1843) (Adelgidae, Aphidoidea)', Zoologica, vol. 109 (1961), pp. 1-112.

——, 'Die Artenkreise der Gattung *Sacchiphantes* (Adelgidae, Apidoidea)', Proc. XI Int. Congr. Entomol. Vienna, vol. 1 (1962), pp. 57-63.

Stephan, W. P. and Cheldelin, I. H. 'Esterase Polymorphism in Select Clones of the Parthenogenetic Cockroach *Pycnoscelus* (Orthoptera: Blatteridae)', in R. Robert (ed.), *Genetics Lectures*, vol. 4 (Oregon State University Press, Corvallis, Oregon, 1975), pp. 205-21.

Stephens, R. T. and Bergmann, A. A. 'The B-chromosome System of the Grasshopper *Melanoplus femur-rubrum*', Chromosoma, vol. 38 (1972), pp. 297-311.

Stephenson, J. 'On Some Scottish Oligochaeta, with a Note on encystment in a Common Freshwater Oligochaete, *Lumbriculus variegans*', Trans. Roy. Soc. Edinburgh, vol. 53 (1922), pp. 277-95.

—, *The Oligochaeta* (Oxford University Press, 1930).

Stephenson, T. A. *The British Sea Anemones*, vol. II (The Ray Society, London, 1935).

Stern, C. 'An Effect of Temperature and Age on Crossing Over in the First Chromosome of *Drosophila melanogaster*', Proc. Nat. Acad. Sci. US, vol. 12 (1926), pp. 530-2.

Stern, V. M. and Bowen, W. R. 'Further Evidence of a Uniparental Race of *Trichogamma semifumatum* at Bishop, California', Ann. Entomol. Soc. Amer., vol. 61 (1968), pp. 1032-3.

Sterrer, W. 'On the Biology of Gnathostomulida', Vie et Milieu, suppl. 22 (1971), pp. 493-508.

—, 'Gnathostomulida', in A. C. Giese and J. S. Pearse (eds.), *Reproduction of Marine Invertebrates*, vol. 1 (Academic, New York, 1974), pp. 345-57.

Sterrer, W. and Rieger, R. 'Retronectidae – A New Cosmopolitan Marine Family of Catenulida (Turbellaria), in N. W. Riser and M. P. Morse (eds.), *Biology of the Turbellaria* (McGraw-Hill, New York, 1974), pp. 63-92.

Stevens, N. M. 'A Study of the Germ Cells of *Aphis rosae* and *Aphis oenotherae*', J. Exp. Zool., vol. 2 (1905), pp. 313-33.

— 'Further Studies on Reproduction in *Sagitta*', J. Morph., vol. 21 (1910), pp. 279-319.

Stiven, A. E. 'Concerning the Survival Curve of *Hydra*', Ecology, vol. 43 (1962a), pp. 173-4.

—, 'The Effect of Temperature and Feeding on the Intrinsic Rate of Increase of Three Species of *Hydra*', Ecology, vol. 43 (1962b), pp. 325-8.

Stotz, W. B. 'Functional Morphology and Zonation of Three Species of Sea Anemones from Rocky Shores in Southern Chile', Mar. Biol., vol. 50 (1979), pp. 181-8.

Strand, A. 'The Norwegian Species of *Amischa* Thoms.', Norsk. Entomol. Tidsskr., vol. 8 (1951), pp. 219-24.

Strasburger, E. 'Über Periodische Reduction der Chromosomenzahl im Entwicklungsgang der Organismen', Biol. Zentralblatt, vol. 14 (1894), pp. 817-38, 849-66.

Strobeck, C. 'Sufficient Conditions for Polymorphism with N Niches and M Mating Groups', Amer. Natur., vol. 108 (1974), pp. 152-6.

Strobeck, C., Maynard Smith, J. and Charlesworth, B. 'The Effects of Hitchhiking on a Gene for Recombination', Genetics, vol. 82 (1976), pp. 547-58.

Stunkard, H. W. 'The Life History and Systematic Relations of the Mesozoa', Quart. Rev. Biol., vol. 29 (1954), pp. 230-44.

Sturtevant, A. H. 'The Probable Occurrence of Parthenogenesis in *Ochthiphila polystigma* (Diptera)', Psyche, vol. 30 (1923).

Sturtevant, A. H. and Dobzhansky, T. H. 'Geographical Distribution and Cytology of "sex ratio" in *Drosophila pseudoobscura* and Related Species', Genetics, vol. 21 (1936), pp. 473-90.

Subba Rao, M. V. and Pantulu, J. V. 'The Effects of Derived B-chromosomes on Meiosis in Pearl Millett *Pennisetum typhoides*', Chromosoma, vol. 69 (1978), pp. 121-30.

Suomaleinen, E. 'Beitrage zur Zytologie der Parthenogenetischen Insekten. I. Coleoptera', Ann. Acad. Sci. Fenn., Ser. A, vol. 54 (1940a), pp. 1-144.

——, 'Polyploidy in Parthenogenetic Curculionidae', Hereditas, vol. 26 (1940b), pp. 51-64.

—— 'Beitrage zur Zytologie der Parthenogenetischen Insekten. II. *Lecanium hemisphaericum* (Coccidae)', Ann. Acad. Sci. Fenn., Ser. A, vol. 57 (1940c), pp. 1-30.

——, 'Parthenogenese und Polyploidie bei Rüsselkäfern (Curculionidae)', Hereditas, vol. 33 (1947), pp. 425-56.

——, 'Parthenogenesis and Polyploidy in the Weevils', Ann. Entomol. Fenn., vol. 14 (suppl.) (1948), pp. 206-12.

——, 'Parthenogenesis in Animals', Adv. Genet., vol. 3 (1950), pp. 193-253.

——, 'The Kinetochore and the Bivalent Structure in the Lepidoptera', Hereditas, vol. 39 (1953), pp. 88-96.

——, 'Zur Zytologie der Parthenogenetischen Curculioniden der Schweiz', Chromosoma, vol. 6 (1954), pp. 627-55.

——, 'A Further Instance of Geographical Parthenogenesis and Polyploidy in the Weevils, Curculionidae', Arch. Soc. 'Vanamo', vol. 9 (suppl) (1955), pp. 350-4.

——, 'On Morphological Differences and Evolution of Different Polyploid Parthenogenetic Weevil Populations', Hereditas, vol. 47 (1961), pp. 309-41.

——, 'Significance of Parthenogenesis in the Evolution of Insects', Ann. Rev. Entomol., vol. 7 (1962), pp. 349-66.

——, 'Die Polyploidie bei dem Parthenogenetische Blattkafer *Adoxus obscurus* L. (Coleoptera, Chrysomelidae)', Zool. Jahrb. Syst., vol. 92 (1965a), pp. 183-92.

——, 'On the Chromosomes of the Geometrid Moth Genus *Cidaria*', Chromosoma, vol. 16 (1965b), pp. 166-84.

——, 'The First Known Case of Polyploidy in a Parthenogenetic Curculionid Native of America', Hereditas, vol. 56 (1966a), pp. 213-16.

——, 'Achiasmatische Oogenese bei Trichopteren', Chromosoma, vol. 18 (1966b), pp. 201-7.

——, 'Evolution in Parthenogenetic Curculionidae', in T. Dobhzansky, M. Hecht and W. Steere (eds.), *Evolutionary Biology*, vol. 3 (Appleton-Century-Crofts, New York, 1969), pp. 261-96.

Suomaleinen, E. and Saura, A. 'Genetic Polymorphism and Evolution in Parthenogenetic Animals. I Polyploid Curculionidae', Genetics, vol. 74 (1973), pp. 489-508.

Suomaleinen, E., Saura, A. and Lokki, J. 'Evolution of Parthenogenetic Insects', Evol. Biol., vol. 9 (1976), pp. 209-57.

Sutton, S. L. 'The Population Dynamics of *Trichoniscus pusillus* and *Philoscia muscorum* (Crustacea, Oniscoidea) in Limestone Grassland', J. Anim. Ecol., vol. 37 (1968), pp. 425-44.

——, *Woodlice* (Hutchinson, London, 1972).

Svedalius, N. 'Alternation of Generations in Relation to Reduction Division', Bot. Gaz., vol. 83 (1927), pp. 362-84.

Taberly, G. 'La Cytologie de la Parthénogénèse chez *Platynothrus peltifer* (Koch) (Acarien Oribate)', CR, vol. 247 (1958), pp. 1655-7.

——, 'La Regulation Chromosomique chez *Trpochthonius tectorum* (Berl.), Espèce Parthénogénêtique d'Oribate (Acarien): Un Nouvel Example de Mixocinèse', CR, vol. 250 (1960), pp. 4200-1.

Takenouchi, Y. 'Polyploidy in Some Parthenogenetic Weevils: A Preliminary Report', Ann. Zool. Japan, vol. 30 (1957a), pp. 38-41.

——, 'On a Parthenogenetic Weevil, *Catapionus gracilicornis* Roelofs', Zool. Mag., vol. 66 (1957b), pp. 198-205.

——, 'Some Oecological Observations of Three Species of Curculionid Weevils, with Special Reference to Parthenogenetic Reproduction', J. Hokkaido Gakugei Univ., vol. 10 (1959), pp. 297-339.

——, 'The Cytology of Bisexual and Parthenogenetic Races of *Scepticus griseus* Roelof (Curculionidae: Coleoptera)', Can. J. Genet. Cytol., vol. 3 (1961), pp. 237-41.

——, 'A Preliminary Note on the Chromosomes of Four Parthenogenetic Weevils (Brachyrhininae) in Canada', Jap. J. Genet., vol. 39 (1964), pp. 74-9.

——, 'Chromosome Survey in Thirty-four Species of Bisexual and Partheno-genetic Weevils of Canada', Canad. J. Genet. Cytol., vol. 7 (1965), pp. 663-87.

——, 'Tetraploid and Pentaploid Races of the Japanese Parthenogenetic Weevil, *Catapionus gracilicornis* Roelofs (Curculionidae, Coleoptera)', Annot. Zool. Japan, vol. 39 (1966), pp. 47-54.

——, 'A Chromosome Study on Bisexual and Parthenogenetic Races of *Scepticus insulans* Roelofs (Curculionidae: Coleoptera)', Can. J. Genet. Cytol., vol. 10 (1968), pp. 945-50.

——, 'A Further Study on the Chromosomes of the Parthenogenetic Weevil, *Listroderes costirostris* Schönherr, from Japan', Cytologia, vol. 34 (1969), pp. 360-8.

——, 'A Further Chromosome Study in Bisexual and Parthenogenetic Races of the Weevil *Catapionus gracilicornis* Roelofs (Curculionidae: Coleoptera)', Japan J. Genet., vol. 45 (1970a), pp. 457-66.

——, 'Three Further Studies of the Chromosomes of Japanese Weevils (Coleoptera: Curculionidae)', Can. J. Genet. Cytol., vol. 12 (1970b), pp. 273-7.

Takenouchi, Y. and Takagi, K. 'A Chromosome Study of two Parthenogenetic Scolytid Beetles', Ann. Zool. Japan, vol. 40 (1967), pp. 105-10.

Taliev, D. N. 'On the Unisexual Reproduction of *Comephorus* (Pisces, Come-phoridae)', Dokl. Akad. Nauk. SSSR, vol. 6 (1950), pp. 105-8.

Tanaka, H. 'Remark on the Viviparous Character of *Coeloplana*', Ann. Zool. Japan, vol. 13 (1932), pp. 399-403.

Tanner, J. E. '*Oecodema cephalotes*', Trinidad Field Naturalists' Club, vol. 1 (1892), pp. 123-7.

Tannreuther, G. 'History of the Germ Cells and Early Embryology of Certain Aphids', Zool. Jahrb., vol. 24 (1907), pp. 609–42.

—, 'The Development of *Asplanchna ebbesbornii* (Rotifer)', J. Morph., vol. 33 (1920), pp. 389–437.

Tappa, D. W. 'The Dynamics of the Association of Six Limnetic Species of *Daphnia* in Aziscoos Lake, Maine', Ecol. Monogr., vol. 35 (1965), pp. 395–423.

Tauson, A. O. 'Wirkung des Mediums auf das Geschlecht des Rotators *Asplanchna intermedia* Huds.', Int. Rev. Ges. Hydrobiol. Hydrogr., vol. 13 (1925), pp. 70, 282–325.

Taylor, F. '*Paludestrina jenkinsoni* at Droylsden, Lancashire', J. Conchol., vol. 9 (1900), p. 340.

Taylor, H. L. and Medica, P. A. 'Natural Hybridization of the Bisexual Teiid Lizard *Cnemidophorus inornatus* and the Unisexual *Cnemidophorus perplexus* in Southern New Mexico', Univ. Colorado Stud. Biol., vol. 21 (1966), pp. 1–27.

Taylor, H. L., Walker, J. M. and Medica, P. A. 'Males of Three Normally Parthenogenetic Species of Teiid Lizards (Genus *Cnemidophorus*)', Copeia (1967), pp. 737–43.

Taylor, J. S. 'An Unusual Moth from Pearston', African Wildlife, vol. 20 (1966), pp. 23–7.

Taylor, M. G., Amin, M. B. A. and Nelson, G. S. ' "Parthenogenesis" in *Schistosoma mattheei*', J. Helminthol., vol. 43 (1969), pp. 197–206.

Taylor, P. D. 'An Analytical Model for a Short-term Advantage for Sex', J. Theoret. Biol., vol. 81 (1979), pp. 407–21.

Tease, C. and Jones, G. H. 'Chromosome-specific Control of Chiasma Formation in *Crepis capillaris*', Chromosoma, vol. 57 (1976), pp. 33–49.

Telford, E. R. and Campbell, H. W. 'Ecological Observations on an All-female Population of the Lizard *Lepidophyma flavimaculatum* (Xantusiidae) in Panama', Copeia (1970), pp. 379–81.

Templeton, A. R. and Rothman, E. D. 'The Population Genetics of Parthenogenetic Strains of *Drosophila mercatorum*', Theor. Appl. Genetics, vol. 43 (1973), pp. 204–12.

Thane, A. 'Rotifera', in A. C. Giese and J. S. Pearse (eds.), *Reproduction of Marine Invertebrates*, vol. 1 (Academic, New York, 1974), pp. 485–505.

Thiesen, B. F. 'Life History of Seven Species of Ostracods from a Danish Brackish-water Locality', Medd. f Dan. Fisk. Dg Hav., vol. 4, no. 8 (1966), pp. 215–70.

Thomas, R. 'The Smaller Teiid Lizards (*Gymnophthalmus* and *Bachia*) of the Southeastern Caribbean', Proc. Biol. Soc. Washington, vol. 78 (1965), pp. 141–54.

Thompson, V. 'Does Sex Accelerate Evolution?', Evol. Theory, vol. 1 (1976), pp. 131–56.

Thomsen, M. 'Studien über die Parthenogenese bei Einigen Cocciden und Aleurodiden', Z. Zellforsch., vol. 5 (1927), pp. 1–116.

—, 'Sex-determination in *Lecanium*' in *Transcripts of Fourth International Congress on Entomology* (1929), pp. 18-24.

Thomson, G. and Feldman, M. W. 'Population Genetics of Modifiers of Meiotic Drive. III. Linkage Modification in the Segregation Distorter System', Theoret. Pop. Biol., vol. 5 (1974), pp. 155-62.

Thomson, G. and Feldman, M. W. 'Population Genetics of Modifiers of Meiotic Drive. IV. Equilibrium Analysis for the Genetic Control of Segregation Distortion', Theoret. Pop. Biol., vol. 10 (1976), pp. 10-25.

Thornton, I. W. B. and Broadhead, E. 'The British Species of *Elipsocus* Hagen (Corrodentia, Mesopsocidae)', J. Bri. Soc. Entomol., vol. 5 (1954), pp. 47-64.

Thornton, I. W. B. and Wong, S. K. 'The Peripsocid Fauna (Psocoptera) of the Oriental Region and the Pacific', Pacific Inst. Monogr., vol. 19 (1968), pp. 1-158.

Thorp, J. H. and Bartholemew, G. T. 'Effects of Crowding on Growth-rate and Symbiosis in Green Hydra', Ecology, vol. 56 (1975), pp. 206-12.

Thorson, G. 'The Larval Development, Growth and Metabolism of Arctic Marine Bottom Invertebrates', Medd. Grønland, vol. 100 (1936), pp. 1-155.

Tinkle, D. W. 'Observations on the Lizards *Cnemidophorus tigris*, *Cnemidophorus tesselatus* and *Crotaphytus wislizeni*', Southwest Natur., vol. 4 (1959), pp. 195-200.

Trautmann, W. 'Parthenogenesis in der Familie der Psychidae', Entomol. Z., vol. 3 (1909), pp. 267-8.

Treisman, M. 'The Evolution of Sexual Reproduction: A Model Which Assumes Individual Selection', J. Theoret. Biol., vol. 60 (1976), pp. 421-31.

Treisman, M. and Dawkins, R. 'The "Cost of Meiosis": Is There Any?', J. Theoret. Biol., vol. 63 (1976), pp. 479-84.

Trembley, A. *Mémoires pour Servir à l'Histoire d'un Genre de Polypes d'Eau Douce* (Paris and Leiden, 1744). (See also Phil Trans. Roy. Soc. London, vol. 43, p. 474). Translation by S. G. Lenhoff and H. M. Lenhoff (University of Miami Press).

Tressler, W. 'Ostracoda' in W. T. Edmondson (ed.), *Freshwater Biology*, 2nd edn (Wiley, New York, 1959), pp. 657-734.

Triantaphyllou, A. C. 'Polyploidy and Reproductive Patterns in the Rootknot Nematode *Meloidogyne hapla*', J. Morphol., vol. 118 (1966), pp. 403-14.

—, 'Environmental Sex Differentiation of Nematodes in Relation to Pest Management', Ann. Rev. Phytopathol., vol. 11 (1973), pp. 441-64.

Triantaphyllou, A. C. and Hirschmann, H. 'Reproduction in Plant and Soil Nematodes', Ann. Rev. Phytopathol., vol. 2 (1964), pp. 57-80.

Trudgill, D. L. 'The Effect of Environment on Sex Determination in *Heterodera rostochiensis*', Nematologica, vol. 13 (1967), pp. 263-72.

Tucker, K. W. 'Automictic Parthenogenesis in the Honey Bee', Genetics, vol. 43 (1958), pp. 299-316.

Tuomikoski, R. 'Ein Vermutlicher Fall von Geographischer Parthenogenesis bei der Gattung *Tachydromia* (Dipt., Emphididae)', Ann. Entomol. Fenn., vol. 1 (1935), pp. 38-43.

Turner, J. R. 'Why Does the Genome Not Congeal?', Evolution, vol. 21 (1967), pp. 645-56.

Tyler, A. 'Artificial Parthenogenesis', Biol. Rev., vol. 16 (1941), pp. 291-335.

Tyler, J. 'Reproduction without Males in Aseptic Root Cultures of the Root-knot Nematode', Hilgardia, vol. 7 (1933), pp. 373-88.

Uéno, M. 'The Freshwater Branchiopoda of Japan. III. Genus *Daphnia* of Japan. 1. Seasonal Succession, Cyclomorphosis and Reproduction', Mem. Coll. Sci. Kyoto Univ. B, vol. 9 (1934), pp. 289-320.

Ullerich, F. H. 'Achiasmatische Spermatogenese bei der Skorpionsfliege *Panorpa* (Mecoptera)', Chromosoma, vol. 12 (1961), pp. 215-32.

Ullyot, P. and Beauchamp, R. S. A. 'Mechanisms for the Prevention of Self-fertilization in Some Species of Freshwater Triclads', Quart. J. Microscop. Sci., vol. 74 (1931), pp. 477-89.

Ulrich, H. 'Experimentelle Untersuchungen uber den Generationswechsel der heterogonen Cecidomyide *Oligarces paradoxus*', Z. Indukt. Abstamm-u. Vererb., vol. 71 (1936), pp. 1-60.

—, 'Generationswechsel und Geslechtsbestimmung einer Gallmucke mit Viviparen Larven', Verh. Deutsch. Zool. Ges. (Wien) (1962), pp. 139-59.

Uvarov, B. *Locusts and grasshoppers* (London, 1928).

Uzel, H. *Monographie der Ordnung Thysanoptera* (Konigratz, Berlin, 1895).

Uzzell, T. M. 'Natural Triploidy in Salamanders Related to *Ambystoma jeffersonianum*', Science, vol. 139 (1963), pp. 113-15.

—, 'Relations of the Diploid and Triploid Species of the *Ambystoma jeffersonianum* Complex (Amphibia, Caudata)', Copeia (1964), pp. 257-300.

—, 'Meiotic Mechanisms of Naturally Occurring Unisexual Vertebrates', Amer. Natur., vol. 104 (1970), pp. 433-45.

Uzzell, T. M. and Berger, L. 'Electrophoretic Phenotypes of *Rana ridibunda*, *Rana lessonae*, and Their Hybridogenetic Associate *Rana esculenta*', Proc. Nat. Acad. Sci. US, vol. 127 (1975), pp. 13-24.

Uzzell, T. M. and Darevsky, I. S. 'Biochemical Evidence for the Hybrid Origin of the Parthenogenetic Species of the *Lacerta saxicola* Complex (Sauria: Lacertidae), with a Discussion of Some Ecological and Evolutionary Implications', Copeia (1975), pp. 204-22.

Uzzell, T. M. and Goldblatt, S. M. 'Serum Proteins of Salamanders of the *Ambystoma jeffersonianum* Complex and the Origin of Triploid Species of This Group', Evolution, vol. 21 (1967), pp. 345-54.

Vandel, A. 'La Parthénogénèse Géographique: Contribution à l'Étude Biologique et Cytologique de la Parthénogénèse Naturelle', Bull. Biol. France Belgique, vol. 62 (1928), pp. 164-281.

—, *La Parthénogénèse* (G. Doin, Paris, 1931).

—, 'La Parthénogénèse Géographique. II. Les Mâles Triploides d'Origine Parthénogénétique de *Trichoniscus (Spiloniscus) elizabethae* Herold, Bull. Biol. France Belgique, vol. 68 (1934), pp. 419-63.

—, 'Recherches sur la Sexualité des Isopodes. III. La Déterminisme du Sexe et de la Monogénie chez *Trichoniscus* (*Spiloniscus*) *provisorius* Racovitza', Bull. Biol. France Belgique, vol. 72 (1938), pp. 121-46.

——, 'Monogénie chez les Oniscoides', Bull. Biol. France Belgique, vol. 75 (1941), pp. 316-63.

——, *Biospeleology. The Biology of Cavernicolous Animals* (Pergamon, Oxford, 1965).

Vannini, E. 'Studi sulla sessualità e Sui Poteri Rigenerativi nel Polichele Ermafrodita *Salmacina incrustans* Clap. I. Osservazioni sul Ciclo Riproduttivo Sessuale e Assessuale', Pubbl. Sta. Zool. Napoli, vol. 22 (1950), pp. 211-56.

Vanzolini, D. E. 'Unisexual *Cnemidophorus lemniscatus* in the Amazonas Valley: A Preliminary Note (Sauria, Teiidae)', Papéis Avulsos de Zool., S. Paulo, vol. 23 (1970), pp. 63-8.

Vanzolini, D. E. 'Parthenogenetic lizards', Science, vol. 201 (1978), p. 1152.

Ved Brat, S. 'Genetic Systems in *Allium*. II. Sex Differences in Meiosis', Chromosomes Today, vol. 1 (1966), pp. 31-40.

Ved Brat, S. and Dhingia, B. 'Genetic Systems in *Allium*. II. Breakdown of Classical System in *Allium cepa*', Nucleus, vol. 6 (1973), pp. 11-19.

Ved Brat, S. and Rai, K. S. 'Chiasma Frequency and Crossing Over in Translocation Heterozygotes of *Aedes aegypti*', Genetics, vol. 77 (1974), pp. 567.

Verner, S. 'Selection for Sex Ratio', Amer. Natur., vol. 99 (1965), pp. 419-21.

Verrill, A. E. 'Description of New American Actinians, with Critical Notes on Other Species, 1', Amer. J. Sci., vol. 6 (1898), pp. 493-8.

Viguier, C. 'L'Hermaphroditisme et la Parthénogénèse chez les Echinodermes', CR Acad. Sci. Paris, vol. 131 (1900).

Viinika, Y. 'The Effect of B-chromosomes on Meiosis', Hereditas, vol. 78 (1974), p. 331.

Vorstmann, A. 'Some Fresh-water Bryozoa of West Java', Treubia, vol. 10 (1928), pp. 1-14.

Vosa, C. G. 'Two-track Heredity: Differentiation of Male and Female Meiosis in *Tulbaghia*', Caryologia, vol. 25 (1972), pp. 275-81.

Vosa, C. G. and Barlow, P. W. 'Meiosis and B-chromosomes in *Listera ovata* (Orchidaceae)', Caryologia, vol. 25 (1972), pp. 1-8.

Vrijenhoek, R. C. 'Genetic Relationships of Unisexual-hybrid Fish to Their Progenitors Using Lactate Dehydrogenase Isozymes as Gene Markers (*Poeciliopsis*, Poeciliidae)', Amer. Natur., vol. 106 (1972), pp. 754-66.

Vrijenhoek, R. C. and Schultz, R. J. 'Evolution of a Trihybrid Unisexual Fish (*Poeciliopsis*, Poeciliidae)', Evolution, vol. 28 (1974), pp. 306-19.

Vrijenhoek, R. C., Angus, R. A. and Schultz, R. J. 'Variation and Heterozygosity in Sexually vs. Clonally Reproducing Populations of *Poeciliopsis*', Evolution, vol. 31 (1977), pp. 767-81.

——, 'Variation and Clonal Structure in a Unisexual Fish', Amer. Natur., vol. 112 (1978), pp. 41-55.

Wager, A. 'Some Observations on *Convoluta*' in *Report of Tenth Annual Meeting of South African Association for the Advancement of Science* (1913).

Wagler, E. 'Faunistische und Biologische Studien an Freischwimmenden Cladoceren Sachsens', Zoologica (Stuttgart), vol. 67 (1912), pp. 305-66.

Wagner, E. 'Über *Campyloneura virgula* Herrich-Schäffer (Hem. Het. Miridae)', Mitt. Deutsch. Entomol. Ges., vol. 27 (1968), pp. 46-7.

von Wagner, F. 'Beiträge zur Kentniss der Reparationsprocesse bei *Lumbricus variegatus*', Zool. Jahrbuch, Anat., vol. 13 (1900), pp. 603-82.

Walton, C. L. and Jones, N. W. 'Further Observations on the Life-history of *Limnaea truncatúla*', Parasitology, vol. 18 (1926), pp. 144-7.

Warwick, T. 'Strains in the Mollusc *Potamopyrgus jenkinsoni* (Smith)', Nature, vol. 169 (1952), pp. 551-2.

Watson, I. D. and Callan, H. G. 'The Form of Bivalent Chromosomes in Newt Oocytes at First Metaphase of Meiosis', Quart. J. Microscop. Sci., vol. 104 (1963), pp. 281-94.

Weismann, A. *Beiträge zur Naturgeschichte der Daphniden* (Leipzig, 1876-9).

——, 'Parthenogenese bei den Ostracoden', Zool. Anz., vol. 3 (1880), pp. 82-4.

——, *Essays upon Heredity and Kindred Biological Problems*, translated by E. B. Poulton, S. Schonland and A. E. Shipley (Clarendon Press, Oxford, 1889).

Weiss, M. J. and Levy, D. P. 'Sperm in "Parthenogenetic" Freshwater Gastrotrichs', Science, vol. 205 (1979), pp. 302-3.

Weissman, D. B. 'Geographical Variability in the Pericentric Inversion System of the Grasshopper *Trimerotropis pseudofasciata*', Chromosoma, vol. 55 (1976), pp. 325-47.

Welch, P. S. and Loomis, H. A. 'A Limnological Study of *Hydra oligactis* in Douglas Lake, Michigan', Trans. Amer. Microscop. Soc., vol. 43 (1924), pp. 203-35.

Wells, H. W., Wells, M. J. and Gray, I. E. 'Ecology of Sponges in Hatteras Harbor, North Carolina', Ecology, vol. 45 (1964), pp. 752-67.

Wells, J. W. 'Evolutionary Development in the Scleractinian Family Fungiidae', Symp. Zool. Soc. London, vol. 16 (1966), pp. 223-46.

Wells, L. 'Seasonal Abundance and Vertical Movement of Planktonic Crustacea in Lake Michigan', Fish. Bull. Fish Wildlife Serv. US, vol. 60 (1960), pp. 343-69.

Werner, B. 'On the Development ∠nd Reproduction of the Anthomedusan *Margelopsis haeckeli* Hartlaus', Ann. New York Acad. Sci., vol. 62 (1955), pp. 1-29.

——, 'Effect of Some Environmental Factors on Differentiation and Determination in Marine Hydrozoa, with a Role on their Evolutionary Significance', Ann. New York Acad. Sci., vol. 105 (1963), pp. 461-88.

Wesenberg-Lund, C. 'On the occurrence of *Fredericella sultana* and *Padudicella ehrenbergi* in Greenland', Medd. Grönland, vol. 34 (1907), pp. 61-75.

——, *Plankton Investigations of the Danish Lakes. General Part: The Baltic Freshwater Plankton, Its Origin and Variation* (Gyldendalske Boghandel, Copenhagen, 1908).

—, 'Contributions to the Biology of the Rotifera. I. The Males of the Rotifera', D. Kgl. Danske Vidensk. Selsk. Skrifter Naturwidensk. og Mathem. Afd. 8, vol. 4, no. 3 (1923), pp. 191-345.

—, 'Contributions to the Biology and Morphology of the Genus *Daphnia*', D. Kgl. Danske Vidensk Selsk. Skrifter Naturwidensk og Mathem. Afd. 8, vol. 11, no. 2 (1926), pp. 92-250.

—, 'Contributions to the Biology of the Rotifera. Part II. The Periodicity and Sexual Periods', D. Kgl. Danske Vedensk. Selsk. Skrifter Naturwidensk. og Mathem. Afd. 9, vol. 2, no. 1 (1930), pp. 3-230.

Wessenberg, H. 'Studies on the Life-cycle and Morphogenesis of *Opalina*', Univ. Calif. Publ. Zool., vol. 61 (1961), pp. 315-70.

Westerman, M. 'The Effect of X-Irradiation on Male Meiosis in *Schistocerca gregaria* (Forskal). I. Chiasma Frequency Response, Chromosoma, vol. 22 (1967), pp. 401-16.

—, 'Population Cytology of the Genus *Phaulacridium*. II. *Phaulacridium marginale* (Walker); Chiasma Frequency Studies from South Island, New Zealand', Chromosoma, vol. 46 (1974), pp. 207-16.

—, 'Population Cytology of the Genus *Phaulacridium*. IV. *Phaulacridium marginale* (Walker) – The North Island Populations', Heredity, vol. 53 (1975), pp. 165-72.

White, M. J. D. 'The Effect of Temperature on Chiasma Frequency', J. Genetics, vol. 29 (1934), pp. 203-15.

—, 'A New and Anamalous Type of Meiosis in a Mantid, *Callimantis antillarum* Saussure', Proc. Roy. Soc. London, vol. B125 (1938), pp. 516-23.

—, 'The Cytology of the Cecidomyidae (Diptera). II. The Chromosome Cycle and Anomalous Spermatogenesis of *Miastor*.', J. Morph., vol. 79 (1946a), pp. 323-70.

—, The Evidence Against Polyploidy in Sexually Reproducing Animals', Amer. Natur., vol. 80 (1946b), pp. 610-19.

—, 'The Chromosomes of the Parthenogenetic Mantid *Brunneria borealis*', Evolution, vol. 2 (1948), pp. 90-3.

—, 'Cytological Studies on Gall Midges', Univ. Texas Publ., vol. 5007 (1950), pp. 1-80.

—, 'Cytogenetics of Orthopteroid Insects', Adv. Genet., vol. 4 (1951a), pp. 267-330.

—, Bull. Soc. Roy. Entomol. Egypte (1951b), pp. 104-48.

—, 'Cytogenetic Mechanisms in Insect Reproduction', in K. C. Highnam (ed.), *Inseet Reproduction* (Roy. Entomol. Soc. London Symp. 2, 1964).

—, 'Chiasmatic and Achiasmatic Meiosis in African Eumastacid Grasshoppers', Chromosoma, vol. 16 (1965), pp. 271-307.

—, 'Further Studies on the Cytology and Distribution of the Australian Parthenogenetic Grasshopper *Morabo virgo*', Rev. Suisse Zool., vol. 73 (1966), pp. 383-98.

—, 'Heterozygosity and Genetic Polymorphism in Parthenogenetic Animals', in M. K. Hecht and W. C. Steere (eds.), *Essays in Evolution and Genetics in Honor of Theodosius Dobzhansky* (Elsevier North-Holland, Amsterdam, 1970), pp. 237-62.

——, *Animal Cytology and Evolution*, 3rd edn (Cambridge University Press, 1973).

——, 'The Karyotype of the Parthenogenetic Grasshopper *Xiphidiopsis lita* (Orthoptera: Tettigoniidae)', Caryologia, vol. 31 (1978), pp. 291-7.

White, M. J. D. and Contreras, N. 'Cytogenetics of the Parthenogenetic Grasshopper *Warramaba* (Formerly *Morabo*) *virgo* and Its Bisexual Relatives. V. Interaction of *W. virgo* and a Bisexual Species in Geographic Contact', Evolution, vol. 33 (1979), pp. 85-94.

White, M. J. D. and Webb, G. C. 'Origin and Evolution of Parthenogenetic Reproduction in the Grasshopper *Morabo virgo* (Eumastacidae; Morabinae)', Aust. J. Zool., vol. 16 (1968), pp. 647-71.

White, M. J. D., Cheney, J. and Key, K. H. L. 'A Parthenogenetic Species of Grasshopper with Complex Structural Heterozygosity (Orthoptera: Acridoidea)', Australian J. Zool., vol. 11 (1963), pp. 1-19.

White, M. J. D., Contreras, N., Cheney, J. and Webb, G. C. 'Cytogenetics of the Parthenogenetic Grasshopper *Warramaba* (Formerly *Morabo*) *virgo* and Its Bisexual Relatives. II. Hybridization Studies', Chromosoma, vol. 61 (1977), pp. 127-48.

Whiting, P. W. 'The Evolution of Male Haploidy', Quart. Rev. Biol., vol. 24 (1949), pp. 231-60.

Whitney, D. D. 'The Relative Influence of Food and Oxygen in Controlling Sex in Rotifers', J. Exp. Zool., vol. 24 (1917), pp. 101-38.

Wiese, L. 'Genetic Aspects of Sexuality in Volvocales', in R. A. Lewin (ed.), *The Genetics of Algae* (Blackwell, Oxford, 1976), pp. 174-97.

Wildish, D. J. 'Adaptive Significance of a Biased Sex Ratio in *Orchestia*', Nature, vol. 233 (1971), pp. 54-5.

Wilhelmi, J. 'Tricladen', Fauna u. Flora des Golfes von Neapel, vol. 32 (1909), pp. 104-5.

Williams, C. B. 'Some Problems of Sex Ratios and Parthenogenesis', J. Genetics, vol. 6 (1917), pp. 255-67.

——, 'Aggregation during Settlement as a Factor in the Establishment of Coelenterate Colonies', Ophelia, vol. 15 (1976), pp. 57-64.

Williams, G. C. *Adaptation and Natural Selection. A Critique of Some Current Evolutionary Thought* (Princeton University Press, Princeton, NJ, 1966).

——, *Group Selection* (Aldine, New York, 1968).

——, *Sex and Evolution* (Princeton University Press, Princeton, NJ, 1975).

——, 'The Question of Adaptive Sex Ratio in Outcrossed Vertebrates', Proc. Roy. Soc. London, B, vol. 205 (1979), pp. 567-80.

Williams, G. C. and Mitton, J. B. 'Why Reproduce Sexually?' J. Theoret. Biol., vol. 39 (1973), pp. 545-54.

Williams, R. W. 'Parthenogenesis and Autogeny in *Culicoides bermudensis* Williams', Mosquito News, vol. 21 (1961), pp. 116-17.

Williams, S. A. 'The British Species of the Genus *Amischa* Thomson (Col., Staphylinidae), Including *A. soror* Kraatz, an Addition to the List', Entomol. Monthly Mag., vol. 105 (1969), pp. 38-43.

Williamson, M. H. 'Studies on the Colour and Genetics of the Black Slug', Proc. Roy. Phys. Soc. Edinburgh, vol. 27 (1959), pp. 87-93.

Wilson, D. P. 'The Settlement Behaviour of the Larvae of *Sabellaria alveolata*', J. Mar. Biol. Ass. UK, vol. 48 (1968), pp. 387-435.

Wilson, D. S. 'A Theory of Group Selection', Proc. Nat. Acad. Sci. US, vol. 72 (1975), pp. 143-6.

——, 'Structured Demes and the Evolution of Group-advantageous Trails', Amer. Natur., vol. 111 (1977), pp. 157-85.

——, *The Natural Selection of Populations and Communities* (Benjamin/Cummings, Menlo Park, Ca., 1980).

Wilson, E. O. *The Insect Societies* (Belknap Press of Harvard University Press, Cambridge, Mass., 1971).

Wilson, F. 'Some Experiments on the Influence of Environment Upon the Forms of *Aphis chloris* Koch', Trans. Roy. Entomol. Soc. London, vol. 87 (1938), pp. 165-80.

Wilson, F. and Woolcock, L. T. 'Temperature Determination of Sex in a Parthenogenetic Parasite, *Ooencyrtus submetallicus* (Howard) (Hymenoptera: Encyrtidae)', Aust. J. Zool., vol. 8 (1960), pp. 153-69.

Wilson, H. V. 'Notes on the Development of Some Sponges', J. Morphol., vol. 5 (1891), pp. 511-19.

——, 'Observations on the Gemmule and Egg Development of Marine Sponges', J. Morphol., vol. 9 (1894), pp. 277-408.

——, 'On the Asexual Origin of the Sponge Larvae', Amer. Natur., vol. 36 (1902), pp. 451-9.

Wilson, J. Y. 'Chiasma Frequency in Relation to Temperature', Genetica, vol. 29 (1959), pp. 290-303.

——, 'Chiasma Frequency in the Two Sex Mother Cells of an Hermaphrodite', Chromosoma, vol. 11 (1960), pp. 433-40.

Winkler, H. *Verbreitung und Ursache der Parthenogenesis im Pflanzen und Tierreiche* (Jena, 1920).

Winslow, R. D. 'Some Aspects of the Ecology of Free-living and Plant-parasitic Nematodes', in J. N. Sasser and W. R. Jenkins (eds.), *Nematology* (University of North Carolina Press, Chapel Hill, 1900), pp. 341-415.

Winterbourn, M. 'The New Zealand Species of *Potamopyrgus* (Gastropoda: Hydrobiidae)', Malacologia, vol. 10 (1970), pp. 283-321.

Wohlgemuth, R. 'Beobachtungen und Untersuchungen über die Biologie der Süsswasserostracoden, Ihr Vorkommen in Sachsen und Bohmen, Ihre Lebensweise und Ihr Forpflanzung', Int. Rev. Ges. Hydrobiol., (suppl. 6), vol. 20 (1914), pp. 1-72.

Wolf, E. 'Die Chromosomen im der Spermatogenese Einiger Nematoceren', Chromosoma, vol. 2 (1941), pp. 192-246.

—, 'Chromosomen Untersuchungen auf Insekten', Z. Naturf., vol. 1 (1946), pp. 108-9.

—, 'Die Chromosomen in der Spermatogenese der Dipteren *Phryne* und *Mycetobia*', Chromosoma, vol. 4 (1950), pp. 148-204.

Wood, T. 'Colony Development in Species of *Plumatella* and *Fredericella* (Ectoprocta: Phylactolaemata)', in R. S. Boardman, A. H. Cheetam and W. A. Oliver (eds.), *Animal Colonies* (Dowden, Hutchinson and Ross, Stroudsburg, Pa., 1973).

Wood, T. R. 'Resting Eggs That Fail to Rest', Amer. Natur., vol. 66 (1932), pp. 277-81.

Wright, J. W. 'Parthenogenetic Lizards', Science, vol. 201 (1978), pp. 1152-4.

Wright, J. W. and Lowe, C. H. 'Evolution of the Alloploid Parthenospecies *Cnemidophorus tesselatus* (Say)', Mammal Chromosome Newsletter, vol. 8 (1967), pp. 95-6.

Wright, S. 'Evolution in Mendelian Populations', Genetics, vol. 16 (1931), pp. 97-159.

Wyatt, I. J. 'Pupal Paedogenesis in the Cecidomyiidae (Diptera). I', Proc. Roy. Entomol. Soc. London, vol. 36 (1961), pp. 133-43.

Wynne-Edwards, V. C. *Animal Dispersion in Relation to Social Behaviour* (Oliver and Boyd, London, 1962).

Yamazi, I. 'Autotomy and Regeneration in Japanese Sea-stars', Ann. Zool. Japan, vol. 23 (1950).

Yonge, C. 'Observations on *Hipponix antiquatus*', Proc. California Acad. Sci., ser. 4, vol. 28 (1953).

Young, P. W. 'Enzyme Polymorphisms and Reproduction in *Daphnia magna*', PhD Thesis (Cambridge University, 1975).

Zaffagnini, F. 'Rudimentary Hermaphroditism and Automictic Parthenogenesis in *Limnadia lenticularis* (Phyllopoda, Conchostraca)', Experientia, vol. 25 (1969), pp. 650-1.

Zaman, M. A., Patwary, U. and Matin, A. 'Karyomorphology and Meiotic Behaviour of *Galphimia gracilis* Barth (Malpighiaceae)', Caryologia, vol. 30 (1977), pp. 429-34.

Zarchi, Y., Simchen, G., Hillel, J. and Schaap, T. 'Chiasmata and the Breeding System in Wild Populations of Diploid Wheats', Chromosoma, vol. 38 (1972), pp. 77-94.

Zecevic, L. and Paunovic, D. 'The Effect of B-chromosomes on Chiasma Frequency in Wild Populations of Rye', Chromosoma, vol. 27 (1969), pp. 198-200.

von Zeppelin, M. G. 'Ueber den Bau und die Teilungsvorgänge des *Ctenodrilus monostylos* n. sp.', Z. Wiss. Zool., vol. 39 (1883), pp. 615-52.

Zweifel, R. G. 'Variation and Distribution of the Unisexual Lizard *Cnemidophorus tesselatus*', Amer. Mus. Novit., vol. 2235 (1965), pp. 1-49.

TAXONOMIC INDEX